6G to Build a Sustainable Future

6G to Build a Sustainable Future

Edited by

Mikko A. Uusitalo
Nokia Bell Labs
Espoo
Finland

Patrik Rugeland
Ericsson Research
Stockholm
Sweden

Mauro R. Boldi
Telecom Italia
Turin
Italy

Ahmad Nimr
Technische Universität Dresden
Dresden
Germany

WILEY

This edition first published 2026.

Registered Offices
John Wiley & Sons, Inc., 111 River Street, Hoboken, NJ 07030, USA
John Wiley & Sons Ltd, The Atrium, Southern Gate, Chichester, West Sussex, PO19 8SQ, UK

For details of our global editorial offices, customer services, and more information about Wiley products visit us at www.wiley.com.

The manufacturer's authorized representative according to the EU General Product Safety Regulation is Wiley-VCH GmbH, Boschstr. 12, 69469 Weinheim, Germany, e-mail: Product_Safety@wiley.com.

Wiley also publishes its books in a variety of electronic formats and by print-on-demand. Some content that appears in standard print versions of this book may not be available in other formats.

Library of Congress Cataloging-in-Publication Data

Names: Uusitalo, Mikko A, editor.
Title: 6G to build a sustainable future / edited by Mikko A. Uusitalo, head of a research department, Radio Systems, Patrik Rugeland, Mauro R. Boldi, Wireless Innovation, Telecom Italiavia Reiss Romoli·274, Italy, Ahmad Nimr.
Description: Hoboken, NJ, USA : Wiley, 2026. | Includes bibliographical references and index.
Identifiers: LCCN 2025044987 (print) | LCCN 2025044988 (ebook) | ISBN 9781394363575 (cloth) | ISBN 9781394363599 (adobe pdf) | ISBN 9781394363582 (epub)
Subjects: LCSH: 6G mobile communication systems. | Sustainable engineering.
Classification: LCC TK5103.252 .A186 2026 (print) | LCC TK5103.252 (ebook)
LC record available at https://lccn.loc.gov/2025044987
LC ebook record available at https://lccn.loc.gov/2025044988

Cover Design: Wiley
Cover Image: © Eoneren/Getty Images

Set in 9.5/12.5pt STIXTwoText by Straive, Chennai, India

Printed and bound by CPI Group (UK) Ltd, Croydon, CR0 4YY

C9781394363575_050226

Contents

List of Contributors

Daniel Adanza
Centre Tecnològic de Telecomunicacions de Catalunya (CTTC-CERCA)
Castelldefels (Barcelona)
Spain

Ozgur Umut Akgul
Nokia
Espoo
Finland

Pol Alemany
Centre Tecnològic de Telecomunicacions de Catalunya (CTTC-CERCA)
Castelldefels (Barcelona)
Spain

George C. Alexandropoulos
National and Kapodistrian University of Athens
Athens
Greece

Angeliki Alexiou
University of Piraeus Research Center
Piraues
Greece

Muhsin Ali
LeapWave Technologies
Leganés, Madrid
Spain

Bassem Arar
Technische Universität Dresden
Dresden
Germany

Italo Atzeni
University of Oulu & Keysight
Oulu
Finland

Amirhossein Azarbahram
University of Oulu & Keysight
Oulu
Finland

Sokratis Barmpounakis
WINGS ICT Solutions
Athens
Greece

Arturo Basaure
University of Oulu
Oulu
Finland

Riccardo Bassoli
Technische Universität Dresden
Dresden
Germany

Gilberto Berardinelli
Aalborg University
Aalborg
Denmark

Rafael Berkvens
IMEC
Antwerp
Belgium

Nabeel Nisar Bhat
University of Antwerp-IMEC
Antwerp
Belgium

Elham Dehghan Biyar
Ericsson Research
Istanbul
Türkiye

Henry Blue
VTT Technical Research Centre of Finland
Espoo
Finland

Mauro R. Boldi
Telecom Italia
Turin
Italy

Panagiotis Botsinis
Apple
Munich
Germany

Xie Boxuan
Aalto University
Espoo
Finland

Pere Garau Burguera
Aalto University
Espoo
Finland

Maurizio Cecchi
Institute PIIU
Rome
Italy

Hui Chen
Chalmers University of Technology
Gothenburg
Sweden

Nicolas Chuberre
Thales Alenia Space
Cannes
France

Josué Castañeda Cisneros
Ericsson Research
Stockholm
Sweden

Francesca Costanzo
Université Grenoble Alpes and
CEA-Leti
Grenoble
France

Davide Dardari
University of Bologna
Bologna
Italy

Panagiotis Demestichas
WINGS ICT Solutions
Athens
Greece

Benoît Denis
Université Grenoble Alpes and
CEA-Leti
Grenoble
France

Claude Desset
IMEC
Antwerp
Belgium

Sotiris Droulias
University of Piraeus Research Center
Piraues
Greece

Sameh Eldessoki
Apple
Munich
Germany

Mårten Ericson
Ericsson Research
Luleå
Sweden

Yigit Ertugrul
IMEC
Antwerp
Belgium

Jeroen Famaey
IMEC
Antwerp
Belgium

Hamed Farhadi
Ericsson Research
Stockholm
Sweden

Frank H.P. Fitzek
Technische Universität Dresden
Dresden
Germany

Luis G. Uzeda Garcia
Nokia Networks
Massy
France

Yu Ge
Chalmers University of Technology
Gothenburg
Sweden

Pietro G. Giardina
Nextworks
Pisa
Italy

Luis González
Universidad Carlos III de Madrid
Leganés, Madrid
Spain

Bikshapathi Gouda
University of Oulu & Keysight
Oulu
Finland

Alperen Gundogan
Apple
Munich
Germany

Hao Guo
Chalmers University of Technology
Gothenburg
Sweden

Sebastian Haas
Barkhausen Institut
Dresden
Germany

Christian Häger
Chalmers University of Technology
Gothenburg
Sweden

Rreze Halili
IMEC
Antwerp
Belgium

Hanne-Stine Hallingby
Telenor
Oslo
Norway

Kawon Han
University College London
London
UK

Katsuyuki Haneda
Aalto University
Espoo
Finland

Hasanin Harkous
Nokia
Germany

János Harmatos
Ericsson Research
Budapest
Hungary

Iman Hmedoush
Nokia
Massy
France

George-Roberto Hotopan
Nokia
Germany

Akshay Jain
Nokia Bell Labs
Espoo
Finland

Riku Jäntti
Aalto University
Espoo
Finland

Ling Jie
Ericsson
Finland

Markku Juntti
University of Oulu
Oulu
Finland

Grigorios Kakkavas
Institute of Communication and Computer Systems
National Technical University of Athens
Athens
Greece

Alexandros Kaloxilos
GIA
Brussels
Belgium

Leyli Karacay
Ericsson Research
Istanbul
Türkiye

Efstathios Katranaras
Sequans Communications
Paris
France

Sylvaine Kerboeuf
Nokia Bell Labs
Paris, Saclay
France

Gunes Kesik
Ericsson Research
Istanbul
Türkiye

Musa Furkan Keskin
Chalmers University of Technology
Gothenburg
Sweden

Bilal Khan
University of Oulu & Keysight
Oulu
Finalnd

Bahare M. Khorsandi
Nokia
Munich
Germany

Dani Korpi
Nokia Bell Labs
Espoo
Finland

Kalle Koskinen
Aalto University
Espoo
Finland

Karol Kuczyński
Orange Innovation Poland
Warsaw
Poland

and

Warsaw University of Technology
Warsaw
Poland

Pekka Kyosti
University of Oulu & Keysight
Oulu
Finland

and

Keysight
Oulu
Finland

Vasiliki Lamprousi
WINGS ICT Solutions
Athens
Greece

Per Hjalmar Lehne
Telenor
Oslo
Norway

Meng Li
IMEC
Antwerp
Belgium

Jingyi Liao
Aalto University
Espoo
Finland

Diego López
Telefónica
Seville
Spain

Onel L. A. López
University of Oulu & Keysight
Oulu
Finland

Charitha Madapatha
Chalmers University of Technology
Gothenburg
Sweden

Giyyarpuram Madhusudan
Orange
Meylan
France

Nurul Huda Mahmood
University of Oulu & Keysight
Oulu
Finland

Pilar Andres Maldonado
Nokia
Aalborg
Denmark

Diamanti Maria
Institute of Communication and Computer Systems (ICCS)
Athens
Greece

Christos Masouros
University College London
London
UK

Marja Matinmikko-Blue
University of Oulu
Oulu
Finland

Emil Matus
Technische Universität Dresden
Dresden
Germany

Nafiseh Mazloum
Sony
Nordic
Sweden

Sami Mekki
Nokia Networks France
Massy
France

Ishita Mishra
Institute PIIU
Rome
Italy

Ahmad Mohammad
LioniX International
Enschede
The Netherlands

Raul Muñoz
Centre Tecnològic de Telecomunicacions de Catalunya (CTTC-CERCA)
Castelldefels (Barcelona)
Spain

Samer Nasser
IMEC
Antwerp
Belgium

Simon Nellen
Fraunhofer Institute for Telecommunications
Heinrich Hertz Institute
Berlin
Germany

Nhan Thanh Nguyen
University of Oulu
Oulu
Finland

Ahmad Nimr
Technische Universität Dresden
Dresden
Germany

Toon Norp
TNO
The Hague
The Netherlands

Behnam Ojaghi
Centre Tecnològic de Telecomunicacions de Catalunya (CTTC-CERCA)
Castelldefels (Barcelona)
Spain

Antonio de la Oliva
University Carlos III of Madrid
Leganés, Madrid
Spain

Akshay Vayal Parambath
Chalmers University of Technology
Gothenburg
Sweden

Noelia Perez Palma
University of Murcia
Murcia
Spain

Dorin Panaitopol
Thales SIX GTS France
Gennevilliers
France

Ana Pereira
Ubiwhere
Aveiro
Portugal

Gregorio Martínez Pérez
University of Murcia
Murcia
Spain

Manuel Gil Pérez
University of Murcia
Murcia
Spain

Dinh-Thuy Phan-Huy
Orange
Chatillon
France

Pablo Picazo-Martínez
University Carlos III of Madrid
Leganés, Madrid
Spain

Rafael Pires
VTT Technical Research Centre of Finland
Oulu
Finland

Pawani Porambage
VTT Technical Research Centre of Finland
Oulu
Finland

Priyesh Pappinisseri Puluckul
IMEC
Antwerp
Belgium

Tianwen Qian
Fraunhofer Institute for Telecommunications
Heinrich Hertz Institute
Berlin
Germany

Vignesh Raman
Technische Universität Dresden
Dresden
Germany

José Miguel Mateos Ramos
Chalmers University of Technology
Gothenburg
Sweden

Philippe Ratajczak
Orange
Valbonne
France

Chris Roeloffzen
LioniX International
Enschede
The Netherlands

Osmel Martínez Rosabal
University of Oulu & Keysight
Oulu
Finland

Patrik Rugeland
Ericsson Research
Stockholm
Sweden

Juha-Matti Runtti
Keysight
Oulu
Finland

Joachim Sachs
Ericsson Research
Bremen
Germany

Merve Saimler
Ericsson Research
Istanbul
Turkey

Sharief Saleh
Chalmers University of Technology
Gothenburg
Sweden

Luís Pedro Santos
Ubiwhere
Aveiro
Portugal

Bikramjit Singh
Ericsson Research
Jorvas
Finland

Ritesh Kumar Singh
IMEC
Antwerp
Belgium

Antonio Skarmeta
University of Murcia
Murcia
Spain

Xosé R. Sousa
Optare Solutions
Vigo
Spain

Athanasios Stavridis
Ericsson Research
Lund
Sweden

Tommy Svensson
Chalmers University of Technology
Gothenburg
Sweden

Halina Tarasiuk
Orange Innovation Poland
Warsaw
Poland

and

Warsaw University of Technology
Warsaw
Poland

Venkatesh Tentu
Chalmers University of Technology
Güteborg
Schweden

Nuutti Tervo
University of Oulu & Keysight
Oulu
Finland

Huy Q. Tran
Nokia Bell Labs
Paris, Saclay
France

Vasileios Tsekenis
WINGS ICT Solutions
Athens
Greece

Christos Tsokos
National Technical University of Athens
Athens
Greece

Ioannis Tzanettis
Institute of Communication and Computer Systems
National Technical University of Athens
Athens
Greece

Mikko A. Uusitalo
Nokia Bell Labs
Espoo
Finland

José María Jorquera Valero
University of Murcia
Murcia
Spain

Antonio Varvara
TIM
Turin TO
Italy

Ricard Vilalta
Centre Tecnològic de Telecomunicacions de Catalunya (CTTC-CERCA)
Castelldefels (Barcelona)
Spain

Usman Virk
Keysight
Oulu
Finland

Enrico Maria Vitucci
University of Bologna
Bologna
Italy

Stefan Wänstedt
Ericsson Research
Luleå
Sweden

Stefan Wendt
Orange
Chatillon
France

Henk Wymeersch
Chalmers University of Technology
Gothenburg
Sweden

Anastasia Yagafarova
TNO
The Hague
The Netherlands

Anastasios Zafeiropoulos
Institute of Communication and Computer Systems
National Technical University of Athens
Athens
Greece

Fengchun Zhang
Aalborg University
Aalborg
Denmark

Peize Zhang
University of Oulu & Keysight
Oulu
Finland

Marcin Ziółkowski
Orange Innovation Poland
Warsaw
Poland

and

Warsaw University of Technology
Warsaw
Poland

Milan Zivkovic
Apple
Munich
Germany

Cristóbal Vinagre Zúñiga
TNO
The Hague
The Netherlands

Preface

As we enter a new era in digital technology, including connectivity, Europe renews its commitment to innovation, sustainability, and global collaboration, aiming to lead in the future of communication networks.

Building on the achievements of 5G, 6G represents a paradigm shift that promises unprecedented capabilities, including ultra-high data rates, low-latency communication, and seamless integration of artificial intelligence. It is not merely an evolution but a revolution that will redefine how society interacts with the digital and physical worlds. From holographic communication and extended reality to remote surgery and smart cities, the applications of 6G may touch every aspect of our lives, with the potential to serve as humanity's 'sixth sense'.

The European Union has embraced this challenge in particular through the Smart Networks and Services Joint Undertaking (SNS JU). With €1.8 billion in funding allocated for 2021–2027, the EU is fostering technological sovereignty while addressing critical challenges such as privacy, cybersecurity, and sustainability. These efforts are complemented by international partnerships with other countries, while Europe seeks to contribute to global standards while strengthening its position in innovation.

Among more than 70 other projects funded by the SNS JU so far, the 6G flagship European research project Hexa-X-II has a cornerstone in laying the foundation for future mobile networks, building on research building blocks with a focus on systemization and pre-standardization.

As we delve deeper into the pages of this book, we are reminded that 6G is not just about technological advancement; it is about shaping a future that aligns with our values. The integration of sustainability, security, resilience, and climate-neutral objectives into network infrastructures demonstrates Europe's commitment to the United Nations' Sustainable Development Goals.

This book also reflects Europe's quest for digital sovereignty. By investing in cutting-edge research, fostering collaboration across industries and nations, and addressing societal challenges head-on, Europe is paving the way for a connected future that benefits society as a whole.

I invite readers – scientists, policymakers, industry players, and interested citizens – to explore the possibilities of 6G within these pages. Together, they can seize the opportunity to lead responsibly in creating a future where connectivity empowers innovation, inclusiveness, and sustainability.

Head of Unit, European Commission *Miguel González-Sancho*
19 April 2025

Modern telecommunication networks play a critical role in all aspects of everyday life. During the late-2000s, the world witnessed a dramatic improvement in telecommunication services with the arrival of 4G networks. Since the beginning of the following decade, scientists and organizations worldwide worked on the design and deployment of 5G networks. Their target was to address the growing traffic needs and to provide an advanced set of new services that would lay the foundations for the flexible support of various vertical markets (e.g. smart industry, energy, automotive, transportation and logistics, health, media, gaming, etc.). Although 5G networks will remain the key mobile network technology for many years to come, because of the long lead time and complexity involved, the R&D activities of 6G systems are now well advanced and its standardization is starting, with the expectation that the first 6G systems will be deployed in the 2030 timeframe.

This book contains the latest consolidated results from the European Smart Networks and Services Joint Undertaking (SNS-JU) research programme foremost from the European 6G Flagship project Hexa-X-II as well some other projects. The approach is strongly based on European values, with sustainability as a key focus, including all dimensions, environmental, social, and economic sustainability. This book provides a cutting-edge insight into what 6G will be, from many of the experts who have created the technology.

This book provides a unique insight into not just the solutions but also the rational behind their selection. The 6G system will be designed as a platform, which provides services including and beyond communication, and this book provides the end-to-end blueprint of this system. This book has specific chapters on the different enablers and components of the 6G system, related to, for example, transceiver and radio design, architecture, artificial intelligence (AI)/machine learning (ML), and integrated sensing and communication (ISAC). The book also provides an overview of the demonstrators and proof-of-concepts, developed within Hexa-X-II, that showcase some of the expected performance of 6G.

As the technical standardization for 6G is set to begin in 2025, the book also outlines the potential paths from the project results towards standardization and further towards introduction to the markets. Overall, I expect that this book, written by many of the top professional of the European ICT sector, will be a reference point for the future research, standardization and development activities for 6G networks and, thus, extremely useful for professionals and academics.

6G IA Board Chair
March 26th, 2025

Colin Willcock

1

Introduction

Patrik Rugeland[1], Mikko A. Uusitalo[2], Sylvaine Kerboeuf[3], Jeroen Famaey[4], Stefan Wendt[5], Henk Wymeersch[6], Sokratis Barmpounakis[7], Hamed Farhadi[1], and Mauro R. Boldi[8]

[1]*Ericsson Research, Stockholm, Sweden*
[2]*Nokia Bell Labs, Espoo, Finland*
[3]*Nokia Bell Labs, Paris, Saclay, France*
[4]*IMEC, Antwerp, Belgium*
[5]*Orange, Châtillon, France*
[6]*Chalmers University of Technology, Gothenburg, Sweden*
[7]*WINGS ICT Solutions, Athens, Greece*
[8]*Telecom Italia, Rome, Italy*

The mobile networks have developed in the past half-century with a new generation every decade. The first generation (1G) was introduced in 1979 and relied on analogue radio to provide voice communication. The second generation (2G) was launched in 1991 and relied on digital communication enabling more efficient use of the radio spectrum. It also introduced mobile data services, including short messaging services (SMS) and general packet radio service (GPRS). In the third generation (3G), a global standard was introduced by the 3rd Generation Partnership Project (3GPP) in 2001, aiming to provide faster data transfer, but still relying on circuit-switched networking. The fourth generation (4G) was developed by 3GPP and was deployed in 2010, adopting a packet-switched network and introducing enhanced radio features such as carrier aggregation (CA) and multiple input multiple output (MIMO), significantly increasing the data rates and capacities. The fifth generation (5G) was also developed by 3GPP and was first deployed in 2019, further enhancing the mobile broadband (MBB) performance partly through the introduction of new frequency bands along with a multitude of other features and components. For instance, 5G was also targeting use cases related to machine-type communication (MTC) as well as ultra-reliable low-latency communication (URLLC). However, 5G was initially deployed as a non-standalone (NSA) variant, relying on 4G radio access network (RAN) for the control signalling, as well as the 4G core network (CN), known as the evolved packet core (EPC). It was only several years later that standalone (SA) 5G, connected to the 5G core network (5GC), began to be deployed, and in many regions, the NSA deployment is still the dominant deployment option. This has fragmented the market, delaying the adoption of SA 5G, and many of the envisioned new functionalities of 5G have yet to be realized. For the sixth generation (6G), a similar timeline (i.e. a decade after the previous generation) is expected, i.e. the first 6G deployments to arrive around 2030.

Although both 4G and 5G were initially designed to be forward compatible, allowing for continuous development and improvements, certain design choices made in the progressive design of those systems limited this ambition, necessitating non-backward compatible changes. For instance, 4G employed transmission of frequent reference signals, which were transmitted every 5 ms in every

6G to Build a Sustainable Future, First Edition. Edited by Mikko A. Uusitalo, Patrik Rugeland, Mauro R. Boldi, and Ahmad Nimr.

cell, regardless of the instantaneous traffic load. In 5G, this was modified to allow a much sparser transmission, up to every 160 ms, enabling significant power savings at low loads. Furthermore, 5G introduced concepts such as network slicing, allowing a network operator to provide guaranteed performance for a particular service in a specific area, by sectioning parts of the network off, which was not compatible with the 4G air interface. At the same time, it is equally important to leverage on the existing technology to enable as much reuse as well as smooth integration and interworking with previous generations as possible. For 6G, the generational shift provides a similar opportunity to remove and redesign features and signalling from previous generations to further improve performance and ease of use, both for end-user devices and Communicaion service providers (CSPs).

This book details the work and results from the European flagship 6G research project Hexa-X-II, describing the benefits and requirements posed by the envisioned 6G use cases as well as the design for the 6G end-to-end system architecture, taking into account lessons learned from previous generations. In addition, key technological enablers related to radio design and network architecture, as well as enablers related to novel beyond-communication features such as artificial intelligence (AI)/machine learning (ML) and integrated sensing and communication (ISAC), are explored. To complement the contributions from the Hexa-X-II project, specific sections are provided by other European SNS-JU projects describing, e.g. sensing by 6G-DISAC (2025) and 6G-MUSICAL (2025), non-terrestrial networks (NTNs) by 6G-NTN (2025), sub-networks by 6G-SHINE (2025), deterministic networking by DETERMINISTIC6G (2025), security by RIGOROUS (2025), data-intensive E2E security management by ROBUST-6G (2025), MIMO architectures and transmission schemes as well as photonic heterodyne sub-THz transceivers by TERA6G (2025), and ISAC by TERRAMETA (2025).

1.1 Why Do We Need 6G?

The current mobile networks are very capable of delivering the existing services such as MBB, videoconferencing/streaming, position-based mapping services as well as plain old telephone service (POTS). There are scant few plausible scenarios that would require much higher bitrates than the peak bitrates offered by 5G. For instance, a single stream of 4K video requires approximately 25 Mbps, which can be compared to the 20 Gbps peak data rate of 5G, an 800-fold difference. Of course, there is hardly anyone, outside of a dedicated speed test, whoever experiences the peak data rate, but even the average performance is well equipped to handle these scenarios. However, the 5G coverage is not ubiquitous, and the data traffic is provided on a best effort basis, i.e. there are no guarantees that the bit rate may not suddenly drop due to, e.g. poor coverage or congestion in the network. Here, 6G offers an opportunity to introduce enhanced reliability in the services, both in terms of deterministic performance and coverage extension by incorporating NTNs with global coverage. There are also predictions showing significant data traffic growth via a combination of new services, device expansion, and a surge in AI usage. We should be prepared to address this and do it in a sustainable way.

Furthermore, there are several nascent technologies being considered for 6G, e.g. ISAC, which can fundamentally change how we interact with the digital landscape by collecting and offering a digital representation of the surrounding environment. Another example is artificial AI/ML, which has leapfrogged its advances in recent years, promising unprecedented opportunities in network optimization and automation as well as new services to the end users.

Similarly, technological advances, e.g. in the hardware, can provide performance improvements, e.g. incorporating distributed MIMO (D-MIMO), a technique to transmit and receive to and from multiple devices from multiple distributed radio antenna elements at the same time. Another promising technology is the so-called energy neutral devices, which are unplugged, battery-less devices that harvest energy from the surrounding environment, e.g. from radio frequency (RF) wireless power transfer (WPT) on the same frequencies used to carry the data.

As the digital and mobile services have become enmeshed in our societies, the push has so far been to keep increasing the bit rate to support more and more services. At the same time, societies face increasing environmental, ecological, and societal challenges, with climate change and societal unrest that need to be addressed, where mobile digital services can offer significant benefits for sustainable development. However, to address these sustainability challenges, the networks would need to take this into account at the onset, instead of adding it on top of the existing networks, and this is where 6G offers an advantage. By designing 6G with efficient technologies, integrating sensing, NTN and AI/ML from the beginning, it will be possible to develop a mobile system offering fundamentally new services, while operating more autonomously and more efficiently, reducing the overall power consumption and operational costs. In particular, if the 6G air interface were simply an extended version of 5G new radio (NR), it would be encumbered by significant backward compatibility requirements, e.g. supporting NSA, higher-layer split (HLS), and ever-growing extensions to the configuration messages. By allowing a clean slate for the 6G air interface, the best aspects of the legacy systems can be adopted, while inefficient or unnecessary features can be pruned. As a contrast, 5GC introduced a so-called service-based architecture, which already allows modular extensions to the network and can easily be extended to simultaneously support both 5G and 6G RAN, allowing a smooth migration from 5G to 6G.

1.2 Global View on Development Towards 6G

The work towards 6G began in 2019, with initiatives from the Finnish 6GFlagship initiative, releasing a number of white papers (Latva-Aho and Leppänen 2019). In addition, several researchers and industrial players published white papers or articles outlining 6G research ambitions (Rappaport et al. 2019; NTT Docomo 2020; Wikström et al. 2020; Ziegler et al. 2020). In 2021, the European 6G flagship project Hexa-X was launched, developing a large number of technical enablers, ranging from radio aspects of, e.g. ISAC, architectural aspects related to network flexibility, efficiency, and intelligence, AI/ML as well as management and orchestration (M&O). In addition, the Hexa-X project introduced the concept of *Key Value Indicators* (KVIs), which should be seen as a potential means to quantify the impact that the 6G system can deliver to be balanced against the traditional *Key Performance Indicators* (KPIs). Beginning of 2023 and following the footsteps of Hexa-X, the Hexa-X-II project took over as the European 6G flagship project, continuing the research on technical components, as well as creating an end-to-end (E2E) system view. In the Hexa-X project, it has been envisioned that 6G will be able to intertwine the physical, digital, and human world, empowering individuals, industries, and societies with digital tools while ensuring environmental, social, and economic sustainability. This is illustrated in Figure 1.1.

In standardization, the International Telecommunication Union (ITU) Radiocommunication Sector (ITU-R) outlined the framework for 6G, starting with trends, usage scenarios, and required capabilities (ITU-R 2023).

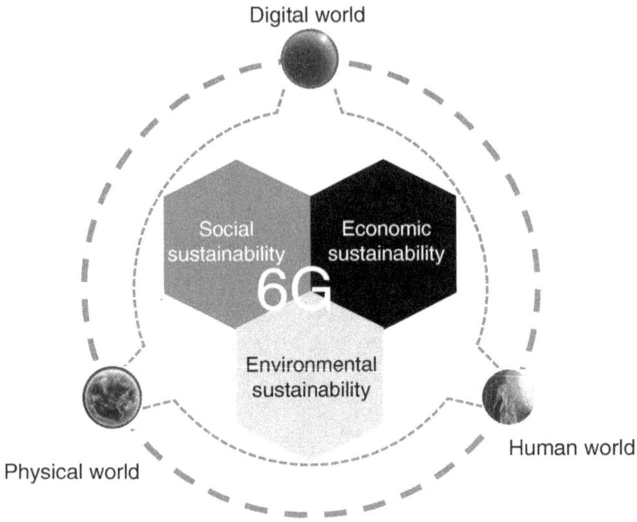

Figure 1.1 Hexa-X-II vision for 6G to intertwine the Human, Physical, and Digital worlds.

The usage scenarios expand upon the 5G usage scenarios of enhanced MBB, MTC, and URLLC into immersive communication, massive communication, and hyper reliable low-latency communication. In addition, three new usage scenarios were introduced, namely ISAC, ubiquitous connectivity, and AI and communication, capturing the envisioned novel beyond-communication functionalities.

The requirements included those from IMT-2020 (specifying the requirements for 5G) but allowed extensions of the values to more extreme values for 6G. In addition, the requirements also introduced new categories to capture the new usage scenarios.

In 3GPP, the Service and System Aspects working group 1 (SA1) began its work towards 6G with a workshop on use cases in May 2024, gathering input from initiatives outside the regular 3GPP members. For the RAN groups, RAN1 working on the physical layer will begin the first 6G study item of release 20 (Rel-20) during quarter 3 (Q3) of 2025, and RAN2 working on Layer 2 and Layer 3 will begin working on 6G in Rel-20 in Q4 2025. It is planned that the technical specification for 6G will be completed in Rel-21 in 2028, with the initial 6G deployments ready in 2030.

The 6G will need a combination of frequency ranges to meet the coverage and capacity requirements and to serve new emerging IMT-2030/6G use cases. It has been estimated that at least 500 MHz of new wide-area spectrum is needed per network, in addition to the reuse of existing spectrum. The ITU World Radiocommunication Conference 2023 (WRC23) defined a number of frequency ranges to be studied. These bands within WRC-27 agenda include 4.4–4.8, 7.125–8.4, and 14.8–15.35 GHz.

Globally, there have been several initiatives towards 6G. In Europe, the 5G Infrastructure Public Private Partnership (5G-PPP), which funded the Hexa-X project, also funded several other, smaller 6G projects. The follow-up programme, the smart networks and services joint undertaking (SNS-JU) has, apart from funding Hexa-X-II, also funded over 60 other 6G projects. There are also a few larger national 6G initiatives in Europe, e.g. the 6G Platform Germany, the Dutch Future Network Services (FNS), or the Finnish 6GBridge to name a few. Further details on the initiatives towards 6G are presented in Chapter 9.

1.3 Structure of the Rest of the Book

Chapter 2 introduces sustainability, emphasizing its three pillars: economic, social, and environmental. It defines 'value' in relation to 6G and outlines a method for analysing and assessing this value. It identifies various stakeholders within the 6G ecosystem, highlighting their roles and interactions. An overview of the 6G use case families and specific use cases is presented, as defined by the Hexa-X-II European project. The discussion then focuses on value analysis, using the example of cooperating mobile robots. Finally, it addresses business modelling for the 6G system, continuing with the cooperating mobile robots example. Chapter 3 focuses on the design of a sustainable 6G platform, prioritizing key values-based design principles, including simplification, enhanced security and privacy, and environmental sustainability. It presents a comprehensive blueprint for a sustainable 6G platform, outlining a multi-layered framework for the 6G E2E system incorporating pervasive functionalities across all layers to enable full-scale automation, seamless service management, and sophisticated AI capabilities. The chapter also addresses the growing security challenges posed by advanced technologies, including AI/ML, evolving cyber-attacks, and the threat of quantum computing to break current encryption, introducing new concepts for security by design and their integration into the 6G platform blueprint. Chapter 4 presents technical enablers for the 6G radio transceiver design that will make it possible to achieve the stringent requirements for 6G. It offers insights into the general radio design process, framework, architecture, and deployment. Subsequent sections cover specific aspects of 6G radio design, such as transceivers and antennas (covering sub-THz and reconfigurable intelligent surfaces (RIS) hardware design), channel and hardware non-ideality models, MIMO architectures and transmission schemes, as well as 6G devices (covering system-on-chip integration and energy-neutral device design). Chapter 5 describes the architectural enablers that will enhance the performance of the network and incorporate the new connectivity options as well as novel beyond-communication functions. Chapter 6 explores the role of AI/ML in advancing 6G networks, emphasizing their impact across multiple layers of the 6G system. It discusses the motivations for AI/ML adoption within 6G, including data-driven architecture, enhancements in physical layer performance, and AI-driven M&O. The chapter also addresses trustworthiness, highlighting AI/ML's role in security, privacy, and reliability. Additionally, it covers key enablers such as AI-driven radio air interface, network optimization strategies, and intent-based service management, alongside relevant standardization efforts within 3GPP and O-RAN Alliance to ensure interoperability and scalability in future networks. Chapter 7 explores ISAC as a key 6G enabler, covering its use cases, architectural design, physical layer, signal processing, and hardware challenges. It examines sensing configurations, AI-driven enhancements, data representation, and self-interference suppression, highlighting opportunities and challenges in security, privacy, and standardization. Chapter 8 explores the early validation of 6G concepts through proof-of-concept implementations carried out in the context of the Hexa-X-II project. The chapter focuses on enablers such as advanced M&O, beyond-communication functions, and flexible topologies while showcasing real-world use cases leveraging collaborative robot-powered industrial operations. Chapter 9 outlines the path forward towards the development and deployment of 6G, presenting also the status of the standardization activities for 6G and a snapshot of the initiatives to research around the main directions towards the new system. Finally, Chapter 10 describes the conclusions that can be drawn from this book.

Disclaimer

Funded by the European Union. The views and opinions expressed are however those of the author(s) only and do not necessarily reflect the views of *the project Consortiums and their participating Partners involved in establishing this book*, nor those of the European Union or Horizon Europe SNS JU. Neither the European Union nor the granting authority can be held responsible for them.

Acronyms and Abbreviations

1G/2G/3G/4G/5G/6G	First/second/third/fourth/fifth/sixth generation
3GPP	3rd Generation Partnership Project
5GC	5G core network
5G-PPP	5G Infrastructure Public Private Partnership
AI	Artificial intelligence
CA	Carrier aggregation
CN	Core network
CSP	Communicaion service provider
E2E	End to end
FNS	Future network services
GPRS	General packet radio service
HLS	Higher-layer split
ISAC	Integrated sensing and communication
KPI	Key performance indicator
KVI	Key value indicator
M&O	Management and orchestration
MBB	Mobile broadband
MIMO	Multiple input multiple output
ML	Machine learning
MTC	Machine-type communication
NSA	Non-standalone
NTN	Non-terrestrial network
POTS	Plain old telephone service
Q3/Q4	Quarter 3/4
RAN	Radio access network
RF	Radio frequency
RIS	Reconfigurable intelligent surface
SA	Standalone
SA1	System aspects working group 1
SMS	Short messaging services
SNS-JU	Smart networks and services joint undertaking
URLLC	Ultra-reliable low-latency communication
WPT	Wireless power transfer
WRC	ITU World Radiocommunication Conference

References

6G-DISAC (Distributed Intelligent Sensing and Communication for 6G) (2025). Grant ID: 101139130. [Online]. Available at: www.6gdisac-project.eu (accessed 25 August 2025).

6G-MUSICAL (6G-Multiband Wireless and Optical Signalling for Integrated Communication, Sensing and Localisation) (2025). Grant ID: 101139176. [Online]. Available at: www.6gmusical.eu (accessed 25 August 2025).

6G-NTN (6G Non-Terrestrial Networks) (2025). Grant ID: 101096479. [Online]. Available at: www.6g-ntn.eu (accessed 25 August 2025).

6G-SHINE (6G Short Range Extreme Communication in Entities) (2025). Grant ID: 101095738. [Online]. Available at: www.6gshine.eu (accessed 25 August 2025).

DETERMINISTIC6G (Deterministic End-to-End Communication with 6G) (2025). Grant ID: 101096504. [Online]. Available at: www.deterministic6g.eu (accessed 25 August 2025).

ITU-R (International Telecommunication Union, Radio Communication Sector) (2023). Framework and overall objective of the future development of IMT for 2030 and beyond. Recommendation ITU-R M.2160-0. [Online]. Available at: https://www.itu.int/rec/R-REC-M.2160/en (accessed 25 August 2025).

Latva-Aho, M. and Leppänen, K. (eds.) (2019). Key drivers and research challenges for 6G ubiquitous wireless intelligence. 6G Flagship White Paper. [Online]. Available at: https://urn.fi/URN:ISBN:9789526223544 (accessed 25 August 2025).

NTT Docomo (2020). 5G Evolution and 6G. NTT DOCOMO White Paper. [Online]. Available at: https://www.docomo.ne.jp/english/binary/pdf/corporate/technology/whitepaper_6g/DOCOMO_6G_White_PaperEN_20200124.pdf (accessed 31 March 2025).

Rappaport, T.S., Xing, Y., Kanhere, O. et al. (2019). Wireless communications and applications above 100 GHz: opportunities and challenges for 6G and beyond. *IEEE Access* 7: 78729–78757.

RIGOROUS (Secure Design and Deployment of Trustworthy Continuum Computing 6G Services) (2025). Grant ID: 101095933. [Online]. Available at: https://rigourous.eu/ (accessed 25 August 2025).

ROBUST-6G (Smart Automated, and Reliable Security Service Platform for 6G) (2025). Grant ID: 101139068. [Online]. Available at: https://robust-6g.eu/ (accessed 25 August 2025).

TERA6G (TERAhertz Integrated Systems Enabling 6G Terabit-Per-Second) (2025). Grant ID: 101096949. [Online]. Available at: https://www.uc3m.es/research/TERA6G (accessed 25 August 2025).

TERRAMETA (TERahertz ReconfigurAble METAsurfaces for Ultra-High Rate Wireless Communications) (2025). Grant ID: 101097101. [Online]. Available at: https://terrameta-project.eu/ (accessed 25 August 2025).

Wikström, G., Peisa, J., Rugeland, P. et al. (2020). Challenges and technologies for 6G. *Proceedings of the 2020 2nd 6G Wireless Summit (6G SUMMIT).*

Ziegler, V., Viswanathan, H., Flinck, H. et al. (2020). 6G architecture to connect the worlds. *IEEE Access* 8: 173508–173520.

2

Value of 6G

Stefan Wendt[1], Hanne-Stine Hallingby[2], Cristóbal Vinagre Zúñiga[3], Marja Matinmikko-Blue[4], Arturo Basaure[4], Maurizio Cecchi[5], Ishita Mishra[5], Ana Pereira[6], and Anastasia Yagafarova[3]

[1] *Orange Innovation, Châtillon, France*
[2] *Telenor, Oslo, Norway*
[3] *TNO, The Hague, The Netherlands*
[4] *University of Oulu, Oulu, Finland*
[5] *Institute PIIU Roma, Italy*
[6] *Ubiwhere, Aveiro, Portugal*

As we transition from previous generations of mobile communication to the anticipated sixth generation (6G), it is essential to understand the newly rising value concept that underpins this new technology. Historically, mobile networks have primarily focused on performance principles, emphasizing speed and capacity. Each generation, from 1G to 5G, has aimed to enhance these performance metrics, enabling faster data transfer rates and increased connectivity for a growing number of devices. The International Telecommunication Union (ITU) has played a pivotal role in this evolution, defining standards through frameworks like IMT-2000 for 3G (International Telecommunication Union 2020), IMT-Advanced for 4G (International Telecommunication Union 2008), IMT-2020 for 5G (International Telecommunication Union 2015), and the ongoing IMT-2030 for 6G (International Telecommunication Union 2023).

Before 5G, the main criteria for mobile networks were centred around increasing data rates, primarily focusing on mobile broadband services for private customers. 5G introduced a paradigm shift by emphasizing the need to improve reliability and latency, as well as addressing Internet of Things (IoT) requirements, thereby targeting business customers. This multifaceted approach is often referred to as the '5G triangle', which balances enhanced mobile broadband (eMBB), ultra-reliable low latency communications (URLLC), and massive machine-type communications (mMTC).

Notably, IMT-2030 marks a significant shift by mentioning sustainability for the first time, recognizing its importance in the development of future networks. However, integrating sustainability into the framework has proven challenging due to the varying perspectives and priorities across different regions of the world. This shift acknowledges that future communication networks must not only deliver superior performance but also address environmental and societal needs. By prioritizing these broader values, 6G aims to create a more interconnected and equitable digital landscape, ultimately transforming how individuals and communities interact with technology and with one another.

6G to Build a Sustainable Future, First Edition. Edited by Mikko A. Uusitalo, Patrik Rugeland, Mauro R. Boldi, and Ahmad Nimr.

The remainder of this chapter is structured as follows:

- **Section 2.1: Sustainability and values**
  This section introduces the concept of sustainability, outlining its three pillars, and defines the term *value* in relation to 6G technology. It also discusses a method for analysing and assessing value.
- **Section 2.2: Stakeholders in the 6G system**
  This section outlines the various stakeholders involved in the 6G ecosystem.
- **Section 2.3: 6G use-case families and use cases**
  Here an overview of the 6G use-case families and specific use cases is provided, as defined by the Hexa-X-II European project (Hexa-X-II Project Consortium 2023–2025).
- **Section 2.4: 6G use case and value design | cooperating mobile robots**
  This section delves into value analysis, applying it to the use case of cooperating mobile robots, as an example.
- **Section 2.5: Business models**
  This section addresses business modelling for the 6G system in general, and using the cooperating mobile robots use case still as an example.
- **Section 2.6: Conclusions**
  Chapter 2 wraps up with its conclusions.

2.1 Sustainability and Values

Sustainability is often perceived solely in terms of environmental issues, particularly related to energy consumption. However, it encompasses much more. The Hexa-X-II project defines sustainability through three interconnected pillars, environmental, social, and economic sustainability, which form the basis for sustainable development. For a legitimate and successful development and implementation of 6G technologies, it is essential to respect and balance all three pillars while addressing the expectations of stakeholders.

Environmental sustainability aims to preserve natural resources and ecosystems by promoting responsible use and protecting biodiversity. It seeks to minimize waste and pollution, ensuring human activities do not harm the environment. Key strategies include adopting renewable energy, implementing eco-friendly practices and supportive policies, and developing innovative production methods potentially enabled by 6G technology.

Social sustainability focuses on creating inclusive, non-discriminatory structures and services that promote equal access to opportunities and resources. It fosters social cohesion through trust, cooperation, and shared values, emphasizing digital inclusion. Achieving social sustainability requires policies addressing social inequalities and is essential for the well-being of current and future generations.

Economic sustainability focuses on building resilient, adaptable, and inclusive economies that promote growth without compromising the environmental integrity or social equity. It aims for long-term stability by fostering conditions for businesses and communities to thrive. Achieving economic sustainability requires balancing economic, social, and environmental concerns in decision-making and integrating sustainable practices into economic policies.

Moreover, sustainability can be divided into two dimensions, namely, 6G for sustainability and sustainable 6G. 6G for Sustainability emphasizes the potential of 6G to enable or help various sectors enhance their sustainability efforts. Sustainable 6G refers to ensuring that the sixth generation of mobile communications is sustainable. It refers to a sustainable construction of the system itself, including all components. For each of the two dimensions, the Hexa-X-II (Hexa-X-II Project Consortium 2023–2025) project has considered the sustainability impact in the mentioned three pillars: environmental, social, and economic. The overall objective is to minimize the environmental impact and secure long-term economic growth with respect to environmental and social needs.

While the assessment of sustainable 6G throughout its life cycle (LCA) is well-established, albeit complex, the evaluation of 6G for sustainability, and in general of Information and Communication Technology (ICT) for sustainability, is a less mature area. The challenge lies in linking desired sustainability outcomes of both dimensions to the design of next-generation mobile technologies. The performance-based approach, which has been the primary focus so far, is to be complemented by a value-based approach, which reflects sustainability considerations in the mentioned three pillars. Indeed, the existing approach used in standardization to translate requirements into measurable objectives focuses mainly on performance through the establishment and utilization of key performance indicators (KPIs). Currently, KPIs are primarily derived from technical and business goals and are quantifiable in an objective manner. However, values may be subjective, context-dependent, and very challenging to quantify.

To address this challenge, the Hexa-X-II project has defined a set of values, named high-level human and planetary goals (H&PG), spanning over the three sustainability pillars. These values as goals align with the United Nations Sustainable Development Goals (UN SDGs) (United Nations 2015) or a subset of them.

These goals should be connected to specific use cases. The impact of 6G use cases on these goals across the three sustainability pillars must then be evaluated. The outcome, i.e. the impact of the implementation of the use case with the 6G technology and its usage, on the high-level H&PG is defined as a key value (KV). The overall aim is to maximize positive impacts on these KVs and minimize negative ones.

To assess a potential positive or a potential negative impact of these KVs, indicators need to be defined. Key value indicators (KVIs) are then a set of indicators, which can, directly or indirectly, measure a particular KV. A KVI is a qualitative or quantitative indicator for assessing a KV. The purpose of KVIs is to gauge the impact from the execution of a use case on the 6G technology in all three sustainability pillars.

Finally, as a use case is designed to interact within an ecosystem, many KVs (and consequently their indicators) fall beyond the reach of what can be influenced by 6G design. These are called use-case key value indicators (UCKVIs) and assess the impact of the application of a use case. A UCKVI is a quantitative or qualitative indicator. On the other hand, those KVIs that can be influenced by the technical design of 6G are called enabler key value indicators (EKVIs). EKVIs assess the impact of the technical enabler, i.e. the 6G technology applied to a use case. They are also quantitative or qualitative indicators. Given that EKVIs are linked to the 6G system, it is possible that EKVIs can be mapped to an existing technical KPI, a new technical KPI or that it remains classified as EKVI.

Figure 2.1 depicts the transition from goals to key performance and KVs, leading to the key performance and KV indicators. It also demonstrates how the value domain completes the performance domain.

The Hexa-X-II project introduced an innovative approach to use case and sustainability analysis, focusing on a use case-oriented methodology. This involves evaluating use cases to determine their contributions towards achieving H&PG or their impact on them. Use cases may be specifically designed to address these goals. Conversely, existing use cases may be retrospectively associated with particular H&PG. The primary objective in both scenarios is to conduct a sustainability analysis. The approach is depicted in Figure 2.2 and will be explained below.

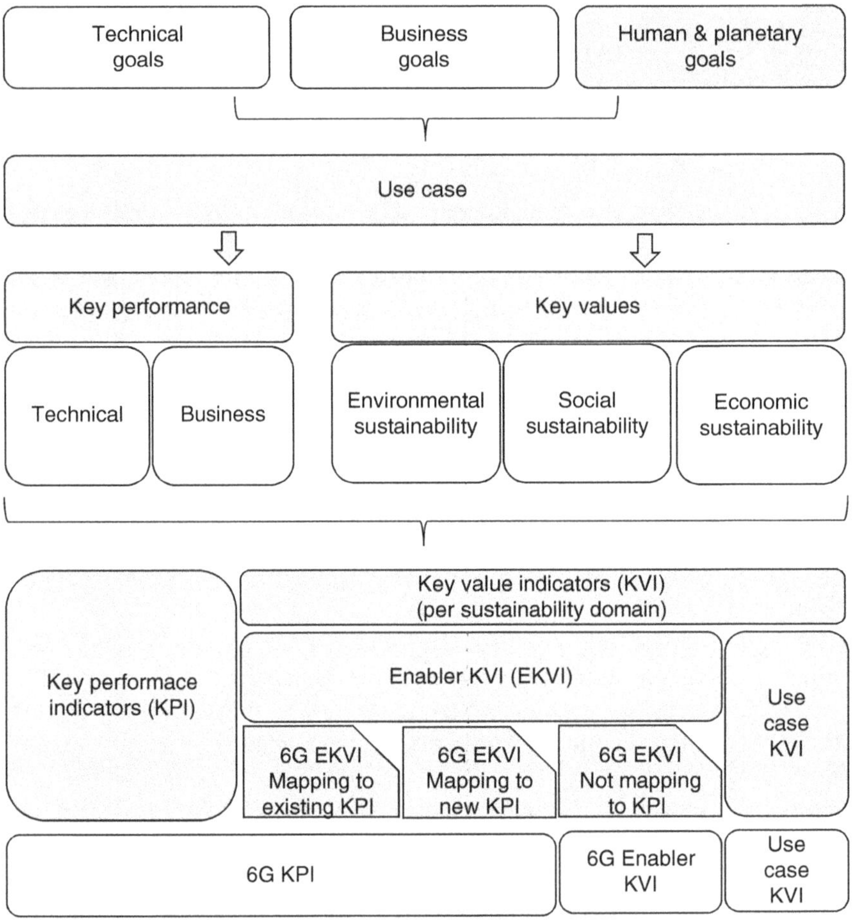

Figure 2.1 Goals, key values, and indicators.

In a first step, it is important to link the H&PG to a given use case. Hexa-X-II proposes a list of 15 items listed in the following.

1) **Foster Resilience**
   'Resilience is the capacity of a system, be it an individual, a forest, a city or an economy, to deal with change and continue to develop' (Stockholm Resilience Centre 2015). It is linked to the environmental, social, and economic sustainability pillars and connected to all UN SDGs.
2) **Protect Biodiversity (land, water, and air)**
   Biodiversity encompasses the variety within species, among different species, and across ecosystems, highlighting the range of plants, animals, and ecological systems. It is linked to the environmental, social, and economic sustainability pillars and connected to UN SDG #14 and #15.
3) **Curb Climate Change**
   Climate change is affected by human activities, particularly those linked to the rise in greenhouse gas (GHG) emissions, leading to an increase in global temperatures. This goal belongs to the environmental sustainability pillar and is linked to UN SDG #13.

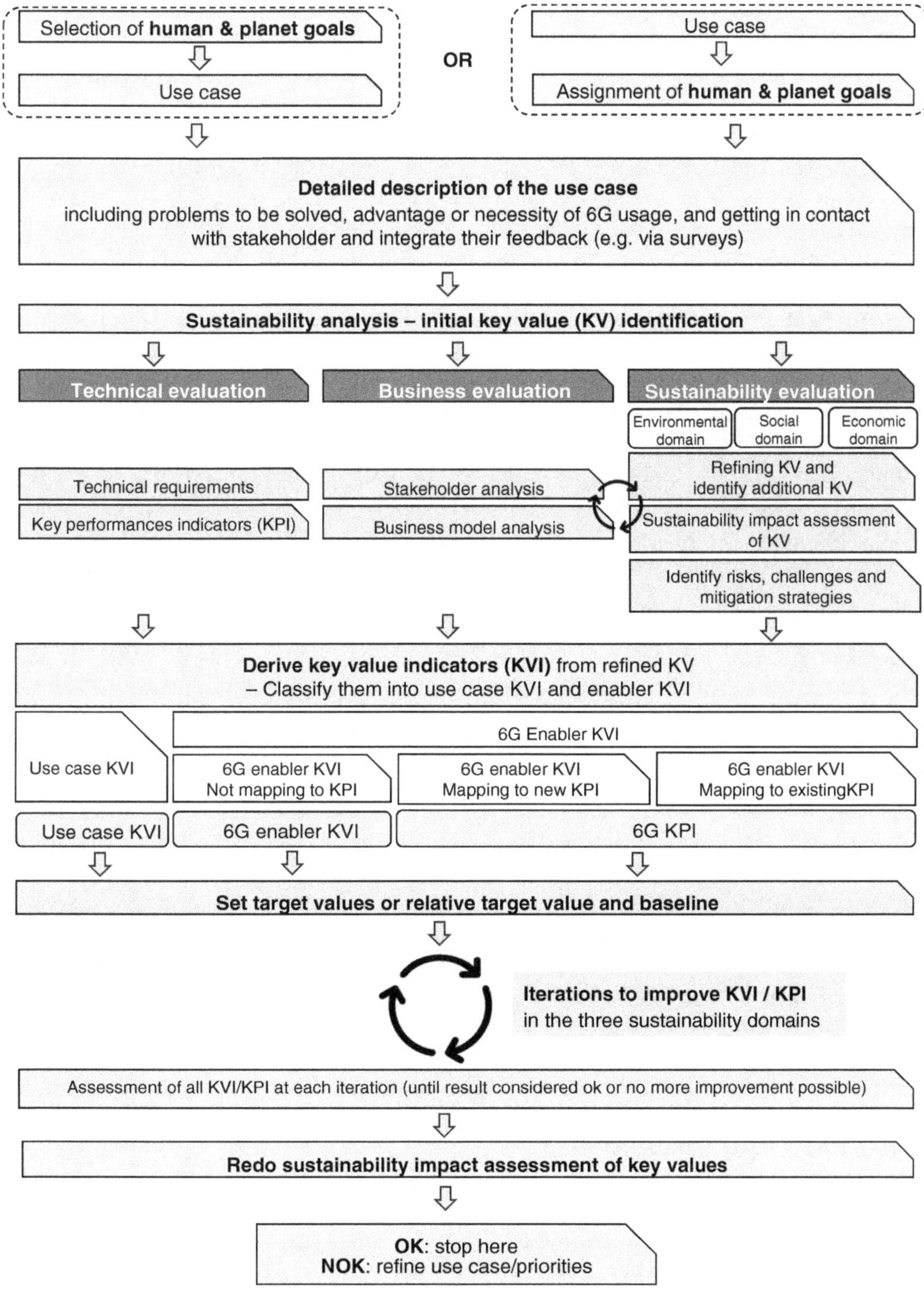

Figure 2.2 Use case and sustainability analysis process.

4) **Preserve Natural Resources**
   Natural resources are raw materials located on or beneath the Earth's surface. They are formed without any human intervention and cannot be artificially created (e.g. gold). This goal belongs to the environmental sustainability pillar and is linked to UN SDG #6 and #12.
5) **Minimize Waste**
   In conjunction with goal number four, there is an essential need to minimize waste. An example of this is the proper management of electronic waste generated in ICT and other sectors. This goal belongs to the environmental sustainability pillar and is linked to UN SDG #12, #14, and #15.
6) **Minimize Emissions and Disposals (to Water, Air, and Soil)**
   Minimizing all forms of emissions and disposals as well as their interactions with the atmosphere, biosphere, and geosphere emphasizes the importance of the processes throughout the entire life cycle of ICT products and services. This goal belongs to the environmental sustainability pillar and is linked to UN SDG #13, #14, and #15.
7) **Protect Human Rights**
   'Human rights are rights inherent to all human beings, regardless of race, sex, nationality, ethnicity, language, religion, or any other status. Human rights include the right to life and liberty, freedom from slavery and torture, freedom of opinion and expression, the right to work and education, and many more. Everyone is entitled to these rights, without discrimination' (United Nations n.d.). This goal belongs to the social sustainability pillar and is linked to UN SDG #2, #3, #4, #5, #8, #11, and #16.
8) **Increase Digital Inclusion**
   Digital inclusion means ensuring that all individuals have access and the ability to effectively use ICT related services. This goal belongs to the social sustainability pillar and is linked to UN SDG #4, #5, #8, #9, and #10.
9) **Ensure ICT Trustworthiness**
   Trustworthiness encompasses security and privacy. It is based on the confidence that individuals, data, and devices will operate and behave as anticipated. This goal belongs to the social and economic sustainability pillars and is linked to UN SDG #9 and #16.
10) **Enhance Health and Well-Being**
    Health is meant as a state of complete physical, mental, and social well-being, rather than just the absence of disease or illness. This goal belongs to the social and environmental sustainability pillars and is linked to UN SDG #2, #3, and #6.
11) **Foster Decent Employment**
    Decent employment refers to the availability of opportunities for all individuals to engage in productive work that provides a fair income, job security, and social protection for families. This goal belongs to the social sustainability pillar and is linked to UN SDG #8.
12) **Supporting Quality Education**
    Education is a fundamental necessity for individuals to enhance their lives and contribute to communities, societies, and economies. This goal belongs to the social sustainability pillar and is linked to UN SDG #4.
13) **Build Long-Term Sustainable Economic Growth**
    Economic growth is sustainable when it does not negatively impact social equity and the environment, both in the short term and long term. This goal belongs to the economic sustainability pillar and is linked to UN SDG #8, #9, #10, and #12.
14) **Ensure Transparency**
    Transparency is the process of making information about current conditions, decisions, and actions accessible, visible, and understandable. It is essential for the governance of both

companies and the public sector, as well as for technological solutions that foster stability, reliability, mutual trust, and public support. This goal belongs to the social and economic sustainability pillars and is linked to UN SDG #8, #9, #10, #11, #12, and #16.

15) **Ensure Ethical Conduct**
ICT ethics aims to explore the application of ethical principles to the development and (mis)use of systems that incorporate digital and electronic components. This goal belongs to the social and economic sustainability pillars. It is linked to UN SDG #12 and #16.

This 15-item list is based on sustainability focus areas outlined in the Hexa-X-II project deliverables D1.1 (Wendt 2023) and D1.2 (Almodóvar and Vinagre 2023) and establishing a connection to the UN SDGs. It is described in project deliverable D1.4 (Ramos 2025).

In a second step, it is then important to provide a detailed description of the use case (cf. Figure 2.2). The description should also address the problems to be tackled, the challenges to be faced, and detail how the 6G technology can bring advantage to this situation and why 6G should be used. The description should be complemented with surveys and feedback from external stakeholders, including civil society. Surveys and feedback from external stakeholders will help to obtain a first broad picture and to identify the needs of stakeholders that the use case is supposed to address.

Surveys and feedback from stakeholders also help to identify, in a third step, an initial set of KVs (cf. Figure 2.2). As mentioned before, KVs represent the impact of the application of the use case on the High-level H&PG. Objective is to reach the goals and thus improve the KVs in this direction.

In a fourth step, in addition to a technical evaluation, which represents the well-known analysis for a use case (based usually on KPIs), Hexa-X-II carried out a business evaluation (based on the developed business model and a stakeholder analysis) and a sustainability analysis of the given use case (taking into account the environmental, social, and economic sustainability domains) as depicted in Figure 2.2.

The establishment of KVs is an iterative process, and it is undertaken in parallel with the stakeholders and the business model analysis. As a result, the refined list of KVs can be compiled. This is not an exhaustive list but captures the most expected values. It is therefore to mention that all stakeholders (as identified in the stakeholder analysis) should be involved in brainstorming sessions to identify possible KVs. Furthermore, it is important that a high number of people of each stakeholder participate in these sessions to come up with the broadest possible vision on KVs.

Given the possibly large list of KVs, it is important to assess the impact of each KV. The overall objective is detecting KVs on which focusing first, having a high or higher impact. Hexa-X-II proposed an evaluation methodology described in the deliverable D1.4 (Ramos 2025), but any other assessment methodology targeting the same purpose could also be used. Subsequently, the most prominent risks can be identified, and mitigation strategies can be suggested to guarantee sustainability requirements could be met.

In a fifth step, for each KV (out of the refined list), one or more KVIs have then to be worked out for each KV. These indicators are used to measure, qualitatively or quantitatively, the KV. The indicators can subsequently be grouped into UCKVIs (which are, as stated earlier, a quantitative or qualitative indicator used to assess the impact of a use case or its application) and EKVIs (being a quantitative or qualitative indicator to assess the impact of the technical enabler, i.e. the technology applied to a use case). Finally, the EKVIs themselves may be classified in three categories: either they could be mapped to an existing KPI, to a new KPI, or they stay as EKVIs. Please refer to the illustration in Figure 2.2.

In a sixth step, to optimize the system, all KVIs and KPIs should be checked against a target value that is to be defined. Then, each KVI and KPI should be improved in an iterative process (cf. Figure 2.2). Please note that especially an improvement for KVIs is targeted, but that KVIs improvement may also impact on KPIs and consequently KVIs and KPIs have to be taken into account. It is very likely that compromises are the best solution possible.

Running through continuous improvements towards the target values of KVIs and KPIs, a new assessment of the associated KVs should be triggered at each iteration. Indeed, improvements in KVIs should consequently lead to an improvement of the KVs and thus in an improvement of the overall situation. At each iteration, a decision is to be done if further improvement is possible or wanted or if the obtained situation is acceptable. This iterative process should also be done regularly throughout the life cycle of the use case on the 6G system. The iterative process was out of scope within the Hexa-X-II project and further work must be dedicated to it.

In Section 2.3, the different use-case families and 6G use cases identified in the Hexa-X-II project (and identified as having a significant impact on at least one sustainability pillar) are presented. In Section 2.4, 6G use case and value design | cooperating mobile robots (CMRs), an application of the described use case and sustainability analysis to one example use case is provided.

2.2 Stakeholders in the 6G System

As 6G aims to create a more interconnected and equitable digital landscape, it will ultimately transform how businesses and communities work and interact with technology and among each other. Therefore, it becomes crucial to understand how the potential stakeholders in the 6G system view this process and what their understanding of it is. These stakeholders can be divided into different profiles or groups that have direct interest in or will be affected by the development and implementation of this digital landscape. These stakeholders play a key role in decision-making and use, therefore having the power to influence the success or failure of wider adoption of a particular technological development.

The stakeholders in the 6G system can be divided into market ecosystem stakeholders, technology ecosystem stakeholders, spectrum ecosystem stakeholders, policy ecosystem stakeholders, and public ecosystem stakeholders. Hexa-X-II suggested potential business ecosystems for the six 6G use cases, and associated stakeholders (Ramos 2025). The ecosystems for each use case and the stakeholders therein are not mutually exclusive.

The **6G use-case business ecosystems** concentrate on the development of 6G use-case value propositions and business models centred around these stakeholders – also see Section 2.5 on business modelling. Even though the business modelling perspective was the focus, sustainability aspects and risks and challenges brought forward new stakeholders as provided in Hexa-X-II deliverable D1.3 (Bantouna 2024).

Market ecosystem stakeholders encompass the different entities that develop, deploy, and wish to commercialize and benefit from the implementation of 6G technology. These include entities such as network providers and operators, providers of equipment, and applications or components that will leverage the technology.

Technology ecosystem stakeholders represent the developers, providers, and operators of the 6G networks that bring the necessary expertise to develop, implement, and maintain the technology, contributing to standards' definition and, more recently, adjusting contributions and work plans considering sustainability ambitions and goals. Additionally, standardization associations and clusters can also be considered as an important part of the technology ecosystem stakeholders, due to their role in defining, coordination and even disseminating new technological solutions.

Spectrum ecosystem stakeholders are quite fundamental, as Spectrum is scarce and a crucially necessary resource for 6G, which is subject to regulations, licensing, and competition related to division or sharing. Therefore, entities like the EU Radio Spectrum Policy Group and Working Groups on Spectrum from different associations are key stakeholders for these discussions and definitions. Moreover, market, technology, and policy ecosystem stakeholders will play a role in the evolution of spectrum allocation.

The stakeholder goals for spectrum are suggested in the Hexa-X-II D1.1 project deliverable (Wendt 2023).

Policy ecosystem stakeholders are intrinsically connected to the standardization contributors of the technology ecosystem and with the Spectrum ecosystem stakeholders, as they are involved in creating and establishing regulations and policies on how to implement and maintain the solution, necessary infrastructure, and other related topics. Within this ecosystem, some of the considered stakeholders include BEREC, SNS-JU, 6G-IA, and the United Nation's frameworks and organizations.

Finally, the **public ecosystem stakeholders** are often perceived as referring only to the general-public and individuals with interest in utilizing technological solutions. However, in the context of technological adoption, different groups such as public authorities or government bodies, industry representatives, academia representatives, and civil society are involved.

All these stakeholders have different perceptions and perspectives on technology, its advantages, and shortcomings, but all are crucial to ensure a smooth implementation and to increase the rate of adoption.

The stakeholder analysis revealed that the true value of 6G may extend far beyond speed and connectivity. Through the exchanges with diverse stakeholders, 6G development is being shaped to ensure that it is not just another technological upgrade but a transformative force that aligns with social needs, creating a future where digital advancements serve humanity, rather than the other way around. By enabling hyper-efficient, intelligent, and sustainable communication networks, 6G may unlock economic opportunities through advanced automation, new business models, and enhanced industrial efficiency. Socially, it will bridge the digital divide, support next-generation healthcare and education, and foster inclusive growth. Environmentally, 6G will drive energy-efficient infrastructure, reduce carbon footprints, and support real-time environmental monitoring.

Moreover, the role of stakeholders in shaping 6G development is pivotal in ensuring that the technology is aligned with needs and social priorities. The Hexa-X-II project exemplifies how meaningful stakeholder engagement can drive innovation that benefits all, ensuring that 6G contributes to a more inclusive, equitable, and resilient digital future. The learnings from the analysis of different ecosystems are elaborated below.

2.2.1 6G Use-Case Business Ecosystem Stakeholders

Hexa-X-II envisions the future by six 6G use-case families, with six representative use cases (Almodóvar and Vinagre 2023). The use cases have been the main tool to identify all potential stakeholder roles, both existing and new. Stakeholders' role in an ecosystem, their struggles, and potential benefits from being part of the 6G business ecosystem were systematically elaborated. The roles of stakeholders were further analysed when providing assessments of sustainability challenges, risks, and mitigation strategies, in addition to their business concerns. When identifying key stakeholders, we assumed roles that may hold a stake in a future **6G ecosystem**, and where actors such as firms, governmental bodies, or citizens may populate the role. Thus, the term key stakeholder does not refer to a few significant stakeholder roles but embraces most roles that are relevant for 6G use cases i.e. overlapping with e.g. stakeholders in the spectrum ecosystem.

Stakeholders can be part of a business ecosystem, with emphasis on providers and customers. The sustainability aspects of 6G development and deployment add new stakeholder roles to the ecosystem, such as people who are unintentionally affected by 6G or agents who can decide or act to affect the 6G evolution. Not the least, future generations and our planet Earth should be seen as holding huge stakes in our 6G use cases. In the elaboration of stakeholders below, we will see that they can be found using different ecosystem perspectives, and that e.g. the 6G use case is one perspective to use to identify them for further analyses.

The summary of the key stakeholders' roles based on the earlier categorization is presented in Tables 2.1 and 2.2.

2.2.1.1 Business Ecosystems' Expansion with 6G

The *business modelling* of potential future 6G use-case ecosystems filtered out the most important stakeholder roles. Ecosystem stakeholder roles that have already been mentioned in, e.g. the 5G ecosystem (Hallingby et al. 2021; 5GPPP 2023) are the provider of infrastructure equipment and

Table 2.1 Key 6G business ecosystem stakeholder roles.

Continuation of existing:	Emerging or reinforced:
Provider of network infrastructure, network operator, end user, customer, system integrator, cloud provider, regulators	Provider of equipment/applications/components on the user/customer side, providers of sensors, providers of Industrial 6G devices, modem chipset manufacturer and provider, providers of sensing capabilities, provider of domain-specific services, provider of indoor network, provider of local network, building owners, Non-governmental organizations, provider of energy and energy infrastructures, innovators, financial party

Table 2.2 Key 6G ecosystem stakeholder roles for sustainability.

Stakeholder roles and examples of specific actors *affected* by sustainability risks:	To be mobilized for strategic risk *mitigation*:
Provider of network infrastructure, network operators, end-user equipment providers (e.g. smartphones, robots, industrial 6G devices), 6G service providers End user (e.g. consumer, patient, doctor, earth-quake survivor, the rescuer, the security staffer, content creator, child, elderly, caretaker, employee, pedestrian, future generations) Customer (e.g. consumer, employer, employee, factory, insurance companies, local/central politicians, domain-specific entities) Entities in the environment where use case is deployed, not involved in use case (e.g. pedestrians, employees) Entities not in 6G use cases at all (e.g. other users of power or material, cities affected by waste) Nations, economies, markets, industries	Provider of network infrastructure, network operator, user equipment provider, other parties in the supply chain or ecosystem (raw material, components, logistic, content, power supply), solution (domain specific) providers, technology developers Global and local regulatory and competition bodies (industry, climate), governmental bodies, non-governmental organizations, vertical industry alliances, end-user alliances, telecommunication and vertical certification bodies Customers (large and influential private and public customers) Telecommunication industry associations, standard setting organizations Investors, research and academic institutions

network operator, the consumer as end user and also subscriber. However, professional entities, such as employers, factory owners, or health care institutions, more often populate the subscriber or customer stakeholder roles. The system integrator, cloud provider, and not at least innovators are also well-established stakeholder roles; however, they may become even more prominent in a modular 6G architecture.

Stakeholder roles that are reinforced in the 6G use-case ecosystems are the providers of equipment to users and customers and the accompanying applications, e.g. equipment for immersive communication experiences, robots, and moving vehicles. Providers of components that enable all kinds of equipment to connect are also becoming more significant. Not the least, new stakeholder roles and characteristics are emerging in the use cases where sensing capabilities are relevant, e.g. providers of sensors, infrastructure with sensing capabilities, and operators of sensing capabilities.

Several of the 6G use cases are very localized, indoor, or in areas currently with low coverage. To address such circumstances, it can be assumed that the network operator stakeholder role provides both local/private and wide-area networks. However, the 6G use cases illustrate a situation with more differentiated needs for networks, which in turn could lead to a split of the network operator stakeholder role into sub-roles. Moreover, calls for more differentiated networks introduce new ways to finance them. Thus, new stakeholder roles have emerged, such as building owners and new financing actors populate the subscriber role, such as non-governmental organizations. Finally, in all 6G use cases, energy efficiency is imperative, and providers of energy and energy infrastructures will probably emerge as a prominent 6G stakeholder role.

How, and if, the 6G ecosystem and 6G use-case ecosystems will emerge and evolve is uncertain. The uncertainty is partly due to challenges and risks in the hands of different stakeholders. In the literature on technology evolution (Bergek et al. 2008) and business ecosystems (Gawer and Cusumano 2014), the positions of, and tensions between, stakeholder roles and actors are a major source for both failure and success. It follows that, to reach the imagined 6G future, which is guided normatively by sustainability values, it is important to identify and address those stakeholder roles and actors who can contribute to mitigate the risks.

2.2.1.2 Sustainability Risk Assessment of the 6G Use Cases

The *full risk assessment* of the 6G use cases considers stakeholder roles that are *affected* by risks, and those that may be influential *mitigating* them. The stakeholder roles affected by 6G risks overlap with the more familiar roles from the 6G ecosystem, such as providers of network operations, users, and customers. For instance, network operators investing in new 6G use cases are hugely affected by market failures. Again, the actors that populate the roles can be more varied. This variation is, and should be, an expected consequence of a focus on sustainability, e.g. focus on inclusion or protection of certain groups, should expand our understanding of who a user and customer may be.

2.2.1.3 The Envisioned 6G User

The user may be a human, or a human overseeing a machine using 6G services. Thus, the potential *user* of 6G-enabled equipment and services can be, e.g. a private or professional. He/she can be the patient or the doctor, the farmer or the application owner giving agricultural recommendations, an earth-quake survivor or the rescuer, poor or affluent, the human in a crowd or the security staffer, content user or producer, or someone who operates/owns or configures a robot. He/she can be in rural or central locations, and may be a child, an elderly person, or a caretaker. In many situations, a user is supposed to use some form of 6G equipment for a period; in other situations, the equipment may be embedded in a body or a thing. One person can be many types of users in parallel, and in turn a user is affected by many 6G sustainability risks.

2.2.1.4 Indirect 6G User Impact

In several of the representative 6G use cases, people are affected, even though they are not actively taking part in 6G use-case scenarios. This is an implication of the rich use of sensors and sensing capabilities in both public and more private areas. For instance, pedestrians will unintentionally be part of the environment that is sensed in public areas where mobility is network-assisted. If 6G sensing and monitoring capabilities are used to capture, e.g. earthquakes or flooding in rural areas, people may by chance be monitored. On a factory floor, or other indoor locations where collaborative robots and digital twins are used, employees, clients, and customers could be on the floor and somehow affected by risk events. Moreover, in some risk scenarios, non-6G users may be affected if 6G services consume a disproportionate share of available energy or 6G waste is not handled in sustainable ways. For example, heating, lighting, and cooking in homes could suffer from expensive energy caused by energy demanding 6G deployments.

2.2.1.5 Customers and Innovators as Stakeholders

The *customer*, or the actor that fills the role of paying for a service, may vary in 6G use cases and thus be affected by risks. For instance, customers such as factories and hospitals could perceive lower benefits because of risks associated with energy consumption or low demand; for instance, those that finance future 6G solutions may be insurance companies, local politicians, central domain-specific entities within health, or central governments via national budgets. Private companies may find it beneficial to invest in 6G use cases, e.g. for educational or health purposes. Thus, such customers with initial willingness to pay for a service could suffer from 6G risks, such as dysfunctional market mechanisms or non-collaborative ecosystem players. The 6G use cases are addressing the functioning and well-being of local and global communities, and thus these stakeholders could also suffer from the realization of 6G use-case risks. As a part of the more varied presence of stakeholders in 6G use cases, innovators are also important. Innovators are large and small developers of hardware and software, but also entities that will use 6G to achieve sustainability goals and who would suffer from risks such as delayed take off or privacy breaches. Thus, there are many examples of future actors populating the user and customer stakeholder roles; specific 6G use cases will need to uncover not only those who can extract the benefits but also those who would suffer in the event of risks.

2.2.1.6 Risk Mitigation

The stakeholder roles who are important to mobilize to *mitigate* risks go beyond those already introduced as relevant in the 6G ecosystem business modelling. Still, it is also important to mobilize the well-established stakeholder roles such as providers of infrastructure, networks, and equipment to address, e.g. risks from competition and lack of standardization. In turn, investors need to consider how to consolidate and support the necessary scale in future 6G deployments. Sustainability is a normative and political topic, and governmental and political institutions on all levels may also have a say. Regulating and competition bodies should also be convinced to use their means to shape a sustainable 6G evolution. Non-governmental organizations may play a role alongside governmental bodies, in particular for environmental sustainability. In a technological evolutionary process such as for 6G use cases, vertical industry alliances and organized user groups can be influential both to confirm their needs and ability to allocate budgets, and to work precautionary to avoid negative effects, e.g. 5G-ACIA or 5G-AA. 6G use cases can be domain specific, for instance within health, mobility, or manufacturing, and domain-specific entities should be approached to mitigate risks. This could be the certifier of health equipment and solutions, urban planners, or road authorities.

2.2.1.7 Building a Resilient 6G

Providers of network infrastructure devices and network operations, and other competing providers, may be tempted to capture market shares or do not have sufficient legitimacy to be taken seriously in the market. In this case, larger private and public customers can act as pioneers and help building trust and enable a market take-off in a societal-critical domain. The telecommunication industry associations also have an important role to play to leverage the playing field, for instance GSMA, NGMN, and TM-forum. These associations have also a role in promoting harmonization, interoperability, standards; even certification can be advanced by for instance GCF (Global Certification Forum). Standard setting organizations will have a role to play – e.g. ITU, 3GPP, ETSI, and IEEE – and 6G use cases will require an even stronger focus on the interplay between the standard focus in these organizations. Not the least, starting as soon as possible, it is important to mobilize research and academic institutions in developing and building belief in the realization of 6G use cases, as technological, social, and business innovations.

The Hexa-X-II project has carried out a systematic identification of key 6G stakeholder roles. One observation is that some are new or reinforced with 6G. A next step in the analyses will be to stabilize the 6G stakeholder roles building on those already known from the 4G and 5G era.

2.2.2 Spectrum Ecosystem Stakeholders

The spectrum ecosystem stakeholders in Hexa-X-II were also identified using the lenses of the 6G use cases. The use cases capture a customer need, which motivates key 6G capabilities and in turn leads to demand for spectrum. As spectrum is a scarce resource that may benefit stakeholders in different ways, it is therefore subject to negotiations and regulations. Across stakeholders, the below goals for spectrum use were identified.

- **Provide additional capacity:** With ever-growing use of mobile communication, 6G needs to provide more capacity along all dimensions. Ideally, the spectrum to be made available will be enough to accommodate the required use cases using the existing grid of base stations, reducing the cost of identifying, acquiring, and deploying additional base station sites by operators.
- **Support high data rate services:** New services such as extended reality (XR) and holographic presence require the support of high data rates. For high data rates, large bandwidths are needed. These large bandwidths will be needed while still enabling mobility, meaning that frequency ranges that could offer a good balance between capacity and coverage will be relevant.
- **Support wide-area coverage:** The goal to make 6G inclusive implies that 6G should be available everywhere, and with service continuity (when user/device is moving). For wide-area coverage, lower-frequency ranges (i.e. up to ~8.5 GHz) are more suitable than higher bands due to radio wave propagation characteristics. In addition, the deployment of non-terrestrial networks (NTN) might be a supplement to the coverage provided by terrestrial networks (TN), in particular in rural areas that are out of reach for TNs.
- **Provide indoor coverage:** New high data rate applications like mobile augmented reality (AR)/virtual reality (VR) communication and gaming are expected to also be used indoors, increasing the volumes of indoor mobile data. Indoor coverage can be delivered in different ways:
  - **Outdoor to indoor coverage:** Here, indoor coverage is provided from outdoor base stations. Lower-frequency ranges are preferred, as the higher the frequency range is, the more walls and windows are attenuating the radio signals.
  - **Indoor to indoor coverage:** Here, indoor base stations are used that, e.g. are connected to fixed fibre networks. It is to be noted that this alternative may not be economically feasible in all cases.

- **Service continuity:** Mobile data applications increasingly require seamless continuation of connectivity travelling across, e.g. outdoor–indoor, urban–rural, and private–public situations. Currently, the device is configured to decide the handover between, e.g. cellular and Wi-Fi networks.
- **Enable positioning and sensing:** 6G will have to support ever more accurate positioning. This positioning can be based on base stations or on dedicated beacons, or on combinations of all available methods, including satellite based, triangulation, observed time difference of arrival, etc. Joint sensing and communication may be new technology advancements in 6G. For sensing, both base stations and mobile devices can be transmitters and/or receivers. In general, higher-frequency ranges imply more accurate positioning and sensing.

The stakeholders in the spectrum ecosystem will have different interests in the above goals. The most relevant stakeholders are:

- **Telecom operators** will continue to play a key role in mobile networks during the 6G timeframe. Existing networks in frequency ranges below 6 GHz are expected to migrate to 6G in the future and thus, similar to previous generations, 6G is expected to be deployed in the same frequency bands as used for earlier generations. Nationwide coverage can be achieved using sub-GHz spectrum and the frequency range 1–7 GHz can be used for combined coverage and capacity. New advanced 6G radio technology should make it possible to also use spectrum in the 7–15 GHz range, in particular enabling 7–8.4 GHz on existing base station grids. This will make this frequency range attractive for use by telecom operators to provide 6G coverage and capacity.
- **Stakeholders deploying private networks and specific purpose networks**, starting with 4G and continuing with 5G, include telecom operators and others. For example, 3.4–3.8 GHz is the core band for 5G public networks in many markets. Public-network-integrated non-public-networks (PNI-NPN) might share RAN and spectrum with the public network. Stand-alone NPN (S-NPN) is often deployed in dedicated spectrum (3.8–4.2 GHz). It can be expected that further spectrum made available for 6G will also be of interest for such use. For spectrum above 100 GHz, the very short communication ranges potentially allow an efficient reuse of the same spectrum, hence could be of interest in this case.
- **Network vendors, software vendors, and/or system integrators:** New 6G hardware and software elements need to be added to enable 6G services. Deployment of 6G in new frequency bands does not require backwards compatibility to earlier system generations. For existing frequency bands, migration solutions to ensure soft re-farming to 6G may be required. Network deployments may be 'global' or local, and a variety of operator models can emerge.
- **End-user equipment manufacturers:** 6G will play an integral role in many end user devices. Still, devices are likely to include a multitude of wireless technologies enabling a large variety of applications. Many frequency ranges are already allocated to mobile services in certain geographical regions. Regional or global frequency band harmonization and allocation will benefit the 6G ecosystem as scale will be bigger, bringing down cost.
- **Incumbent users:** There is a variety of primary incumbent services in the frequency range 7–15 GHz. Parts of this range are envisaged for 6G. These primary incumbents will need to be considered accordingly.
- **Regulators** will play a crucial role in enabling 6G by making spectrum available for 6G and assigning spectrum access rights to those requesting them.
- **Building owners** demand indoor coverage in their home, office, factory, or other building even where coverage from outside public mobile networks is a challenge. Because they see indoor coverage as crucial, they are also prepared to invest in Wi-Fi networks, private networks, or indoor

cellular solutions. Which of the available solutions these building owners chose determines the deployed frequency range and spectrum access mechanisms. Note that building owners can be both those that carry out a specific activity in the building, e.g. hospitals and manufacturer, and professional building owners renting premises.

- **Neutral hosts'** idea is to build and operate an (indoor) radio network and rent capacity to network operators. Neutral hosts deployed solutions would determine spectrum use. This is a role that might add on to the other role but is strictly speaking not mandatory.

It is important to acknowledge that there are different goals, and that different stakeholders have different interests. Thus, the further negotiations are better informed and can also add both goals and stakeholders which are relevant.

2.2.3 Public Ecosystem Stakeholders

Technological advancements and evolution are received and perceived by all those targeted or affected by them in different ways and contexts. Therefore, it becomes essential to evaluate and understand public perceptions, attitudes, and concerns about them.

An analysis was performed with different stakeholders within the public ecosystem through surveys and expert discussions, with the goal to understand citizen attitudes towards advanced telecommunications, user expectations, and the social implications of connectivity advancements, laying the groundwork for developing inclusive and trustworthy 6G services.

A structured approach was adopted in 2024 to analyse public perceptions of 6G through engagements in France, Portugal, and Greece, along with insights from the Expert Group on 6G Acceptance within the Hexa-X-II project. These efforts aimed to gather diverse stakeholder perspectives on expectations, concerns, and opportunities related to 6G development.

The Expert Group on 6G Acceptance, composed of external members from academia, civil society, and research institutions, as well as internal representatives from Hexa-X-II consortium partners, facilitated dialogue on critical themes such as sustainability, trust, and digital inclusion. Through expert discussions and surveys, the group provided recommendations on transparent policies, ethical AI governance, and minimizing environmental impact to align 6G development with societal needs.

In France, engagements with municipalities and civil society organizations involved workshops, focus groups, and surveys to understand public attitudes toward digital infrastructure. A total of 487 participants from civil society and businesses took part in these discussions, highlighting concerns over digital dependency, cybersecurity, and sustainability, and emphasizing the need for responsible governance in deploying future networks.

In Portugal, discussions with local governments focused on the role of municipalities in 6G adoption. Structured interviews and questionnaires helped assess priorities such as regulatory frameworks, security enhancements, and digital literacy initiatives, particularly in remote areas where inequalities in digital access remain a challenge.

In Greece, engagement with policymakers, businesses, and citizens utilized surveys and stakeholder consultations to evaluate expectations for 6G applications in sectors like education, healthcare, and public administration. A total of 37 participants responded to the questionnaires, which included a mix of multiple-choice, checkbox, and Likert scale (1–5) questions. The analysis focused on the importance of affordability, user control over data privacy, and digital inclusion.

Findings from these engagements were analysed through qualitative and quantitative methods, combining thematic analysis of stakeholder interviews with statistical assessments of survey responses. This multi-dimensional approach provided a comprehensive understanding of stakeholder priorities, shaping 6G as a technology that is inclusive, secure, and aligned with societal expectations.

Each stakeholder group contributed a unique perspective on what they expect from next-generation networks. Public authorities, e.g., emphasized the importance of equitable access to digital resources, robust data privacy measures, and governance frameworks that enhance digital sovereignty. Their input focused on addressing the digital divide and creating infrastructure that ensures equal opportunities for all, regardless of geography or socioeconomic status. Academia, on the other hand, highlighted the need for interdisciplinary research to address the social, environmental, and economic implications of 6G. Researchers stressed the need for long-term collaborations to embed sustainability at the core of 6G system design, ensuring that technological advancements align with human and environmental well-being.

Civil society organizations and municipalities brought forward concerns about transparency in technological advancements, emphasizing the need for a participatory approach that prioritizes public engagement from the earliest stages of development. Many participants voiced fears of the increasing digital divide and the environmental degradation that could result from unchecked technological progress. Industry representatives contributed insights on the scalability, security, and economic viability of 6G networks. They emphasized the need for innovative capabilities that support enterprise-level applications, drive economic growth, and ensure seamless integration with emerging technologies such as AI, automation, and the Internet of Things (IoT).

Several core themes have emerged from public stakeholder discussions, highlighting the value of 6G in transforming the digital landscape. Accessibility and inclusion emerged as primary concerns, with many stakeholders stressing the importance of addressing the needs of underserved regions and marginalized communities. Stakeholders emphasized the necessity of affordable infrastructure that ensures widespread network coverage. Ensuring equitable access will be essential in maximizing 6G's social impact. Furthermore, sustainability was a recurring theme, with demands for energy-efficient networks, reduced electronic waste, and eco-friendly deployment practices. These concerns reflect a growing social awareness of climate change and resource limitations. Green infrastructure, low-energy transmission technologies, and responsible resource management were viewed as integral to the long-term success of 6G.

Trustworthiness and data privacy were identified as essential pillars for social acceptance. Stakeholders called for transparent policies regarding data usage and enhanced measures to safeguard user privacy. Trust was seen as critical for ensuring widespread adoption of 6G, especially in an era where digital interactions are increasingly integral to daily life. Health and well-being were also prominent concerns, with participants advocating for networks that support mental and physical wellness by minimizing the risks of over-reliance on technology.

Another major insight from public stakeholder engagement was the importance of digital literacy and skills adaptation. Many participants highlighted the need for training programmes to equip individuals with the skills to navigate the complexities of next-generation networks. This was particularly emphasized by civil society and academia, who advocated for initiatives that promote lifelong learning and ensure no one is left behind in the digital transition.

Finally, ethical AI and automation were central considerations for 6G development. Stakeholders expressed concerns about dehumanization and over-reliance on automated systems, calling for AI frameworks that complement human capabilities rather than replace them. They stressed the importance of transparency and accountability in AI-driven decision-making to build trust and ensure ethical practices.

2.3 6G Use-Case Families and Use Cases

The 20 use cases identified by Hexa-X-II are displayed in Figure 2.3. These were grouped into 6 use-case families based on a series of commonalities such as industries targeted, main aim for the use case, or types of capabilities required. Within each use-case family, a 'representative use case' was selected to reflect the main features of the family. For each of the representative use cases, several future scenarios were envisaged and are presented below.

2.3.1 Immersive Experience

The Immersive Experience use-case family comprises use cases based on XR technologies. The immersion is built through the combination of immersive/sensorial technologies (e.g. 3D visual perception, spatial hearing, haptics) and the synchronization of data streams so that multiple participants have simultaneous, consistent, and collaborative experiences.

The representative use case for this family is *Seamless Immersive Reality* (*SIR*). Two example scenarios are (i) **Immersive collaboration meeting, immersive classroom**, where this use case describes the possibility of displaying a particular content in an AR/VR/Mixed Reality environment while interacting with remote and co-located participants; and (ii) **Immersive Experiences 'on-the-go'**, where information from the real world is incorporated and used to provide a better user experience (e.g. sight-seeing landmarks) but also can be shared with others (e.g. joint virtual city tour).

To achieve this superior Quality of Experience and seamless service continuity, data rates beyond what 5G delivers are required, plus strict end-to-end (E2E) latency and reliability considerations. Additionally, new 6G capabilities such as sensing, positioning, and artificial intelligence (AI)/machine learning (ML) are key technology enablers for *SIR*.

Finally, privacy and security requirements are crucial for the proper functioning of these use cases, as vast amounts of data will be employed. Technical and administrative/legislative means

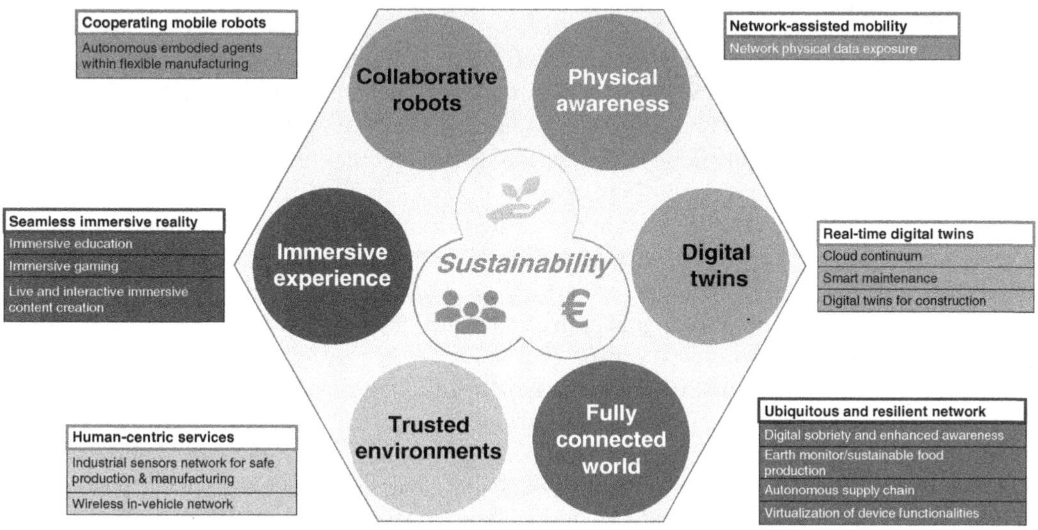

Figure 2.3 Hexa-X-II's use cases and use case's families.

are required to protect privacy in a hierarchical and heterogeneous 6G system, i.e. across wide-area networks, public and private local networks, as well as subnetworks.

2.3.2 Physical Awareness

The *Physical Awareness* family refers to use cases where sensing and positioning are used together with communication and network intelligence to enable for instance physical scene analysis, tracking, context awareness, trajectory prediction, navigational support, and collision avoidance. The network sensing can be enhanced by embedded sensors and derivative information delivered by network nodes and surrounding devices. Example scenarios refer to cars, automated guided vehicles, and drones as well as pedestrians and bikes.

The representative use case for this family is *Network-Assisted Mobility*. Here, networks measure the physical environment in traffic scenarios and analyse to detect and capture information on objects (e.g. size, speed, trajectories, positioning). From this large data set, information is extracted and relayed to vehicles subject to the vehicle's capabilities. Thereby, the network can support vehicles with different levels of autonomy and modes of operation and also enable smart transport in urban areas. Example scenarios described comprise (i) **Autonomous drone transport**, whereupon carrying goods in urban areas, drones can be updated in real-time on the fastest and most energy-efficient traffic routes; (ii) **Smart intersections**, where vehicles are informed about the conditions at an intersection to advert collisions; and (iii) **Assisted vehicles** where the benefits of autonomous driving are expanded into the ecosystem, benefiting from the contextual data resulting through the fusion of the wide-area sensor information provided by the network, vehicle onboard sensors, and/or even sensors embedded in the transport infrastructure.

As this use case aims at preventing undesired outcomes such as collisions, the system is required to be highly reliable, both in communication aspects (Reliability, Coverage, Service availability) and data quality/accuracy (Positioning, Sensing, AI/ML) in a 3D environment. Additionally, the availability of reliable compute capabilities offered by the network, and privacy and security requirements are also essential to this use case.

2.3.3 Digital Twins

The *Digital Twins* family collects a set of use cases where digital equivalents of the real world are created and displayed for interaction, control, and maintenance, as well as process and component management. Digital twins can be part of managing manufacturing plants, construction sites, city infrastructure, or communication networks with or without real-time needs. Also, specific use cases around distributing the software between operators and network locations are included.

The representative use case for this family is *Real-Time Digital Twins*. As the name suggests, along with the creation of an accurate digital representation of a given object/process/person, this use case expands the potential offered by these representations by benefiting from the real-time aspect, which allows extending the digital twin also towards the direct control of ongoing physical processes. An example scenario is (i) control over a process in real-time in a **Production plant** but also building an (ii) **Overarching smart city digital twin** to support its full management, analysis, planning and predictions, and response to emergencies.

To achieve the latter, AI/ML, positioning, and sensing capabilities are required to deliver a seamless user interaction. Moreover, network and device interoperability are key to the success of the working ecosystem. For the latter, using open interfaces and relying on strong system integrators is essential. Finally, other relevant requirements are 3D coverage, compute capabilities, and privacy/trustworthiness of the network.

2.3.4 Fully Connected World

The *Fully Connected World* family unites use cases demonstrating the need for ubiquitous network access and service coverage across the Earth's population. Ubiquitous access is enabled both through TN and NTN and amended by infrastructure-less network extensions to wide-area network deployments.

The representative use case for this family is *Ubiquitous and Resilient Network*. This use case focuses on delivering Mobile Broadband connectivity to every human on Earth, leaving no 'white zones'. Here, three scenarios are described: (i) **Connectivity at remote locations** describes the possibility of a user in a mountainous area now having a wide range of options (e.g. remote health consultations, voting, or emergency calls if being lost in the mountains); (ii) **Improved connectivity in developing countries** enables small-scale farmers to apply the benefits of connectivity to their crops (e.g. precision farming); and (iii) **Connectivity during natural disasters and emergencies** builds on the role NTN can play in case of a terrestrial natural disaster (e.g. earthquake, fire, flood) for emergency services and disaster relief.

Ubiquitous and Resilient Network not only has a big impact on social (e.g. digital inclusion, education, disaster relief), economic (e.g. deployment cost, new business opportunities), and environmental (e.g. earth monitoring, less invasive deployment) issues but also increases network resilience significantly. Therefore, the biggest challenge lies in delivering a tight and reliable integration of different networks from the initial phases of 6G.

2.3.5 Trusted Environments

The Trusted Environments use-case family encompasses use cases with high reliability and privacy requirements as key characteristics. These use cases build primarily on AI, sensing, and computing capabilities to create spatial and situation awareness to enable context-driven interventions.

The representative use case for this family is *Human-centric services*. This use case describes three scenarios displaying how Hexa-X-II envisions putting humans at the centre of a wide range of 6G services. These are (i) **Precision healthcare**, which delivers personalized diagnosis and treatment; (ii) **Safe environments**, which is based on spatial- and situation awareness to prevent accidents or trigger changes in the environment (e.g. ramps deployment); and (iii) **Public safety services during big events** aimed at predicting disorder and delivering solutions to aid the people in moments of need.

Delivering these services imposes very high privacy requirements and consequently the mechanisms to ensure the data is held secure and the system is trustworthy (e.g. anonymization, advanced information coding, additive homomorphic technology). For the latter, availability and reliability also play a key role when dealing with health and well-being situations. Both the new technical requirements/capabilities and the considerations emerging from the use case (secure by design) are topics for which 6G can provide an answer.

2.3.6 Collaborative Robots

The ***Collaborative Robots*** family includes several use cases for intelligent, collaborative, and mobile robots with the ability to move, sense their environment, perform a task, and cooperate with humans or with one another to achieve their goals. Application domains include home robots, robots for facility management, robots in day care or hospitals, as well as robots in flexible shop floor manufacturing.

The representative use case for this family is ***Cooperating Mobile Robots (CMRs)***. Further details on this use case are presented in the following section, where it is used as an example for applying the value design methodology.

2.4 6G Use Case and Value Design | Cooperating Mobile Robots

As the name suggests, this use case revolves around robots being able to move, sense their environment, and cooperate safely, enabling them to perform tasks beyond their individual capabilities and perceive their environment beyond their local capabilities. Furthermore, it focuses primarily on local ad hoc connectivity embedded in private networks, and the challenges imposed by the mobility aspects.

In this context, the network of the future is envisioned to enable local cooperation among robots, support autonomous task-solving by the robots, enhance the safety of human-machine interaction, and ensure that only authorized machines and humans can participate in task-solving.

Scenarios within this use-case family revolve around the need to complete a task that surpasses the capabilities of a single robot, but potentially also helping people to perform tasks irrespective of their current capabilities. The solution lies in robot cooperation, where multiple robots form a cluster to collaboratively accomplish the task. These robots utilize local ad hoc connectivity to establish a subnetwork for the following purposes:

- Coordinating and controlling their actions for collaborative successful task execution.
- Exchanging raw or processed sensor data,
- Sharing AI/ML models, weights, and federated learning.
- Ensuring safety for humans nearby in the event of an emergency.
- Potentially connect to other local capacities such as edge computing, building sensor data, digital twins, and AI servers.

A subnetwork consists of a group of devices that communicate with each other for a specific task in a confined space temporarily. Its creation is initiated by the network or by a device, in this case, a robot. The subnetwork utilizes a portion of the communication resources from the private network within which it is embedded. Once the task is completed, the subnetwork is disbanded, and the communication resources are subsequently freed. Depending on the scenario, the subnetwork might be used by the collaborative robots for multiple tasks.

2.4.1 Human and Planetary Goals

As shown previously, the connection between a use case and the H&PGs can be done either by designing a UC as a solution to contribute to one or several of these high-level goals, or it can be designed as a response to a business need or opportunity. This second scenario is what describes the situation for the introduction of CMRs.

Here, the introduction of a new generation of robots demands energy and materials, which can be directly linked to H&PGs #3: Curb climate change, #4: Preserve Natural Resources, and #5: Minimize waste.

From a social perspective, the introduction of CMRs will have an impact primarily on the work environments, in terms of jobs, safety, inclusion, to name a few. Thus, the main H&PGs impacted are #7: Protect human rights, #8: Increase digital inclusion, #9: Ensure ICT trustworthiness, #10: Enhance Health and Well-being, and #11: Foster decent employment.

Finally, from an economic perspective, the CMRs will favour goal #13: build long-term sustainable economic growth, but it should also abide by goal #15: Ensure ethical conduct.

This initial exploratory approach is further expanded in later stages of the process and can be found in Section 2.4.9.

2.4.2 Problems to Be Solved and Challenges

The CMRs use case presents several problems to be solved, but also identifies several challenges related to the proper deployment of the latter.

The first challenge is **understanding and addressing the communication requirements of machines in the future**. This includes existing industrial automation systems, as well as the emerging type of *autonomous embodied agents* that encompasses industrial robots, service robots, and cobots, i.e. collaborative, cooperating, or cognizant robots. These robots interact with the physical world through sensors and actuators, use advanced ML for decision-making, and employ wireless connectivity to collaboratively perform complex tasks and adapt to changing environments.

Then, **using limited resources efficiently (materials, energy, and human effort)** is a crucial challenge. For a sustainable deployment, it is essential to further automate well-defined, structured, and repetitive tasks in the physical world. This will require advances in connectivity for smart tools and machines that will allow mobile robots to collaborate, thereby optimizing productivity, resource efficiency, and costs.

A third challenge is to adapt to the **dynamic requirements of the market**. In the future, the ability to address real-time fluctuations in demand and personalized customer specifications will be essential in industries such as manufacturing, transportation, construction, and logistics. A key advantage will be the ability to use machines in different configurations. Within a collaborative network, the potential of a single machine to leverage the capabilities of other machines amplifies the versatility of the fleet. Combined with enhanced automation, this allows for more agile responses to real-time market requirements.

Moreover, it enables **end-user access to custom manufacturing**. Enhancing the capabilities of end-users, including tradespeople and individuals without technical expertise, to manufacture products such as mechanical components, electronic devices, and architectural structures, is a step towards self-sufficiency. As CMRs make manufacturing processes more accessible and decentralized, goods can be produced as and when needed at the specific locations where they are needed. New possibilities for customization will emerge. This shift will transform business relationships and will lead to improved transparency and accountability in the manufacturing sector.

Finally, a crucial challenge relates **to safe and trustworthy interactions with tools that can make decisions**. As robots gain decision-making and collaboration capabilities, one major concern is the assurance of human safety and the establishment of trust when these autonomous machines operate in human-inhabited spaces.

2.4.3 Why 6G Is Needed

The exchange of locally relevant information between robots will be crucial for cooperation. Key building blocks are the evolution of embedded device-to-device communication towards subnetworks supporting highly reliable and low-latency industrial communication, together with mechanisms for local reservation of communication resources. Mobile robots may need the capability to request and establish these reservations when local connectivity is required.

Current 5G deployments as of 2025 do not meet the latency, reliability, and scalability requirements in several demanding industrial and vertical scenarios such as motion control, wired-2-wireless link replacement, mobile robots, or mobile operation panels, especially when looking at scalability with larger numbers of URLLC connections and high reliability in operational settings. Increasing the adoption of existing millimetre wave communication technologies and the development of new sub-THz connectivity may provide additional mechanisms for achieving the latency and reliability targets and managing interference.

Cooperating robots significantly benefit from advances in ML. Network-side support for efficient exchange of training data, weights, and models, along with federated learning, will likely become relevant. Complementing the onboard computation capabilities of robots, offloading computation to nearby edge nodes may add another layer of efficiency. Moreover, future systems are expected to exploit context information derived from sensing, the application, or a shared knowledge base, to improve both application and communication performance.

The integration of sensing and positioning capabilities in the radio interface may aid in the collection of data that is needed for advanced ML applications. Trust management is crucial for safe machine–machine and human–machine interactions and will require reliable, deterministic, and low-latency E2E message transport mechanisms along with secure sensing results and manipulation-protected positioning.

2.4.4 Example Scenarios

The representative use case CMRs contain several scenarios, focusing on specific characteristics in different industrial settings. Figure 2.4 illustrates two of the example scenarios.

2.4.4.1 Cooperative Carrying with Mobile Robots

Cooperative carrying with mobile robots is a concept where multiple autonomous mobile robots (AMRs) work together to transport an object that exceeds the carrying capacity of a single robot, as shown in Figure 2.4.

- **Industrial communication:** Collaborative robots exchange sensor data and industrial control information with each other for cooperative carrying. Time synchronization and deterministic highly reliable and low-latency communication between cooperating robots are needed.

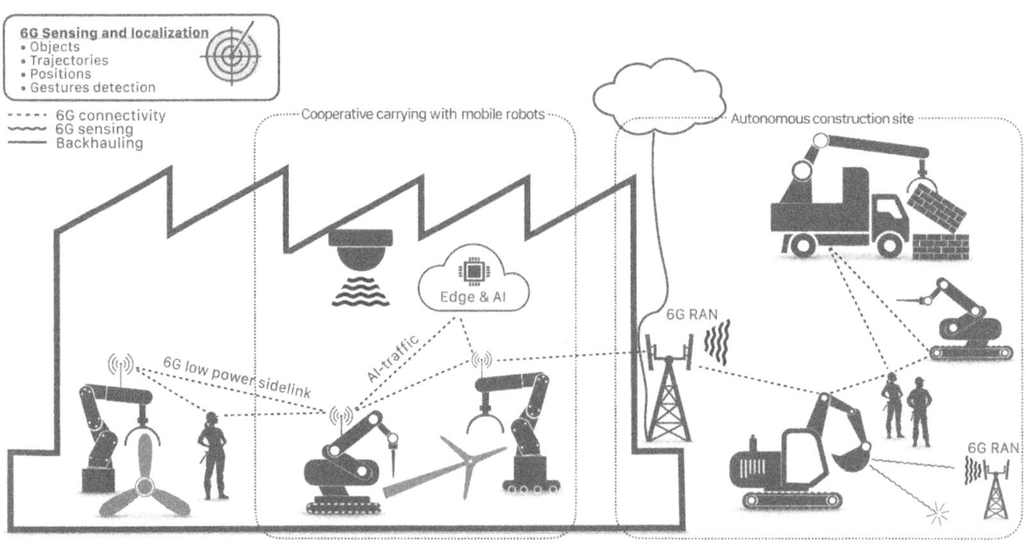

Figure 2.4 Cooperating mobile robots example scenarios. *Source:* © by Geir Egeland, Telenor.

- **Digital twinning:** Sensor data and sensing information are sent to edge nodes, digital twins, and potentially to AI applications.
- **Artificial intelligence:** The AMRs have AI capabilities for task planning and autonomous operation. AI model transfer occurs in case of decentralized learning or from a local AI/ML compute server.

The collaborative mobile robots are in close proximity. They are mobile and can move in unison as a group. They have sensing capabilities as well as AI capabilities for task planning and autonomous operation. The collaborative mobile robots for cooperative carrying can be found indoors on the factory floor and outdoors in the yard of a factory, a construction site, a campus, and similar places.

The special focus of this scenario is on collaboration between several AMRs, and deterministic industrial communication between AMRs in a moving subnetwork.

2.4.4.2 Lot-Size-1 Production

Lot-size-1 production is a manufacturing approach in which each product is manufactured individually and uniquely, rather than in batches. This method is facilitated by the collaboration of robots with different capabilities within a production cell, as well as the cooperation between logistics robots and production cells.

- **Industrial communication:** Collaborative robots exchange industrial control information with each other for cooperative manufacturing. Time synchronization and deterministic URLLC communication between cooperating robots are needed.
- **Logistics AMRs:** AMRs are used for logistics purposes such as providing the necessary materials to the production cell. These are usually single AMRs.
- **Digital twinning**: The task management system will send data of the product digital twin to the collaborative manufacturing robots for task planning. Product status, production process information, sensor data, and sensing information are sent to edge nodes, digital twins.
- **Artificial intelligence:** The AMRs have AI capabilities for task planning, process planning, collaboration, and autonomous operation. AI data can be transferred to, e.g. a local AI/ML compute server. AI model transfer occurs in case of, e.g. decentralized learning.

The collaborating robots are in close proximity within the same production cell. They can be considered static, and there might be only small movement in order to get to a better position within the production cell. Interaction with logistics AMRs is only necessary if they are close to the production cell. The collaborating robots have sensing capabilities as well as AI capabilities for task planning and autonomous operation.

The collaborative mobile robots for lot-size-1 production can be found indoors on the factory floor. Logistics AMRs interact with the collaborative robots in the production cell. For specific products/materials, the logistics task might be done by a group of cooperatively carrying AMRs as described in the previous scenario.

The special focus of this scenario is on collaboration between robots and dynamic task and process planning.

2.4.4.3 Automated Industrial Tasks

In warehouses or logistic sites, cobots perform automated industrial tasks (e.g. quality check at the production line, movement of goods between different workstations, from storage to shipping, or from quality check to repair), autonomous system diagnostics, and functionality (re-)allocation of the participating robots (e.g. for charging or repairment) without disrupting the warehouse process.

- **Logistics AMRs:** AMRs are used for logistics purposes such as moving goods between different workstations, from storage to shipping, or from quality check to repair. These are usually individual AMRs.
- **Machine vision:** Robots perform quality checks at the production line. Depending on the sensing type and the video resolution, high data throughput may be required.
- **Artificial intelligence and digital twinning:** Artificial intelligence is used in quality checks during normal operation and system diagnostics and functionality re-allocation. Digital Twinning supports normal operation and mitigation of unexpected situations such as low batteries and repairs. AI data transfer to, e.g. a local AI/ML compute server. Product status, production process information, task status, AMR status, sensor data, and sensing information are sent to local edge nodes and local digital twins.

The collaborative mobile robots are mobile and move indoors on the factory floor. They have sensing capabilities as well as AI capabilities for quality checking, autonomous operation, system diagnostics, and functionality re-allocation.

The special focus of this scenario is on automated industrial tasks of collaborating robots such as machine vision services and autonomous incident mitigation, for instance, by functionality re-allocation.

2.4.4.4 Autonomous Farming

Autonomous farming refers to the use of self-driving, smart machines that work collaboratively in agricultural operations such as weed picking or harvesting.

- **Communication:** Local machine-to-machine communication for the exchange of status information, sensing information, as well as other information relevant to the collaborative agricultural operation. Wide-area human-to-machine communication supports remote monitoring, remote control, and human intervention.
- **Sensing and artificial intelligence:** Artificial intelligence is used in agricultural operations such as detecting weeds and animal safety. The necessary sensing information may come from various sources such as the 6G devices with ISAC or from HD video cameras. The latter requires high uplink data throughput for transferring AI data and sensing information to, e.g. a local AI/ML compute server.
- **Positioning:** Positional awareness and location are required, both relative to other collaborating robots and absolute. The location service is used for coordination between the collaborating self-driving smart machines and other services such as route management.

The collaborative self-driving smart agricultural machines are in close proximity and can move in unison as a group. They operate outdoors in urban areas (fields, meadows). They have location, sensing, and AI capabilities for autonomous, collaborative operation, and safety.

The special focus of this scenario is on location capabilities in a moving subnetwork.

2.4.4.5 Autonomous Construction Site

On an autonomous construction site, different types of robots work collaboratively in unison to construct a building. The transportation of materials, such as bricks, must seamlessly integrate with advanced construction machinery like 3D printers. Other robots are responsible for post-processing tasks, such as drilling holes autonomously.

- **Industrial communication:** Collaborative construction machines exchange industrial control information, location information, and sensing information with each other for cooperative construction work. Time synchronization and deterministic URLLC communication between cooperating machines are needed. Furthermore, this includes human-to-machine communication for remote monitoring, remote control of construction machines, and human intervention.

- **Sensing and localization:** Sensing and localization capabilities for positioning, location services, detection of human presence, quality monitoring, collision avoidance, etc. The sensing information may come from various sources such as the 6G devices with ISAC or from HD video cameras (requires high uplink data throughput). The location service, both absolute and relative, is used for coordination between the collaborating construction machines and other services such as route management.
- **Digital twinning:** Digital twins of the building provide detailed information about all project stages. The digital twin information is used for inspections, previews of complex construction aspects, documentation, etc. Sensors and advanced IoT integration provide the necessary current data to the digital twin.
- **Immersive extended reality:** High-bandwidth and low-latency communication for local and outdoor walkthroughs of future construction stages at the construction site but also in remote offices. Virtual guidance supports previews and better understanding of complex construction aspects, allowing for remote participation of stakeholders.
- **Artificial intelligence:** The autonomous construction machinery has AI capabilities for task planning, process planning, collaboration, and autonomous operation. AI data can be transferred to, e.g. a local AI/ML compute server. AI model transfer occurs in case of, e.g. decentralized learning.

The collaborative construction machinery is operated in a local private network on a construction site outdoors but becoming more and more indoors as the construction advances. They have location, sensing, and AI capabilities for autonomous, collaborative operation, and safety.

The special focus of this scenario is on the dynamics of the collaboration and the large variety of 6G services such as industrial communication, positioning, AI/ML, digital twinning, and immersive XR.

2.4.4.6 Smart Workshop

In a smart workshop, cooperating robots seamlessly connect material flow between various manufacturing stations. A high level of integration and autonomy of the machines enables tradespeople and hobbyists to produce items that were previously only possible in a factory setting.

- **Industrial communication and logistics AMRs:** Cooperating AMRs exchange sensor data and industrial control information with each other for handover of material and workpieces. Human-to-machine communication is used for real-time monitoring and control. Time synchronization and deterministic highly reliable and low-latency communication between cooperating robots and between HMI devices and manufacturing stations are needed.
- **Sensing and localization:** Sensing and localization capabilities support positioning, detection of human presence, collision avoidance, optimized task planning, etc.
- **Digital twinning:** Digital models of the items to be produced.
- **Artificial intelligence:** The AMRs and smart workshop manufacturing stations have AI capabilities for task planning and autonomous operation.

The cooperating AMRs move indoors within the smart workshop. They have sensing and positioning capabilities as well as AI capabilities for autonomous operation and dynamic task planning. The manufacturing stations include AI capabilities for dynamic task planning, too.

The special focus of this scenario is on dynamic task planning and dynamic material flow.

2.4.5 Deployment Aspects

2.4.5.1 Environment

CMRs will be present in many environments. Several considered scenarios are mainly **indoor** – some closer to consumers with smaller areas (smart workshop), others closer to industrial manufacturing with potentially larger areas (cooperative carrying, lot-size-1 production, automated industrial tasks). Some are **outdoor** within a local well-confined geographical area such as a factory yard (collaborative carrying), a field (autonomous farming), or a construction site (autonomous construction site). In these environments, line-of-sight links may be scarce for macro cells. Thus, femto- or pico-cells should be preferred for beam-based coverage. Propagation models used for network planning should include the appropriate multipath effects.

2.4.5.2 Type of Deployment

- **Private networks:** It is assumed that the environment in which the network is deployed is dedicated to a specific purpose and to private use, e.g. manufacturing, construction, farming, living, and similar. Trade-offs among coverage, data rate, latency, reliability, etc. can be tailored according to specific application needs inside the private, local networks.
- **Local networks:** The collaborating mobile robots use case deploys local wireless networks in well-confined geographical areas such as factories, industrial sites, enterprises, construction sites, or fields. The local networks are primarily private networks, in several scenarios with functionality and data storage on-premises.
- **Subnetwork:** Support for (very) low latency and high reliability is enabled by peer-to-peer and mesh networking of co-located collaborative robots (especially in cooperative carrying scenarios).
- **Connection to the local network:** Subnetworks usually have a (very) low-latency/high-reliability connection of one or multiple subnetwork nodes to the local wireless network in order to reach local edge nodes, local AI/ML compute server, and other centralized local functionalities.
- **Infrastructure option:** Support for (very) low latency and high reliability is enabled by 6G communication between collaborating manufacturing robots and between collaborating robots and functionalities located in the local 6G network such as local edge nodes, AI/ML compute servers, and other local centralized functionalities.
- **Wide-area connectivity:** Especially outdoor scenarios (autonomous farming, autonomous construction site) benefit from wireless 6G communication for wide-area connectivity.

2.4.5.3 Users and Devices

- **The majority of users are machines**, with a comparatively smaller number of human users.
- **The device class is differentiated based on the data traffic demands of the robots.** The traffic demands are influenced by the level of autonomy of the robots. Machines with low autonomy, requiring frequent exchange of deterministic coordination messages, could potentially expand the 'Highly reliable low-latency' device class. Conversely, machines with a high level of autonomy, reliant on sensor data exchange and AI/ML traffic, fall into the 'Reliable high data rate with bounded latency' device class.

2.4.5.4 Constraints and Challenges

- **Power is generally available.** But power efficiency is nonetheless required for battery-driven devices like mobile robots and as a general requirement supporting sustainability.

- **The management and orchestration must fit a single operator, private network.** Resources must be properly allocated for providing the CMR service with the required level of QoS. In particular, the resource allocation is executed at the radio access level, either on macro cell or femto- or pico-cells and at the operator network level. Depending on the network architecture, the resources available on different segments of the network (e.g. transport, edge, core, subnetwork) must be properly allocated.

2.4.6 Requirements

The main technical requirements are listed below:

- **Local ad hoc connectivity:** A collaborative task is characterized by its localized nature. Direct communication between connected devices allows this to be exploited, leading to the formation of subnetworks. These subnetworks, envisioned to be task-specific, temporary, and localized, are embedded within broader campus networks. This structure aids in meeting stringent latency and reliability targets within each subnetwork.
- **Support of private networks:** The network environment is usually dedicated to a specific purpose and private use in geographically limited areas. Trade-offs among coverage, data rate, latency, reliability, etc. can be tailored according to specific application needs inside the private, local networks and to on-premises functionalities and data storage.
- **Extremely reliable and low latency communications:** Interruptions in industrial manufacturing operations typically entail huge financial and material losses. Safety-critical applications are intended to protect human lives. As such, applications in this domain have among the strictest service-level reliability and E2E latency requirements. The service area may be highly localized to subnetworks and increased autonomy in machines may accommodate lower reliability and higher latency.
- **Mobility:** The localized and ad hoc nature of subnetworks in a 6G environment may induce frequent handovers as machines join and leave. These subnetworks, while logically embedded in a campus network, may exhibit nomadic behaviour within it, and the roaming of subnetworks between different campus networks may also occur.
- **Sensing, positioning, and AI/ML:** Integrated sensing capabilities within the 6G network and devices (ISAC) can potentially enhance a robot's perception of its environment. The introduction of AI/ML traffic types and AI/ML execution in edge nodes can further enhance robot coordination. For maximum accuracy, it is likely that data from device-based sensors will be utilized in fusion with information from the network.
- **Energy efficiency:** Energy efficiency is a general requirement for wireless communication. However, energy saving in the 6G network must not endanger successful, efficient, and high-quality industrial manufacturing.

2.4.7 Key Performance Indicators

Table 2.3 shows the main KPIs identified for the CMR use case. Please note, as the IMT-2030 capability-related KPI values are formulated at an early point in time and derived from theoretical analysis and simulations, real deployments may not always fulfil these requirements. In the end, use cases are only meaningful if they can be delivered with acceptable cost and sustainability footprint. The feasibility of such use cases in an environmentally, socially, and economically sustainable way still needs to be further evaluated.

Table 2.3 Cooperating mobile robots KPIs.

	KPI	Target range	Comments/justification
Communication	Application data rate – industrial Communi-cation/control traffic (Mb/s)	≤2.5 DL and UL	Symmetric DL/UL data rates: UL = DL **Example 1:** 500 B payload at min. transfer interval 1.66 (ms) = 2.41 (Mb/s) The data rate might be rather low depending on the transfer interval and payload size
	Application data rate – between robots and network (Mb/s)	<10 <250	**Critical:** timely transfer of (raw) sensor data, video, immersive XR, AI/ML traffic, Digital Twin data, etc. The data rate might be significantly higher depending on the type of data exchanged with the campus network (e.g. video stream).
	Connection density (devices/m^2)	≤0.5	Local concentration of mobile robots: depends on the scenario, e.g. cooperative carrying assuming 8 robots in a space of 4 m × 4 m carrying an object
	Mobility (km/h)	<20	Slow vehicular Linear movement of subnetwork → ≈ no relative mobility between devices of subnetwork
	6G E2E latency (one way) – D2D latency (ms)	<5 ≥ 0.8	**Critical:** exchange of coordination messages and control messages up to 600 times per second. Note, this results in a transfer interval of ca. 1.66 (ms). E2E latency limit is set to at most half that interval [22.104]. This provides enough margin for ARQ.
	6G E2E latency (one way) – between robots and network (ms)	1–10	**Critical**, if timely response expected (e.g. digital twin data) Not critical, if no real-time response is expected (e.g. AI/ML model exchange)
	Max/min transfer interval (ms)	≤10 ≥1.66	**Critical:** deterministic periodic communication with very low latencies in collaborative robotic use cases
	Message error rate	$<10^{-5}$–10^{-7}	**Critical:** Message losses lead to interruptions in the collaboration process and may entail financial and material loss (damaged carried object). Application-side safety net mechanisms like 'survival time' or 'grace period' are employed to compensate for occasional packet losses and delays at the link level. Selected applications may have an even more strict packet reliability requirement of up to 99.999999% (= message error rate of 10^{-9}).
	Message size (Byte)	250–500	Applications are putting more and more information into the control data, 500 Bytes with localization information
	Coverage (%)	100 (localized)	The localized nature of a joint task makes local ad hoc connectivity favourable.

Table 2.3 (Continued)

	KPI	Target range	Comments/justification
Positioning and new capabilities	Positioning accuracy – general (m)	<0.1 <1	Collaborative tasks such as handing over material and objects or robot navigation in close distance to other objects require more accurate positioning. Tasks like robot localization can work with less accurate positioning.
	Positioning accuracy – human presence detection (m)	0.4	Identifying human presence is very important: 40 cm resolution within 1 m of the machine, velocity. (Scenarios autonomous construction, smart workshop.)
	Sensing-related capabilities (Y/N)	Yes	Robots and cobots depend on capturing the environmental context. Network-integrated sensing may complement or replace dedicated onboard sensors. Efficient transport of data/information from connected external sensors is likely needed.
	AI/ML-related capabilities (Y/N)	Yes	Robots and cobots depend on advanced ML. Execution may be embedded in the device and/or offloaded at a local edge and/or provided by the network as an over-the-top service.

2.4.8 Key Values and Key Value Indicators

Table 2.4 presents the results of the KV analysis for the CMRs use case, represented by the respective sustainability handprints (positive impacts) and sustainability footprints (negative impacts). This analysis aimed to capture the main impacts of the use case on the ecosystem. However, note that this list is not exhaustive and must not be considered as such. Then, as KVs are subjective in nature, to understand whether one is progressing towards achieving a value, it needs to be measurable. This is where KVIs provide a measurable reference point, looking to quantify the respective KV, either directly or indirectly.

As KVIs are designed to capture the value that the use case brings to the ecosystem, this newly created value should be evaluated against a reference or baseline, which, in the context of technological innovation, corresponds to the most recent technology currently in use. Specifically, for CMRs, static scenarios (e.g. machine cell or fixed production line) or cooperation with human control (e.g. construction site or farming) could be considered as a reference.

Given the pervasive nature of mobile technologies, their impact on the ecosystem spans a wide range of areas. Most of the KVs and KVIs fall outside the scope of what can be influenced by 6G, as they are directly related to the implementation of the use case. These are called use-case KVIs (UCKVI). However, other KVIs can be influenced by the design and implementation of 6G. These are called enabler KVIs (EKVI) and are highlighted in **bold** in Table 2.4. It is important to underline that the influence that 6G has on the selected enabler KVIs also varies depending on the indicator. Additionally, Table 2.4 shows the associated H&PG as defined in Section 2.1, as well as the UN SDGs linked to these KVs.

The results of the sustainability analysis and subsequent KVI determination indicate which aspects need to be considered within the design of 6G (Enabler KVIs), for the application of this particular use case.

Table 2.4 Cooperating mobile robots KV/KVIs.

SP	H/F	KV	KVIs (EKVIs & UCKVIs)	H&PG	SDG
Environmental	Sustainability handprint	Resource efficiency: Functionalities may be provided by machines with less materials, energy, and waste generated	1. Overall GHG footprint of the materials involved (tCO_2e) 2. Virgin materials (in) (tons)/Material used in products (out) (tons) 3. % of produced products remaining in stock/not possible to sell/scrapped 4. **Energy consumption per process/overall (MWh)**	4, 5	6, 12, 14, 15
		Function integration eliminates the need for multiple dedicated machines with individual functions	1. # of different dedicated machines in a production plant 2. Downtime of robot while switching functionality 3. # of functions a robot can undertake	4	6, 12
	Sustainability footprint	Energy is consumed and materials are used to manufacture, deploy, and operate robots and associated services	1. Energy usage related to cobots through their life cycle (MWh) 1a Energy used in manufacturing components (MWh) 1b Energy used in transportation processes (MWh) 1c **Energy used in operation phase (MWh)** 1d **Energy used in data transfer (optimizing packets and volume of data) (MWh)** 1e Energy used in end-of-life phase (MWh) 2. Emissions related to energy usage in all previous KVIs (tCO_2e) 3. Environmental footprint related to materials usage, across the life cycle 4. Material usage of cobots, across their life cycle (tons, total footprint) 5. **'Data efficiency' on storage** TeraByte **(TB)**	3, 4	6, 12, 13
		The manufacturing, including material extraction and industrial processes, and transportation of robots generate GHG emissions	1. **Total life-cycle emissions (tCO_2e)** 2. **Share of recycled or reused materials used (%)** 3. **Materials used after/materials used before (%)**	3	13
		The disposal of machines and devices results in increased electronic waste	1. **Life expectancy of robots (unit of time)** 2. % of recyclability of robots 3. **# of virtualized functionalities** 4. Weight of e-waste (tons) 5. Weight of e-waste properly recovered/reused/recycled (tons)	5	12, 14, 15

Social	Sustainability handprint	Help people perform tasks irrespective of their current capabilities (e.g. heavy lifting)	1. # of newly enabled people to perform a given task, previously restricted by personal abilities 2. Share (%) of 'newly introduced' working force performing these tasks	11	8
		Safer work environment leading to fewer injuries.	1. **# of injuries at work** 2. Level of severity of work-related injuries 3. # of days on sick leave (e.g. stress, severity) 4. Perception of risk (survey 1–10 rank) | 5. Accident insurance premium (cost, rates)	7, 11	2, 3, 4, 5, 8, 11, 16
		Increased trustworthiness of on-time delivery of the expected outcomes	1. % of processes completed in time 2. Time required for completing each process 3. # of times where no resources were available (e.g. cobots busy)	9	9, 16
		Relieving human participation robots could fulfil assistance roles, such as supporting the elderly at home (M2H support)	1. Level of acceptance of cobots by humans (feeling comfortable. Survey) (rating) 2. # of humans who use cobots for assistance roles 3. # of accidents at home, from more accurate and speedy reaction by cobots 4. # of positions required in the healthcare sector, not filled (closing the gap)	10	2, 3, 6
		Increased autonomy by robots supporting people with disabilities	1. Level of acceptance of cobots by humans (feeling comfortable. Survey) (rating) 2. # of humans who use cobots for assistance roles 3. Level of autonomy (according to autonomy scale) 4. Level of safety perception (Survey) (rating)	7	2, 3, 4, 5, 8, 11, 16
	Sustainability footprint	Rendering roles obsolete: may eliminate job roles involving manual, linear, and repetitive tasks	1. # of 'manual, linear and repetitive' jobs eliminated/absorbed by the introduction of cobots 2. # of workers losing their jobs ('manual, linear and repetitive') 3. Ratio of jobs eliminated/jobs created 4. Ratio of workers losing their jobs without getting retrained (upskilled)/workers retrained (upskilled)	11	8
		Humans may be resistant to obtain new skills w.r.t the method of using maintenance of using collaborating robots	1. # of jobs eliminated/absorbed by the introduction of cobots 2. # of new jobs created by the use of cobots 3. # of workers retrained/upskilled 4. # of workers losing their jobs		8

(continued)

Table 2.4 (Continued)

SP	H/F	KV	KVIs (EKVIs & UCKVIs)	H&PG	SDG
Economic	Sustainability handprint	Benefits of robots and cobots might not be distributed evenly among society	1. # of industries or stakeholders that cannot access cobots for a given reason/# total industries or stakeholders, per domain/vertical and/or per geographical area	8	4, 5, 8, 9, 10
		People's privacy may be breached by unauthorized use of robots' and cobots' sensors	1. # of data leaks/breaches/cyberattacks, with personal information compromised 2. # of data Points of personal data gathered by cobots 3. Data volume of personal data gathered by cobots (TB) 4. Total days for which personal data gathered by cobots is retained	9, 15	9, 16, 12, 16
		Social isolation due to robots used in healthcare	1. Level of perceived lonelinessl Assessment from healthcare personnel, regular checks (rating)	10	2, 3, 6
		Increased local and global productivity, cost efficiency and enhanced competitiveness from the use of the collaborative robots	1. Revenue (€) 2. New revenue streams (€) 3. Production time (hours) 4. **Downtimes (hours)** 5. **Non-conformance cost (€)** 6. Other production KPIs (distinctive per factory)	13	8, 9, 10, 13
		New business opportunities may emerge from technology leadership in autonomous robotics from the development of these solutions	1. New revenue or revenue streams in the cobots'-related value chain (€) 2. # of new types of businesses that enter the value chain	13	8, 9, 10, 13
		New business opportunities for old and new businesses from the use of collaborative robots (new markets around robots beyond the factory)	1. **Average price of cobots (€)**	13	8, 9, 10, 13
		New job opportunities in factories for people working together and handling robots requiring new skill sets	1. # of people taking on new roles 2. Various productivity measures to assess higher value jobs/better human capital	13	8, 9, 10, 13

Sustainability footprint	Cobots leading to increased safety including fewer work-related injuries and deaths and consequently savings in healthcare costs, fewer lawsuits etc.	1. # of accidents 2. Average cost per accident (healthcare, productivity, lawsuits) (€)	10, 11, 13	2, 3, 6, 8, 9,10, 13
	High initial CAPEX/investment as a hurdle for starting a business with use of cobots, including building skill sets, buying equipment, etc.	1. # of Businesses using cobots 2. **Initial investment required for setting up Collaborative Robots (€)**	13	8, 9, 10, 13
	Increased reliance on robots/cobots can pose a risk in case of failures or cyberattacks	1. # of cyberattacks 2. Additional costs imposed by failures/cyberattacks per year (€)	9, 15	9, 12, 16
	Monopolization risks due to winner-take-all dynamics if e.g. 6G-connected devices use proprietary specifications or operational systems which start to dominate markets (communication still relies on open 6G standards)	1. # of enterprises involved in the industry of creating cobots 2. # of platforms exploiting cobots 3. # of open APIs included in the solution for integration 4. # of reported incompatibilities for interoperability aspects	13	8, 9, 10, 13
	The use of collaborative robots that are autonomous, intelligent, and collect data can raise new regulatory and legal issues	1. Costs for legal consultation related to *CMR* (€) 2. Indemnity costs (€)	13	8, 9, 10, 13
	Higher maintenance costs compared to traditional robots	1. **OPEX from equipment, maintenance, operation, service, training, etc. of *CMR* (€)** 2. **Running CAPEX (€)**	13	8, 9, 10, 13
	Cost for society from lost jobs and re-skilling people	1. Costs for people's retraining/upskilling spent per year by the company (€) 2. Costs for people's retraining/upskilling spent per year per country (€)	13	8, 9, 10, 13

Table 2.5 Value requirements mapping to technical requirements.

Categories	Value requirements	Example EKVIs	Technical requirements
Energy	Minimize the energy usage for the correct operation of robots, both on communication links and in the data processing, data storage, and data collection aspects.	• Energy consumption per process/overall (MWh) • Energy used in operation phase (MWh) • Energy used in data transfer (optimizing packets and volume of data) (MWh) • 'Data efficiency' on storage (TB)	• Energy-efficient Network operation • Energy neutral devices • Energy-efficient AI/ML training and inference
Materials/ waste	Reduce material dependency, foster longer-lasting materials/products, use materials where it is really necessary, choose materials with lower environmental impact, etc. The latter spans from infrastructure-related materials to operational aspects, where 6G-related services can have an impact.	• Total life-cycle emissions (tCO2e) • Share of recycled or reused materials used (%) • Materials used after/Materials used before (%) • Life expectancy of robots (unit of time) • # of virtualized functionalities	• Modularization • Virtualization • Softwarization • Compute offloading • Resilience • Predictable low latency
Safety	Ensure that the systems are reliable, both in their ability to meet the required communication aspects always and accurately interpreting the information coming from different sources.	• # of injuries at work	• Ubiquitous connectivity • Resilience • Explainability
Trust-worthiness/ privacy/ security	Ensuring the operation of CMRs is reliable, data is kept secure, and outcomes are properly understood.	• # of data leaks/ breaches/ cyberattacks, with personal information compromised	• Reliability • Trust and security • Explainability
Costs	Ensuring that adoption and deployment of CMR is affordable. Focus on the 6G-related aspects that contribute to the affordability of the solution.	• Average price of cobots (€) • Initial investment required for setting up Collaborative Robots (€) • OPEX from equipment, maintenance, operation, service, training, etc. of CMR (€) • Running CAPEX (€)	• Low CAPEX • Closed-loop Control • Reliability • Monitoring and telemetry • Low energy/energy neutral • Zero-touch • Service exposure

Table 2.5 (Continued)

Categories	Value requirements	Example EKVIs	Technical requirements
Productivity/ efficiency	Ensure that 6G delivers the right conditions to enhance the several productivity and efficiency targets, including reliability and resiliency related to the communication aspects.	• Downtimes (h) • Non-conformance cost (€)	• Ubiquitous connectivity • Low latency • AI-native • Reliability • Resiliency • Service exposure • Ensuring SLAs

First, to lower the environmental impact in the application of CMRs, energy consumption and efficiency are key for the communication links, but also in the data processing, data storage, and data collection aspects.

Then, another crucial topic for environmental aspects relates to materials. This encompasses, e.g. reducing the material dependency, fostering longer-lasting materials, using materials where it is really necessary, choosing materials with lower environmental impact, and including circularity principles in the design of infrastructure and devices (e.g. modularity, recyclability, upgradeability).

On social aspects, safety is key. This means ensuring that the systems are reliable, both in their ability to meet the required communication aspects always, as well as accurately interpreting the information coming from different sources.

Then, it is a must to integrate privacy and security mechanisms, together with considerations on what data is being collected (e.g. type, volume, source). These will contribute to the overall trustworthiness of 6G-enabled systems.

Additionally, from an economic perspective, the implementation of CMRs must be affordable. Thus, ensuring that the communication and coverage aspects in the infrastructure deployment are not a bottleneck for the implementation of CMR, i.e. it is easy and not expensive to deploy the required communication environment for a proper application of CMRs.

Finally, CMR are expected to reach the desired outcomes in an efficient and reliable manner. Consequently, it is crucial to ensure that 6G is reliable and resilient, and so, no downtimes in a given production scenario are due to failures in the communication-related aspects.

2.4.9 Feedback into Technical Design

After the KVIs have been extracted from the KVs of the corresponding use case, these have been labelled as either EKVI or UCKVI. EKVIs are related to technical aspects that enable the delivery of a use case, while UCKVIs relate to the broader impacts that the use case creates in the environment.

The next step in the process is to look further into these EKVIs and determine how these can be influenced by the design of the technology. The results of this process are summarized in Table 2.5. First, the KVIs are grouped into categories, to then describe what are the main requirement(s) sought from a value perspective, accompanied by some of the KVIs identified for this requirement. The latter is then analysed to identify how these requirements can be mapped to specific technical requirements (Tiermann and Fichter 2016).

As the main technical mechanisms and/or requirements connected to the desired value requirements have been identified, the process continues by analysing, tweaking, and incorporating these requirements in the 6G design. An example of such corresponds to the technical enabler selection, where the application of this process allowed the identification and selection of previously discarded enablers, thus, strengthening the overall value proposition and contributing to the sustainability of the 6G blueprint.

2.5 Business Models

The business model concept is rooted in strategic management, focusing on a firm's competitive landscape, including, e.g. opportunity, value, and advantage. It serves as a framework for designing, planning, and executing business, especially in technology contexts, where it connects technical potential with economic value. A classical visual tool for business models is the Business Model Canvas (BMC) (Osterwalder 2004), which outlines nine components: customer segments, value propositions, channels, customer relationships, revenue streams, key resources, key activities, key partnerships, and cost structure.

The development of sustainable business models for new technologies requires a long-term business vision. This implies an integration of economic, environmental, and social sustainability perspectives into the value proposition, expanding the number of relevant stakeholders. New tools have emerged to address these aspects of sustainability. Sustainable business models such as Sustainable Business (Tiermann and Fichter 2016) add to previous tools sustainability elements such as environmental and social costs and benefits, a circular design perspective, and the need to define a safe space for business that avoids harming people or the planet. Other tools focus on the innovation ecosystem, like the Ecosystem Pie (Talmar et al. 2020) bringing into the analysis the stakeholder value, the related resources, and culture. Taking into account the different elements of existing business modelling tools, the Hexa-X-II project developed its own ecosystem-level business modelling approach for analysing the ongoing technical 6G developments.

2.5.1 Business Modelling for 6G Ecosystem

Hexa-X-II project has developed a new ecosystem-level business modelling approach for 6G ecosystem that is applicable to use case-specific business modelling at the ecosystem level. It builds on the approaches presented in D1.2, 1.3, and 1.4 (Almodóvar and Vinagre 2023; Bantouna 2024; Ramos 2025). The developed business modelling methodology consisted of three steps for developing the ecosystem-level business model for the use cases: (i) ecosystem BMC, (ii) stakeholder analysis, and (iii) ecosystem pie visualization.

The new ecosystem-level business modelling approach for 6G ecosystem is built upon the existing models and expands them by taking elements to adjust for sustainability at the ecosystem level. At the same time, it also caters to firms' business models. The following steps were taken for each use case:

- Step 1: Develop a generic ecosystem BMC that consists of the following main elements:
  - Ecosystem Value Propositions including Value Proposition
  - Ecosystem Co-value Creation Process including Value Co-creations and Value Co-destructions,
  - Ecosystem Outcomes/Value Capturing including Benefits, Costs, Co-pricing and Revenue Sharing.
- Step 2: Conduct a Stakeholder Analysis where key stakeholders for the use cases are identified and analysed by presenting the following elements:
  - Stakeholder's description, Value proposition, Value co-creation towards other stakeholders, Benefits, Costs, Revenues and pricing, Activities and Resources.

Table 2.6 Cooperating mobile robots business model canvas.

	Ecosystem value propositions
Value proposition (the intended value arising from a customer-oriented solution characterized by a system-level goal)	**Value proposition: Improved efficiency, quality, safety, security, flexibility, and reliability from collaborative mobile robots conducting complex tasks in a coordinated manner within pre-defined geographical spaces.** This can result, e.g., in less injuries for workers in dangerous environment, higher efficiency of conducting tasks without breaks, reduced likelihood for collisions, enabling of large-scale automatization, introduction of intelligence to adapt to changes, new 'robots as a service business models', high consistency in quality, and avoiding dangerous and monotonous work. Furthermore, this can increase circularity in operations using robots to sort materials, reuse components and allow flexible reaction to customer requests, creating innovative ways and a paradigm shift beyond automatization for manufacturing/managing/construction in specific environments.
	Ecosystem co-value creation process
Value co-creations (the value that is co-created in the ecosystem between the stakeholders for the end user, depending on but not controlling each other)	Co-creating and enabling a total solution for robots, machines and humans to conduct complex tasks jointly, ensuring that the robots and components of different providers cooperate for the improved efficiency, quality, security, flexibility, and reliability.
Value co-destruction (the value that is destroyed including obstacles for creating value and hindering realization of the use case)	Actions that hinder other stakeholders in the ecosystem from together creating value for users and customers including, e.g.: • potential lack of collaboration due to interoperability challenges between components from different providers, hindering innovation and the ability to achieve economies of scale • lack of interoperability between hardware and software hindering robots' ability to cooperate • high initial investment for some stakeholder • holding back information from relevant stakeholders • actions to dominate or control the developing and provisioning of the solution, limiting easy access to and use of own components • monopolization risks due to winner-take-all dynamics
	Ecosystem outcomes/value co-capturing
Benefits (benefits related to environmental, social, and economic aspects from ecosystem value co-creation describing what kind and how much value created by the ecosystem is captured through different means)	Environmental, social, and economic benefits from, e.g.: • higher resource efficiency and productivity of processes • improved reuse of resources • enhanced competitiveness • cost savings from same solution used in multiple locations • high precision leading to less waste • improved safety for workers • new business from developing and manufacturing robot/cobots and services, • improved flexibility • opportunity for automizing production and continuous operations • allowing remote operations • enabling setting up factory rapidly and producing closer to customer with potential environmental and social benefits from bringing production back to Europe.

(Continued)

Table 2.6 (Continued)

Costs (costs related to environmental, social, and economic aspects covering different costs from delivering the value proposition)	Environmental, social, and economic costs from, e.g.: • manufacturing and operating of robots • replacing people with robots and the need to learn new skills • coordination of robots for their resource usage including energy/charging • back up for robots in case of failures • high upfront investment from deployment of whole system, calling for a phased approach.
Co-pricing and revenue sharing (ways of pricing and sharing revenues in the ecosystem)	Revenues could be collected with different pricing models from, e.g.: • solution as a service, where the whole solution is from one major player that partners with others in the ecosystem • payments as you go per robot • fixed price per component for 'do-it-yourself' • pricing based on customers' improved efficiency or other values • building owner investing in building with the robots and rents the facility as a service/contractual agreement • manufacturing (construction, logistics) as a service.

- Step 3: Visualize the ecosystem using the Ecosystem Pie for each use case consisting of the following elements:
  - The ecosystem's value proposition, value co-creation, value co-capture and key stakeholders' roles.

In Step 1, the value analysis for the use case is done at the ecosystem level considering the ecosystem as a whole. The value proposition refers to the intended value arising from a customer-oriented solution characterized by a system-level goal. Value co-creations refer to the value that is co-created in the ecosystem between the stakeholders for the end-user, depending on but not controlling each other. Value co-destructions refer to the value that is destroyed including the obstacles for creating value and hindering realization of the use case. Benefits refer to the environmental, social, and economic benefits from ecosystem value co-creation describing what kind and how much value created by the ecosystem is captured through different means. Costs related to environmental, social, and economic aspects cover different costs from delivering the value proposition. Co-pricing and revenue sharing refer to the ways of pricing and sharing the revenues in the ecosystem.

In Step 2, the value analysis is done at the stakeholder level by first identifying the key stakeholders of the ecosystem in the given use case and then analysing the stakeholders in more detail with respect to each stakeholder's value proposition, value co-creation towards other stakeholders, benefits, costs, revenues and pricing, activities, and resources. The elements analysed for each stakeholder in Step 2 are similar to Step 1 ecosystem-level elements, while activities and resources are additional elements. Activities are the mechanisms to generate productive contributions to the ecosystem. Resources are the key tangible and non-tangible resources including assets, skills, capabilities, etc. needed for value creation.

In Step 3, the value analysis from Steps 1 and 2 is presented in a graphical illustration that brings together key stakeholders' analysed elements in one figure. The selected elements for the ecosystem pie visualization include ecosystem value proposition in the middle and stakeholder specific sectors that include stakeholder's value proposition, value co-creation, and value co-capture that is a merge of benefits and pricing.

Table 2.7 Stakeholder analysis modelling of cooperating mobile robots use case.

Stakeholder	Description	Value proposition to 6G use case user/customer	Value co-creation towards other stakeholders in ecosystem	Benefits	Costs	Revenues and pricing	Activities	Resources
Mobile network operator	Provider of local/wide-area connectivity services, in particular for the local area of robots. Can be public, or local private network	To offer and operate the network infrastructure for running collaborative robots in the target area, where investment and costs in network can be justified by the use and users, and what they pay for deploying and running the network	Ability to optimize and easily adjust network compatibility and real-time processing and low-latency network solutions to use	New business opportunities and markets (e.g. connectivity as a service)	Infrastructure initial investments, maintenance, energy and material consumption	Connectivity-as-a-service, pay-per-user, network slicing as a service as a service, edge computing services, data monetization, revenue sharing	Operate local/wide-area network. Handle relationship with network owner/financer - and users, access rights of network. Authentication	Local/wide-area active (and passive network). Human operation and support
Network infrastructure provider	Vendor of communication network infrastructure	To deploy and optimize use of active infrastructure for different purposes and customers, extract synergy effects, keep installed base maintained and up to date	Optimized network compatibility and real-time processing and low-latency network solutions. Providing network equipment which constitute a predictable foundation for mobile network operation, with flexibility to enable adjustment to varying use case and customer requirements	New business opportunities and markets (e.g. infrastructure as a service)	Infrastructure initial investments, maintenance, energy and material consumption	Subscriptions, pay-as-you-go, revenue sharing, bundled offerings, infrastructure-as-a-service, data monetization, marketplace, licensing	Optimizing use of servers and installations, maintenance, sales and customer relationships, energy optimizing activities	Servers, cooling installations, contracts for green and cost-efficient energy, licenses for platform SW

Provider of robots and cobots	Providers of robots and cobots	To provide a device made ready to autonomously carry out collaborative tasks according to the demand from the customer in a specific context	To provide robots/cobots that are easy to integrate, set-up, and use, and maintain	Increased productivity, resource efficiency, new business opportunities and markets (e.g. cobots as a service)	Infrastructure initial investments, maintenance, energy and material consumption, training costs, elimination of job roles involving manual, linear, and repetitive tasks	Cobot-as-a-service, subscriptions, revenue sharing, bundled offerings, training-as-a-service, data monetization, predictive maintenance as a service, pay-per-task	Research, design, manufacturing (outsourced), sale/distribution	Design processes. IPR, knowledge. People. Own factories - or outsourcing expertise
Modem chipset manufacturer	Manufacturers of modem chipsets that are integrated into different devices/equipment	To provide components for the robots/cobots equipment to connect to networks	To provide compatible modems which are easily integrated into devices/equipment	New business opportunities and markets (e.g. reduced price per quantity)	Investments in manufacturing modems	Price per quantity	Design and development, sales and distribution, manufacturing (outsourced)	Design processes. IPR, knowledge. People. Own factories – or outsourcing expertise

(Continued)

Table 2.7 (Continued)

Stakeholder	Description	Value proposition to 6G use case user/customer	Value co-creation towards other stakeholders in ecosystem	Benefits	Costs	Revenues and pricing	Activities	Resources
Provider of software for collaborative robots	Provider of software and applications needed for the robots to function	To provide software for robots to collaborative conduct tasks enabling, e.g. efficient and automatic choice of network 'paths'	Easy to integrate, set up and use of applications	New business opportunities and markets (e.g. subscription software-as-a-service)	Develop, maintain solution/service; customer service	Subscriptions software as a service, licensing, pay per user, custom software development, revenue sharing, software bundling, training, analytics as a service, open-source platforms, predictive maintenance, cloud integration	Design and development, sales and distribution, management of access/licenses, etc.	Algorithms, application, IPR and copyright, experts, vertical domain knowledge
System integrator	Integrator of components from different providers	To quick, easily, and cost-efficiently recombine, reconfigure, and handle many types of components into offering	Develop and maintain insight and skills about different solution components and their capabilities and how to best handle them and the requirements from users. Seek information from qualified sources	New business opportunities and markets (e.g. cobot-as-a-service)	Labour costs, licensing costs	Subscriptions, revenue sharing, outcome-based pricing, managed services model, cobot as a service, analytics and insights as a service, training, pay per model	Handle standard development and open-source community processes from start to result. Mobilize contributors. Recruit and manage financing institutions. Manage 'brand' and trust	Experts. Trust, legitimacy. Level of deployed results (standards). Accessibility. Processes and systems for developing and publishing

Manufacturer	Company that operates the manufacturing site where the collaborative robots are located	To improve manufacturing and process efficiency by means of collaborative robots	Open and precise about problems to be solved and achievements wanted in short and long term	Increased productivity, resource efficiency, new business opportunities and markets (e.g. training as a service), cost savings	Training costs, May eliminate job roles involving manual, linear, and repetitive tasks The disposal of machines and devices results in increased electronic waste	Cobot-as-a-service, customization and co-development, subscriptions, revenue sharing, training-as-a-service, data monetization, predictive maintenance as a service, outcome based	Define processes and manufacturing activities, define requirements, operate robots	Site, infrastructure, other manufacturing machinery and devices
Co-workers of robots	People working with robots	To improve process efficiency in collaboration with robots	Collaboration with the collaborating robots to conduct tasks	Safer work environment, ability to perform tasks beyond human capabilities, relieving human participation by using robots	Switching costs, training, social isolation due to using robots	Not applicable	Operate robots and work with robots	Skills and know-how

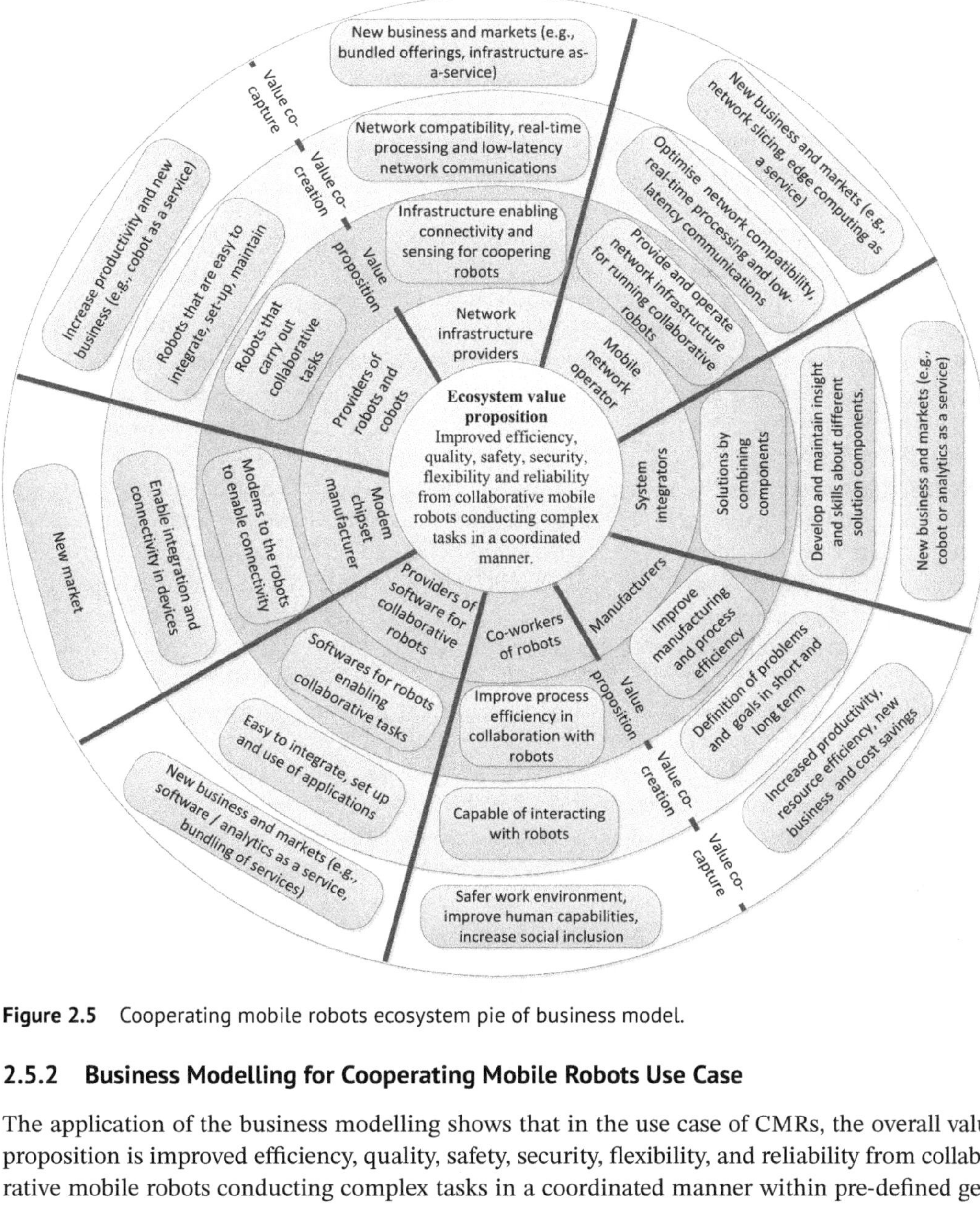

Figure 2.5 Cooperating mobile robots ecosystem pie of business model.

2.5.2 Business Modelling for Cooperating Mobile Robots Use Case

The application of the business modelling shows that in the use case of CMRs, the overall value proposition is improved efficiency, quality, safety, security, flexibility, and reliability from collaborative mobile robots conducting complex tasks in a coordinated manner within pre-defined geographical spaces. Table 2.6 summarizes the output of Step 1 presenting the ecosystem-level BMC of the CMRs use case.

The stakeholder analysis presenting Step 2 of the business modelling of CMRs use case is shown in Table 2.7.

Finally, the ecosystem pie visualization presenting the output of Step 3 of business modelling is shown in Figure 2.5.

2.6 Conclusions

Sustainability is referred to as environmental, social, and economic sustainability, which are three interconnected pillars that form the basis for sustainable development. A legitimate and successful development and implementation of 6G technologies must respect all three pillars as well as cater to and balance the expectations of stakeholders.

Sustainability encompasses two aspects, namely, '6G for Sustainability' that focuses on the opportunity for 6G to support solutions that will promote different sectors of society in getting more environmentally, socially, and economically sustainable and 'Sustainable 6G' that targets a 6G system that is environmentally, socially, and economically sustainable.

This chapter provided a new way of exploring a sustainability analysis when it comes to use cases. This use case and sustainability analysis process was presented in this section. It helps bringing together specific KVs and their indicators for each of the three sustainability domains. Moreover, it can help identify some KVs, considered to be more important and on which a focus should be put at first.

The chapter reminded the 20 identified use cases by Hexa-X-II, all of which were enabled by the innovative capabilities that 6G will provide. These were then grouped into six families, and subsequently, one use case per family was chosen for further exploration. This detailed analysis not only encompassed the technical aspects and requirements but also approached the potential impact that the deployment of such use case can have on the larger ecosystem. This analysis enables a series of consequent actions, which span from aspects that can be taken into consideration upon the design of the technology itself, all the way to highlighting key points to acknowledge upon deployment and which lay beyond what the ICT industry can control (e.g. policy, behaviour). An example of such process was carried out for the CMRs use case and presented in the present document.

Finally, business modelling was used as tool to study the economic value of 6G through the developed use cases. The business modelling approach consisted of ecosystem BMC, stakeholder analysis, and ecosystem pie visualization, providing a new tool for values-driven technology development while considering sustainability aspects. The extended stakeholder analysis demonstrated how their varying interests and capabilities together will shape the future of 6G use cases and systems. Hexa-X-II exchanged views with many stakeholders to ensure that new insights contribute to informed decision-making. Furthermore, the engagement with stakeholders uncovered public perceptions, attitudes, and concerns about technological advancements, including perspectives from public authorities, industry representatives, and civil society.

Acknowledgement

This chapter is based on the valuable research and findings from the Hexa-X-II project in its Work Package 1 (WP1) – Value, Requirements, and Ecosystem. The editors and contributors would like to express their sincere gratitude to all individuals who have participated in the work. Additionally, we would like to extend special thanks to (in an alphabetic order):

Karin Ahltin, José Luis Almodóvar Chico, Michael Bahr, Aimilia Bantouna, Sokratis Barmpounakis, Rishikesh Chakraborty, Stéphanie Char, Giovanna d'Aria, Panagiotis Demestichas, Geir Egeland, Patrick Everaert, Nelly Giannopoulou, Marie-Hélène Hamon, Samira Jaber, Maria Kounalaki, Lisa Lessing, Nafiseh Mazloum, Ali Rezaki and Tommy Svensson for their valuable contributions, having played a key role in shaping the insights of the sustainability work within the Hexa-X-II project and, consequently, also in the development of this chapter.

Acronyms and Abbreviations

AI	Artificial intelligence
AMR	Autonomous mobile robot
AR	Augmented reality
BMC	Business Model Canvas
CMR	Cooperating mobile robot
E2E	End to end
EKVI	Enabler key value indicator
eMBB	Enhanced mobile broadband
H&PG	Human and planetary goals
ICT	Information and communication technology
IMT	International Mobile Telecommunications
ITU	International Telecommunication Union
KPI	Key performance indicator
KV	Key value
KVI	Key value indicator
LCA	Life cycle analysis
ML	Machine learning
mMTC	Massive machine-type communications
NTN	Non-terrestrial networks
SGD	Sustainable development goal
SIR	Seamless immersive reality
TN	Terrestrial networks
UC	Use case
UCKVI	Use-case key value indicator
UN	United Nations
URLLC	Ultra-reliable low latency communications
VR	Virtual reality
XR	Extended reality

References

5GPPP (2023). New edition of 5G PPP stakeholders pictures and glossary. [Online]. Available at: https://5g-ppp.eu/update-of-the-5g-ppp-stakeholders-pictures-and-glossary/ (accessed 1 February 2024).

Almodóvar, J. and Vinagre, C. (2023). Deliverable D1.2 – 6G use cases and requirements. [Online]. Available at: https://hexa-x-ii.eu/results/ (accessed 11 March 2025), also at https://doi.org/10.5281/zenodo.16779960.

Bantouna, A. (2024). Deliverable D1.3 – environmental and social view on 6G. [Online]. Available at: https://hexa-x-ii.eu/results/ (accessed 13 March 2025), also at https://doi.org/10.5281/zenodo.16780232.

Bergek, A., Jacobsson, S., Carlsson, B. et al. (2008). Analyzing the functional dynamics of technological innovation systems: a scheme of analysis. *Research Policy* 37: 407–429.

Gawer, A. and Cusumano, M.A. (2014). Industry platforms and ecosystem innovation. *Journal of Product Innovation Management* 31: 417–433.

Hallingby, H.K., Fletcher, S., Frascolla, V. et al. (2021). *5G Ecosystems*. Brussels: 5G Infrastructure Association.

Hexa-X-II Project Consortium (2023–2025). Hexa-X-II – European level 6G flagship project. [Online] Available at: https://hexa-x-ii.eu/ (accessed 4 March 2025).
International Telecommunication Union (2008). REPORT ITU-R M.2134 – requirements related to technical performance, s.l.: s.n.
International Telecommunication Union (2015). Recommendation ITU-R M.2083-0 – IMT vision – framework and overall objectives of the future development of IMT for 2020 and beyond.
International Telecommunication Union (2020). Recommendation ITU-R M.1457-15 – detailed specifications of the terrestrial radio interfaces of International Mobile Telecommunications-2000 (IMT-2000), s.l.: s.n
International Telecommunication Union (2023). Recommendation ITU-R M.2160 – framework and overall objectives of the future development of IMT for 2030 and beyond, s.l.: s.n.
Osterwalder, Alexander. (2004). The Business Model Ontology – A Proposition in a Design Science Approach.
Ramos, E. (2025). Deliverable D1.4 – 6G value, requirements and ecosystem. [Online]. Available at: https://hexa-x-ii.eu/results/ (accessed 6 May 2025), also at https://doi.org/10.5281/zenodo.15772633.
Stockholm Resilience Centre (2015). What is resilience?[Online]. Available at: https://www .stockholmresilience.org/research/research-news/2015-02-19-what-is-resilience.html (accessed 3 April 2025).
Talmar, M., Walrave, B., Podoynitsyna, K.S. et al. (2020). Mapping, analyzing and designing innovation ecosystems: The Ecosystem Pie Model. *Long Range Planning* 53 (4): 101850.
Tiermann, I. and Fichter, K. (2016). Developing business models with the sustainable business canvas. Manual for Conducting Workshops. Oldenburg and Berlin, s.n.
United Nations (2015). Sustainable development goals. [Online]. Available at: https://sdgs.un.org/goals (accessed 14 March 2025).
United Nations (n.d.). Human rights. [Online]. Available at: https://www.un.org/en/global-issues/human-rights (accessed 3 April 2025).
Wendt, S. (2023). Deliverable D1.1 – environmental, social and economic drivers and goals for 6G. [Online]. Available at: https://hexa-x-ii.eu/results/ (accessed 4 March 2025), also at https://doi.org/10.5281/zenodo.16778706.

3

Sustainable 6G Platform

Sylvaine Kerboeuf[1], *Akshay Jain*[2], *Diego Lopez*[3], *Raul Munoz*[4], *Pol Alemany*[4], *Behnam Ojaghi*[4], *Luís Pedro Santos*[5], *José María Jorquera Valero*[6], *Manuel Gil Pérez*[6], *Gregorio Martínez Pérez*[6], *Pietro G. Giardina*[7], *Huy Q. Tran*[1], *Gunes Kesik*[8], *Noelia Perez Palma*[6], *Antonio Skarmeta*[6], *Pawani Porambage*[9], *Josué Castañeda Cisneros*[13], *Elham Dehghan Biyar*[10], *Ioannis Tzanettis*[11], *Grigorios Kakkavas*[11], *Anastasios Zafeiropoulos*[11], *Xosé R. Sousa*[12], *and Patrik Rugeland*[10]

[1] *Nokia Bell Labs, Paris, Saclay, France*
[2] *Nokia Bell Labs, Espoo, Finland*
[3] *Telefonica, Seville, Spain*
[4] *Centre Tecnològic de Telecomunicacions de Catalunya (CTTC-CERCA), Castelldefels (Barcelona), Spain*
[5] *Ubiwhere, Aveiro, Portugal*
[6] *University of Murcia, Murcia, Spain*
[7] *Nextworks, Pisa, Italy*
[8] *Ericsson Research, Istambul, Turkey*
[9] *VTT Technical Research Center of Finland, Oulu, Finland*
[10] *Ericsson Research Istanbu, Türkiye*
[11] *Institute of Communication and Computer Systems, National Technical University of Athens, Athens, Greece*
[12] *Optare Solutions, Vigo, Spain*
[13] *Ericsson Research, Stockholm, Sweden*

Unlike its predecessors, the sixth generation (6G) transcends the traditional role of simply connecting users and applications. It must not only provide connectivity and data delivery but also process data, generate insights, and offer additional valuable services like compute services, artificial intelligence (AI) services, and spatial/temporal data services. 6G aims to become a comprehensive platform, offering a flexible and adaptable set of services tailored to the ever-changing needs of diverse applications and verticals. This platform should create an open and accessible environment for a myriad of application providers and developers to build and deploy innovative solutions. The platform should enable seamless and adaptable access for diverse users and devices, and integration of new service types and their exposures. The expansion of 6G, as a platform encompassing new services and an evolving ecosystem, should be achieved in a manner that considers and prioritizes sustainability across social, economic, and environmental dimensions.

To guide the evolution towards a sustainable platform, the establishment of an end-to-end (E2E) system blueprint is crucial for ensuring a cohesive system that can adapt to future needs responsibly. Specifically, the design of the blueprint must prioritize key values-based design, ensuring a 6G system with long-term viability, minimizing negative impacts, and having a positive impact on our world. This includes incorporating fundamental architectural design principles grounded in the sustainability pillars that shape the sustainable 6G platform. These principles are outlined in Section 3.1.1 and prioritize simplification via streamlining operations and procedures, enhanced security and privacy, and environmental sustainability, ensuring a robust and responsible system.

6G to Build a Sustainable Future, First Edition. Edited by Mikko A. Uusitalo, Patrik Rugeland, Mauro R. Boldi, and Ahmad Nimr.

Then the 6G platform must support the 6G use case families, such as immersive experiences, physical awareness, collaborative robots, and a fully connected world, which require compute offloading, data-driven architecture, sensing and positioning, ubiquitous connectivity, high bit rates, predictable low-latency, etc. The chapter further analyses the key requirements of the system derived from all the use cases, both in terms of required functional capabilities and expected key performance indicators (KPIs) and key value indicators (KVIs).

To accommodate the unique requirements and design principles of 6G, a detailed blueprint of the E2E system is needed, integrating infrastructure, network, application enablement platform, and application layers. This blueprint incorporates pervasive functionalities across all layers, including management and orchestration (M&O) for realizing full-scale automation and seamless management of E2E services across the cloud continuum, security, privacy, and resilience, and an AI framework and data management framework enabling real-time intelligence with data-centric workflows and sophisticated AI capabilities. In addition to the system blueprint, a structured design process guides the development and integration of technical enablers for adhering to the design principles and for achieving the 6G E2E objectives, both in terms of KPI and KVIs.

Furthermore, to adapt to the 6G complex ecosystem, the use of Intent-Based Management (IBM) towards E2E services M&O is introduced, especially for the foreseen multi-stakeholder scenarios. To this end, Section 3.3 describes the expected functionalities and functions of an IBM solution managed by a digital service provider (DSP) such as a network operator (NOP). It presents a set of four enabler-based use cases for E2E services management that demonstrate how IBM can streamline the M&O operations by removing human-dependent actions and making the M&O solutions more autonomous and efficient.

The chapter further reflects upon the paramount importance of trustworthiness in 6G, which extends the security and privacy, encompassing resilience, availability, dependability, and ethical compliance. This chapter addresses, in Section 3.4, the growing security challenges posed by the increased complexity and reliance on advanced technologies, including the new vulnerabilities introduced by the widespread use of AI, the evolution of cyberattacks, and the potential for quantum computing to break current encryption. New concepts to ensure security by design for 6G are presented, as well as their integration in the 6G system blueprint.

3.1 Sustainable 6G System: Principles and Requirements

The evolution of advanced technologies, including extended reality, real-time environment sensing, digital twins (DTs), and AI, is placing increasing demands on computing power, data storage, network infrastructure, and energy consumption. 6G must embrace a new paradigm where technological advancement goes hand-in-hand with environmental, economic, and social sustainability. In addition to improving the traditional performance, the 6G platform must prioritize energy efficiency, cost optimization, and responsible resource utilization, ensuring a sustainable and responsible system. As 6G expands into new frontiers with technologies like integrated sensing and communication (ISAC) capability for large-scale monitoring, also known as Joint (or Integrated) Communication and Sensing (JCAS/ICAS) and pervasive AI-driven network operations, concerns regarding trustworthiness become paramount. A robust blueprint must address these concerns. 6G must bridge the digital divide, ensuring access to networks and devices for all, regardless of geographical or economic constraints.

Therefore, in order to comply with the mentioned needs, the design of the 6G system should be guided by architecture design principles that prioritize environmental, economic and social sustainability. These principles are presented in Section 3.1.1, also considering their impact on the

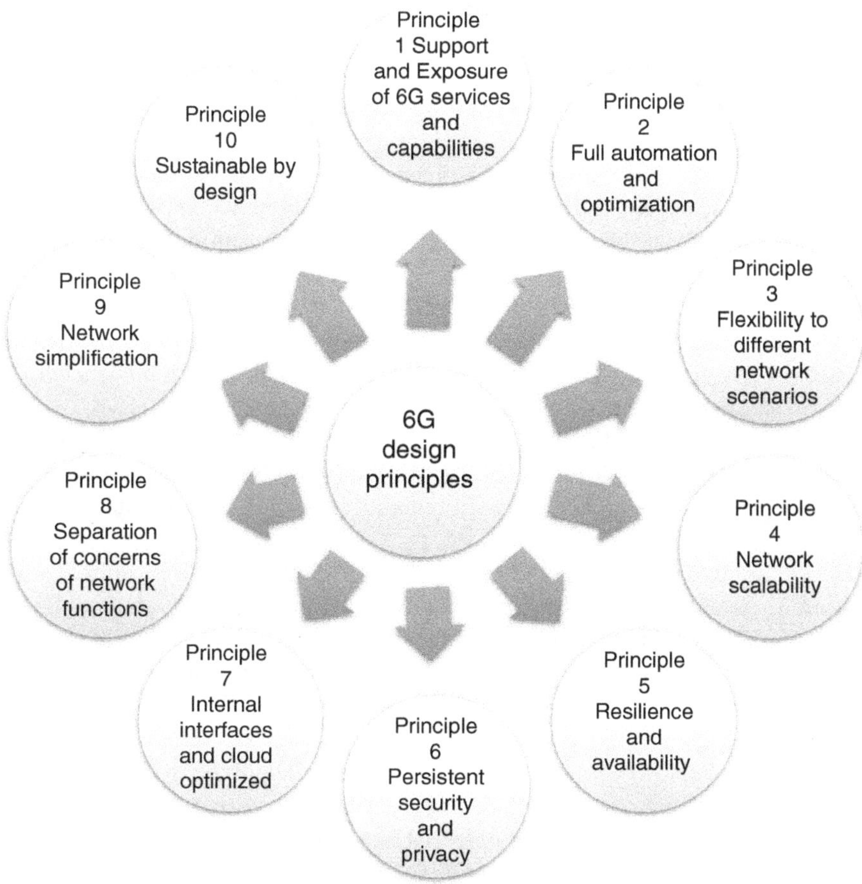

Figure 3.1 6G design principles.

6G E2E system. Future 6G systems will have to cope with stringent requirements from use cases comprising immersive experiences, collaborative robots (cobots), and DTs. These use cases, outlined in Chapter 2, impose strong functional and operational requirements in terms of modularity and scalability, resilience and security, overall sustainability goals and intent-based automation. These requirements, including the considered KPIs and KVIs, are described in Section 3.1.2.

3.1.1 Design Principles

The design principles for the 6G system, as introduced in the European Union (EU) 6G flagship projects, Hexa-X (Ericson et al. 2021), and Hexa-X-II (Porambage et al. 2023a), provide fundamental guidelines for the design and development of the 6G E2E system. These design principles, illustrated in Figure 3.1, are summarized below.

Principle 1 – Support and exposure of 6G services and capabilities: The 6G system should be able to provide comparable or superior user experience for all existing services, for example, voice and mobile broadband, while allowing gradual network deployment without disrupting legacy services. Additionally, it should enable efficient incorporation of novel 6G digital services, such as sensing and computational offloading, and be able to expose new or existing capabilities to E2E applications.

Principle 2 – Full automation and optimization: The 6G system should be AI-native, i.e. integrate AI agents and functionalities wherever beneficial to enhance performance and efficiency in network and service management automation, as well as to support new functionalities. This will include both observability as well as IBM. The system should also support continuous orchestration over multiple stakeholders and domains, in an E2E approach.

Principle 3 – Flexibility to different network scenarios: The 6G system should be flexible to address different types of scenarios and topologies, including local networks, wide area networks, non-terrestrial networks (NTNs), public and private networks, and utilization of new spectrum in an efficient way.

Principle 4 – Network scalability: The 6G system should be scalable to different networks and deployments, from small to large wide area networks, low-power wide area, fixed wireless access, energy neutral access, or private networks and should be capable of dynamically scaling network resources to minimize energy consumption while fulfilling the requirements.

Principle 5 – Resilience and availability: The architecture should allow network deployments with high resilience and availability, which is matched to the service needs. This includes support of separation of control plane and user plane, resilient mobility solutions, as well as subnetwork resilience.

Principle 6 – Persistent security and privacy: The 6G system should be able to address current as well as future threats in a resilient manner and incorporate security fundamentals in its design. Furthermore, the 6G system should inherently support the preservation of privacy and allow different levels of anonymity, also for future services such as sensing, offloading, or AI-as-a-Service.

Principle 7 – Internal interfaces are cloud optimized: Network interfaces should be designed to be used in a cloud-native environment, utilizing state-of-the-art cloud technologies, design patterns, platforms, and IT tools in a coherent and consistent manner. Care should be taken to design proper service separation to maximize the potential for service reuse and allow for service innovation with minimal integration efforts (plug-and-play).

Principle 8 – Separation of concerns of network functions: The network functions should have bounded context ("*separation of concerns*") and all dependencies among services should be through their application programming interfaces (APIs), with minimal dependency on other network functions. This allows network functions to be developed and replaced more independently from each other and allows efficient scaling of network resources. The functionality of network functions should not be duplicated.

Principle 9 – Network simplification in comparison to previous generations: The network architecture should be streamlined to reduce the complexity of radio access network (RAN) and core network (CN) functions with fewer (well-motivated) parameters to configure and fewer external interfaces, to maximize innovation, and to reduce time to market. It will be important that the new 6G services do not introduce detrimental complexities for the legacy services.

Principle 10 – Minimizing environmental footprint and enabling sustainable networks: The design of the 6G system should ensure that the environmental footprints can be justified, and any increased footprint should be clearly balanced with added value, cost efficiency, and societal benefits. The architecture design should make it possible to disable or deactivate functions that are not in use, either in the short-term or the long-term. As an example, to ensure zero power at zero load, one should be able to turn off the radio frequency transceiver chains without any loss in performance when there is network load again. It should also facilitate circular practices through higher modularity of components to limit the replacement to only the degraded part, or compatibility with software updates to extend the lifetime.

3.1.2 Requirements

The requirements from the 6G system can be categorized into two aspects:

- What the 6G system should do: These are functional requirements, which define the 6G system capabilities that it should achieve.
- How the 6G system should behave or the constraints under which it must operate: These are non-functional requirements, which specify the system quality characteristics, including the performance, security, reliability, and scalability.

Building upon the technical requirements for the collaborative robots use case and the other 6G use cases presented in Chapter 2, this sub-section provides a comprehensive overview of functional and non-functional requirements for the 6G E2E system, as defined in Porambage et al. (2023a, b) and Zou et al. (2024).

3.1.2.1 Functional Requirements

These requirements, as identified by Hexa-X-II and other Smart Networks and Services Joint Undertaking (SNS-JU) projects, are briefly outlined, with key requirements highlighted in Figure 3.2.

Ubiquitous connectivity and indoor coverage: As more and more mobile services become imperative to everyday life as well as indispensable for efficient industries, the expectation for connectivity wherever you are is continuing to rise. At least basic 6G services (e.g. video streaming and simpler low-tier extended reality [XR] services) should be available everywhere, to facilitate truly global services, and to provide connectivity to unconnected or under-connected areas and regions.

6G systems will have both terrestrial networks (TNs) and NTNs as possible deployment solutions. Hence, the 6G system should consider convergence between TN-NTN systems with respect to how the spectrum resources should be shared. Furthermore, it should also consider the interworking between Third-Generation Partnership Programme (3GPP) and non-3GPP radio access technologies (RAT).

Furthermore, a large share of the current data traffic is consumed indoors, either at home or in offices, whereas the majority of mobile connectivity is provided via outdoor base stations

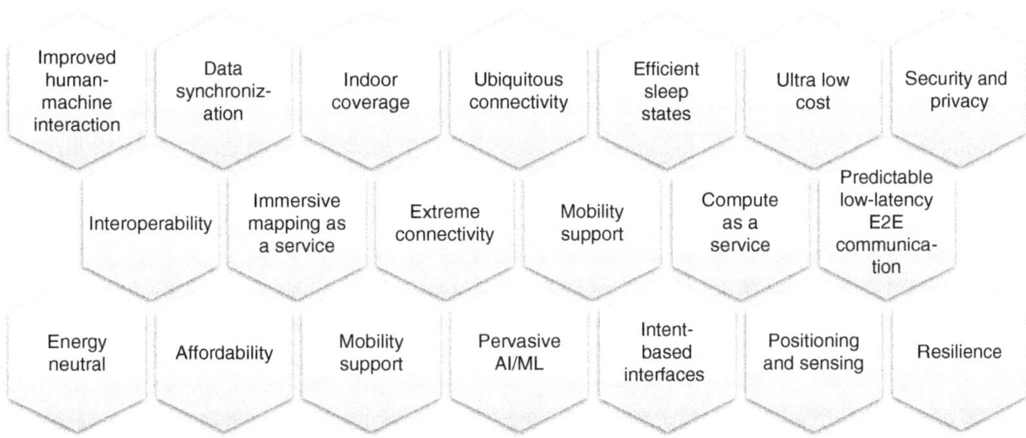

Figure 3.2 Main 6G system requirements.

(Ericsson Mobility Report 2021). Although the propagation characteristics and penetration losses from outdoor-to-indoor, which are further pronounced at higher carrier frequencies, limit the performance, the users will expect full 6G service indoor as well as outdoors.

Extreme connectivity with high bit rates and predictable low-latency: Although most of the use cases require modest experienced bit rates, it is expected that a few niche applications, such as fully immersive merged reality, will push the envelope towards extreme performance beyond what is possible in current generations.

On the other hand, even though fifth-generation (5G) systems are capable of delivering millisecond air interface latencies in specific deployments, in most cases it is not the best-case or even average latency which is relevant to provide the quality-of-experience (QoE) or to fulfil the delay and timing requirements for machine type communication, but instead the bounded worst-case or a predictable low-jitter latency would be more relevant to avoid processing chokeholds or sudden spikes in delay.

Multi-connectivity: The 6G system should support aggregation of contiguous and non-contiguous bands in all available frequency bands, from single and multiple transmission points for communication and beyond-communication services (e.g. sensing) to increase instantaneous bandwidth and reliability. The solution should avoid complex user equipment (UE) configuration schemes.

Mobility support and service continuity: The distinctive feature of mobile networks is, of course, the mobility, which will naturally continue to be of prime relevance. This accounts for connectivity at high speeds, where Doppler effects begin to affect the radio channels, as well as the expectation for service continuity during mobility.

Moreover, certain services and use cases can operate with relatively long service interruptions, e.g. file downloading or cached video streaming, where several seconds of interruption can go unnoticed. However, for other use cases, even a few milliseconds of interruption can have dire consequences, e.g. inducing nausea in an XR application or losing control of a remote-controlled vehicle. The service continuity is particularly important during mobility, when a device moves out of coverage of its current connection, the network needs to seamlessly hand over the connection to another node with sufficiently low service interruption.

Pervasive AI: The development of AI is progressing exceedingly fast, which is likely to impress expectations from the end-users on the availability of AI service anywhere. This entails access to AI services from any device, to optimize performance and enhance applications, as well as the availability of the AI services, wherever there is network coverage. The 6G network should have AI built in to the functionalities of the network. The network should also support learned and/or customized waveforms in the form of predefined look-up tables or via other means of agreement between the transmitter and the receiver. Additionally, AI will also be applied as an essential tool in M&O, security, and privacy frameworks and in data analytics to improve the performance of the network as well as to serve the third-party applications/network applications (Nw Apps). Moreover, with the implementation and integration of AI models in the 6G system, it will be vital to monitor and evaluate different AI models' complexity in both training and inferencing. In addition, the scalability of the AI models will depend on how easily it will be to generalize the AI models (Simeone and Sifaou 2024).

Compute and immersive mapping as a service: As one of the beyond-communication key services, compute as a service (CaaS) from the end-user devices to the network is expected to enable more advanced services for longer durations on power- and computationally constrained devices, where the bulk of the processing is handled in the network.

Furthermore, as the 6G network will collect vast amounts of sensing data, it will be important for the network to be able to compile, analyse, and distribute this sensing data as an immersive mapping service to different consumers of the service.

Positioning/sensing, data synchronization, and improved human–machine interaction: By leveraging the radar properties of radio propagation, 6G is poised to broaden the utility of mobile networks, to provide cost-efficient detailed positioning and sensing capabilities to end-users.

Also, as more users experience and consume common digital services, e.g. a joint immersive experience, it will be important to be able to ensure accurate time synchronization between them. If multiple users are to interact simultaneously with the same digital object, they need to be accurately time-synchronized from an end-user point of view and might also require accurate positioning.

It is further expected that 6G, utilizing positioning/sensing and time synchronization alongside incorporating new devices with unparalleled interfaces, will seek to improve human–machine interaction, e.g. immersive audio/visual mobile experiences as well as haptic.

Intent-based interfaces: With networks becoming more and more heterogeneous and complex, end-users will expect the network to discern the users' preferred connectivity option (e.g. necessary bandwidth, or utilization of ISAC or offloading) without any overt commands from the user.

New 6G capabilities exposure: The 6G system should provision a generic and dynamic exposure functionality beyond communication services, comprising sensing, enhanced localization and tracking, and CaaS. Furthermore, it should be able to expose new and existing capabilities via APIs to developers of applications. Such APIs should be simplified (abstracted) so they can be used by application developers without the need to be network experts. Those APIs can provide services that third-party application developers can consume with their familiar tools and processes while still providing them access to a wide set of network capabilities (e.g. application flows steering). Additionally, it should also enable controllability by the NOP of the capabilities that are exposed to third parties (tenants).

Privacy protection: Privacy, like previous generations, encompasses confidentiality of data transmission and storage, data minimization and explicit support for consent and transparency. Enhanced privacy capabilities should be developed to support the novel ISAC capabilities and the evolution of trustworthy AI in 6G. Avoid any malicious disruption (tampering, interception, impersonation, etc.) of networks, devices, and services. The network should support the definition and verification of privacy requirements, incorporating them into service-level requests and agreements.

Interoperability: To provide service continuity, it will be important that the 6G devices can interwork with other wireless technologies, especially supporting fallback to, e.g. 5G. Standardized interfaces will continue to be relevant to keep specific open interfaces, e.g. lower-layer split (i.e. between medium access control (MAC) and physical (PHY) or within PHY) should be considered to make the network more flexible and more manageable.

3.1.2.2 Non-functional Requirements

The 6G system is expected to provide functionality, which will not be directly visible to the end-users but is required from the NOP's perspective to efficiently fulfil the use case requirements.

Scalability: Network scalability has been identified as a design principle, as presented in Section 3.1.1. However, to implement this, it will be important to define specific requirements to achieve efficient and sustainable networks. To provide this, it will be important that the network can scale from a small number of nodes or users to the largest number of nodes and users expected for the use case with similar performance (Lopez Da Silva 2023).

Reliability and resilience: With increasing reliance on mobile services for personal, commercial, and industrial applications, the expectation for service reliability will be much more pronounced, and users will assume that services will remain available even during mobility or at unforeseen events.

Additionally, in order to ensure fulfilment of the service level agreement (SLA), it is important to ensure continuous operation and service continuity, even in the case of unexpected disruptions. In particular, the resilience is the networks' ability to mitigate or quickly recover from disruptions caused either by attacks impacting security or privacy, as stated above, or any other causes, e.g. in terms of natural disasters, power outages, or mechanical failures.

Security: Similar to previous generations, security will continue to be crucial in 6G to prevent malicious disruption (tampering, interception, impersonation, abuse, etc.) of network, devices, and services. 6G security faces threats from legacy vulnerabilities, increased complexity due to AI integration, and evolving cyberattacks leveraging AI and the future quantum computing. The network should support security automation that encompasses dynamic monitoring, attack detection, decision-making, and remediation actions to orchestrate security and enforce policies. Furthermore, the network should support the definition and verification of security requirements, incorporating them into service-level requests and agreements. A comprehensive set of security requirements can be found in the SNS-JU projects RIGOUROUS (ICT-FI 2023), HORSE (Siachos et al. 2023), PRIVATEER (Pastor et al. 2023), and ROBUST-6G (Valero and Perez 2024).

Flexible radio protocols: For the 6G system to achieve its potential, one of the methods is to introduce a flexible protocol stack in the RAN. This should consider approaches such as radio resource control (RRC) simplification, improving/optimizing the data recovery and reordering mechanisms, making security features more optimized, and in general, ensuring as much flexibility as reasonably possible.

Unified mobility procedures: For the mobility aspects, the 6G system should support a solution wherein the mobility procedures provisioned at multiple layers, such as layer-1 mobility and layer-3 mobility, may be unified. Furthermore, it should be expected that the 6G system mobility procedures are agnostic to network architecture/deployment options. Moreover, prevalent methods should be further optimized for 6G, including support for frequency range 3 (FR3) centimetre-wave (7–15 GHz). Additionally, the inclusion of AI and ISAC towards improving mobility procedures should also be considered.

Intent-based management: Automation in 6G network operations should leverage a controller/manager architecture based on IBM and closed-loop automation. IBM can be applied at the different management layers. Within each management layer, intents are used to define the goal state, which the closed loop (CL)-based controller then tracks. AI methods are employed within each layer to perform interpretation of the intent, modelling the network's state, and making decisions to ensure the desired goal is achieved. Status information is provided as feedback to intents that are issued.

Seamless orchestration across the compute continuum: The 6G system should support the deployment of 6G services over heterogeneous resources across the device-to-edge-to-cloud continuum. It should also support intent-driven orchestration for deployment of cloud-native applications with strict quality-of-service (QoS) requirements (e.g. latency-sensitive application components) across resources in the Internet of things (IoT)-to-edge-to-cloud continuum.

Furthermore, the 6G system should support interaction between network providers and over-the-top (OTT) players (edge/cloud providers) for optimal deployment of applications.

Support of 6G service delivery across multiple digital service providers: Managing 6G E2E services requires the involvement of several service providers (e.g. network service providers and cloud service providers) to compose an E2E service. For example, an application may require 6G network resources such as RAN, transport, and CN services from one NOP, transport network resources from another NOP, and cloud computing resources from hyperscalers. Hence, the 6G system should support the composition and orchestration of multiple services coming from different DSPs to provision and assure the 6G E2E service. It should also provide aggregator means for interfacing with different service providers and for exposing their services to third parties, such as verticals.

Extreme transport capacity: Any mobile network will need to be able to efficiently transport the data from the base station to the CN, and further to the remote servers. To handle the increased capacity of 6G, the FLEX-SCALE project (Uzinidis et al. 2023) identified a need for transport capacity beyond 100 Terabit per second (Tbps).

Efficient sleep states and energy neutral: With sustainability and energy-efficiency as key driving forces for the development of 6G, it will be imperative to efficiently conserve energy when it is not needed and that devices and equipment can enter partial or fully power-saving sleep-modes to extend the operation, as well as minimizing the cost and environmental impact of the service. Next, a large detriment to current machine-type communication devices is their limited lifetime. Even if the battery lifetime is ten years, the lifetime of the equipment and machinery it supports is often several decades (e.g. in a factory), which will necessitate semi-regular battery replacements. By enabling energy-neutral operation, e.g. through energy harvesting, the device lifetime could be extended beyond the lifetime of the systems they are embedded in. This could enable deployments in harsher environments, where access for battery replacements can be omitted.

Ultra-low-cost and affordability: While certain 6G use cases predict unprecedented performance, others lie at the other extreme end of the range, enabling near-zero-cost devices to communicate via the network. These devices would be possible to distribute and embed nearly anywhere and provide simple connectivity to nearly anything.

Moreover, for the sake of social sustainability, it will be important that at least basic levels of 6G services and devices should be affordable to be able to bridge the digital divide. For instance, the 6G system should be designed to cost-efficiently expand the service coverage with flexible topologies and integration of NTNs, as well as fully leverage AI capabilities for automation to reduce the operating expenses (OpEx), enabling cheaper services to be offered.

Key value and key performance requirements: While the last described requirements are crucial for the 6G system's sustainability, the sustainability aspects have been further emphasized through the consideration of KVIs introduced in Chapter 2. For example, the value requirements and enabler KVIs, such as energy consumption, energy usage, total life-cycle emissions, and recycled/reused material percentage, are presented in Chapter 2, for the collaborative robots use case.

The performance requirements for 6G systems are diverse and complex, reflecting the various use cases identified in Chapter 2. These requirements are not always met simultaneously or by the same network deployment.

For immersive experiences like virtual reality and gaming, 6G needs to achieve low-latency (under 10 ms for split rendering, 50 ms for voice, and 150 ms for collaboration), high data rates (up to 250 Mbps uplink and downlink), high area traffic capacity (up to 250 Mbps/m^2), high reliability (99.999%), and accurate positioning (under 10 cm horizontally and vertically).

Collaborative robots in warehouses require data rates up to 10 Mbps, positioning accuracy up to 1 m, low-latency (under 0.8 ms), high reliability (99.99999%), and support for mobility up to 20 km/h.
Network-assisted mobility, a physical awareness use case, demands data rates up to 100 Mbps, latency under 20 ms, high reliability and availability (99.99%), wide coverage (99.9%), and support for high device connection density (up to 104 devices/km^2).
DTs for real-time applications like industrial processes or smart cities require high reliability (99.99999%), data rates up to 100 Mbps, wide coverage (99.99%), accurate location (under 10 cm), and low latency (a few milliseconds).
The fully connected world envisioned for 6G requires latency under 100 ms, high reliability (99.999%), availability of 98.5%, data rates up to 25 Mbps, and wide coverage (99.9%), including indoor coverage.
Finally, trustworthy environments for human-centric services require latency under one second, high reliability (99.999%), location accuracy up to 10 m, positioning accuracy up to 1 m, and relative positioning accuracy of 0.1 m.

3.2 Blueprint of the Sustainable 6G Platform

6G era represents a significant evolution in mobile networks, moving away from its traditional role of simply connecting users and applications. 6G aspires to become a comprehensive platform, offering a flexible and adaptable set of services tailored to the ever-changing needs of diverse applications and verticals. In this dynamic ecosystem, the 6G platform will create an open and accessible environment for application providers and developers to build and deploy innovative solutions.

To guide this transformative evolution and ensure a cohesive system that can adapt to future needs, a robust blueprint is essential (Kerboeuf et al. 2024). As a technical realization of the 6G platform, the 6G E2E system blueprint is designed according to the architectural principles driven by the key values of environmental, economic, and social sustainability, as outlined in Section 3.1. This blueprint should provide an efficient framework for integrating the new functionalities and novel technologies, ensuring they meet the requirements of 6G. Furthermore, recognizing 6G's role as a platform serving applications, it is also important to demonstrate how services and functionality will be exposed to applications and users. The 6G E2E system blueprint encompassing the novel aspects is built through an iterative process. This process begins by analysing 6G use case requirements, the architecture design principles, and the ecosystem environment to identify key capabilities. These capabilities are then evaluated against system performance and key values (social, economic, and environmental sustainability) to ensure a positive impact. This iterative design approach, incorporating technology enablers for various aspects of the 6G system, aims to create a blueprint that aligns with real-world needs and expectations, fostering a more adaptive and responsive design. The resulting blueprint developed in Hexa-X-II (2023) is presented in Section 3.1.1. The iterative design approach, along with the methodology to select the appropriate 6G enablers that will realize the blueprint, is further elaborated in Section 3.1.2.

3.2.1 E2E System Architecture

The 6G E2E system blueprint, illustrated in Figure 3.3, introduces several novel aspects (Le Sauze and Kerboeuf 2025). Compared to previous mobile generations, the 6G system is cloud native via the cloud continuum and able to expose service and capabilities via the application enablement

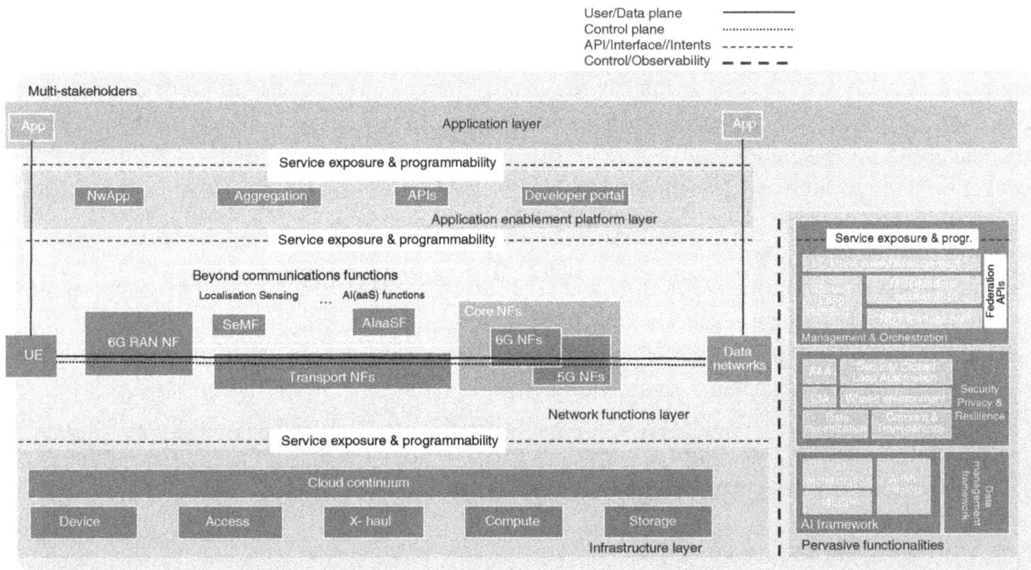

Figure 3.3 6G E2E system blueprint.

platform and application layers, providing a unified platform for developers and third-party applications A sustainable and flexible network functions layer leverages existing 5G core functions, enabling a cost-effective and streamlined migration to 6G, while introducing new 6G capabilities. 6G devices connect via a single 6G RAT, further simplifying the transition. Pervasive functionalities address massive usage of AI in-network and are exposed to the end-user. They also streamline M&O of multiple technologies and distributed infrastructures, involving diverse stakeholders, and address E2E security and privacy of 6G. These aspects are further elaborated in the following paragraphs.

3.2.1.1 Infrastructure Layer

The layer sits at the lowest level of the 6G E2E system blueprint, encompassing the physical components that handle most 6G network transmission. It provides essential resources, including **Device** hardware, **Access**, **X-haul**, **Compute** and **Storage**, to support the various functionalities of the 6G system. **Device** hardware encompasses not only end-user equipment, such as smartphones, laptops, and other wireless/wired equipment, but also IoT devices and other intermediary devices that access the internet and process data. 6G introduces new device categories tailored for specific applications (Balakrishnan et al. 2023), ranging from low-data rate energy harvesting devices to high-data rate devices designed for immersive experiences like extended reality (Vega et al. 2020).

Access technologies connect the user devices to the CN; including the RAN enabling today's 4G and 5G connections (composed of physical infrastructures such as base stations, small cells, and antennas) and soon 6G connections, the wired access distributed through fibre optics and other cables that handle broadband connections, and Wi-Fi connections through access points and routers. 6G access should leverage diverse frequency spectrum, employing lower frequency bands for broad coverage and higher frequency bands for ultra-high data rates and enhanced radio sensing. The access layer of the 6G E2E system will integrate a wide range of deployment types and radio architectures, each tailored to specific environmental characteristics, coverage requirements, mobility conditions, and connection densities. The 6G access infrastructure will be optimized for multiple coverage options, ranging from localized networks in personal devices or body area networks to wide area coverage. It also comprises the access infrastructure of the integration of NTN/TN for supporting seamless and ubiquitous connectivity (Harounabadi and Heyn 2023). Moreover, 6G access will cater to temporary deployments, where mobile infrastructures, such as flying base stations, are critical (Mishra et al. 2021). Altogether, these features of the access contribute to increasing the network flexibility. More details are provided in the flexible network-specific view of the system blueprint outlined in Chapter 5.

The interconnected resources in the infrastructure layer enable the E2E data transmission. Stretching across the whole infrastructure layer, the **Cloud continuum** is a platform that abstracts and aggregates resources across public and private cloud services, on-premises, and edge infrastructures. While considering resources from the end devices to the central cloud, a continuum of those resources between adjacent organizations could be foreseen. With the development based on numerous virtualisation technologies, these resources can be abstracted from their physical representation, enabling the Cloud continuum to easily and flexibly partition the underlying resources and optimize resource utilization for the Network functions layer.

3.2.1.2 Network Functions Layer

The layer consists of the logical network functions. The **UEs** comprise the logical network functions residing in the devices used directly by end-users to communicate. The protocol stack of the UEs is controlled by the network through signalling procedures, such as those for application deployment or resource allocation. These UEs can collaborate among themselves for more efficient communications, allowing for device-to-device connections through dedicated subnetworks, like personal networks or unmanned aerial vehicle (UAV)-assisted networks.

6G RAN network functions cover the different RAN deployment scenarios for various stakeholder needs, through centralized, distributed, cloud-based, etc., envisioning an architecture where network functions related to the RAN will consist of physical network functions and virtualized network functions. The increased softwarization enables programmability, autonomous orchestration, and analytics-based CL resource control of these network functions. The blueprint depicts 6G RAN network functions only connected to the evolved Core, highlighting the streamlining of the architecture by eliminating the possibilities of having multiple options, compared to 5G. More details are briefly introduced but not represented on the blueprint figure for space considerations. A low-layer split architecture, with baseband functions in central processing and physical layer functions in radio unit, is crucial for building scalable and effective distributed massive input, massive output (MIMO), as outlined in Chapter 4. Multi-connectivity will be achieved through a single, enhanced carrier aggregation architecture. To facilitate a single-step migration to 6G standalone, multi-RAT spectrum sharing will enable dynamic 5G-6G spectrum sharing, more efficiently than in 4G-5G sharing solutions, see in Chapter 9. 6G radio protocols streamlining will prioritize improved control coordination, enhanced mobility management, and adaptiveness, building upon 5G and supporting new capabilities like ISAC, NTN, and AI/machine learning (ML). Details are further developed in Chapter 5.

CN functions are based on an evolution of the previous 5G core and comprise both 6G and 5G Network functions. The recent advancements in softwarization techniques and the service-based architecture of the 5G CN enable modular CN functions to be flexibly deployed, involving the user plane to functions close to the network edge and the control plane to functions in more centralized locations. The separation between control plane and user plane is further supported by the existence and integration of both edge and central cloud infrastructures. Detail on modular network functions is outlined in Chapter 5.

Beyond-communication functions extend 6G capabilities to a new level and provide novel services. Network-based sensing data can be collected from connected IoT sensors and further augmented by AI, giving the 6G network new functionalities and services such as the sensing management function (SeMF), to facilitate efficient coordination of sensing procedures, or the Artificial Intelligence as a Service Function (AIaaSF), responsible for providing accessible AI capabilities, like pre-built AI models, datasets, algorithms and tools into applications. Additional details of the necessary beyond-communication functionalities are outlined in Chapter 5 with a specific view of the novel services on the 6G E2E system blueprint.

The evolution of **Transport network functions** is driven by the development of software-defined networking (SDN), enabling the programmability and dynamic management of system elements in the network infrastructure. This allows for virtualized transport network functions to be dynamically designed, deployed, activated, and removed with minimal or non-existent service disruption, according to the evolution of transport infrastructures as well as the changes of transport service demands. For example, a control and management solution for programmable and context-aware transport enables QoS guarantees for specific network slices (Akgul et al. 2025).

The **Data Network (DN)**, encompassing in standardized 3GPP architecture the functionalities related to services provided by the operator through Internet access or third-party services, is going to be more integrated in 6G. While DNs are not usually considered part of mobile networks, the increased usage of edge and cloud infrastructures for current and future network deployments considers them relevant for the scope of the 6G E2E system blueprint. Stronger cooperation is needed between the DN (including edge computing), CN functions, and beyond communication functions related to compute offloading (see in Chapter 5). This collaboration is crucial for meeting the E2E QoS requirements of time-critical edge application.

3.2.1.3 Application Enablement Platform Layer

The layer acts as a catalyst for new 6G use cases by providing applications an easy access to the full range of 6G services. The capabilities of the layer are detailed below and further explored with Figure 3.4, which develops an exposure-centric view of the 6G E2E system blueprint.

In this layer, the **APIs** are abstracted from the underlying network APIs, M&O APIs as well as **Nw App** APIs, simplifying their use for non-experts. These APIs may be standardized, for example, through CAMARA (Linux Foundation 2024), to further enable an easier and seamless access to the 6G capabilities. Nw Apps are centred around network capabilities and are under network control. For example, using data collection analytics and AI models, an Nw App can enable additional intelligent operations and optimizations across the network. Another example would be an application providing enabling services like device location management. The NW Apps can expose their APIs possibly externally via the service exposure of the application enablement platform layer, so they can also be leveraged by applications. In addition, supporting the complexity of the 6G ecosystem involving multiple stakeholders, the platform enables the **Aggregation** of services from multiple 6G service providers, over multiple networks, facilitated by federation APIs.

However, to ensure that the applications can build upon the 6G platform APIs, their providers need to be supported with additional functionalities (① in Figure 3.4). The support includes operational and management (**OAM**) services for API consumers to integrate the APIs from service perspectives (such as available APIs discovery, API access ordering, etc.). The platform provides a **Developer portal**, offering resources like specifications, documentation, software development kits (SDKs), and a test environment to facilitate development. The layer must also address the challenges associated with exposing APIs externally for commercial use. This includes concerns related to monetization, authorization, and compliance with legal regulations, such as privacy

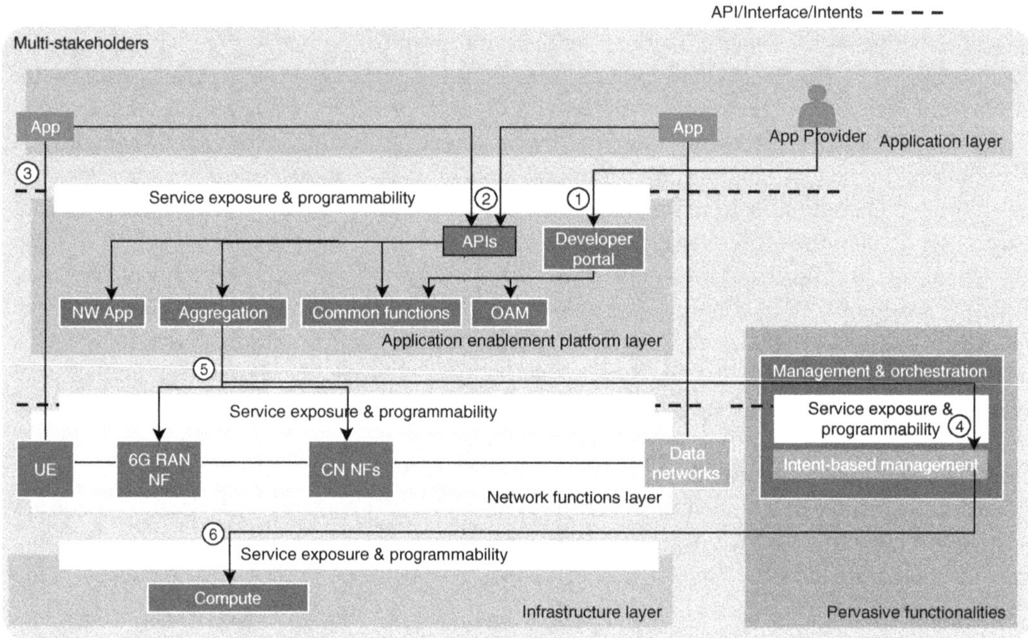

Figure 3.4 Exposure-centric view of the 6G E2E system blueprint.

laws. To facilitate this, **Common functions** leverage existing frameworks like the Common API Framework (CAPIF) Core Function, defined by 3GPP for northbound APIs (3GPP, 29.222 2022), which provides services to address these challenges. By offering common functionalities for monetization, authorization, and compliance, the layer can simplify the implementation and management of commercially exposed APIs.

Given the diversity of 6G capabilities, the service exposure framework must be adaptable to the requirements of each capability. It should support both out-of-band, over the top (② in Figure 3.4) or through the user–network interface (③ in Figure 3.4), and in-band (through transport protocols) access channels. Each of these access methods presents its unique requirements for the service exposure framework, including protocols to use, consumer technical capabilities, and trust in the consumer. By accommodating these various access and their requirements, the application enablement platform layer ensures a flexible and adaptable service exposure for the 6G platform.

The introduction of new types of services in 6G presents new requirements across the layers for enabling their exposure. The first relates to technical delivery requirements. While simple representational state transfer (REST) APIs may suffice for many services, some require specialized delivery mechanisms to meet stringent performance demands. For example, a sensing service might require the delivery of a large stream of spatial data in real-time (Akgul et al. 2025). The second kind of requirements is the integration with the underlying network capabilities. Many 6G APIs aim to provide programmatic control over 6G services, enabling applications to manage or influence their behaviour. For example, a 6G API will allow configuring the desired QoS for a specific UE. This relies on a high degree of programmability of the underlying 6G capabilities. The increased programmability introduces new demands on the management capabilities exposure framework (④ in Figure 3.4), which must be able to expose management services that were previously used only internally, potentially requiring enhancement in areas like automation, scalability and security. For instance, service management automation with an intent-based approach could facilitate the exposure of service lifecycle management requests, possibly up to vertical applications, while simplifying network operations through scalable resource orchestration via cloud continuum APIs (⑤ in Figure 3.4). The management capabilities exposure framework supports the creation of multi-domain services, which combine capabilities from different domains (⑥ in Figure 3.4). For example, a 'quality on demand' service might require integrating capabilities from the RAN, CN, and operational support system (OSS) to configure and monitor connectivity. To enable this, the service exposure framework needs to facilitate the integration of services from multiple domains, and the application enablement platform layer should provide tools like workflow engines and information translation services to enable seamless composition.

3.2.1.4 Application Layer

The application layer sits atop the 6G E2E system blueprint, representing how the applications might have some service expectations from the network itself. These expectations concern application requirements such as bandwidth, latency, security, and scalability.

In the 6G system blueprint, the network functions layer and application enablement platform layer expose both network information and network control APIs through the service exposure & programmability, towards the application layer. This allows developers to specify their application requirements to the network, as well as adjust according to current network conditions and capabilities.

M&O operations on the services can also be exposed to the application ecosystem, such as developers, verticals, and service providers, through developer-friendly intent-based abstractions and human-oriented intent-based APIs.

3.2.1.5 Pervasive Functionalities

The pervasive functionalities facilitate the four layers of the 6G E2E system blueprint to realize the full potential of the 6G platform. This facilitation is achieved independently or through joint operation, as these functionalities may permeate the entire system blueprint.

The **data management framework** is designed to support the complex data requirements of 6G system. A comprehensive data collection is essential for gathering diverse data from various network domains and layers. This framework should support cloud-native observability, enabling the collection and processing of various signals to generate valuable insights. These signals encompass a wide range of monitoring data, including QoS metrics, resource consumption, virtual function and microservice-specific metrics, distributed traces, and logging information. Data fusion techniques are employed to combine these semantically interlinked data, leveraging data representation schemes. This allows for the development of open and extensible analysis processes, extracting valuable insights that can be fed into operational or business support systems of network providers. The framework adheres to open telemetry specifications, ensuring compatibility and interoperability with industry standards. At the core of the data management framework, data operations (DataOps) automate the lifecycle management of distributed data across the cloud continuum. It enables the availability of high-quality data, ensuring its accuracy and consistency. It establishes robust data governance, addressing stringent data security and privacy, and ensuring compliance with regulations.

The **AI framework** is essential for realizing the full potential of 6G, enabling a data-driven architecture with distributed intelligence. This framework will be deeply integrated into the core of 6G, unlike in 5G, driven by the need to leverage 6G infrastructure flexibility for dynamic resource allocation and optimization, cope with the increasing complexity of 6G networks and services, and unlock new revenue streams through novel AI-enabled services and applications. This systemic application of advanced AI techniques requires a support framework spanning all layers of the 6G E2E system blueprint. This framework encompasses both domain-specific and cross-domain AI, ranging from specialized AI for specific tasks to AI spanning multiple layers and functionalities. It also includes AI embedded into network functions and devices. For instance, in infrastructure, AI will drive cost-efficient radio transceiver design that adapts to channel changes and mitigates hardware impairments (Letaief et al. 2022). In management & orchestration, AI techniques will enable zero-touch operations (Kokkonen et al. 2022), including multi-agent systems for resource orchestration across the computing continuum, from the edge to the cloud (Zafeiropoulos et al. 2022). Central to this pervasive application of AI, the AI framework provides the tools and components for advanced AI workflows across the entire 6G ecosystem. The AI Orchestrator dynamically selects and deploys the optimal AI model for specific tasks, ensuring both AI performance and overall network efficiency. The AI/ML Catalogue serves as a repository of pre-trained AI services, models, algorithms, and metadata, using standardized interfaces like AIaaS functions for exposing and accessing AI models as services, enabling seamless integration of AI models into network functions and applications. Finally, the MLOps Orchestrator manages the entire lifecycle of AI models, including retraining and adaptation, using automated pipelines to maintain model accuracy and relevance over time. More details on the AI framework are provided in Chapter 6.

A new **M&O framework** that goes beyond 5G approaches is required to address the 6G's complex ecosystem. This framework prioritizes simplified operations, intuitive interfaces, automated processes, and intelligent decision-making. IBM is a key pillar toward simplification, enabling user-friendly service requests and supporting diverse vocabulary for service deployment, including sustainability targets. However, managing intents from multiple tenants with conflicting service requirements and resource competition necessitates a robust intent-based conflict resolution solution. 6G will face significantly more dynamic traffic at the edge, relying on a vast continuum of heterogeneous resources from extreme edge devices to cloud resources. This distributed and interconnected environment necessitates a shift from centralized management to a distributed but coordinated approach, jointly managing both network and compute resources. Multi-platform orchestration provides unified M&O of network services and applications across this cloud continuum, ensuring seamless management across diverse network environments. Closed-loop control, potentially powered by AI, is crucial for automating and achieving autonomy in 6G network operations. AI will be extensively used for proactive network management, including energy efficiency optimization. Network digital twins (NDTs) management offers a solution for safe AI-based orchestration and control by providing digital replicas of the network infrastructure, enabling 'what-if' scenarios and testing before applying actions to the real network. The system collects data from the real network to ensure accurate feedback, enabling informed decisions about applying actions to the live network. Monitoring and telemetry interfaces play a crucial role in gathering real-time data on network element state and energy consumption, enabling sustainable networking through energy-driven decision-making and supporting the creation of DTs for AI-assisted autonomous networks. Details on the E2E intent-based service management are provided in Section 3.3.

The **security, privacy, and resilience pervasive functionality** affects all layers of the 6G E2E system blueprint, from network design to business and operational aspects. AI integration in 6G networks, while promising, introduces new security and privacy challenges. While 5G focused on traditional threats like eavesdropping, spoofing, and jamming, 6G's pervasive intelligence brings new risks, including attacks on ICT infrastructure and data integrity challenges from malicious users. The streamlined application layers, diverse network function topologies, and integration of edge and end-user devices into a cloud continuum in 6G create vulnerabilities. Data exposure through simplified APIs, potential privacy breaches in new network scenarios, and increased vulnerability of user data in less trustworthy edge devices are key concerns. To address these challenges, 6G requires a value-oriented, E2E system design prioritizing trustworthiness. This involves implementing specific security, privacy, and resilience controls that enforce key principles across the network. These controls encompass confidentiality, integrity, and availability (CIA), achieved through cryptography, redundancy, and quantum-safe solutions; authentication, authorization, and accounting (AAA), ensuring secure access and identity management for both human and non-human entities; data minimization, consent, and transparency safeguarding data privacy and confidentiality; automated CL for security operations through dynamic monitoring, attack detection, decision-making, and remediation actions; and a 'What-if' environment leveraging NDTs for assessing network resilience through simulations of various network conditions using synthetic environments and a common interface for seamless integration. By proactively anticipating and mitigating potential risks through adaptive and trust-oriented security measures, 6G can ensure a secure and reliable network environment. Further details on E2E security concepts are elaborated in Section 3.4.

3.2.1.6 Multistakeholder Support

The 6G ecosystem is characterized by increased interactions between diverse stakeholders, shifting from a linear value chain toward a multi-sided one. This evolution necessitates the definition of new roles as outlined in Section 3.3, including DSPs, capability operators (COPs), and resource providers, alongside various digital service customers (business to business(B2B), Application service providers, aggregators). The multi-stakeholder dimension is particularly relevant in the 6G cloud continuum, where interactions between multiple DSPs offering digital services to customers, and collaboration between COPs and diverse resource providers (network, cloud, AI, applications) for the M&O of E2E services are crucial. Furthermore, integrating resources provided by vertical industries and extreme-edge resources belonging to end-users is essential for aggregating heterogeneous resources and realizing the cloud continuum. This ecosystem involves the multiplication of exposed interfaces from multiple domains, including north-bound and east-west interfaces. Fundamental concepts underpinning this ecosystem are depicted in Figure 3.5. They include controllable capability exposure, where M&O capabilities are offered as services accessible via APIs with access and security controls, and a federation model and federation APIs that enable the aggregation of resources and services across different administrative domains, employing either peer-to-peer or broker-based approaches (Akgul et al. 2025).

3.2.2 Design Process of 6G E2E System

The E2E system design process aims not only to build a 6G system that adheres to the use case performance requirements, but it also aims to address the various sustainability requirements outlined in the design principles and requirements in Section 3.1.

3.2.2.1 Top-Down Versus Bottom-Up System Design

An iterative design process is established for a sustainable E2E 6G system blueprint. This process involves a top-down approach wherein the blueprint is elaborated based on the 6G use-case requirements and the system architecture design principles. As new use cases and associated KPIs/KVIs emerge during the development lifetime of the 6G standard and product phases, the system requirements are frequently revisited, and the considered set of technology innovations is refined or augmented. Conversely, consideration of the technical achievability will give feedback for fine-tuning the use cases and for the consolidation of KPIs and KVIs.

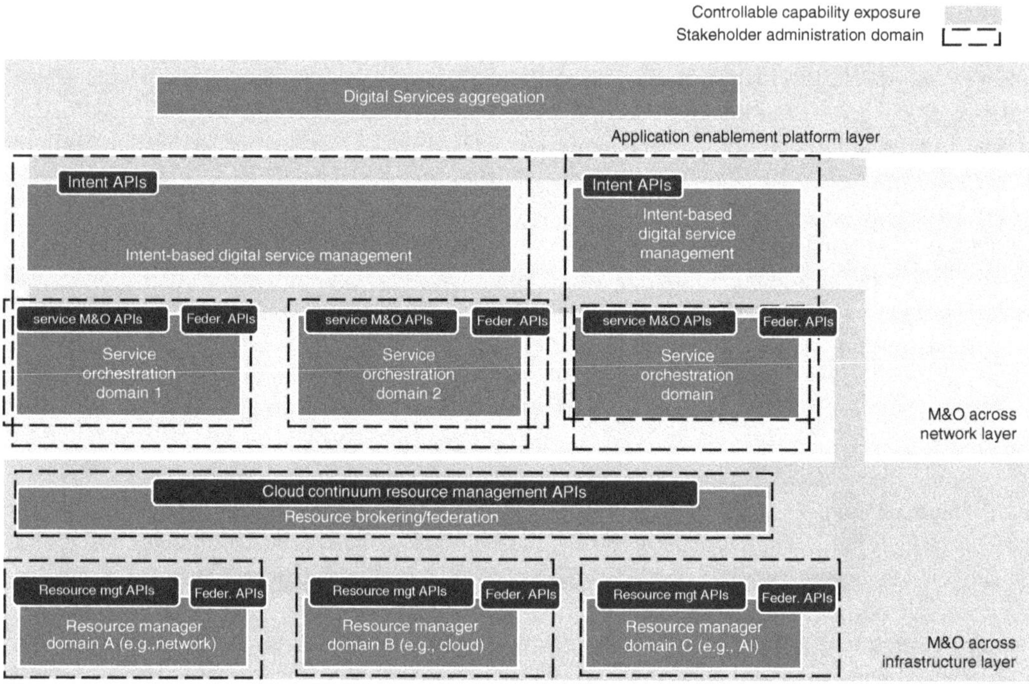

Figure 3.5 Multi-stakeholder centric view of the 6G E2E system blueprint.

A bottom-up approach complements the process, with enablers of the system being developed and selected in separate projects, both in the EU and globally. The enabler selection process evaluates the pros and cons of each enabler, using a checklist to ensure alignment with the E2E performance and operation targets. The E2E alignment process especially carries out several analyses that are further detailed below. This provides feedback for ongoing development of enablers. The analysis results will serve to continually update the 6G E2E system blueprint as well as the enabler design, as enablers become mature during the development cycle of the 6G standard and system.

Figure 3.6 illustrates the iterative E2E system design process, highlighting the reconciliation between the top-down and the bottom-up information flows. Note that the grey part of the figure represents the use case dependency and the KPI/KVI-based iterative system design process. Figure 3.7 outlines the KPI and KVI-based iterative system design process.

The use cases analysis provides the initial requirements, including the targeted KPIs and targeted KVIs, that are expected to be delivered by the 6G system. An initial 6G system design is then developed, integrating the use case requirements in terms of functionalities and potential enablers. Then, this 6G system design is evaluated to estimate KPIs and KVIs. Such an evaluation is usually performed via system-level proof-of-concepts (PoCs), as outlined in Chapter 8.

Next, the E2E alignment process, as depicted in Figure 3.6, analyses enablers in terms of compatibility, commonality, simplification, and unified exposure.

Compatibility analysis: When integrating different enablers together in an E2E system, it is crucial to verify the capability of these enablers to work with each other. For example, whether a proposed protocol/interface provides compatibility to work with other parts of the 6G system. Additionally, the security and privacy implications of the use of these enablers and their interplay must be assessed, incorporating the required controls for addressing them.

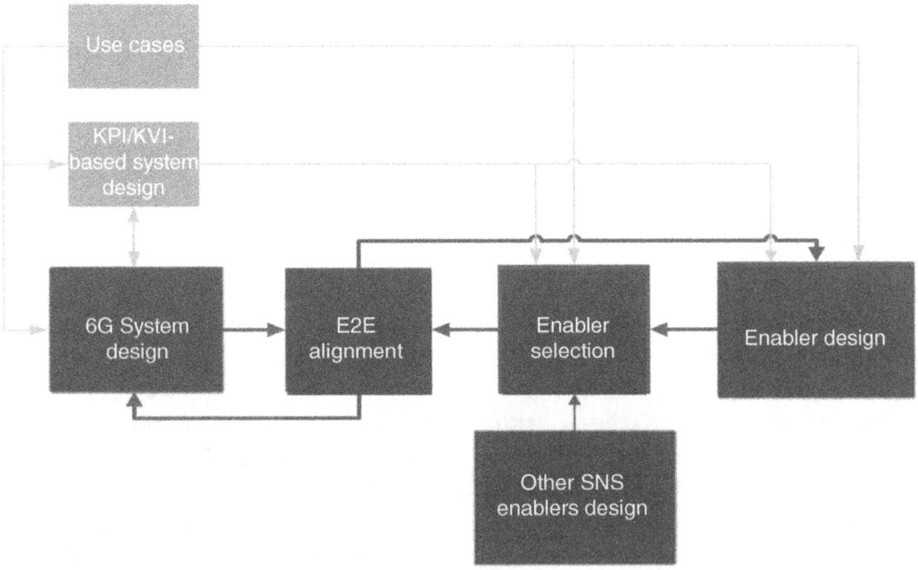

Figure 3.6 Iterative E2E system design process reconciling top-down (light grey arrows and boxes) and bottom-up (dark grey arrows and boxes) approaches.

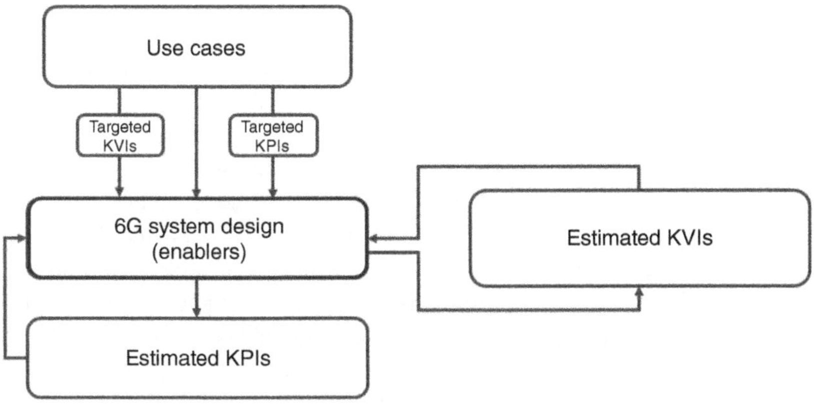

Figure 3.7 Iterative E2E system design process based on KVIs, values, and KPIs.

Commonality analysis: By applying the same design principles, it might happen that similar enablers are elaborated in different projects and studies. The analysis identifies those enabler commonalities and consolidates them into one common enabler. Similarly, different replicas of the same information spread across the system in a distributed manner could possibly be centralized and reused.

Simplification of functionalities: The detection of commonalities could be considered as the first step to simplification (e.g. the same configuration function could be used to configure both the RAN and the Core pub/sub buses). Optimization of the interaction between functions or modules could also be performed to reduce the latency budget (e.g. E2E User Plane latency or Control Plane procedure latency) and meet the requirements. Within this context, the possibility of merging some adjacent enablers, turning their mutual interfaces into internal interfaces, can be considered. Impact evaluation should be provided in this case.

Unified exposure analysis: The analysis checks for uniformization across the various exposures of new interfaces between the network and the vertical service providers. The security exposure functions, provided by the 6G platform for developing new secure and robust services using the API framework, should also be analysed. The exposed APIs can provide analytics, such as network energy measurement, latency, throughput, localization, and sensing information.

This iterative process, with its feedback loops and continuous analysis, ensures the development of a sustainable, high-performing 6G system that meets the evolving needs of users and stakeholders.

3.2.2.2 Enablers Integration in 6G System: A Knowledge Graph-Based Approach

The complexity of 6G system design, involving numerous enablers under exploration during the 6G standardization and system design lifetime, with intricate dependencies and features, necessitates a mechanism that can consider these complex relationships. Building upon the established top-down and bottom-up design process, a knowledge graph (KG) based approach detailed in Zou et al. (2024) has been developed to address this challenge. This approach aims to select a suitable set of enablers for a given set of requirements. By capturing dependencies between recommended enablers, enabler features, design principles, and use case requirements, this approach provides a comprehensive visualization and evaluation framework for myriad enabler combinations, which can then be used to design/deploy the 6G system.

The data-driven KG method, illustrated in Figure 3.8, involves:

Metadata collection: Some metadata is ingested as inputs (e.g. use case, use case required (Req.) KPI/KVI, enablers, functionality provisioned, enabler provisioned (Prv.) KPI/KVI, enabler priority, network statistics that provide insights into enabler performance, etc.) The measured/analysed KPIs and KVIs are collected from the PoCs analysis of Alemany and Ojaghi (2024) and from the use case requirements determined in Almodovar and Vinagre (2023). The relevant KVs and KVIs (qualitative) are collected from the use case requirements analysis in Ramos (2025). Collected metadata related to the technological enablers consists of: maturity level (Technical Readiness Level), requirements for update to standards, criticality towards migration, requirements for new hardware entities, dependencies on other enablers, existence/presence of interfaces, protocols, etc., for communication/management/information exchange with the other dependent enablers, design principles fulfilled, qualitative impact of the technical enablers on the collected KPIs, i.e. positive, neutral, or negative impact on the KPIs.

Noisy KG generation: An initial KG is created, which consists of all nodes, which in the case of the 6G system design are enablers, and edges incorporating the collected enabler data, the use case KPI requirements, and other inferred network statistics, if the network is already operational.

Prioritization step: This step prioritizes the enablers and their dependencies that are essential for the 6G E2E system design. Prioritization is obtained from metadata inputs.

Clustering: Certain inputs might provide the same level of functionality, or they may achieve similar objectives. Hence, those inputs are clustered on the basis of functionality.

Graph pruning: This step eliminates the unessential nodes and edges from the KG. The goal is to identify the most critical enablers and their dependencies for achieving the defined objective while adhering to the constraints. This step involves using a harmonization/combination function F and a quantitative threshold ψ to evaluate the essential nature of the nodes and their corresponding edges. The output of this stage is a pruned KG containing only the selected

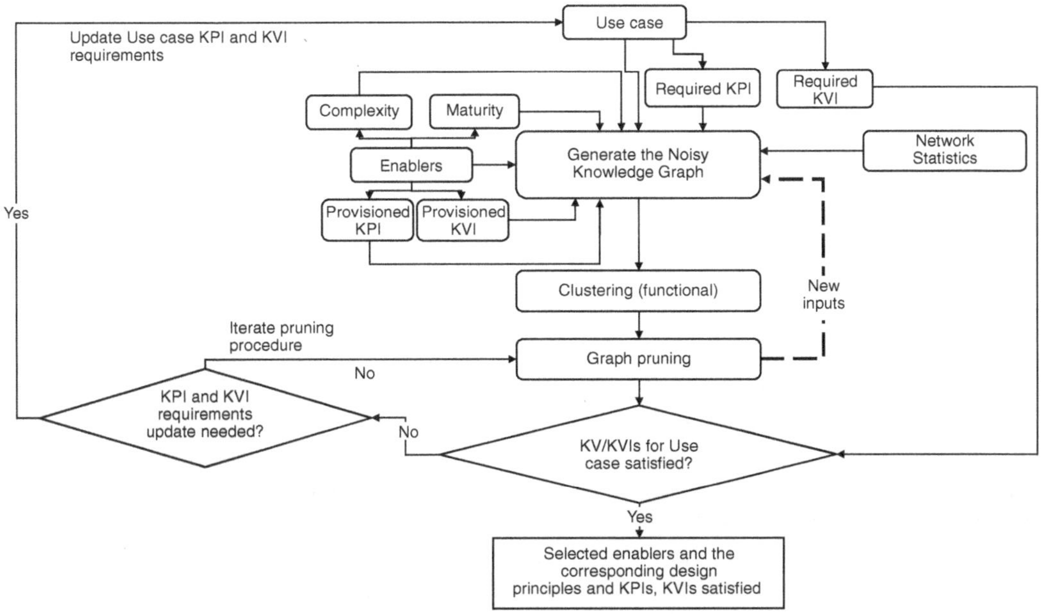

Figure 3.8 Knowledge Graph-based enabler selection method.

enablers. The pruning prioritizes the most positive impact on the KPIs, aiming to select all those enablers and their dependencies that contribute most effectively to achieving the required KPIs.

KVI evaluation: The selected enablers for the E2E system are assessed against the required KVIs for the specific use case. This involves determining the requirements to satisfy/achieve the KVIs, updating the E2E system if the KVI guidelines from WP1 are not met, via re-selection of enablers and updating KVI requirements in case it is realized that the stated requirements are too stringent/loose.

Iteration and refinement: The above process is repeated for all use cases, and the enabler selection is updated as needed based on the results of each iteration.

Note that only a qualitative analysis is performed with respect to the KPIs. This is because, to ensure a particular SLA, an enabler in the radio might have different requirements as compared to another enabler in the M&O domain, while both are necessary for ensuring the SLA. Hence, for a unified analysis and selection process, a qualitative assessment is performed, wherein we finally execute the pruning process via a simple maximum aggregate score function.

As can be seen from Figure 3.8, the enabler selection methodology considers the KVI requirements after a KPI based enabler selection has been performed. This is because of the subjective nature of the KVs and KVIs, as well as the challenges related to analysing enablers from different domains with a specific KPI target. Concretely, since metadata collected for each enabler and their impact on the KPIs is subjective, and the use case KV/KVI requirements are also subjective, the proposed sequential approach provides a less complex enabler selection process. It must be stated that, if the KV/KVI for the use case is not satisfied, then it is checked if there is a requirement to update KPI and KVI targets for the use case. This can typically be avoided if not enough combinations of the enablers have already been explored. In case many enabler combinations have been explored, then a decision to update the KPI and KVI targets needs to be considered. Next, for the iteration of the graph pruning approach, the main objective is to determine another combination of enablers that satisfies the KPI and KVI requirements. This is different from the 'New inputs', wherein the presence of a new enabler triggers the entire process again.

3.3 Multi-Stakeholder Intent-Based Service Management

The 6G transforms the traditional 'Telco' model to a Technology Company or 'TechCo' ecosystem by redefining roles and expanding service capabilities. NOPs manage network and cloud resources, whereas communication service providers (CSPs) create and offer services. Tier-1 operators often mix both tasks, with NOPs focusing on infrastructure and CSPs on commercial matters. This ecosystem includes digital services provided by DSPs via APIs, in addition to traditional communication channels. To fulfil the diverse demands of these services, the NOP function is expanding into specialized COPs and resource providers.

In a multi-stakeholder scenario, given the existence of various types of tenants, two models for intent-based service provisioning were identified: (i) the aggregation model, in which a single tenant assumes full responsibility for requesting the necessary intents and maintains direct communication with the Intent-based digital service manager (DSM) of involved administrative domains, and (ii) the federation model, in which this responsibility is transferred to the Intent-based DSM of a single administrative domain (SAD) that has received the tenant's request. In the federation model, this DSM subsequently communicates with other DSMs in distinct administrative domains to request the necessary intents.

3.3.1 End-to-End Multi-DSP Service Management

The high-level architecture (Figure 3.9) for automating E2E intent-based service management defines the interactions among key actors: tenants (Digital Service Consumers(DSCs)), DSPs, COPs, and resource providers. Three types of tenants (i.e. aggregators, verticals, and application service providers) interact with DSPs to begin and maintain service intents. The DSP, which is equipped with an Intent-based DSM, assesses these requests and translates them into specific needs for COPs, who manage domain-specific resources using enablers such as the Integration Fabric (IF). Resource providers operate across several areas, including network, AI, and cloud resources. This multi-layered architecture allows two possible multi-DSP intent-based service management: aggregation and federation, which offers cross-domain collaboration for full-service delivery via the Intent-DSC and the Intent-DSP APIs. Further details can be found in Le Sauze and Kerboeuf (2025).

3.3.1.1 Multi-DSP Aggregation Service Provisioning

The aggregation model for multi-DSP service provisioning is designed to manage intent-based services by delegating the responsibility of requesting multiple required intents to the Tenant-1 Type (i.e. Aggregator) as illustrated on top of Figure 3.9. The Aggregator takes on the responsibility of obtaining the necessary intents from the most appropriate SADs. As an example, if a request specifies "users in domain B", the Aggregator will check with domain B for the services needed. However, the Aggregator initially lacks insight into service availability; hence, it queries all known SADs to determine their available capabilities to identify which domain can best meet the expectations in the original request. For instance, if domain B cannot provide the requested vertical service but domain A can, it will focus on that selection. This decision is taken by factors like service availability in each SAD, the ability to meet KPIs/KVIs, and resource-related information shared by the SADs.

After the selection of SADs, the Aggregator issues service-specific intent requests to the involved SAD and their own DSMs, which process and manage the intent requests independently. For example, when the DSM of SAD A receives a request for the vertical service, it validates the intent expectations against the KPIs/KVIs specified in the SLAs between the SAD and the Aggregator. After validation, the DSM evaluates their feasibility against local resources and, if feasible, translates the intent expectations into executable requests to deploy the services. Then, the intent data object is updated with performance outcomes and resource details. This process continues until all conflicts are resolved (if any), and the intent reaches a successful state. Once the provisioning is complete, the Aggregator is notified of the outcome. The same process is repeated with other SADs, such as SAD B, which handles connectivity services between SADs A and B. Finally, after all the intents have been successfully fulfilled (e.g. the SAD A vertical service and the connectivity service linking SADs A and B), the Aggregator informs that the requested services are fully deployed and ready for end users.

3.3.1.2 Multi-DSP Federation Service Provisioning

The federation model aims to eliminate the need for an aggregator role by assigning the responsibility of identifying and selecting participating SADs to the intent-based DSM that receives the initial intent-based request from the tenant (i.e. direct link between tenant and intent-based DSM illustrated in Figure 3.9). Now, tenants directly request services from the intent-based DSM, which then interacts with other intent-based DSMs to deploy the necessary resources.

A key feature of the federation model is the interaction between different domain intent-based DSMs, referred to as 'east-west' interaction through the Intent DSP APIs and the multi-domain integration interfaces in Figure 3.9. If local resources can fulfil the tenant's requirements, the intent-based DSM implements the deployment process. However, if local resources are inadequate, it seeks to engage other known SADs to inquire about their available services and capabilities. The intent-based DSM evaluates the responses to determine the most suitable domains to address unmet expectations based on resource availability as well as the required and offered KPIs/KVIs.

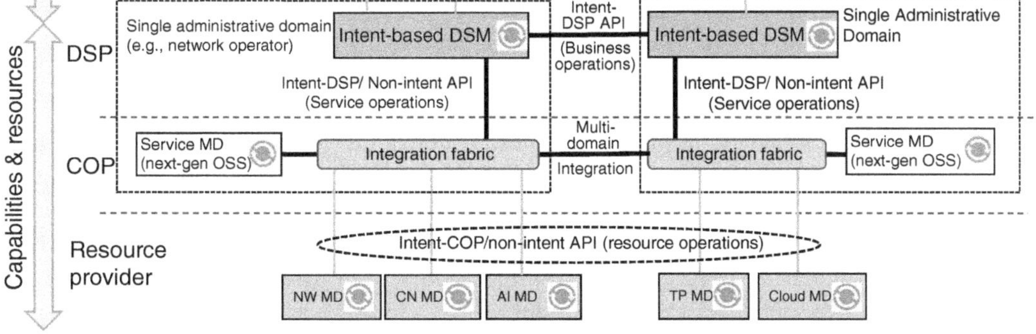

Figure 3.9 E2E digital service management automation framework.

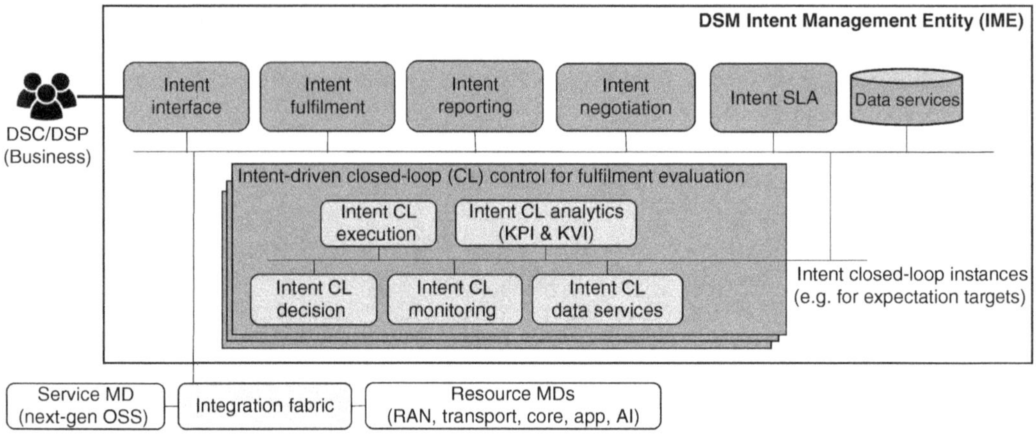

Figure 3.10 Intent-based DSM functional architecture.

Once the appropriate domains for any unmet requirements are identified, an intent-based provisioning process is initiated between the involved intent-based DSMs. Like in the aggregation model, each provisioning process within an intent-based DSM includes the steps involving intent requests, validation, provisioning, and reporting. Finally, the successful fulfilment of all non-local expectations through collaborative efforts, the original intent-based DSM updates the E2E intent data object and informs the tenant about the outcome. The tenant may then proceed with internal processes to provide access information for the newly deployed vertical service to its end users.

3.3.2 Intent-Based Digital Service Manager

Based on the 3GPP guidelines (3GPP TS 28.312 2025), a functional architecture (Figure 3.10) is proposed for the intent-based DSM component previously illustrated in Figure 3.9. This functional architecture, called DSM-Intent Management Entity (IME), interacts with the Service Management Domains (MDs) and the Resources MDs through an IF and is composed of seven functional blocks: (i) the **Intent Interface** allows the users to interact with the whole DSM-IME and it is the architecture core element as it takes care of the intents data objects by managing their intent's life cycle; (ii) the **Intent Fulfilment** contains internal capabilities to accomplish the intent life-cycle phases such as translation, feasibility, or conflict detection/resolution algorithms or the functions to coordinate and manage intent CLs (ETSI ZSM 0091 2025); (iii) the **Intent Reporting** handles the different types of reporting information during the whole intent's life cycle; (iv) the **Intent Negotiation** allows the DSM-IME users to obtain the precise information about the intents they might need before they are even provisioned, such as to explore if all the expectations in an intent are possible to fulfil, to ask for the intent with the most severe requirements possible, or to define preferences as a base for a judgment decision before any action is executed; (v) the **Intent SLA** deals with the management of SLAs for the intents, to provide a tenant full characterization by capturing data on security, trustworthiness, and end-users profiling data based on the contracted services and SLAs, and, finally it offers abstracted information of the available 6G services information and offered SLAs to the tenants; (vi) the **Intent-driven CL Control for fulfilment evaluation** delivers the capabilities to manage the intent CL instances life cycle and ensure they are fulfilled at any time; (vii) the **Data Services** stores the intent data objects and other possible information such as SLAs and policies.

Finally, in terms of AI/ML, two of the modules require the use of AI/ML technologies. On the one hand, the **Intent Fulfilment** on the translation and the conflicts detection/resolution, and on the other hand, the **Intent** Negotiation for predictive capabilities to estimate the effect of an action on the system state. Further details can be found in Le Sauze and Kerboeuf (2025).

3.3.2.1 Intent-Based Interfaces

To support cases where several DSPs fulfil services, the intent-based DSM requires supporting interfaces that enable operability between multiple intent-based DSMs. Thus, such intent-based interfaces should support different intent functionalities to operate intents, manage their lifecycle and requirements/expectations. According to different standardization bodies, these include setting the intent, reporting on intents, negotiating intents, and handling the intent profiles (TMF IG1253 2022; 3GPP TS 28.312 2025; ETSI ZSM 011 2023). Setting intents include creating, modifying, querying, deleting, activating, deactivating, and translating intents. Reporting intents include generating reports for the users. Negotiating intents includes communicating between the DSM and users while expressing their preferences. Finally, the profile handling includes showing the users the services offered by the DSP.

The considered intent-based interfaces to enable the multi-stakeholder scenarios are shown in Figure 3.11. The Intent Interface is the overall block that serves as the access point for users to engage with the DSM. This block is divided into sub-blocks that support the functionalities described in standardization bodies: (i) 'Intent CRUD' enables managing intents by creating, reading, updating, and deleting them; (ii) 'Intent Activation/Deactivation' enables users and DSM to have finer control on when and how an intent should be active or remain on stand-by; (iii) 'Intent Negotiation Management' allows the users and DSM to deliberate which intent and what expectations could be fulfilled by exchanging knowledge between the user and the DSM; (iv) 'Intent Report Management' generates and sends detailed documents about the intent and its expectations fulfilment status; (v) 'Intent Handling Capability' exposure describes what can be

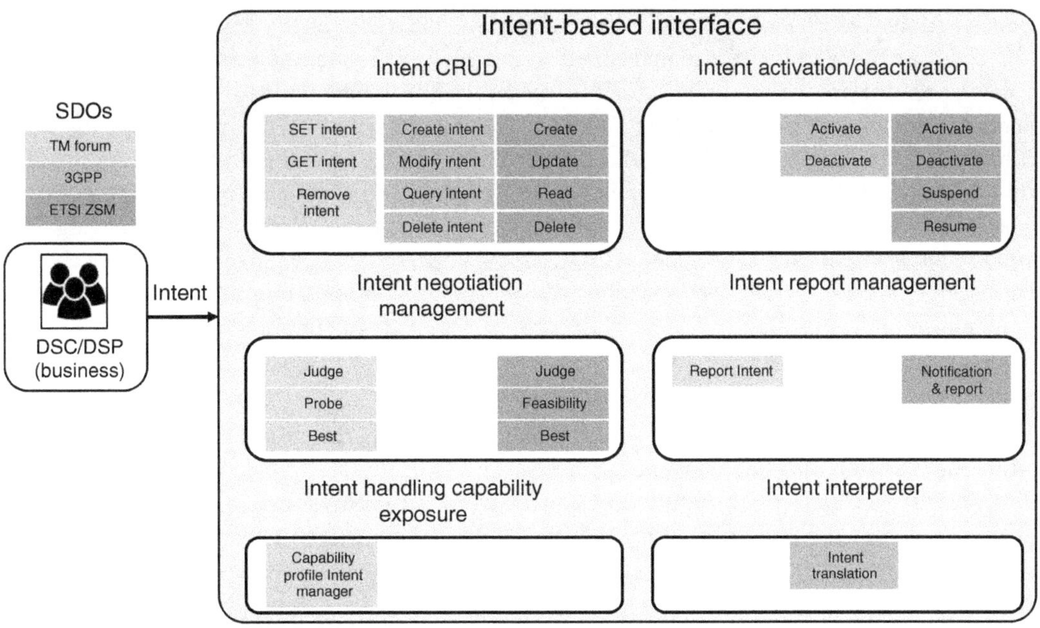

Figure 3.11 The intent interface to accept intents from the users, such as DSC/DSP.

offered and managed through the intents of the DSM; and finally, (vi) 'Intent Interpreter' gets the user's request and enables the DSM to understand the user's expectations.

3.3.3 Intent-Based-Specific Enablers for a Sustainable E2E Service Management

With regards to efficient M&O of intent-based services, it is key to understand how the intents coming from different tenants are properly managed to not let one of them affect the others. To this end, four E2E Service management use cases are presented, with the first two focusing on the deployment of an intent and then its constant evaluation while the intent is alive. In parallel, the other two E2E services are complementary services to accomplish the first two, and their focus is on the intent CLs management and the identification of the most trustworthy elements to deploy a trust-based intent.

3.3.3.1 E2E Intent-Driven Service Fulfilment Management

It extracts end-user expectations and executes a series of intelligent activities, leveraging contextual information, to dynamically design and deploy services using some AI components (language interpretation, service KPI prediction, service design recommendation or intent conflict impact analysis). This approach considers the complete set of capabilities and resources, manageable from an E2E perspective. Intent fulfilment management is achieved through several stages: expectation and target discovery, feasibility assessment of the intent, deployment of the E2E service fulfilling the intent, initiation of CLs activities associated with the intent, and intent lifecycle management.

As previously introduced, several functions within the Intent-based DSM architecture use AI/ML technologies to support different actions in an intent life-cycle such as discovery actions based on natural language processing (NLP) and large language models (LLM); feasibility and deployment actions with smart transition among the desired service levels and resource and capabilities management actions using autonomous AI based systems; analysis actions that enable CL operations; and reporting actions to provide management oversight. The creation of a detailed definition of the customer's service expectations subsequently drives E2E interactions to achieve the desired outcomes. This Intent Model (3GPP TS 28.312 2025) description serves as the interface for E2E fulfilment, which designs and deploys the service whilst striving to maintain the desired service levels. Some intelligent actions using analytics, like cognitive CLs (systems that use feedback to adjust and control inputs, creating a self-regulating loop), need to be considered for managing the conflicts among intents and the Resource Continuum. Then, the service finishes with the feasibility assessment and deployment actions followed by the initiation of CL automation. The status of the intent throughout its lifecycle is reported to the customer via the Intent Reporting.

3.3.3.2 E2E Intent-Driven Service Evaluation Management

It exploits real-time feedback from the services' deployment to enable accurate identification and continuous refinement of the users' actual needs. The COP utilizes translation mechanisms to define specific service level objectives (SLOs) and evaluates services' deployment environment real-time feedback from the users using data fusion and telemetry mechanisms, while trying to resolve failures or performance issues with intent-based service functions or resource management mechanisms.

For tackling intent translation to lower-level service fulfilment evaluation objectives, NLP and LLMs have been proven especially useful, especially due to the abstract nature of real-world requirements that incorporate hidden dimensions. Such algorithms can act as the central communication bus between intent owners and capability providers that traditionally work with programmatic logic. On the other hand, the early identification of failure to respect such objectives is crucial for operators to make fulfilment guarantees. AI-based time series analysis and anomaly detection algorithms have been proven successful to cover such needs since the availability of relevant data is continuously enlarged and allows complex models to function. Furthermore, root cause analysis methodologies are broadly available and can greatly enhance the resolution of system failures or performance shortcomings using advanced reasoning techniques and models. Regarding the continuous operation of established intents, intelligent decision-making is necessary for identifying optimal management actions, such as service placement and resource allocation. While control and optimisation-based methods can leverage domain models to support orchestrators, data-driven models (e.g. ML/reinforcement learning (RL)) can exploit historical information to build adaptable systems that are more efficient and drift-resilient, both crucial qualities for supporting the most daunting user intents.

3.3.3.3 E2E Intent-Driven Closed Loop Coordination

It guarantees intent assurance through CL-based automation processes. Multiple CLs could coexist simultaneously and insist on the same set of resources. CL coordination encompasses specific functions to avoid/mitigate potential conflicts that may lead to undesired/unstable states of the underlying system. The CL Coordination is in the Intent Fulfilment within the DSM architecture (Figure 3.10) that interacts with other internal functions such as the Intent Conflict/Detection Resolution, Closed-Loop Governance, and the Intent Reporting module.

As automation may concern complex tasks, CLs are increasingly integrating AI/ML-driven algorithms, especially for data analysis and decisions. The same rationale can be applied to CL coordination, which usually considers multiple inputs from different sources, environments, and situations: a level of complexity that can make the coordination actions not always modellable in terms of if-then-else rules. Following the European Telecommunications Standards Institute (ETSI) Zero-touch network and Service Management (ZSM) approach (ETSI ZSM 0091 2025), several types of coordination actions can be considered between two or more CLs: (i) identification of interaction (i.e. hierarchical, peer-to-peer), (ii) goal alignment, and (iii) pre-execution and post-execution coordination, i.e. actions to be performed before and after the triggering of the Execution stage, respectively. While all benefit from AI/ML, coordination actions in (iii) are probably the ones that benefit the most. In pre-execution coordination, the CL Coordination performs an analysis of the CLs' action plans to fulfil the CL goal: here, properly trained AI/ML models may efficiently automate the analysis to identify conflicts and to optimize the execution before the Execution stage is triggered. In post-execution coordination, one of the most complex activities is to determine the effects of CL actions when they cannot be done in pre-execution, and implement remediation measures. In both cases, AI/ML models (e.g. neural networks and decision trees) can be used to (i) identify conflictual situations and (ii) suggest remediation/mitigation actions. Moreover, in the multi-DSP federation scenario, where multiple CLs operate among different DSMs, determining the optimal execution sequence becomes increasingly complex. RL offers an effective solution by enabling dynamic and intelligent coordination. An RL agent continuously observes CL information and resource states, and is then trained over time how different execution sequences impact overall performance. Based on these insights, the agent determines the most suitable execution order across all involved DSMs, aiming to fulfil the original intent while optimizing execution time, reducing conflicts, and ensuring system-wide stability and responsiveness.

3.3.3.4 E2E Intent-Based Trust Management

It ensures, via the Level of Trust Assessment Function (LoTAF) (Jorquera Valero et al. 2024), that the end-user to declare trust requirements with trust level agreements (TLAs) and continuously monitor service assurance based on an objective indicator. LoTAF considers functionalities of the intent-based DSM architecture like Intent Interpreter, Intent Handling Capability Exposure, Intent CL Monitoring, and Intent Fulfilment. The LoTAF evaluates the deployment of security technologies and serves as a bidirectional service: It assists trustors (consumers) in making informed and intelligent decisions and provides trustees (network providers) with insights into meeting trust criteria and opportunities for service improvement.

LoTAF is mainly divided into two primary functionalities: (i) the **semi-natural trust-based intent** focuses on semantic interpretation and data representation via KGs. This segment enables users to articulate their trust requirements naturally through a graphical interface. A specialized trust-based interpreter – leveraging a descriptive language and a Named Entity Recognition (NER) model with technologies such as Doccano, Spacy, and transformers (ModernBERT) – extracts key tokens from user input. It also uses a one-touch mechanism by providing a LLM-driven chatbot to validate these tokens, leading to selecting candidate network services, establishing a Trust Level Agreement (TLA), and creating a KG for ongoing monitoring; (ii) the **trust assessment** continuously monitors service assurance (IETF RFC 9417 2023) by evaluating both objective (quantitative) and subjective (qualitative) trust indicators. In this vein, LoTAF updates the LoT via Bayesian methods over time, ensuring that network services consistently meet the TLA terms. Furthermore, trust assessment also introduces a breach prediction component using AI-driven algorithms to forecast potential deviation on TLA parameters by analysing real-time service assurance and historical data. Overall, AI enhances E2E intent-based trust management by enabling natural language interaction, dynamic monitoring, and predictive analytics. This leads to more intelligent, proactive, and trustworthy enforcement aligned with user-defined agreements.

3.4 E2E Security Concepts

The 6G network is required to provide a stable and predictable performance E2E in expected conditions, and *trustworthiness* is one of the crucial key values for 6G. Trustworthiness encompasses multiple facets, ranging from security and privacy to compliance with ethical frameworks, and encompasses availability and resilience, as mentioned in different analyses of the network trustworthiness concept (NGMN 2023; Mahmood et al. 2024). All these considerations on trustworthiness have in common that security and privacy are crucial properties of trustworthy systems. This section focusses on how E2E security and privacy can be achieved in 6G systems, and on the general principles and enablers to achieve a high degree of these properties, thus contributing to a high degree of trustworthiness. This analysis includes not only the technologies considered in the Hexa-X-II project but also other relevant proposals made within the framework of projects in the European 6G research initiative. Further, the topic of E2E trust-awareness in 6G networks is addressed, considering how evidence on properties of the network and its components can be provided, and the construct of a LoTAF, supporting service providers and service users to specify and monitor the security and privacy posture of services and give the user assurance on the trustworthiness of a network service.

When considering E2E security in 6G, it is important to analyse how security features have evolved in the successive generation of mobile networks to identify the *6G Delta*, that is, which are the specific threats posed by the different trends towards 6G, either by the mobile network evolution itself or by the evolution of the security base technologies. The specific E2E security enablers and their characteristics are to be proposed and validated as to how to address this 6G Delta in what relates to architecture, the pervasive use of AI, the evolution of trust infrastructures, and the possibility of enhancing physical layer security (PLS).

The 6G Delta in security comprises three main architectural trends: the Network of Networks (NoN) compositional pattern, the use of a Cloud Continuum as a base infrastructural approach, and the application of radical disaggregation mechanisms, especially in the RAN segment. The NoN concept focuses on the dynamic integration of multiple networks or network segments, and it raises many issues linked with procedures needed for the attachment and separation of the network domains, and procedures related to the proper and optimized operations of the composed network, including short-lived compositions. The Cloud Continuum concept assumes integrating all infrastructure resources and their uniform application exposure, using virtualized, unreliable, and constrained resources at the edge and far-edge environments. Disaggregation may expand the potential attack surface by making the network more vulnerable since each disaggregate element can be a possible entry point for cyberattacks, requiring robust authentication, authorization, and verification mechanisms to prevent unauthorized access and data breaches, and including strong adherence to the zero-trust principles for network, and especially RAN, design. Moreover, as data traverses through various disaggregated elements, ensuring its confidentiality becomes paramount. Encryption and secure data transmission protocols are necessary to protect sensitive information from interception and unauthorized exposure.

AI is expected to play a key role in the realization of the data-driven 6G networks. Potential uses of AI in 6G include network optimization, network automation, and security protection of the network and its components against various attacks. The trustworthiness of AI requires the availability of prevention mechanisms against threats to AI-based systems. These threats would focus on making the AI elements work unexpectedly or make them work in a manner biased by the attacker. Attacks may take place at any place in the AI lifecycle, with special emphasis on the training and inference phases. Furthermore, privacy attacks on AI can be specifically directed to learn sensitive data, such as information about the training data, the parameters of the AI model, or even known patterns from inference queries and their responses. Mitigating these kinds of attacks would require the use of adversarial training and explainable AI (XAI), with the incorporation of transparency mechanisms, and a careful weighing of how to apply federated learning (FL) approaches in support of higher degrees of privacy, looking for the right balance of security and privacy goals.

The capacity of assessing trustworthiness of network services requires addressing trust subjectivity, without a universal standard, beyond current objective and measurable security KPIs. Creating objective, constructive terms regarding trust in the agreement requires agreed standards for trustworthiness assessment. This assessment must also consider indirect/social trust or *reputation*, influenced by third parties. This poses challenges to the monitoring mechanism design regarding real-time feasibility, data integrity, security, and privacy. Furthermore, the technology pillars considered for assurance and trust in 6G networks keep evolving. These include Distributed Ledger Technologies (DLTs), offering transparency, immutability, and non-repudiation, which can be utilized to define agreements and liability of each party, and protect the integrity of data; privacy preserving data sharing, aligning with the shift towards trust-centred frameworks and ensuring confidentiality and privacy; and attestation mechanisms, making identities only bound to the possession of a cryptographic key but to the verification of the overall status of the component. Finally, the impact of the required evolution of crypto techniques must be considered. The use of cryptography in today's networks is pervasive, including all aspects related to infrastructure management and network support systems. In this context, two main trends can be identified. First, and the one most widely discussed recently, we have the threat associated with the vulnerability of current crypto algorithms to quantum computing, becoming much more than a theoretical risk. In addition, the generalization of virtualized and cloud-native solutions requires a comprehensive approach to identity infrastructures and key management procedures, very much focused on static trust fabrics challenged by the virtual nature of current network functions.

To increase the trustworthiness of the mobile network, the current architectural proposals incorporate a range of security measures and controls. While many of these are based on cryptography, it is envisioned that the set of existing security measures will be extended by methods from the domain of PLS leading to an increasing number of design and deployment options regarding achieving desired security goals. But selecting appropriate security controls is not only driven by the security goals, since most (if not all) security controls induce certain costs (e.g. in terms of increased energy consumption, reduced goodput or increased latency) and rely on certain assumptions (e.g., mathematical assumptions or the non-existence of sufficiently powerful quantum computers). Therefore, the overall selection process can be understood as a trade-off or optimization between the desired trustworthiness and other non-functional as well as functional requirements. The relevant decisions must be made not only during the design and development or deployment phase but also during the runtime. For example, methods for PLS in particular are dependent on the current conditions of the transmission channel, demanding context awareness. 6G innovation in ISAC requires an in-depth analysis to understand the security and privacy challenges it implies, in both areas of using sensing for security applications and securing the sensing process itself. The threat of jamming is not new for 6G, but the extended critical nature of the services it supports raises the impact of dealing with this threat, making the classification of jamming signals and the localization of the jamming sources fundamental for initiating countermeasures. Finally, there are new paradigms, like physical layer deception (PLD), that represent an active defence strategy at the physical layer.

3.4.1 Security Controls and Security Enablers

Addressing security and privacy requirements for 6G services does not necessarily require providing new, differential functionality, but rather address specific threats, providing mechanisms to detect them, and to mitigate their impact in E2E system operation and behaviour. While the analysis of general features related to network operational aspects can be structured around well-defined functional elements that make possible to realize or enhance 6G capabilities via a common interface, what is consistently referred as *enablers*, core security functionality requires to be analysed following a more detailed, fine-grained pattern, associated with the threats they address and taking into account that interfaces are not necessarily common, as many of these security, privacy and resilience (SPR) core features may require specific interfaces to be applied to address the specific threat to be considered. A paradigmatic case for this is crypto primitives, which cannot be used through remote interfaces such as REST or remote procedure call (RPC), since they constitute the foundation of the security for such interfaces. This is why the proposal of security features for E2E 6G services as depicted in Figure 3.12, is structured around the concept of *controls*.

A security control corresponds to an essential component providing core security functionality as a safeguard, detector, or countermeasure with respect to specific threats. Controls can be used directly by other network elements via dedicated internal interfaces, or accessed through well-defined functional remote access interfaces, wrapped within enablers, in most cases controlled by orchestration modules and operating under the As-a-Service paradigm. Within the overall E2E system architecture, security controls will appear associated to most (if not all) other architecture

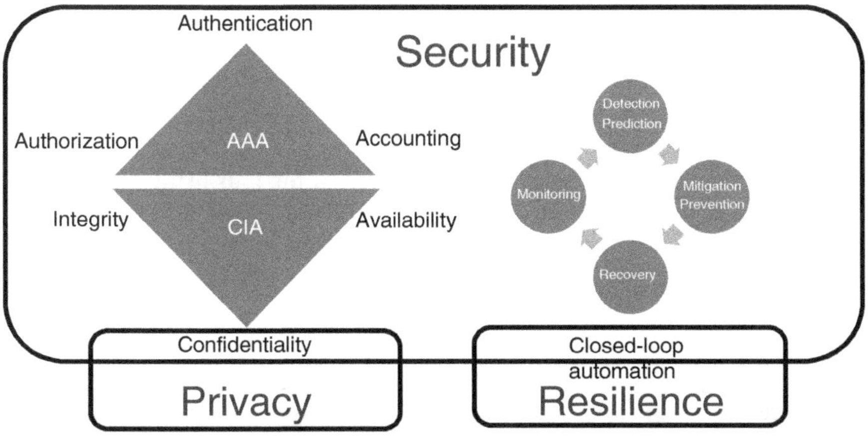

Figure 3.12 Base security features and their connection with the other trustworthiness dimensions.

elements when directly used by these elements, to specific security enablers when they are accessed using the As-a-Service paradigm, or as dedicated functionality to be required in the different phases for the design, deployment, and operation of a network service.

In some sense, security controls act as *security enablers for enablers*, and their use is structured around two basic principles. First, the use of a general loosely coupled model, compatible with the IF described for orchestration and management. Second, the fact that these controls will be accessed through APIs, either local (as in the case of crypto functions) or remote (following general design paradigms like REST).

According to these principles, and following the analysis performed on the threats and potential safeguards, detection mechanisms, or countermeasures, security controls have been identified and classified within the Hexa-X-II project, considering elements proposed within the project and others that have been developed in other European 6G-focused projects within the SNS initiative. These controls are associated with essential security features:

- The CIA triad is achieved by the cryptographic mechanisms and the integration of redundancy mechanisms relating to the evolution of quantum-safe cryptography and the possibilities offered by cloud-native architectural choices.
- An AAA framework, relevant to secure accessibility and identity management for human and non-human identities. Privacy requires the confidentiality of data transferring and storing, data minimization mechanisms and the explicit support for consent and transparency. It is significant to extend these principles to manage as a combination of human and pervasive non-human identities in the 6G ecosystem.
- Automation of security closed-loop operations, including dynamic monitoring, attack detection, decision-making, and remediation actions, leading to security orchestration and the enforcement of security policies and/or intents. This also includes the use of what-if scenarios to assess the impact of different conditions and countermeasures, by using simulation mechanisms and integration with synthetic environments NDTs to emulate network segments.

The controls identified within the Hexa-X-II project address:

1) Aspects related to PLS, including *Physical Context Awareness*, which implies considering the trade-off or optimization between the desired level of security and other non-functional as well as functional requirements. The fundamental concept is that the system 'understands' the

desired (security) goals and the current situation and adapts the security controls accordingly, selecting and configuring them to achieve the goals in an optimal way, which could include graceful degradation. In *Physical Anomaly Detection*, the preliminary step to jamming signal management is the classification of jamming signals, intended to detect physical layer anomalies and discriminate casual interference from malicious jamming. *ISAC Threat Mitigation* is currently focused on threat analysis, addressing the protection of the four main categories of ISAC assets: sensing information, systems, value provided to the operator from the sensing service and value provided by sensing applications. Finally, *PLD* constitutes a novel paradigm for active defence at the physical layer. It is based on deceiving potential eavesdroppers by transmitting compromised information while simultaneously delivering the original data to authorized receivers.

2) The availability of Trustworthy AI, with the implementation of prevention mechanisms against attacks on AI-based systems. This includes the detection and prevention of attacks against AI, applied both during the training and during the inference phase of the AI model lifecycle. These controls will be incorporated into the AI management elements, including the data and action flows at the input and output of the models in use. The trustworthiness dimension also comprises the advanced intrusion detection systems (IDSs) that use ML, providing them with strong security measures in place. Solving this issue requires a comprehensive framework that combines the strengths of ML against adversarial attacks, handling complex attack vectors, and real-time applicability. In addition to regularly updating the systems and training them with a variety of data, incorporating XAI to detect malicious behaviours is promising as an effective strategy for a trustworthy adversarial detection framework.
3) The support of advanced trust fabric mechanisms, related to *Crypto Evolution* and *Infrastructures for Transparency,* in support of the assessment of the *Level of Trust*. The evolution of mobile network infrastructure requires solutions able to support quantum-safe and cloud-aware crypto mechanisms able to satisfy two main requirements: *Agility*, as the possibility to maintain alternate evolution paths as algorithms and technologies evolve, and *Pliability*, as the capacity of being adapted to network management best practices. Hexa-X-II has evaluated the application of a number of relevant trends in the evolution of the cryptography application in mobile networks, namely the urgent transition to quantum-safe schemas to restructure the trust fabrics and their PKI foundations, the use of hybrid certificates (including signatures and keys for different crypto algorithms) and non-PKI mechanisms for establishing session keys, the extension of identity to include matters related to attestation and supply chain verification, and the use of automation procedures suitable to cloud environments.
4) The integration of relevant enablers is considered in other SNS projects. Two of them are included in this section: the use of data-intensive E2E security management driven by holistic AI methods being developed by the ROBUST-6G project (Giardina and Ruta 2024), and the incorporation of security and privacy concerns in the DevOps approach common in cloud-native environments, towards what project RIGOUROUS terms as DevSecOps.

In what follows, the most relevant aspects of these controls, and the specific enablers supported by them, are discussed, with the exception of those dedicated to ISAC threats and AI trustworthiness, discussed in Chapters 7 and 6, respectively.

3.4.1.1 Physical Context Awareness

As an essential part of the 6G security framework, PLS is highly dependent on the physical channel features, and an awareness of such physical context information becomes therefore crucial. For example, the channel measured between two legitimate devices during a time division duplex

(TDD) communication within the channel coherence time, which is reciprocal in nature, can be used in generating symmetrical encryption keys at the respective devices, which is known as the secret key generation (SKG) technique. Such SKG protocol involves randomness extraction, quantization, information reconciliation, and privacy amplification phases to generate the encryption keys. The achievable secret key rate is determined not only by the parameters used during the different stages of the protocol but more critically on the randomness present in the multipath channel, which depends on the channel context, e.g. non-line of sight (NLOS)/line-of-sight (LOS), static, dynamic, and mobility in the channel. Hence, knowing the context beforehand helps in increasing the achievable secret key rates by setting the optimal system parameters for the protocol.

3.4.1.2 Physical Anomaly Detection

The RAN architecture is evolving from a closed, unified infrastructure to an open, disaggregated, and softwarized system, offering flexibility and scalability. However, the distributed data processing and hierarchical formation in disaggregated architectures will cause difficulties for the direct integration of centralized security solutions (e.g. detection of security vulnerabilities) in such networks. Due to disaggregation and other features, the threat surface is increased as compared to previous generations. Furthermore, because of the complex distributed architecture of RAN and RAN components operating in different control loops, it would be rational to deploy anomaly and intrusion detection mechanisms (IDMs) in different locations in RAN. For instance, FL algorithms have great potential in detecting anomalies in such disaggregated RAN environments, enhancing anomaly detection capabilities by allowing distributed model training across edge devices. This approach takes advantage of the distributed nature of disaggregated RAN components, allowing for collaborative model training without the need for centralized data collection.

Moreover, considering the critical use cases anticipated for 6G – such as smart factories that connect cooperative robots – it may be necessary in certain scenarios to go beyond RAN parameter monitoring and cover the wireless spectrum for maintaining a resilient network. In licensed bands, any interfering signal not transmitted by an authorized user is classified as a spectrum anomaly. Detecting such signals is essential, as they can compromise both communication performance and sensing accuracy. On the positive side, the functionalities and interfaces foreseen for 6G enable the extension of NDTs down to the physical layer, including the wireless channel. Information from environment sensing and localization can be bundled with information from resource allocation and combined with real-time ray tracing to generate an accurate estimation of the spectrum occupation, represented, for example, by a spectrogram.

The placement of anomaly detection mechanisms and the design of security solutions need to be followed efficiently, prioritizing cost-aware cybersecurity practices. This necessitates the joint optimization of security with other criteria, such as resource utilization and QoS. The joint security-QoS optimization in RAN involves fine-tuning security mechanisms to achieve the best performance without compromising protection. This includes balancing the load on network elements for security operations, ensuring efficient resource use, and minimizing latency introduced by security processes. Given the disaggregated architecture of RAN, the joint optimization ensures that security measures are tailored to specific components and interfaces, maintaining overall network performance.

Identifying attempts of adversarial attacks at the earliest possible stages will limit the spread of the distribution of unwanted traffic within the network. Discarding such redundant traffic close to the entry point can significantly reduce the total energy consumption in the network. Next-generation RAN, with its flexible and disaggregated architecture, enables the implementation of such strategies more effectively. By using open interfaces and modular components, NOP s can deploy customized IDS/IDM that suit their specific needs. The flexibility of RAN allows for a more adaptive security strategy, where the IDS strategy can be periodically or dynamically adjusted based on the actual state of the network, the required QoS, the expected security level, and the risk level towards the goal of improving energy efficiency.

3.4.1.3 Physical Layer Deception

One common limitation of conventional PLS solutions is that they have generally been developed as passive approaches to defend against possible eavesdropping, without any capability of detecting or actively countering eavesdroppers. Despite the enhancement they provide to the passive security, a notable imbalance still remains in the security model, as eavesdroppers (especially passive ones) can attempt to wiretap with barely any risk of exposure, and significantly lower effort compared to the extensive measures and costs taken by the network and legitimate users to secure data. This imbalance necessitates a strategic pivot towards integrating active defence mechanisms into the wireless security framework.

A possible way to make up for this shortcoming is to introduce deception technologies, which aim to mislead and distract potential eavesdroppers by creating fake data or environments, while keeping the real data and environment secure. Such technologies can even be deployed to lure eavesdroppers into exposing themselves.

Combining the concept of cyber deception and PLS, a new secrecy paradigm called PLD can be developed as a technical enabler for wireless privacy and security. As illustrated in Figure 3.13, the key principle is to cipher each individual message with an independently chosen random key instead of a pre-agreed key. The random key shall then be multiplexed with the ciphertext for a joint transmission. The radio resource and power shall be well allocated between the ciphertext and the key, so that the former is well exposed, but the latter is protected in a PLS sense. If the ciphering algorithm is carefully designed to fulfil the condition that any ciphertext combined with

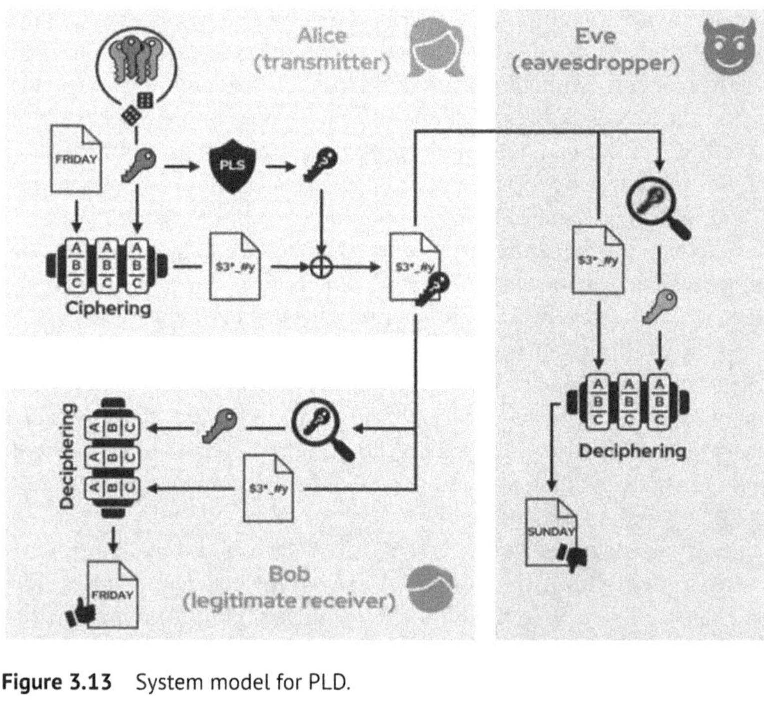

Figure 3.13 System model for PLD.

an incorrect key will generate a misleading false plaintext, the eavesdropper will have a statistically inferior signal to interference-plus-noise ratio (SINR) and face a high risk of being deceived with a falsified message, while the legitimate receiver can still correctly receive the information due to its superior SINR.

More specifically, to reduce the challenge of ciphering algorithm design, a recommended implementation is to randomly activate/deactivate the deceptive ciphering. When deactivated, the plaintext is directly used as the ciphertext and multiplexed with a 'garbage' sequence instead of a key. Thus, the requirement for the ciphering algorithm can be relaxed to that each feasible ciphertext codeword must also be a feasible plaintext codeword.

3.4.1.4 Transparency Services and Level of Trust Assessment

In a nutshell, a transparency service offers an issuer the possibility to register statements about artefacts in such a way that others can cryptographically verify the registration of the signed statement in the service. In more detail, the main notions of this concept are:

- **Artefact:** The entity about which a statement is made.
- **Issuer:** The entity that is making the statement about the artefact. Before registration of a statement, the issuer signs the statement, leading to a signed statement. The signed statement might not include the statement itself, but a hash of the statement and a link to the actual storage of the statement. In this way, the recording of the statement and the recording of the registration can be decoupled.
- **Statement:** Any piece of information, which can be made about an artefact and which should be registered in the transparency service.
- **Client:** The entity registering a signed statement at the transparency service on behalf of the issuer. The client and issuer may be the same entities.
- **Transparency service**: The main task of the transparency service is the registration of new signed statements. If the transparency service receives a signed statement from a client, the transparency service verifies the formal validity of the statement and adds it to the ledger. Thus, the ledger consists of an ordered sequence of all registered signed statements.

  The integrity guarantees of the ledger (append-only, immutability, auditability) are ensured using Merkle binary hash trees. The hashes of registered signed statements are the leaves of the tree. By iteratively calculating the hash of a branch node by hashing the values of its child nodes, the hash value of the root node of the tree is calculated, which contains the complete state of the ledger at a certain time. After having registered a signed statement, the transparency service provides a receipt to the client, which contains a prove that the statement has been added to the ledger. In summary, the proof contains the signed hash value of the root of the Merkle tree, and the hash value of all branch nodes needed to reconstruct the tree root from the hash value of the signed statement.

  Whether the main purpose of the transparency service is to ensure the integrity of the ledger through maintaining the hash tree, or if the transparency service should also provide as a searchable archive for the registered statements, needs to be discussed and depends on the specific requirements of the ecosystem. The transparency service is not implemented by a single node, but by several nodes, which are operating together via a consensus protocol. However, it is not expected that the operation of the transparency service is completely open (like, for instance, in the case of Bitcoin), but that nodes of consortium partners act together based on a set of agreed-upon rules.

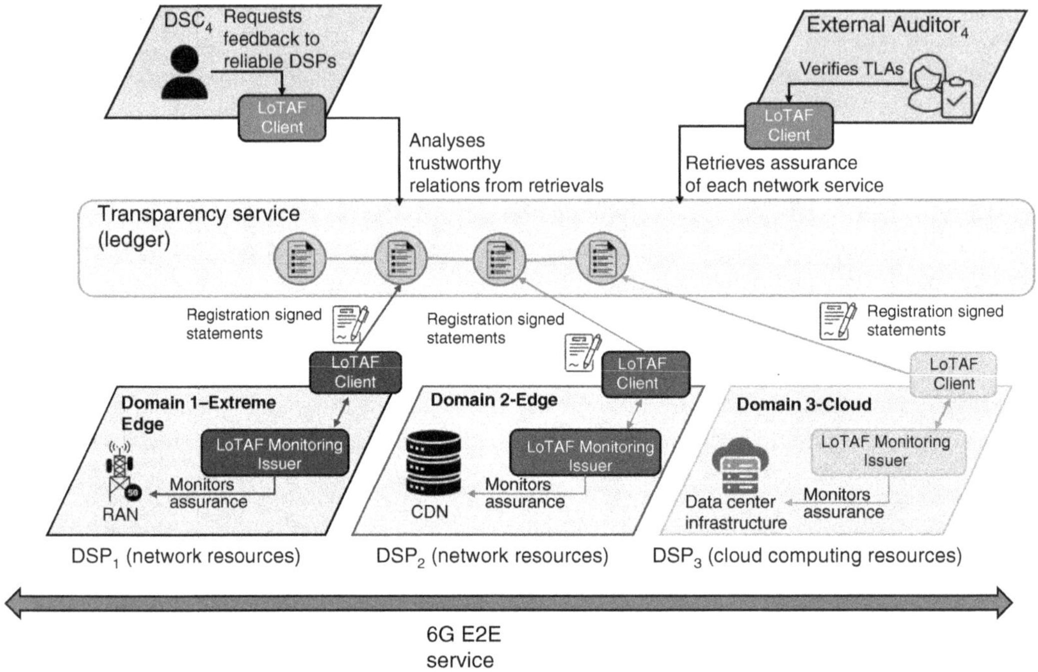

Figure 3.14 Example of a joint Transparency Service to assess LoT in cross-domain DSP relations.

- **Relying parties:** Relying parties are the entities consuming the content of the signed statements. With the help of the proof of inclusion in the receipt, the relying party can verify independently of the transparency service that the signed statement is part of the ledger.

Use cases for the transparency services include, among others, supply chain applications with vendors registering signed statements about their products; the logging of information sensitive to the operation of the network; the tracking of data provenance in complex AI scenarios in which data undergo several different process steps; and the accounting for the utilization of shared resources.

As shown in Figure 3.14, a transparency service can be used for level-of-trust assessment of cross-domain relationships in Cloud Continuum scenarios. In this case, a DSP (e.g. network service provider, and cloud service provider) might register on the transparency service the level-of-trust of its network service regarding a contracted TLA among several DSPs to compose a 6G E2E service. The signed statements may be issued by the level-of-trust assessment function of a given DSP that continuously monitors service assurance and evaluates statements, as described in Section 3.3.3.4.

Furthermore, transparency services may also support a DSC's level-of-trust assessment function to verify the fulfilment of a trust agreement when different DSPs participate in an E2E relationship. Each DSP should log the service assurance outcomes of its network service running under its domain to the joint transparency service. Thus, the transparency service supports the verification (through external audits) of statements of a decomposing E2E service coming from multiple services and different DSPs. Additionally, the transparency service may also help DSCs to analyse forthcoming trustworthy relationships as the level-of-trust assessment function evaluates previous interactions registered in the transparency service.

3.4.1.5 Data-Intensive E2E Security Management

Full automation in the protection of 6G network services and infrastructure against possible cybersecurity risks is an essential feature of the mechanisms to address the 6G threat landscape. A data-extensive E2E smart network and service security management framework is crucial to achieve this full automation of network security. This framework should apply ZSM along with AI techniques. The native integration of AI across all network segments and layers, wherever feasible, will be one of the main differentiators for future 6G networks from the previous generations.

To address the dynamic challenges posed by future 6G services and networks within the cyber-physical continuum, data-driven, AI-based security solutions are essential. Therefore, security operations of 6G networks are anticipated to be data-intensive. These security operations will need advanced data analytics to detect, respond, predict and mitigate threats in real-time while maintaining user privacy and data integrity (Giardina and Ruta 2024).

AI will have a critical role in the automation of decision-making processes, improving network performance, enabling zero-touch automation, and achieving E2E security. To facilitate the use of AI across all network segments and layers, 6G networks must support efficient data management solutions capable of collecting, processing, storing, and exposing the required information for the training of AI algorithms and decision-making processes of smart agents. Hence, AI relies on large amounts of data, which may include sensitive information, ensuring privacy, confidentiality, and security for data is of utmost importance.

By adopting this data-intensive E2E security approach, 6G networks can effectively balance the demands of flexibility, adaptability and programmability with robust security measures, ultimately paving the way for a secure, efficient, and resilient communication infrastructure.

3.4.1.6 DevSecOps

In the rapidly evolving landscape of 6G networks, our approach to DevSecOps transcends traditional security paradigms by weaving continuous privacy and trust evaluations directly into the service lifecycle. At the heart of this methodology lies a robust, declarative privacy manifest – a living document that not only captures the detailed intentions of developers regarding data processing, storage, and communication but also incorporates insights from automated tools analysing data flows and connectivity. In this way, the manifest serves as both a blueprint and a benchmark, ensuring that privacy considerations are integral from the very inception of a service.

As services are designed and deployed, developers articulate their privacy policies within the manifest, detailing which data is handled and how. Concurrently, advanced algorithms assess these practices by examining the service's connectivity with external endpoints and quantifying the overall privacy level into a composite score. This score, which can be tailored to meet specific user preferences or adhere to regulatory standards, offers a dynamic measure of privacy integrity that evolves alongside the service itself. Moreover, recognizing that services are in a state of constant flux, our architecture incorporates a control version mechanism that establishes a baseline privacy benchmark. This benchmark is continuously re-evaluated and refined through what we term 'DevPrivOps', a process of iterative privacy testing and monitoring. In essence, every service update triggers a fresh evaluation, ensuring that the evolving service remains aligned with both emerging privacy standards and the latest security practices. In parallel, the commitment to a secure 6G ecosystem is reinforced through a Zero-Trust security framework that continuously monitors the trustworthiness of network functions, applications, and devices. By scrutinizing behavioural and environmental indicators, this framework proactively identifies any deviations from established trust norms. In a 6G environment – where millions of devices interact across diverse jurisdictions – this level of continuous trust evaluation is crucial for swiftly isolating

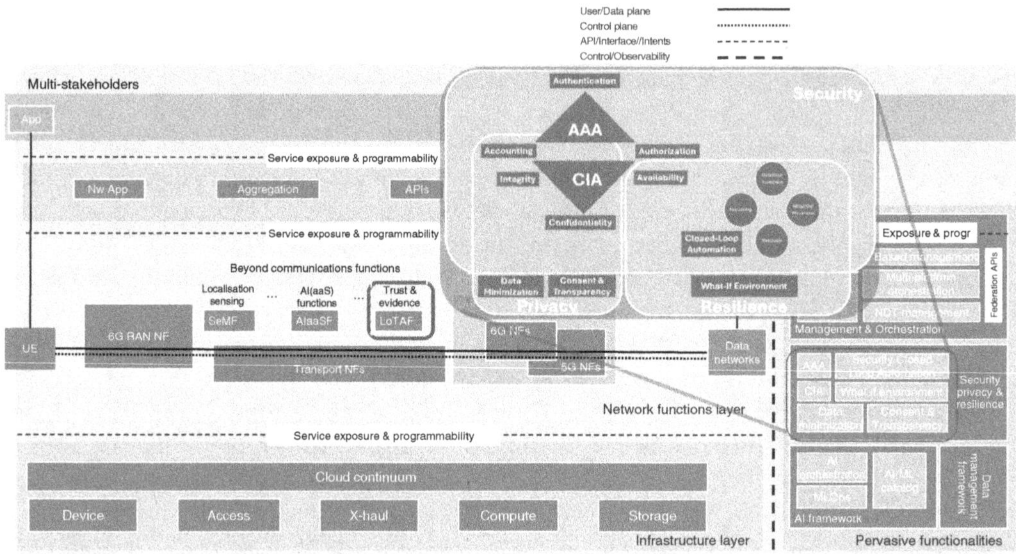

Figure 3.15 Security features within the 6G E2E system blueprint.

potential threats and mitigating risks. Ultimately, this integrated vision of DevSecOps is not merely about meeting technical or regulatory benchmarks. Instead, it represents a holistic philosophy where security and privacy are interwoven into every aspect of service design, deployment, and evolution. By combining a declarative privacy manifest with continuous trust evaluation and adaptive control mechanisms, this approach lays the foundation for a 6G network that is both secure and resilient, ready to meet the challenges of an increasingly interconnected future.

3.4.2 E2E 6G Security

Figure 3.15 illustrates the mapping of the different controls and enablers on the E2E 6G system blueprint defined by Hexa-X-II. The controls providing direct security and privacy support thus become integrated within the pervasive functionalities, as specifically illustrated in the figure by zooming in on the corresponding element, highlighting how the CIA triad, AAA mechanisms, and data protection and transparency have to be applied at all layers. In the following, the use of these controls at the different layers, as pervasive functionalities and/or specific enablers, is analysed.

3.4.2.1 Infrastructure Layer

Cloud and computer service providers need to protect their own infrastructure from attacks and ensure the isolation of workloads by following techniques such as confidential computing to achieve CIA and AAA features, and apply System on Chip (SoC) architectures to prevent data leaks, ensure resilience against attacks, and provide robust foundations for tiny ML applications.

3.4.2.2 Network Functions Layer

Security and privacy should be carefully considered in this layer, considering multi-level security solutions. For instance, when a mobile network operator (MNO) relies on DNs deployed by other compute resource providers, AAA features should be protected. The associated security controls in this layer ensure secure communication through ciphering and integrity protection and provide robust mobility management to support seamless transitions and uninterrupted service. PLS mechanisms aim to secure RAN protocols.

3.4.2.3 Application Enablement Platform Layer

The paradigmatic scenario, when the UE acts as a CaaS consumer, considers the DN managed by a compute control node (CCN), which is the CaaS provider. When the CCN manages compute resources at the DN and interfaces with the 6G system, it is important to establish a secure communication path between the UE application client and the DN that ensures CIA and AAA features, as well as data privacy properties.

3.4.2.4 Management and Orchestration

As a pervasive functionality in the 6G E2E system blueprint, M&O contribute to advancing security management through closed-loop control and using AI for automating security actions. By orchestrating security services seamlessly together with those CLs, the system will allow the enforcement of remediation policies by automated actions and the application of SecDevOps mechanisms.

3.4.2.5 AI Framework

The integration of AI in the 6G E2E system blueprint is pivotal for ensuring intelligent, adaptive, and efficient operations in all layers. Robust and trustworthy AI mechanisms are to be accompanied by privacy-enhanced data management and mechanisms to ensure CIA and AAA properties across network functions and connectivity options. Security automation and accountability for shared data in federated domains are crucial to maintaining data integrity, reliability, and system trust.

3.4.2.6 Data Framework

The key functionalities performed by the data framework, such as data collection, distribution, and storage, are required to be privacy compliant. This entails not only aligning with the privacy-compliant regulatory requirements but also achieving confidentiality in data processing and storage.

3.4.2.7 Multistakeholder 6G Ecosystem

In a multi-stakeholder 6G ecosystem, it is essential to incorporate continuous monitoring and security assessment to ensure E2E security. This includes developing dynamic and adaptive security solutions for threat detection, mitigation, prediction, and response in a multi-stakeholder environment. Solutions include incorporating cryptographically protected multi-level security enablers or using security policies together with identity and access management schemes. This requires the application of specific enablers, such as the LoTAF and transparency services.

3.4.2.8 Service Exposure and New 6G Services

This allows to securely expose the services and capabilities provided by the 6G platform. The service exposure framework should support a strong AAA framework for secure identity and access management. With the zero-trust security models supported by 6G networks, it is important to maintain dynamic and continuous verification of service exposure, including the application of quantum-safe cryptographic protocols.

In particular, a number of specific enablers are proposed as a specific class of beyond-communications functions labelled Trust & Evidence, as highlighted in Figure 3.15. These functions would include the LoTAF, as exemplified in the figure, helping network users in assessing the trustworthiness of network services, a Transparent Notary Service (TNS) as the common reference for identity (comprising AAA) and crypto (comprising CIA) features, and the possibility of exercising on-demand NDT evaluations of security policies and attack mitigations.

3.5 Conclusion

6G undergoes a fundamental transformation. It will move beyond traditional connectivity to become intelligent, context-aware, and adaptive systems, powered by the integration of cutting-edge technologies like AI, deep-edge computing, NTNs, and ISAC. In addition, it will shift to a 'built for values and performance' approach, prioritizing traditional and new KPIs and KVIs, ensuring environmental, economic and social impact. This requires a complete overhaul of network architecture to ensure ubiquitous, sustainable, and resilient communication across diverse environments, from crowded cities to remote and underserved regions.

A comprehensive methodology has been set up for the design of a 6G E2E system, including principles, a structured design process, and a 6G E2E system blueprint. The architecture design principles prioritize environmental sustainability, digital inclusiveness, and trustworthiness, considering their impact on the 6G E2E system. An iterative design process is defined as a two-way approach, encompassing a key performance and value indicators-based design process and a continuous top-down versus bottom-up alignment process. In the top-down approach, the blueprint is derived from the requirements outlined by the use cases and the underlying architecture design principles. Conversely, in the bottom-up approach, the architectural and technology enablers are developed, tackling various aspects of the 6G system. These enablers can then be analysed for their seamless integration into the 6G E2E system blueprint. Furthermore, the recommendations generated during this process provide valuable feedback to the development of technical work, aiding in the refinement of the enablers and contributing to the iterative enhancement of the overall 6G E2E system blueprint.

The blueprint establishes a multi-layered framework, encompassing infrastructure, network-centric applications, and application layers, along with pervasive functionalities and relevant technological innovations. The infrastructure layer integrates a cloud continuum, exposing in a unified way heterogeneous network infrastructure resources, IoT, extreme edge, and edge cloud computing resources spanning across different administrative and technological domains. It also integrates terrestrial, aerial, and space-based networks for seamless and ubiquitous connectivity. The network functions layer encompasses a software-defined architecture that incorporates 6G RANs, CN functions, and beyond-communication functions. The application enablement platform layer facilitates the exposure of the 6G platform capabilities in a friendly way towards the 6G ecosystem and supports commercial and business aspects. A set of pervasive functionalities permeates all the layers, enabling full-scale automation, real-time intelligence with data-centric workflows and sophisticated AI capabilities, seamless management of E2E services across the cloud continuum, and security against the new and changing surface of attacks.

IBM will play a pivotal role in shaping sustainable and intelligent 6G networks by evolving beyond the capabilities already introduced in 5G. While 5G networks laid the groundwork for IBM by automating certain aspects of network configuration, 6G envisions a far more advanced, autonomous, and context-aware IBM system tailored for dynamic, multi-stakeholder environments. In 6G, IBM will go beyond basic automation by enabling real-time translation of high-level business intents – expressed by diverse stakeholders – into granular, enforceable network policies. This evolution supports smarter resource orchestration across compute, bandwidth, and energy, activating resources only when and where needed to avoid overprovisioning and reduce environmental impact. By continuously monitoring network conditions and autonomously adjusting configurations, 6G-native IBM will reduce operational overhead, minimize human error, and contribute to extending the lifecycle of infrastructure through more efficient usage. With all these aspects in mind, this chapter presented the Hexa-X-II vision for how IBM systems must evolve to address the demands of 6G. It detailed the capabilities IBM must provide to accommodate a wide variety of tenants, ranging from verticals to service providers, and outlined the internal architecture and functional components needed to support this adaptive, scalable, and sustainable approach to IBM in 6G networks.

The integration of advanced enablers in the 6G ecosystem requires rigorous security measures to ensure the reliability and trustworthiness of the overall system. Security is a foundational requirement for system trustworthiness, and this implies a bidirectional relationship between enablers and the security controls discussed in this chapter. The bidirectional relationship implies E2E enablers play a key role in support of security objectives and, conversely, act as the main consumers of the features provided by security controls. From an E2E perspective, enablers must ensure accountability, transparency, and security in processes like intent translation, provisioning, and multi-domain federations. Incorporating security automation and closed-loop coordination is essential to address vulnerabilities and maintain system resilience. Enhanced privacy controls, such as those for threat mitigation, safeguard against unauthorized access and reinforce trustworthiness. Robust and trustworthy AI mechanisms should be integrated, accompanied by privacy-enhanced data management and mechanisms to ensure CIA and AAA properties across network functions and connectivity options. Security automation in edge computing environments and accountability for shared data in federated domains are crucial to maintaining data integrity, reliability, and system trust. These considerations establish a unified framework for enhancing E2E privacy, security, and resilience, aligning with the advanced demands of next-generation network environments.

The blueprint of the 6G sustainable platform serves as a foundation for further exploration of specific system views, offering a starting point for subsequent chapters that delve into various aspects of the 6G ecosystem. By combining a holistic system blueprint with detailed views of specific aspects, this comprehensive approach facilitates a deeper understanding of the intricate workings of the 6G ecosystem and paves the way for a successful implementation of a sustainable 6G future.

Acronyms and Abbreviations

3GPP	Third-Generation Partnership Programme
4/5/6G	Fourth/fifth/sixth generation
AAA	Authentication, authorization, and accounting
AI	Artificial intelligence
AIaaS	AI as a service
API	Application programming interface
CaaS	Compute as a service
CAMARA	A project of The Linux Foundation
CAPIF	Common API framework
CCN	Compute control node
CIA	Confidentiality, integrity, and availability
CL	Closed loop
CN	Core network
COP	Capability operator

CRUD	Create, read, update, delete
DataOps	Data operations
DevSecOps	Development, security, and operations
DLT	Distributed ledger technology
DSC	Digital service consumer
DSP	Digital service provider
DT	Digital twin
E2E	End to end
ETSI	European Telecommunications Standards Institute
FL	Federated Learning
FR3	Frequency Range 3
IBM	Intent-based management
ICT	Information and Communications Technology
IDM	Intrusion detection mechanisms
IDS	Intrusion detection system
IF	Integration fabric
IME	Intent management entity
ISAC	Integrated sensing and communication
JCAS/ICAS	Joint/integrated communication and sensing
KPI	Key performance indicator
KVI	Key value indicator
LLM	Large language model
LoTAF	Level of trust assessment function
M&O	Management and orchestration
MAC	Medium access control
MD	Management domain
ML	Machine learning
MNO	Mobile network operator
NDT	Network digital twin
NER	Named entity recognition
NLP	Natural language processing
NN	Neural network
NTN	Non-terrestrial network
OpEx	Operating expenses
OTT	Over-the-top
PHY	Physical (layer)
PKI	Public key infrastructure
PLD	Physical layer deception
PLS	Physical layer security
QoE	Quality of experience
QoS	Quality of service
RAN	Radio access network
RAT	Radio access technology
RL	Reinforcement learning
SAD	Single administrative domain
SDN	Software-defined networking
SeMF	Sensing management function

SKG	Secret key generation
SLA	Service level agreement
SNS-JU	Smart networks and services joint undertaking
SoC	System on chip
Tbps	Terabit per second
TDD	Time division duplex
TLA	Trust level agreement
TN	Terrestrial network
TNS	Transparent Notary Service
UE	User equipment
URLLC	Ultra-reliable low-latency communication
XAI	Explainable AI

References

3GPP (2022). Technical specification group services and system aspects; common API framework (CAPIF) – framework for northbound APIs; stage 3 (Release 17), v17.5.0, 2022. 3GPP TS 29.222. Available at: https://www.3gpp.org/ftp/Specs/archive/29_series/29.222/ (accessed 29 August 2025).

3GPP (2025). Management and orchestration; Intent driven management services for mobile networks (Release 19). 3GPP TS 28.312, March 2025.

Akgul, O., Ericson, M., Botsinis, P. et al. (2025). Hexa-X-II "Deliverable D3.5 final architectural framework and analysis". Hexa-X-II European Level 6G Flagship Project [Online]. Available at: https://zenodo.org/records/15774984 (accessed 29 August 2025).

Alemany, P. and Ojaghi, B. (2024). Hexa-X-II "Deliverable D2.4: end-to-end system evaluation results from the interim overall 6G system". September 2024. [Online]. Available at: https://zenodo.org/records/15796347 (access 29 August 2025).

Almodovar, J. and Vinagre, C. (2023). Hexa-X-II "Deliverable D1.2: 6G use cases and requirements". Hexa-X-II – European Level 6G Flagship Project. [Online]. Available at: https://zenodo.org/records/16779960 (accessed 29 August 2025).

Balakrishnan, B. (2023). Hexa-X-II "Deliverable D5.2 Characteristics and classification of 6G device classes. [Report] Hexa-X-II European Level 6G Flagship Project. [Online]. Available at: https://zenodo.org/records/16995203 (accessed 29 August 2025).

Ericson, M., Flinck, H., Vlachèas P. et al. (2021). Hexa-X "Deliverable D5.1: initial 6G architectural components and enablers". [Online]. Available at: https://hexa-x.eu/wp-content/uploads/2022/03/Hexa-X_D5.1_full_version_v1.1.pdf (accessed 29 August 29, 2025).

Ericsson Mobility Report (2021). Available at: https://www.ericsson.com/4a03c2/assets/local/reports-papers/mobility-report/documents/2021/june-2021-ericsson-mobility-report.pdf (accessed 29 August 2025).

ETSI (2023). Zero-touch network and Service Management (ZSM); Intent-driven autonomous networks; Generic aspects. ETSI GR ZSM 011, February 2023. [Online]. Available at: https://www.etsi.org/deliver/etsi_gr/ZSM/001_099/011/01.01.01_60/gr_ZSM011v010101p.pdf (accessed 29 August 2025).

ETSI (2025). Zero-touch network and service management (ZSM); closed-loop automation; part 1: enablers. ETSI GS ZSM 009-1, June 2025 [Online]. Available at: https://www.etsi.org/deliver/etsi_gs/ZSM/001_099/00901/01.01.01_60/gs_ZSM00901v010101p.pdf (accessed 29 August 2025).

Giardina, P. and Ruta, M. (2024). ROBUST-6G "Deliverable D4.1: security automation for 6G". [Online]. Available at: https://robust-6g.eu/wp-content/uploads/2025/01/ROBUST-6G-D4.1_Security-Automation-for-6G_v1.0.pdf (accessed 29 August 2025).

Harounabadi, M. and Heyn, T. (2023). Toward integration of 6G-NTN to terrestrial mobile networks: research and standardization aspects. *IEEE Wireless Communications* 30 (6): 20–26.

Hexa-X-II (2023–2025). Hexa-X-II – European Level 6G Flagship Project. [Online]. Available at: https://hexa-x-ii.eu/ (accessed 29 August 2025).

ICT-FI (2023). RIGOUROUS "Deliverable D2.2: RIGOUROUS use-cases". [Online]. Available at: https://zenodo.org/records/10476145 (accessed 29 August 2025).

IETF (2023). IETF RFC 9417 "Service assurance for intent-based networking architecture". July 2023. [Online]. Available at: https://datatracker.ietf.org/doc/rfc9417/ (accessed 29 August 2025).

Jorquera Valero, J.M., Ordóñez Lucena, J.A., Pastor Perales, A.A. et al. (2024). Building trust in the era of 6G: a level of trust assessment function for cloud continuum. *2024 Joint European Conference on Networks and Communications & 6G Summit (EuCNC/6G Summit)*, pp. 812–817. IEEE.

Kerboeuf, S., Porambage, P., Jain, A. et al. (2024). Design methodology for 6G end-to-end system: hexa-X-II perspective. *IEEE Open Journal of the Communications Society.* doi: https://doi.org/10.1109/OJCOMS.2024.3398504.

Kokkonen, H., Lovén, L., Motlagh, N. H. et al. (2022). Autonomy and intelligence in the computing continuum: challenges, enablers, and future directions for orchestration. arXiv preprint arXiv:2205.01423.

Le Sauze, N. and Kerboeuf, S. (2025). Hexa-X-II "Deliverable D2.5 final overall 6G system design". April 2025. [Online]. Available at: https://zenodo.org/records/15796361 (accessed 29 August 2025).

Letaief, K.B., Shi, Y., Lu, J., and Lu, J. (2022). Edge artificial intelligence for 6G: vision, enabling technologies, and applications. *IEEE Journal on Selected Areas in Communications* 40 (1): 5–36.

Linux Foundation (2024). *CAMARA Project*. Retrieved from https://camaraproject.org (accessed November 2024).

Lopez Da Silva, R. (2023) D2.1: definition of use cases, service requirements and KPIs/KVIs. [Report] DESIRE6G. Available at: https://zenodo.org/records/10446047 (accessed 29 August 2025).

Mahmood, N.H., Samarakoon, S., Bennis, M., et al. (2024). Resilient-by-design: a resiliency framework for future wireless networks. arXiv preprint arXiv:2410.23203.

Mishra, D., Vegni, A.M., Loscrí, V., and Natalizio, E. (2021). Drone networking in the 6G era: a technology overview. *IEEE Communications Standards Magazine* 5 (4): 88–95.

NGMN Alliance (2023). 6G trustworthiness considerations. Retrieved from https://www.ngmn.org/wp-content/uploads/NGMN_6G_Trustworthiness.pdf.

Pastor, A., Ramón, H. and Lopez, D. (2023). PRIVATEER "Deliverable D2.2 use cases, requirements and design report". [Online]. Available at: https://zenodo.org/records/10027073 (accessed 29 August 2025).

Porambage, P. et al. (2023a). Hexa-X-II "Deliverable D2.1: draft foundation for 6G system design". [Online]. Available at: https://zenodo.org/records/16995021 (accessed 2 September 2025)

Porambage, P. et al. (2023b). Hexa-X-II Deliverable "D2.2: foundation of overall 6G system design and preliminary evaluation results". [Online]. Available at: https://zenodo.org/records/16995047 (accessed 2 September 2025)

Ramos, E. (2025). Hexa-X-II Deliverable D1.4: 6G value, requirements, and ecosystem. Hexa-X-II – European Level 6G Flagship Project. [Online]. Available at: https://zenodo.org/records/15772634 (accessed 29 August 2025).

Siachos, I., Kontopoulos, G., Cianchini, V., et al. (2023). HORSE: "Deliverable D2.1: HORSE landscape: technologies, state of the art, AI policies and requirements". [Online]. Available at: https://www.horse-6g.eu/?wpdmdl=491&ind=1698663024683 (accessed 29 August 2025).

Simeone, O. and Sifaou, H. (2024) *D2.2 Digital Twin Training Sandbox*. Zenodo. doi: https://doi.org/10.5281/zenodo.12742279.

TMForum, IG1253 (2022). Intent in Autonomous Networks v1.3.0 (IG1253). September 2022. [Online]. Available at: https://www.tmforum.org/resources/introductory-guide/ig1253-intent-in-autonomous-networks-v1-3-0/ (accessed 29 August 2025).

Uzinidis, D., Rivas Moscoso, J. M., Munoz, R. et al. (2023). FLEX-SCALE "Deliverable D2.1: 6G network requirements". [Online]. Available at: https://6g-flexscale.eu/en/outcomes (accessed 29 August 2025).

Valero, J. and Perez, M. (2024). ROBUST-6G "Deliverable D2.2: use cases, requirements, and initial architecture". [Online]. Available at: https://robust-6g.eu/wp-content/uploads/2025/01/ROBUST-6G-D2.2_v1.0-1.pdf (accessed 29 August 2025).

Vega, M.T., Liaskos, C., Abadal, S. et al. (2020). Immersive interconnected virtual and augmented reality: a 5G and IoT perspective. *Journal of Network and Systems Management* 28 (4): 796–826.

Zafeiropoulos, A., Fotopoulou, E., Filinis, N., and Papavassiliou, S. (2022). Reinforcement learning-assisted autoscaling mechanisms for serverless computing platforms. *Simulation Modelling Practice and Theory* 116: 102461.

Zou, Z. and Rugeland, P. et al. (2024). Deliverable D2.3: interim overall 6G system design. [Report]. Hexa-X-II – European Level 6G Flagship Project. [Online]. Available at: https://zenodo.org/records/16995059 (accessed 29 August 2025).

4

6G Transceiver and Radio Design

Ahmad Nimr[1], Luis González[2], Tommy Svensson[3], Italo Atzeni[4], Jeroen Famaey[5], Nurul Huda Mahmood[4], Enrico Maria Vitucci[6], Gilberto Berardinelli[19], Claude Desset[5], Nuutti Tervo[4], George-Roberto Hotopan[7], Philippe Ratajczak[8], Davide Dardari[6], Sotiris Droulias[9], Angeliki Alexiou[9], Bikshapathi Gouda[4], Venkatesh Tentu[3], Charitha Madapatha[3], Akshay Vayal Parambath[3], Hao Guo[3], Sebastian Haas[10], Emil Matus[1], Onel L. A. López[4], Bikramjit Singh[11], Nafiseh Mazloum[12], Samer Nasser[5], Ritesh Kumar Singh[5], Priyesh Pappinisseri Puluckul[5], Riku Jäntti[13], Dinh-Thuy Phan-Huy[8], Efstathios Katranaras[14], Usman Virk[18], Pekka Kyosti[4,18], Katsuyuki Haneda[13], Yigit Ertugrul[5], Meng Li[5], Bilal Khan[4], Christos Tsokos[20], Osmel Martínez Rosabal[4], Amirhossein Azarbahram[4], Ling Jie[11], Peize Zhang[4], Jingyi Liao[13], Kalle Koskinen[13], Xie Boxuan[13], Simon Nellen[15], Tianwen Qian[15], Ahmad Mohammad[16], Chris Roeloffzen[16], Muhsin Ali[17], Juha-Matti Runtti[18] and Fengchun Zhang[19]

[1] *Technische Universität Dresden, Dresden, Germany*
[2] *Universidad Carlos III de Madrid, Leganés, Madrid, Spain*
[3] *Chalmers University of Technology, Göteborg, Schweden*
[4] *University of Oulu, Oulu & Keysight, Finland*
[5] *IMEC, Antwerp, Belgium*
[6] *University of Bologna, Bologna, Italy*
[7] *Nokia, Germany*
[8] *Orange, Valbonne, France*
[9] *University of Piraeus Research Center, Piraeus, Greece*
[10] *Barkhausen Institut, Dresden, Germany*
[11] *Ericsson Research Jorvas, Finland*
[12] *Sony, Nordic, Sweden*
[13] *Aalto University, Espoo, Finland*
[14] *Sequans Communications, Paris, France*
[15] *Fraunhofer Institute for Telecommunications, Heinrich Hertz Institute, Berlin, Germany*
[16] *LioniX International, Enschede, The Netherlands*
[17] *LeapWave Technologies, Leganés, Madrid, Spain*
[18] *Keysight, Oulu, Finland*
[19] *Aalborg University, Aalborg, Denmark*
[20] *NTUA, Athens, Greece*

The sixth generation (6G) of wireless systems is expected to build upon the foundations of fifth generation (5G) by improving key performance indicators (KPIs) such as capacity, latency, and reliability, while also placing greater emphasis on sustainability and adaptability. In this context, 6G radio design adopts a value-based and holistic approach, where technical choices are guided not only by performance but also by environmental, economic, and social sustainability (Mahmood et al. 2025). Central to the radio system is the transceiver, which plays a key role in enabling communication across a wide range of use cases, including industrial automation, connected mobility, and energy-constrained Internet of Things (IoT) deployments. Meeting these varied requirements

6G to Build a Sustainable Future, First Edition. Edited by Mikko A. Uusitalo, Patrik Rugeland, Mauro R. Boldi, and Ahmad Nimr.

calls for radio solutions that are efficient, flexible, and capable of adapting to different operational constraints without compromising on sustainability goals. To fulfil such demands, 6G radio systems must integrate a wide range of technology enablers, including access to new spectrum bands, scalable hardware (HW) architectures, and advanced signal processing.

One of the fundamental drivers of innovation in 6G is spectrum expansion, with particular attention to the frequency range 3 (FR3) band (7.125–24.25 GHz), which offers a promising balance between bandwidth and coverage. While FR3 is currently the main focus for early 6G deployments, research into the sub-terahertz (sub-THz) range (above 100 GHz) is ongoing. Sub-THz is of interest due to its potential to support extremely high data rates and high-resolution sensing capabilities, which may be needed in dense, short-range scenarios or future phases of 6G (Nimr 2023). However, the use of sub-THz bands introduces significant HW challenges, such as power-efficient signal generation, low-phase-noise oscillators, and compact high-gain antenna integration (Desset et al. 2024). To address these, emerging transceiver technologies, including photonic-based solutions (Nellen et al. 2022b), resonant tunnelling diodes (RTDs) (Nishigami et al. 2018), and reconfigurable intelligent surfaces (RIS) (Ratajczak et al. 2024), are being explored for their potential to enable efficient and scalable implementations at these higher frequencies.

Additionally, realistic modelling of the propagation environment and HW impairments is essential for the new spectrum ranges (Mahmood et al. 2025). In particular, higher frequencies increase sensitivity to factors such as mobility, directionality, and interference, requiring updated channel models and abstraction layers. In addition to capturing propagation dynamics, these models must account for HW imperfections, including amplifier non-linearities and phase noise (PN) (Tervo et al. 2023). Such modelling is a critical foundation for both the design and evaluation of transceivers, enabling reliable simulation, performance assessment, and system optimization.

Expanding from transceiver-level design considerations, 6G extends its capabilities to support advanced multi-antenna configurations. Multiple-input multiple-output (MIMO) and distributed MIMO (D-MIMO) systems are key enablers for enhancing spatial resolution, link reliability, and network capacity. Technologies such as massive MIMO and extremely large-scale MIMO (XL-MIMO) (Zhang et al. 2023) that can be integrated with RIS and near-field wavefront engineering, enable refined beam control and integrated sensing and communication (ISAC) (Stratidakis et al. 2024). To support these features efficiently, various transceiver architectures are explored, including low-resolution fully digital designs using simplified analogue-to-digital converters (ADCs) and digital-to-analogue converters (DACs), as well as hybrid analogue–digital implementations (Atzeni et al. 2023). In parallel, novel signal processing aspects, such as 'over-the-air' (OTA) electromagnetic (EM) processing, are investigated to reduce complexity and improve energy efficiency in dense and heterogeneous deployment scenarios (Dardari et al. 2024a).

Considering the diverse needs of 6G applications, radio design should be tailored to a wide range of device types, from high-performance user equipment (UE) to ultra-low-power IoT nodes. This includes accommodating stringent requirements on size, power efficiency, and security. Emerging devices increasingly integrate advanced system-on-chip (SoC) platforms (Weinhold et al. 2023) that combine secure architectures with dedicated accelerators for artificial intelligence (AI), machine learning (ML), and digital signal processing (DSP) (Friedrich et al. 2024). These HW capabilities are essential to support intelligent and adaptive radio functionalities, such as real-time signal optimization, autonomous resource management, and context-aware communication directly at the device level. Moreover, 6G devices may operate under energy-constrained conditions, relying on energy harvesting (EH) or passive technologies such as ambient backscatter (López et al. 2025). Therefore, transceiver designs must ensure both performance and flexibility while enabling integration with modern processing and energy solutions.

This chapter introduces the foundational principles and latest research in 6G radio and transceiver design. Section 4.1 offers a high-level overview of 6G radio design, outlining the main technical components and their interdependencies. The subsequent sections explore specific technical aspects in more detail. Section 4.2 explores various transceiver architectures, with particular attention to HW components and implementation technologies. Section 4.3 focuses on radio channel modelling and HW non-idealities modelling, especially in the FR3 and sub-THz FRs. Section 4.4 highlights recent advances in MIMO architectures, including designs using low-resolution converters, OTA EM signal processing, near-field wavefront engineering, and D-MIMO processing. Finally, Section 4.5 provides an overview of innovation in 6G device design, with emphasis on emerging IoT devices and advanced processing platforms.

4.1 6G Radio Design Overview

The 6G radio design follows a value-based, holistic approach that not only focuses on fulfilling technical specifications but also considers the sustainability aspect in its environmental, social, and economic domains. As illustrated in Figure 4.1, the 6G radio design process takes specific use cases and their associated requirements as inputs, with the goal of satisfying these requirements through innovative solutions under given constraints, such as regulatory aspects for frequency and spectrum usage, and power level. Moreover, 6G radio design emphasizes key value considerations for sustainability, as discussed in Chapter 2, Section 2.2. These constraints impose restrictions on the usage of resources such as energy, material, and software/HW components. Therefore, innovative technologies and solutions are needed to balance performance with value-based requirements. These include the following physical layer (PHY) enablers:

- **New spectrum** to meet the increased demand for capacity and to enable new use cases. For instance, (sub)-THz (above 100 GHz) could be used for short-range communication that requires an extreme data rate and for radio sensing due to the availability of a very large

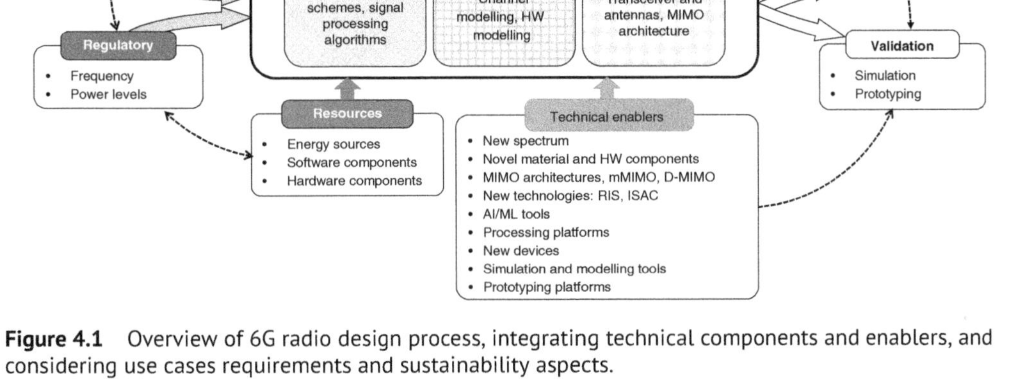

Figure 4.1 Overview of 6G radio design process, integrating technical components and enablers, and considering use cases requirements and sustainability aspects.

bandwidth and the need for narrow beams. Moreover, the cm-Wave (7–15 GHz) is considered a key spectrum that offers both capacity and coverage. Using new spectrum requires exploration of transceiver design (Section 4.2), and the corresponding channel and HW models (Section 4.3).

- **Novel material and HW components** with minimal environmental impact in terms of reduced energy consumption, and improved durability and recyclability, in addition to exploring alternative technologies of transceiver implementation (Section 4.2.2).
- **MIMO architectures**, such as massive MIMO, aim to improve spectral efficiency (SE) and/or energy efficiency by means of developing efficient transmission techniques and optimized HW architecture (Sections 4.4.1 and 4.4.3). Moreover, D-MIMO provides solutions to enhance the coverage and mitigate blockage (Section 4.4.4).
- **New technologies,** such as ISAC, aim to leverage network infrastructure and devices for providing sensing and localization information (Chapter 7). Furthermore, RIS provides low-cost radio solutions that enable controlling the propagation environment (Section 4.2.3), and can be utilized with other technical enablers, such as ISAC (Section 4.4.2) or D-MIMO (Section 4.4.4).
- **AI/ML tools** can be exploited in the design and operation of the radio interface (Chapter 6, Section 6.4). For instance, AI-based modelling can be used for modelling HW non-linearities, HW impairment compensation, resource allocation, channel state information (CSI) acquisition, and for E2E radio optimization.
- **Processing platforms** to meet diverse computational demands and performance metrics, while enhancing energy efficiency and security. These include designing SoC and AI accelerators (Section 4.5.2).
- **New devices,** such as IoT devices, that are optimized for massive deployment, considering sustainability and smart operation (Sections 4.5.1 and 4.5.3).

Other radio enablers include mechanisms for operating the PHY, such as:

- **Flexible topologies** that enable adaptable network configurations to efficiently meet varying use cases demands and environments, such as the integration of terrestrial networks (TNs) and non-terrestrial networks (NTNs) (Chapter 5, Section 5.4).
- **Mechanisms for efficient spectrum sharing** to improve spectrum utilization and reduce interference among different radio access technologies (RATs), including spectrum sharing solutions for NTN spectrum, such as fixed satellite, with TN, as well as multi-RAT spectrum sharing (MRSS) (Chapter 5, Section 5.3.3).

During the radio design, various technologies are explored and initially evaluated using simulation and relying on the mathematical models for channels and HW components (Section 4.3). Thereafter, the technologies are validated using prototyping platforms and experiments (Section 4.2.3).

4.1.1 6G Radio Scenarios

A successful radio design begins with a precise definition of its use cases. These use cases reflect the diverse needs and scenarios in which the radio system operates, from urban high-density connectivity to remote access in rural areas. Over generations, use cases have evolved, with increased demands for bandwidth, low latency, high reliability, and massive connectivity. Defining specific radio scenarios and their requirements sets the foundation for a radio system design.

In a typical process, as illustrated in Figure 4.2, a wide range of application use cases are analysed and grouped into families with similar service requirements (Chapter 2, Section 2.3). For radio

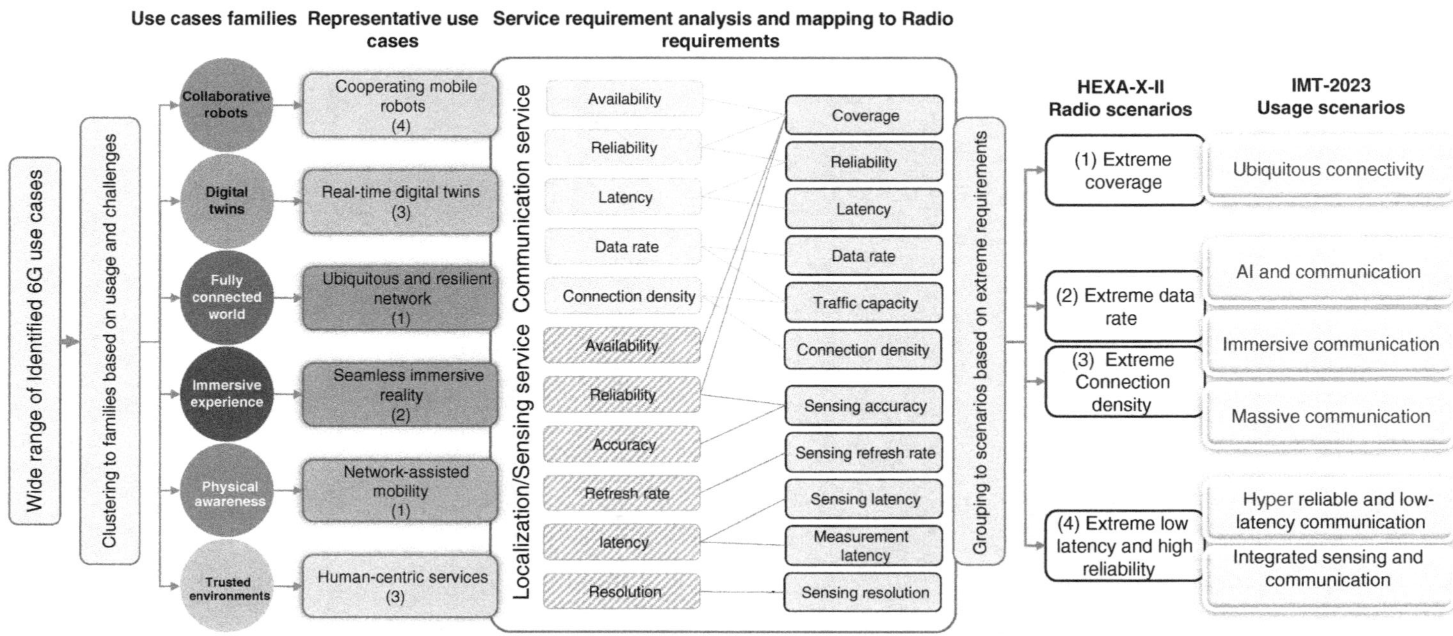

Figure 4.2 Framework for use cases analysis and radio scenario definition.

design, the service requirements are mapped to corresponding radio requirements, and a set of scenarios is derived.

One approach to defining scenarios is to consider extreme requirements, which provide the most challenging and demanding conditions for radio design. This approach ensures that a radio system designed and optimized for an extreme scenario can meet diverse economic and social needs. For instance, HEXA-X-II defines four scenarios (Mahmood et al. 2025), each focusing on one extreme requirement while allowing flexibility in other parameters as flexible. These scenarios are denoted as (i) extreme coverage, (ii) extreme data rate, (iii) extreme connection density, and (iv) extreme low-latency and high reliability. The International Telecommunication Union (ITU) defines six 6G usage scenarios in the IMT-2030 framework (ITU-R M.2160 2023). These scenarios emphasize new technologies (AI and ISAC), while extending the three usage scenarios introduced in 5G (enhanced mobile broadband (eMBB), ultra-reliable low-latency communications (URLLC), and massive machine-type communications (mMTC)).

Along with the requirements, multiple scenarios can be further derived considering further deployment and operation parameters. These are types of *radio devices that* influence transceiver form factor and operational constraints, the *environment* that influences the propagation channel, *mobility* of involved devices that impact the channel and link quality, *deployment options* of the network structure that influence the coverage conditions, and the operating *frequency that has* an impact on the transceiver and antenna design.

Table 4.1 provides a summary of scenarios, including radio requirements and various sub-scenario parameters, with a mapping to the HEXA-X-II use case families (Chapter 2, Section 2.3), and ITU usage scenarios. The extreme radio KPIs are aligned with IMT-2030 recommended capabilities (ITU-R M.2160 2023).

4.1.2 Radio Design Framework

The holistic radio design approach considers the radio system in its entirety, accounting for the interdependencies between its various elements. The radio HW and software are jointly designed to optimize the performance and efficiency of the system. The design methodology can be applied to both infrastructure and user devices, considering the implantation constraints, in size, availability of power supply, and applications, as discussed in Section 4.5. The radio design framework shown in Figure 4.3 illustrates the different modules within the radio system and their interfaces, as well as the group of enablers related to each module.

Propagation environment: It considers physical environment effects on the signal, including obstacles, reflectors, and atmospheric effects. It is influenced by the environment, deployment type, and mobility.

RU: The radio unit (RU) consists of a radio frequency (RF) frontend (FE), frequency conversion, a digital/analogue (D/A) conversion. It interacts with the propagation environment through the antennas and with the PHY via digital samples. From the PHY perspective, RU functionality can be abstracted by means of HW and channel modelling. Various RF transceiver architectures can be implemented in RUs, determined by the number of RF chains and antennas, as well as the resolution of the DACs and ADCs. For example, a fully digital MIMO architecture employs one chain per antenna, whereas hybrid beamforming uses analogue beamforming, with more antennas than RF chains. Low-resolution DAC/ADC can be used in a fully digital architecture, such as 1-bit ADC/DAC massive MIMO, as discussed in Section 4.4.3. Additionally, there are various types of analogue beamforming architectures using fully or partially connected phased arrays,

Table 4.1 Summary of radio scenarios with corresponding radio requirements and parameters. Extreme requirements are highlighted, while boldface KPIs represent relaxed requirements.

	Extreme coverage	Extreme data rate	Extreme connection density	Extreme low latency and high reliability
Radio devices	Enhanced 5G (mMTC, eMBB) devices, Energy-neutral devices	Access points for backhaul, Gateway for sensors	Reliable high data rate with bounded latency devices	High reliability & low latency devices
Environment	Mobile indoor, Public indoor, Outdoor (urban, suburban, rural)	Controlled and semi-controlled indoor and outdoor	Urban indoor/outdoor with high density of users High-rise	Indoor, Embedded network
Mobility	Static, up to 300 km/h	Static, up to 10 km/h, controlled mobility (velocity information available)	Static, up to 100 km/h	Static, up to 20 km/h, controlled mobility (velocity information available)
Deployment options	Long/short range, Fixed/temporary, Mobile infrastructure, TN/NTN integration	Small cell, Fixed access, D2D, Sensor network with a gateway	High density of cells, Macro cell, Micro cell, Cell-free	Small cell, On premises, Infrastructure, Sensor network
Frequency	Sub-GHz, sub-6 GHz, 7–15 GHz, satellite frequency ranges	mmWave, sub-THz, mixed and unlicensed for local connections	Sub-6 GHz, 7–15 GHz, mmWave	Private frequency, sub-GHz, sub-6, GHz, 7–15 GHz, mmWave, sub-THz for sensing
Peak data rate (PHY)	**<1 Gbit/s**	(10–100 Gbit/s)	**<10 Gbit/s**	**<10 Mbit/s**
Link reliability	Variable	Variable	Variable	(99.999–99.999 99%)
Air interface latency	Variable	Variable	Variable	(0.1–10) ms
Connection density	Variable	**<10^6 device/km^2**	(10^6–10^8) device/km^2	Variable
Coverage	Availability (99.99–99.999 999%)	Local	Variable	Local coverage
Sensing capabilities	Variable	Variable	Variable	Positioning accuracy (0.1–1) cm

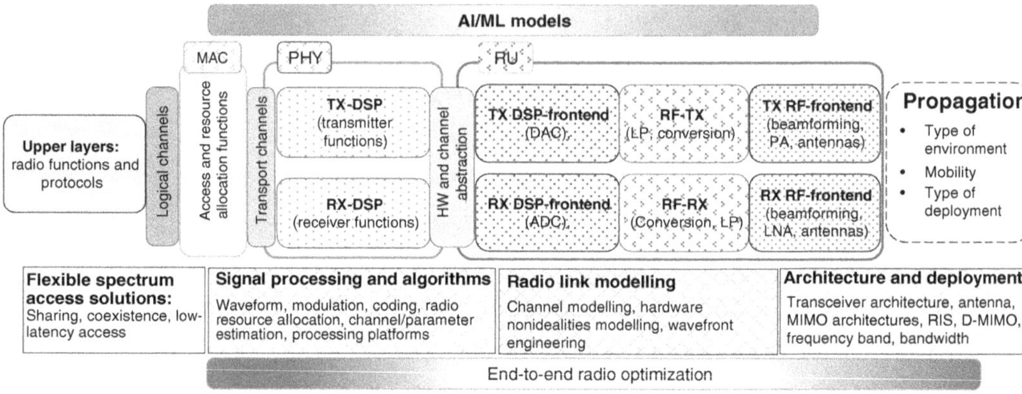

Figure 4.3 Holistic radio design framework.

or switched beam antenna lenses (Desset et al. 2024). Different technologies can be used in the implementation of the transceiver, considering the particularity of the operating frequency, such as sub-THz, as discussed in Section 4.2.2.

Radio link modelling: This is an essential aspect in radio design, providing simulation models for the channel and HW components. Various abstracted channels can be obtained at different levels, as follows:

- Wireless channel: This abstracts the effects of the propagation channel and accounts for the impact of the RF FE. The characteristic of this channel depends on the FR, bandwidth, beamforming, RUs distribution, and deployment scenario. Channel modelling aims to capture the key characteristics and behaviours of wireless channels with respect to the radio signal (3GPP TR 38.901 2024). Modelling at this level also includes aspects like obtaining models for antenna coupling, power amplifier (PA), and low-noise amplifier (LNA). More details are provided in Section 4.3.
- Baseband channel: This extends the wireless channel by considering the effects of frequency converters, including PN and low-pass (LP) filtering.
- Digital baseband channel: This incorporates, in addition to the baseband (BB) channel, the effects of DAC and ADC resolution, and digital frontend (DFE) processing, such as filtering and pre-distortion.

PHY: It encompasses a broad range of functions related to transmission, detection, CSI feedback, and sensing. It provides transport channels to the medium-access control (MAC) layer and interacts with the discrete BB channel. The PHY functions can be implemented using conventional model-based solutions or employing AI/ML data-driven models, as discussed in Chapter 6, Section 6.4, which can address individual functions or a combination of them.

MAC: It is responsible for spectrum access, physical resource allocation, scheduling, and retransmission. It includes functions for flexible spectrum sharing and coexistence, as discussed in Chapter 5, Section 5.3.3.

Upper layers: These layers consist of various radio functions and protocols, such as radio link control (RLC), packet data convergence protocol (PDCP), and radio resource control (RRC), which operate on top of MAC/PHY. Further details on radio protocols are presented in Chapter 5, Section 5.6.

The 6G radio system is integrated in the 6G E2E system (Chapter 3, Section 3.2) at the infrastructure layer through devices and access, and in the network layer via the radio-access network (RAN)

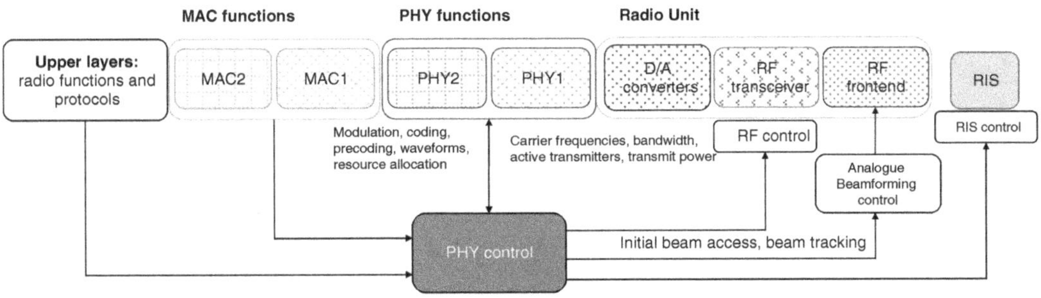

Figure 4.4 Radio logical functional split and configuration parameters.

functions and protocols. Moreover, the basic radio components can be deployed and interconnected in various forms in the realization of the radio network.

4.1.3 Flexible Radio Architecture and Deployment

In the 6G radio system, multiple degrees of flexibility can be achieved in both the implementation and deployment of various functions. At the PHY/MAC level, flexibility is typically realized through variable waveform numerologies, adaptive modulation and coding, and dynamic radio resource allocation. This allows the system to adjust to changing conditions and requirements. In the RU, flexibility can be achieved following a software-defined radio (SDR) approach that enables the configuration of carrier frequencies, bandwidth, power level, the number of active chains, as well as controlling the radiation patterns in the case of an analogue beamforming FE.

Configuration commands can originate from different layers in the system, allowing the radio to be tailored to specific requirements and operating conditions.

Figure 4.4 illustrates a logical functional split of MAC, PHY, and RU functions, along with their corresponding configuration parameters. In typical devices, these functions are often collocated to achieve higher integration and reduce material costs. However, in the RAN infrastructure, implementing various functional splits can offer additional flexibility, enabling a more modular approach. These splits can be beneficial in a range of scenarios, allowing for optimized resource usage and scalability. However, they also affect infrastructure requirements, such as fronthaul capacity and signal processing design. Additionally, supporting multiple options can lead to unnecessary complexity with little benefit. Therefore, only the most suited option should be selected. Figure 4.5 provides examples of the functional splits discussed in this deliverable:

- **Centralized RAN:** In this configuration, the MAC, PHY, and RU functions are collected and controlled at a higher layer. This setup can support, e.g. multi-connectivity at PDCP or carrier aggregation with individual cell- or site-specific MAC schedulers. This corresponds to higher layer split (HLS), as discussed in Chapter 5, Section 5.2.1, which has shown to have several drawbacks in 5G (e.g. complexity, latency to configure, and requiring significant signalling between the centralized unit (CU) and distributed unites (DUs).
- **Distributed MAC:** Here, the MAC functions are split into centralized and distributed entities. Partially centralized MAC functions enable carrier aggregation across different FRs, as discussed in Chapter 5, Section 5.3.2, and offer flexibility in managing resources, such as for filling latency requirements where low-latency tasks are distributed. This split is typically associated with the data plane, and a similar split can be achieved for the control plane. However, a split of the control plane adds latency and complexity to the operations, where the different nodes need to negotiate configurations back-and-forth for each UE.

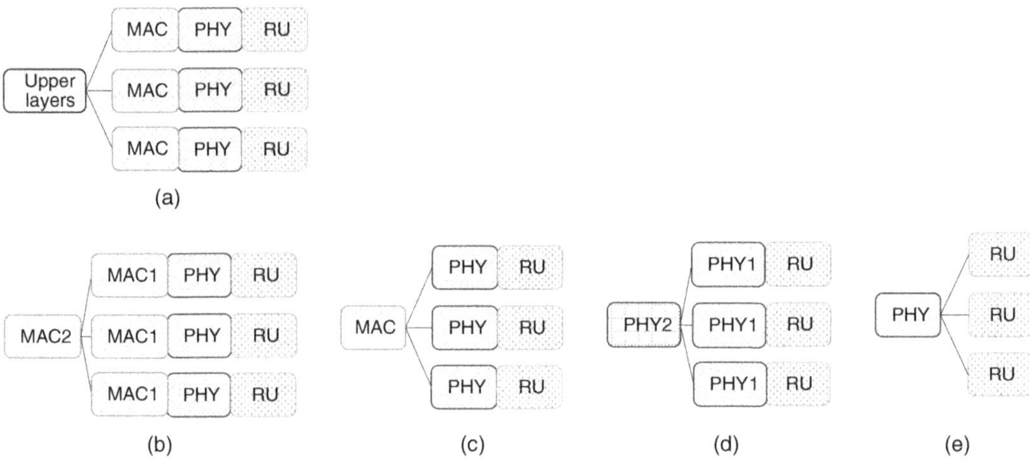

Figure 4.5 RAN deployment options. (a) Centralized RAN, (b) multiple PHYs/distributed MAC, (c) multiple PHYs/centralized MAC, (d) multiple RUs/distributed PHY, and (e) multiple RUs/centralized PHY.

- **Centralized MAC:** In this setup, multiple PHY instances are centrally controlled by MAC. This allows, for example, carrier aggregation in the same FR with a system-wide scheduler. However, latency requirements might require a more distributed topology.
- **Distributed PHY:** This deployment corresponds to D-MIMO with distributed processing, as discussed in Section 4.4.4.1. It reduces fronthaul capacity requirements by offloading part of the PHY processing close to RUs, such as partial channel estimation and precoding. This split is also known as lower layer split (LLS).
- **Centralized PHY:** This is a conventional D-MIMO setup with centralized processing, where all samples from RUs are collected and centrally processed. In some cases, analogue fronthaul may be used, where D/A converters are placed closer to the PHY to reduce digital signal transport.

Further deployment scenarios correspond to the placement of transmitters and receivers (Rxs) as in joint communications and sensing (JCAS) deployment, such as bistatic, monostatic, and multistatic sensing, as discussed in Chapter 7.

4.2 Transceivers and Antennas

While early 6G systems are expected to operate in existing sub-6 GHz bands or potentially in FR3, the growing demand for data may drive later 6G deployments into the sub-THz range (100 GHz to 1 THz), where significantly wider bandwidths are likely to be available compared to current frequency bands. For systems operating in existing bands or FR3, HW transceivers and radio design will not differ substantially from current 5G technology. However, exploring the sub-THz region introduces new challenges due to the much higher carrier frequencies. At these frequencies, signal generation and detection push the limits of current semiconductor technologies. Antenna design also becomes more complex, as wave behaviour transitions between RF and optical domains, and large antenna arrays are required to overcome path loss and meet link budget demands. This section, therefore focuses, on sub-THz technologies. It begins with novel architectures aimed at improving power efficiency, reducing PN, and enabling antenna integration in sub-THz systems using conventional transistor-based technologies. It then explores emerging technologies for sub-THz signal generation and detection, such as RTDs and photomixers. Finally, it introduces RISs as a potential alternative to traditional phased antenna arrays.

4.2.1 Novel Architectures for Transistor-Based Sub-THz Systems

4.2.1.1 Dimensioning

Exploring new frequency bands, especially in the sub-THz region, may lead to very different architectures compared to sub-6 GHz, or to a lesser extent, to FR2 millimetre-wave (mmWave) bands. Carrier frequency and bandwidth are much larger, semiconductor technology and circuit design come with more challenges, wireless propagation differs, and large antenna arrays are required to bridge the link budget. Making new architectures energy-efficient requires a proper dimensioning, considering architecture types (full-digital, analogue, hybrid) or main dimensioning parameters (number of antennas, PA output power). This can be achieved based on a scalable power consumption model (Desset et al. 2020) coupled to link budget assessment such that the target scenario requirements can be met.

Different architectures were optimized in terms of energy efficiency for various hotspot scenarios, serving 10 to 120 users in a range of 10–100 m. A first observation is to recommend just-sufficient architecture dimensioning. For instance, selecting the maximum number of antennas or the highest output power from the available PA technology may lead to significant over-design. Although this will provide the fastest throughput, this can be shown to lead to severe degradation of the system-specific energy consumption, in nJ/bit, by a factor of 10 or more.

A second observation is that the most energy-efficient configuration typically consumes a balanced energy between the PAs and the other components of the analogue FE (Desset et al. 2021). The reason is that for the same link budget performance, it is possible to either increase the number of FE chains and reduce the PA power levels, or, on the contrary, increase the PA power in order to reduce the number of chains. This leads to a parabola-shaped curve of the total power with respect to the number of antennas, where the minimum is reached when the power consumption of both components is similar.

A third interesting observation relates to the relative share of the analogue FE and the digital BB processing. While often believed that digital power consumption will dominate sub-THz systems due to very wide bandwidth and a large number of antenna chains to process, this is generally not the case. As an example, using a hybrid partially connected architecture with 4 BB chains and 64 antennas, the power consumption remains dominated by the analogue FE for a bandwidth of up to 2.3 GHz (see Figure 4.6), even using relatively old (20 nm) digital complementary metal-oxide-semiconductor (CMOS) technology. Extrapolating to more advanced (3 nm) CMOS expands the analogue-dominated region up to above 30 GHz bandwidth.

This analysis has implications for which component to focus on when targeting improved energy efficiency. For instance, DAC/ADC components represent a small share of the power consumption, and they do not dominate power for resolutions around 10-bit that are considered in this study. Given the exponential nature of their power consumption, DAC/ADC would quickly become dominant at 16-bit resolution or more, but on the contrary, they also become negligible with 6 to 8-bit, without necessarily having to push them below that resolution.

4.2.1.2 Phase Noise Mitigation Utilizing Asymmetrical LO Routing

PN is an RF impairment that originates from an imperfect local oscillator (LO) and causes random fluctuations in the phase of a signal. The PN degrades the signal quality, thus impacting the system performance in communications and sensing (Rasekh et al. 2021; Robins 1984; Sarajlić et al. 2023).

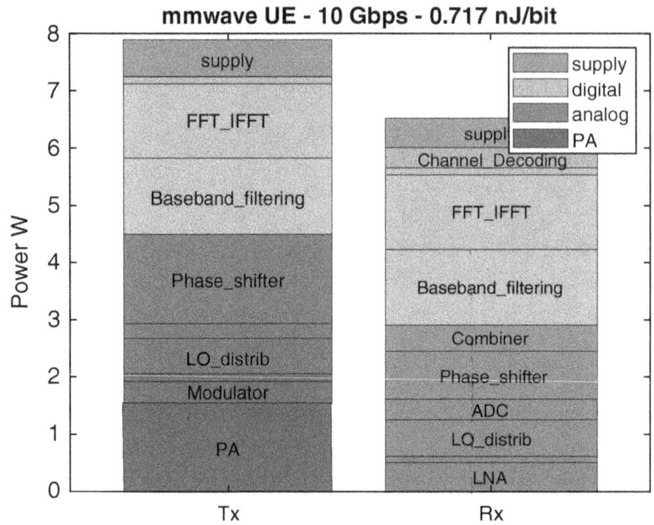

Figure 4.6 Power breakdown for a hybrid partially connected architecture having 4 baseband chains and 64 antennas, where analogue and digital parts consume similar power, i.e. for 2.3 GHz bandwidth. *Source:* Lopez (2025) / Hexa-X-II.

In communication systems, this is experienced as a limit in achievable error vector magnitude (EVM) (Desset et al. 2024), while in radar and sensing applications, the PN causes errors in range and velocity estimation (Bozorgi et al. 2021). The impact of PN increases significantly as a function of centre frequency and bandwidth, making it one of the dominant impairments, especially in wideband sub-THz communications (Dahlman et al. 2020; Nimr 2023). In general, the slow phase variations appearing in low offset frequencies in the PN spectra can be rather effectively mitigated by symbol-level techniques using phase-tracking reference signals (PTRSs) (Qi et al. 2018). However, similar techniques cannot effectively mitigate very fast variations without significant signalling overhead. Therefore, the PN at higher carrier frequencies utilizing wider signal bandwidths poses significant challenges (Chen et al. 2017; Leinonen 2021).

Systems operating at higher carrier frequencies use large antenna arrays with capabilities of steerable and controllable beams through various beamforming architectures. Typically, the larger array is divided into several subarrays, each having a dedicated mixer and digital signal path. Depending on the architecture, each subarray and mixer either shares the same LO or uses individual LOs (Arora et al. 2021; Aminu et al. 2019). To maintain good frequency and phase synchronization across subarrays, which is crucial for beamforming, it is preferred that each mixer shares the LO generated from the same source. The LO is routed to the mixers using different routing architectures depending on the scenario. Desset et al. (2024) highlighted that the routing architecture plays a significant role in array-level PN behaviour across individual paths and can be utilized for compensation. Figure 4.7 shows the subarray-based beamforming system with shared LO architecture. The PN is typically compensated by PTRSs in 5G systems, which tracks and mitigates the common phase error (CPE), that is dominant in smaller offset frequencies from the carrier. However, the PN is constrained by the LO noise floor in wider carrier offset and cannot be mitigated. Utilizing *asymmetrical LO routing* or intentionally introducing controllable delay differences among the LO paths per subarrays, the PN can be mitigated at the wider carrier offset in wideband systems. Due to the shared LO PN, when multiple subarray signals combine with LO delay differences, it creates a notch in the PN spectra when the LO signal combines out of phase (Desset et al. 2024). The notch

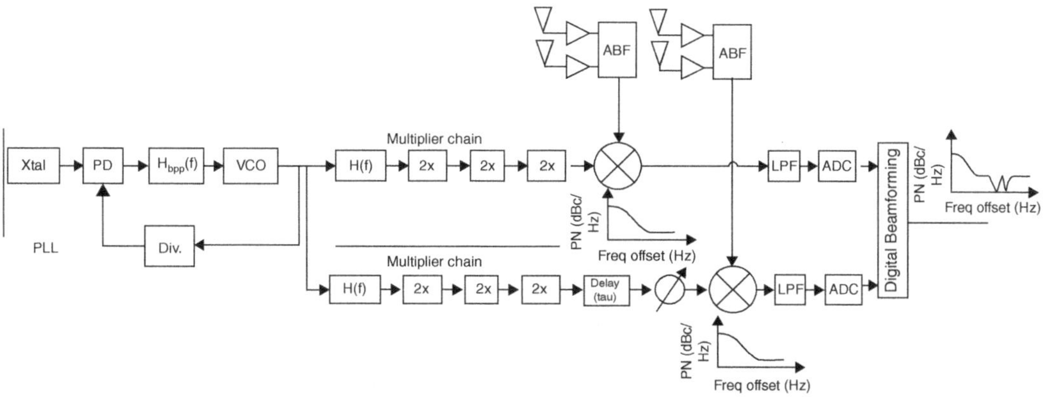

Figure 4.7 Shared LO architecture with delay differences in a subarray-based beamforming system. The LO signal is time delayed to effectively combine the PN with delay differences in the sum node.

location is controlled by tuning the delay difference between the LO signals per subarray. With the proposed technique, combining two paths can introduce a 3 dB gain in the effective PN-limited EVM (Desset et al. 2024). Furthermore, AI-based approaches are also used to compensate for the residual PN after correcting the CPE through the PTRS (Farhadi and Sundberg 2020).

4.2.1.3 Antenna Integration

Phased array technologies are one of the key enablers in sub-THz communication systems, where amplitude and/or phase control per antenna is possible. Half-wavelength spacing between the antenna elements used in conventional phased arrays allows versatile radiation pattern control and avoids grating lobes. However, RF FE and phased array integration become challenging due to the conflict between shrinking wavelength and integrated circuit (IC) size (Sadhu et al. 2019). Hence, irregular array architectures such as clustered and thinned arrays become favourable as they relax the half-wavelength spacing between antenna elements.

Clustered arrays are formed by grouping neighbouring antenna elements and having a single feeding point, while thinned arrays are formed by deactivating a certain number of antenna elements. SE is a widely used metric to assess the communications performance of MIMO systems. Next to that, the peak side lobe level (SLL), a measure of the maximum radiation level in any undesired direction, is a metric that plays a significant role in JCAS, resilience to eavesdroppers and interference minimization. Unlike regular patterns, where grating lobes are responsible for the dominant undesired radiation in specific directions, irregular arrays prevent such specific peaks, justifying the use of SLL as a better metric of unwanted radiation.

Clustered and thinned arrays have been compared in the context of mm-wave single-user MIMO systems and for sub-THz multi-user MIMO systems (Ertuğrul et al. 2024a, b). To that end, clustered array configurations have been generated using Algorithm-X (Knuth 2000) and genetic algorithm-based optimization considering antenna groupings of two and four, also known as *domino*-tiled phased arrays (DTPAs) and *tetromino*-tiled phased arrays (TTPAs), respectively. Thinned array configurations have been generated using a random sample generation considering 50% and 25% thinned arrays, where the former is realized by deactivating half of the antennas and the latter is realized by deactivating 3/4 of the antennas.

Simulation results in Figure 4.8 indicate that the clustered arrays outperform their thinned array counterparts, i.e. DTPA versus a 50% thinned array and TTPA versus a 25% thinned array, thanks to having more radiating elements. The DTPA is an excellent substitute for the conventional fully

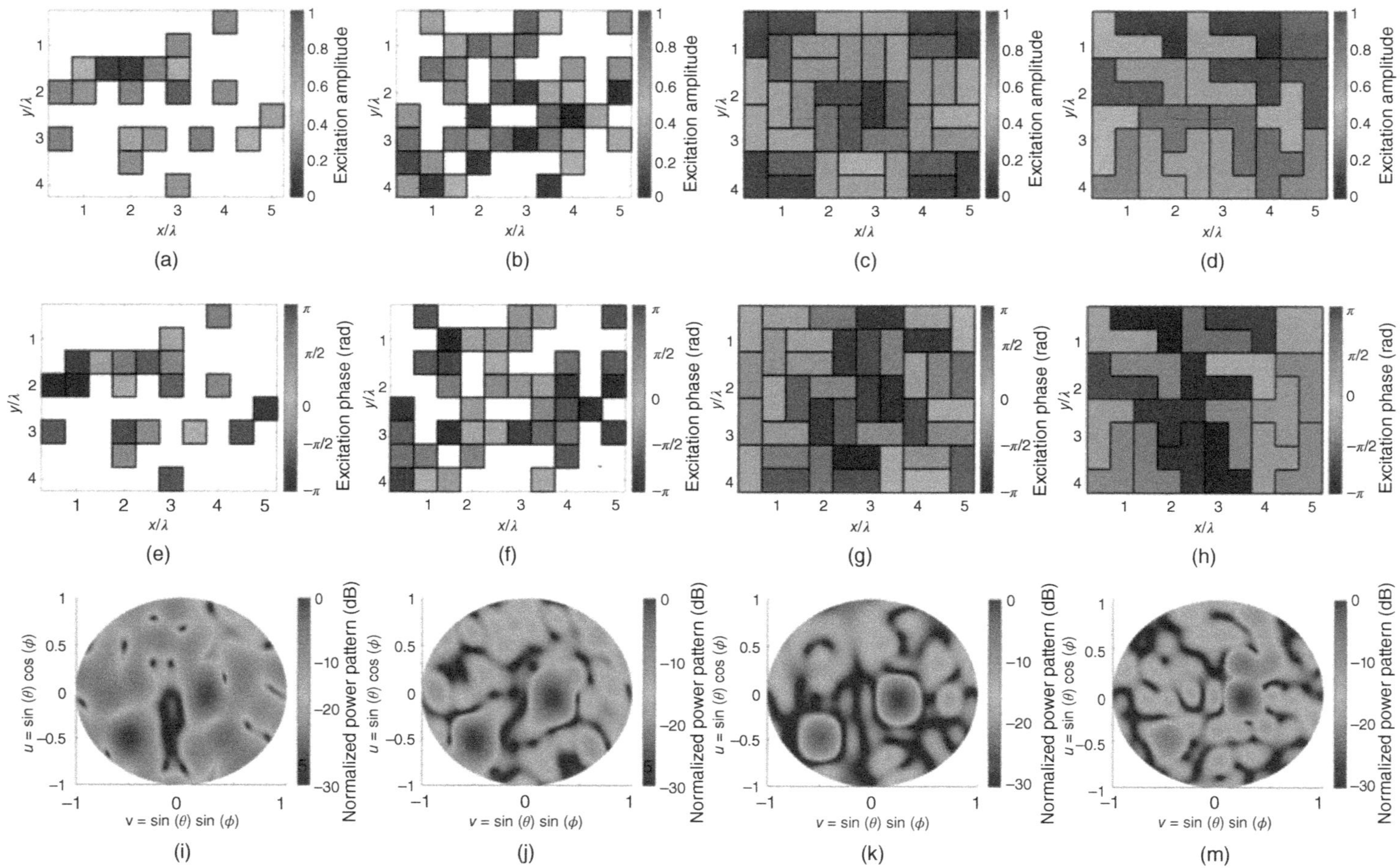

Figure 4.8 Zero-forcing excitation of optimal irregular arrays serving two users located at $(\theta_1, \phi_1) = (-45°, 45°)$ and $(\theta_2, \phi_2) = (15°, -15°)$, simultaneously. (a)–(d) Amplitude and (e)–(h) phase excitations, and (i)–(m) normalized radiation patterns for 50% thinning, 25% thinning, DTPA and TTPA, respectively. *Source:* Ertuğrul et al. (2024b) / IEEE.

populated reference arrays, having good SE and acceptable SLL around −13 dB. Moving to higher degrees of clustering and thinning, TTPA maintains acceptable SLL around −9 dB but loses some control of the main beams towards users. On the other hand, 25% thinned arrays keep strong main beams thanks to amplitude and phase control per antenna but lose SLL control (around −4 dB) due to more than half-wavelength spacing and radiate much more in unwanted directions.

4.2.2 Novel Sub-THz Transceiver Technologies

4.2.2.1 Resonant Tunnelling Diodes

The next-generation cellular technologies, commonly referred to as the 6G, are envisioned to support, among others, a higher system capacity and data rate. RTD-based technology can be a promising candidate solution for future 6G communications, by leveraging its high-frequency operation capability at room temperature with easy circuit realization, increased versatility, and energy-efficient implementation of the RTD-based device.

Currently, the highest fundamental oscillation frequency achieved by an RTD oscillator sits at 1.98 THz (Izumi et al. 2017), which is also the fastest existing purely solid-state device. In addition, RTDs can operate as high-sensitivity Rxs due to the strong nonlinearity in the I-V characteristic. This is achieved when the RTD is biased close to the peak current value on the I-V curve, where the sensitivity of the device is highest. Current state-of-the-art RTD-based transmitters include 2 mW, 15 Gb/s at W-band (Wang et al. 2018) and 1 mW with 110 GHz modulation bandwidth at 260 GHz (Al-Khalidi et al. 2018). The fastest reported wireless data rates using-RTD-based transmitters include 34 Gb/s at 500 GHz using single-channel and 56 Gb/s using dual-channel links (Oshima et al. 2016, 2017). On the other hand, the fastest RTD-based detector was reported recently with error-free data rates of up to 27 Gb/s at 300 GHz (Nishigami et al. 2018). This is the best detector performance in any technology.

The promising potential of RTD-based technology for future 6G communications resides in its simplicity, e.g. a 1 mW 300 GHz source requires only a single RTD device realized using photolithography, while transistor-based technologies such as CMOS require an array of devices, sub-100 nm high-resolution lithography, and advanced circuit design techniques. Moreover, due to RTDs' increased versatility and the fact that most of the time additional transceiver building blocks such as PAs or LNAs can be omitted, the RTD-based FE architecture is simpler than the more conventional transistor-based approach.

To date, research has focused on one hand on introducing the RTD-based architecture (cf., Figure 4.9) and offering a fair comparison with other more established technologies by looking into aspects such as antenna-generated output power, modulation formats, and antenna integration (Lopez 2025). Then, on the other hand, the performance of RTD-based devices was studied via D-band link level simulations (cf., Figure 4.10), where a Tx-Rx link was simulated for

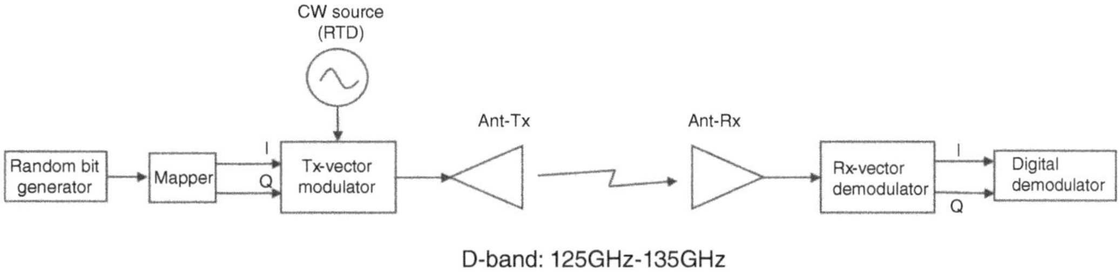

Figure 4.9 SystemVue-simplified schematic of the simulated RTD-based D-band link (125–135 GHz).

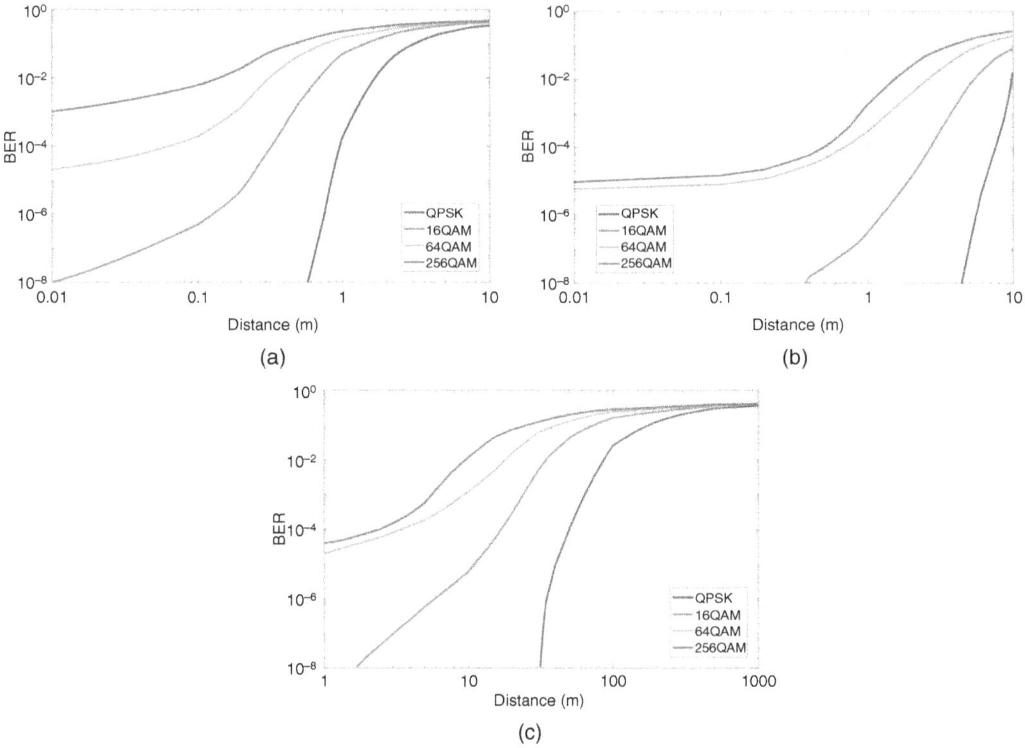

Figure 4.10 Simulated bit error rate (BER) curves for different modulation formats: (a) 1-elem RTD, (b) 2×2 RTD array, and (c) 4×4 RTD array.

several modulation formats (quadrature phase shift keying (QPSK) and M-ray quadrature amplitude modulation (M-QAM) with M∈{16, 64, 256}), distances (10^{-2} m–10^{3} m), and RTD-antenna array configurations (1 element up to 16 elements).

Independent of the number of RTD+antenna elements used (1 up to 16), the main takeaway here is that higher order modulation formats are more sensitive to the nonlinearity effects, and as such provide shorter distance links compared to simpler formats. Note that for a 4×4 RTD array, a 34 m radio link distance can be achieved using QPSK (for BER $= 10^{-6}$) and 10 m radio link with 256QAM (for a BER $= 10^{-2}$). The results presented for a D-band radio link using RTD show that a link distance of more than 30 m can be achieved using a 4×4 RTD antenna array, with a data rate defined by the QPSK modulation and a 10 GHz BW (bandwidth). For shorter distances, the data rate can be increased by using higher-order modulation formats. Moreover, if a 1 element RTD is considered, and therefore lower transmitted power is used, the radio link distance is limited to approximately 0.7 m for QPSK modulation.

4.2.2.2 Photonic Sub-THz Transceivers

Photonic sub-THz transceivers rely on photomixers. In these systems, photomixers are illuminated by a pair of lasers, which interfere, generating an optical beat signal whose beat frequency corresponds to the frequency difference between the two lasers. This optical beat signal is then processed in the photomixer to generate or detect sub-THz radiation. On the transmitter side, the photomixer is typically implemented as a photodiode (PD) (Nellen et al. 2022b). A PD converts the optical beat signal into a signal whose frequency is equal to the beat frequency. To generate sub-THz frequencies, hence, the two lasers must be offset by frequencies within the sub-THz range. Conversely, in Rx architectures, photoconductive antennas (PCAs) are used to downconvert the incoming sub-THz signal to an intermediate frequency (IF) (Deumer et al. 2024a). In PCAs, the optical beat signal modulates the resistivity of the photoconductive gap in the feeding point of an antenna. This serves as the LO for the heterodyne downconversion of the incoming sub-THz signal.

Compared to electronic-based sub-THz generation, photonic transceivers based on PDs and PCAs offer two key advantages: first, they are inherently broadband (i.e. they are not limited to specific frequency windows as CMOS-based sub-THz emitters/Rxs, for example) (Nellen et al. 2020; Deumer et al. 2025), and second, essential data communication functionalities such as modulation and filtering can be implemented using mature optical components, which provide significantly higher bandwidth than their counterparts in microwave electronics.

Two types of PDs are employed for sub-THz applications: positive-intrinsic-negative (PIN) and uni-travelling-carrier (UTC) PDs. Both exhibit extremely broadband operation, spanning from DC to several THz, when integrated with a silicon (Si) lens, and deliver comparable output power levels across this range, with the PIN PD providing higher levels at frequencies below 170 GHz, and the UTC PD outperforming it at higher frequencies (Nellen et al. 2020). To date, the highest output power at sub-THz frequencies from a single PD has been achieved with a coplanar waveguide-integrated UTC-PD, reaching 2 mW of power (measured on chip) at 300 GHz (Ohara et al. 2023). State-of-the-art PIN-PDs, on the other hand, are able to reach radiated output powers higher than 0.5 and 0.05 mW at 100 and 300 GHz, respectively (Deumer et al. 2024b).

For PCAs, recent advancements have focused on enhancing the IF bandwidth and the responsivity (i.e. the ratio of downconverted IF power and the incident sub-THz power) through the development of waveguide-integrated designs, where the light is coupled to the PCA through an optical waveguide instead of a free-space region as in top-illuminated PCAs (Deumer et al. 2025). Waveguide-integrated PCAs are also the key enabler for the photonic integration of photomixer-based sub-THz Rxs. As will be discussed later, photonic integration is crucial for ensuring that these systems remain competitive with alternative solutions for sub-THz generation and detection. When operated as heterodyne Rxs, waveguide-integrated PCAs have demonstrated an order-of-magnitude improvement in responsivity over state-of-the-art top-illuminated PCAs, corresponding to a 20-dB increase in downconversion gain (Deumer et al. 2025).

Figure 4.11 illustrates the typical architecture of a photonic heterodyne sub-THz transceiver with beam-steering capabilities. In the transmitter section, a laser source feeds an optical modulator for data encoding. The modulated optical signal is then amplified and directed to an optical processing unit, which can perform various functions such as optical filtering (e.g. to suppress one of the redundant modulation sidebands) or optical beamforming through an optical beamforming network (OBFN) for beam-steering with a phased array. Following processing, the optical signal is combined with a second laser and delivered to an array of PDs and antennas. On the Rx side, the architecture largely mirrors that of the transmitter, with two differences: the absence of the modulator and the use of PCAs instead of PDs in the array. To operate the phased-array Rx, i.e. to enable beam reception from different directions, one of the Rx lasers must also pass through an OBFN. The outputs from the PCAs are a set of IF signals that are combined to recover the received signal.

To make photomixer-based communication systems competitive with alternative solutions based on CMOS or RTDs in terms of size, weight, and power (SWaP), the optical subsystem must be integrated into a photonic integrated circuit (PIC). Achieving this requires hybrid integration, which allows the combination of different material platforms within a single PIC, enabling the use of the optimum integration platform for each specific function. For example, silicon nitride (SiN) is ideal for the optical signal processing unit due to its ultra-low-loss optical waveguides

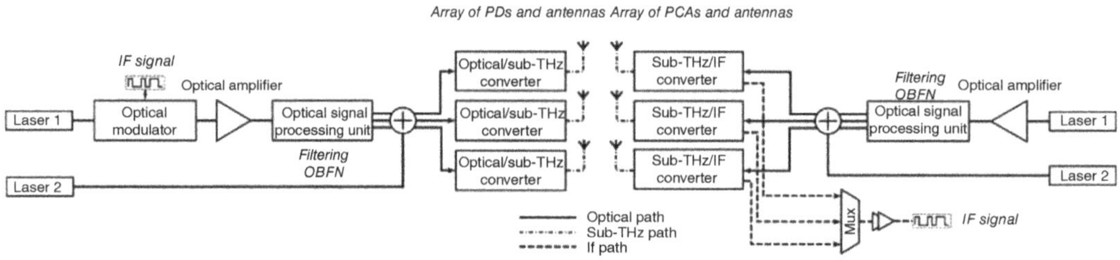

Figure 4.11 Typical architecture of a photonic heterodyne sub-THz transceiver. OBFN: optical beam-forming network.

(Roeloffzen et al. 2018). Polymer-based platforms are advantageous for optical routing, as their 3D structuring capabilities help minimize waveguide crossings (Weigel et al. 2024). Finally, Indium phosphide (InP) is required for active components such as PDs, PCAs, lasers, amplifiers, and high-speed modulators (Gupta et al. 2023).

A demonstration of a photonic sub-THz transmitter is presented in Nellen et al. (2022a), featuring a 1 × 4 phased array. The PDs are integrated with planar bow-tie antennas, and a hyper-hemicylindrical Si lens is mounted atop the planar antennas to enhance coupling efficiency. The array pitch is optimized for operation at 300 GHz. Following the approach of hybrid photonic integration, the OBFN is realized on polymer, while the PDs are fabricated using InP technology. The main limitation of this prototype is that it is not scalable to two-dimensional arrays since the hyper-hemicylindrical lens only enables beam steering in one dimension. To address this, an array of Si rod antennas can be used instead of the lens (Rivera-Lavado et al. 2017; Carpintero et al. 2024).

In Avramopoulos (2024), a sub-THz transmitter and Rx based on photomixers are demonstrated. Figure 4.12(a) depicts the transmitter, which comprises all functional blocks shown in Figure 4.11, including a 1 × 4 Si rod antenna array mounted atop a 1 × 4 array of bow-tie antenna-integrated PDs. The Rx, shown in Figure 4.12(b), does not support beam steering capabilities and therefore consists

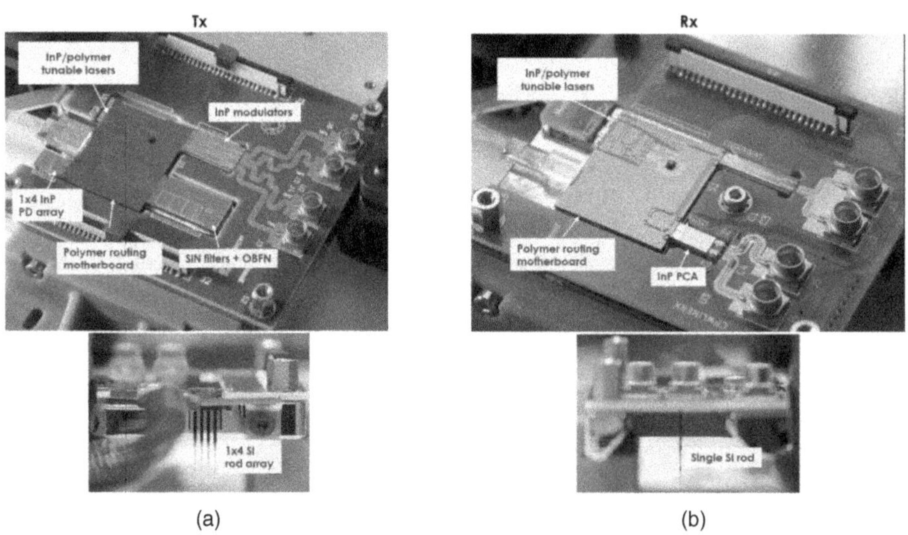

Figure 4.12 Photonic sub-THz, (a) transmitter and (b) receiver. *Source:* Avramopoulos (2024) / European Commission / CC BY 4.0.

of a single Si rod antenna mounted on top of a single bow-tie antenna-integrated PCA, without an OBFN. As shown in Figure 4.12(a) and (b), both modules employ hybrid photonic integration. The transmitter's antenna pitch is 1 mm, optimized for 150 GHz operation. This work represents the first demonstration of a fully integrated sub-THz photomixer-based transceiver system.

4.2.3 RIS Hardware Prototyping and Verification

RISs are planar structures composed of periodically distributed subwavelength scatterers (unit cell) with tuneable responses, capable of reflecting incident waves in desired directions and shaping the wavefront into specific beam profiles. Unlike traditional technologies such as relays or MIMO beamforming, RISs transform the passive environment into a smart one, boosting coverage, localization accuracy, secrecy, and even reducing EM field exposure (Ratajczak et al. 2024). Moreover, they achieve this enhancement without requiring active power sources or complex processing (Basar et al. 2019).

RISs can be realized using different technologies. For example, they can be built using the metasurface technology, i.e. as 2D metamaterials realized in periodic structures composed of several 'electrically small' unit cells (where each unit cell has dimensions much smaller than a wavelength). RIS can also be implemented using conventional reflect arrays or transmit arrays, i.e. arrays of planar antenna elements with half-wavelength spacing, where each antenna element is connected to a tuneable load.

A reflect array-based RIS is shown in Figure 4.13. The proposed surface was designed to work in the frequency band from 24.25 to 27.5 GHz and to continuously control the reflected coefficient by using varactor diodes available on the market (D'Errico 2023). The unit cell size of the reflect array is shown in Figure 4.13(a). Its size has been fixed to 4.4 mm × 4.4 mm, and a varactor diode is placed on each arm of the metallic printed cross. The diode biasing circuit has been included in the design and printed on a multilayered PCB. A prototype of 16 × 16 unit cells (Figure 4.13(b)) has been manufactured and tested.

Figure 4.14 illustrates the overall system concept where the same impinging wave can be reflected either with one configuration of the RIS (Configuration A) or the other (Configuration B), thanks to different settings of the voltages of the varactor, resulting in different phase-shifts of the reflection coefficients of the unit-cells.

For these tests, the same voltage is applied at the centre of each unit cell (cf., Figure 4.15(c)). The calibration of the RIS is done depending on the inverse voltage varying between 0 and 15 V, the incidence (from 0° to 50°), and polarization of the EM field. The reflection coefficient of the RIS is obtained from the transmission between two horns via the RIS (Lopez 2025).

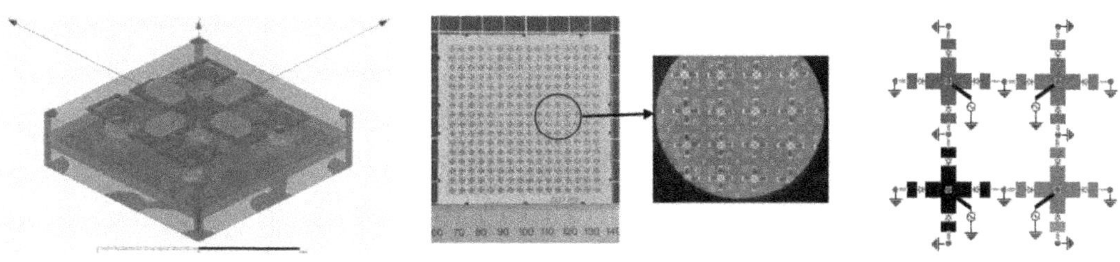

(a) (b) (c)

Figure 4.13 RIS: (a) unit cell RF model, (b) 16 × 16 tile prototype, and (c) varactor diode biasing principle.

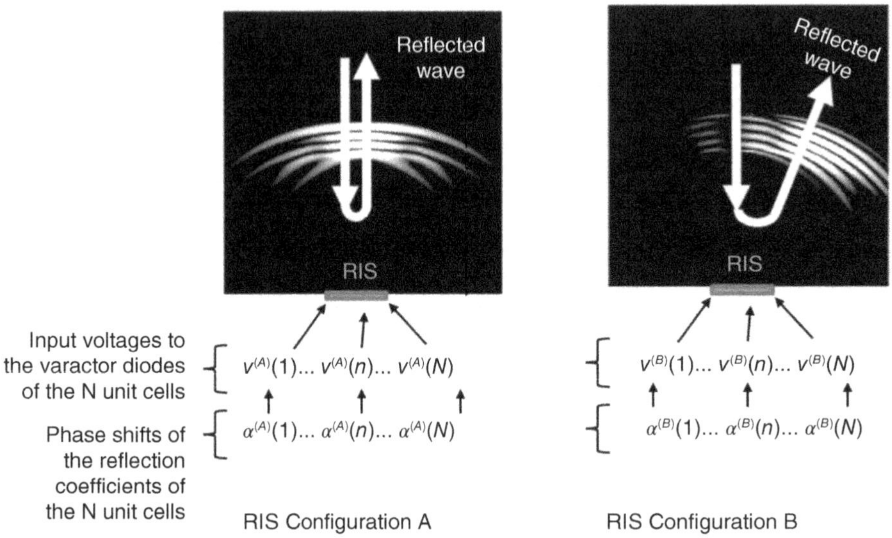

Figure 4.14 Overall system concept where the same impinging wave is reflected by the RIS either without deviation (in Configuration A) or with a wanted direction (in Configuration B), simply by applying the voltages $v^{(A)}(1)$ to $v^{(A)}(N)$ or $v^{(B)}(1)$ to $v^{(B)}(N)$, to the varactor diodes of the N unit-cells of the RIS, to meet the target phase shifts $\alpha^{(A)}(1)$ to $\alpha^{(A)}(N)$ or $\alpha^{(B)}(1)$ to $\alpha^{(B)}(N)$ of the reflection coefficients of the unit cells, in Configuration A or Configuration B, respectively.

The measured reflection coefficients of the prototype were compared to simulation results. For simulation, a normal incidence and for equivalent excursion in capacitance from 0.24 to 0.02 pF was used, while a voltage from 0 to 15 V was used for the experiments. Comparable behaviour was observed for both the amplitude (cf., Figure 4.13(a) and (b)) and phase (cf., Figure 4.13(c) and (d)), with a frequency shift of 1 GHz towards higher frequencies for the experiments.

The RIS characterization is in accordance with the requirements considered for the design: a continuous phase control of at least 300° with a good stability depending on the incidence, up to 50°, and the polarization of the incident field and an isolation between main polarization and cross-polarization above 17 dB. Discrepancies are observed on the amplitude level of the reflection coefficient, lower than expected by simulation, and the frequency shift. This remains reasonable given the unit cell complexity and the number of elements in the tile (256-unit cells with 1024 diodes on 70.4 mm × 70.4 mm).

4.3 Channel and Hardware Modelling

This section reports recent results obtained for radio channel modelling in scenarios of interest for future 6G communications, such as wireless propagation in industrial environment, radio channel characterization in the FR3 frequency band, and RIS-aided communication scenarios, that require realistic and computationally efficient models for channel prediction. The influence of highly directive antennas and different beamforming schemes on the temporal and angular dispersion characteristics of the radio channel at sub-THz frequencies is also discussed. Finally, the last part of the section deals with the modelling of HW non-idealities, with particular reference to PAs at sub-THz and FR3 frequencies.

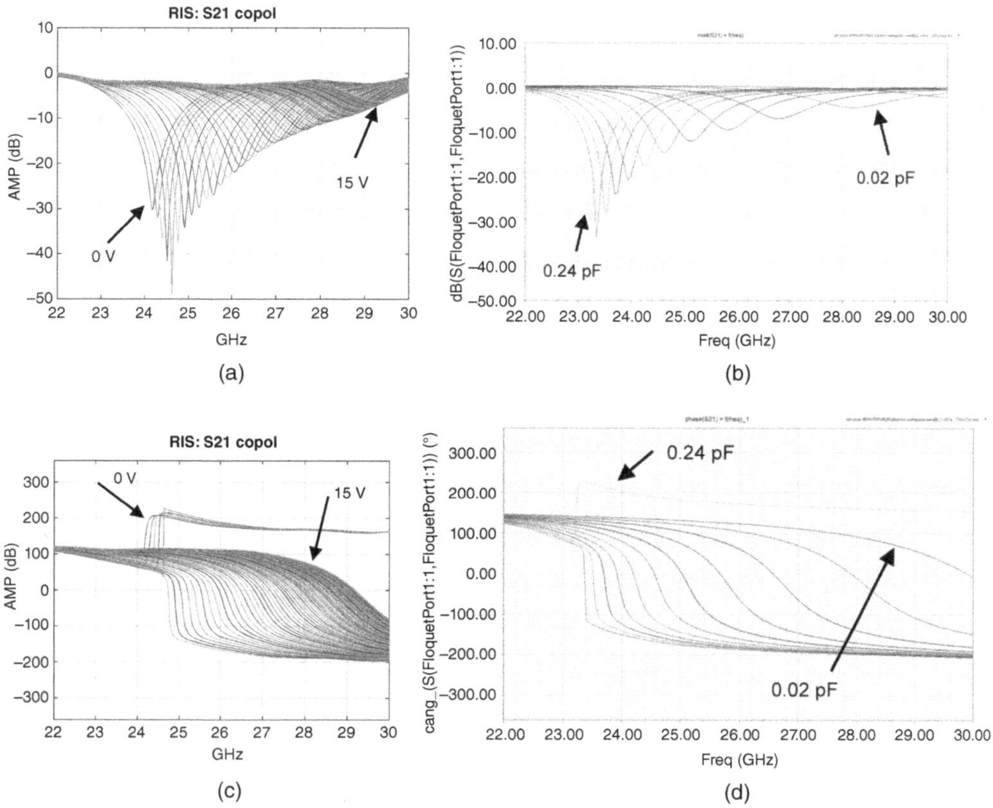

Figure 4.15 RIS reflection phase coefficient: (a) measured amplitude, (b) simulated amplitude, (c) measured phase, and (d) simulated phase.

4.3.1 Short-Range Measurements and Channel Models in Industrial Scenarios

The industry is increasingly interested in developing ultra-fast, low-latency 6G networks for industrial environments to support short-range communications. In addition to the existing FR1 (below 6 GHz) and FR2 (mm-wave) frequency spectra, there is active research into new spectra, particularly the mid-band FR3 FR from 7.125 to 24.25 GHz. The lower end of this spectrum, in particular, has garnered significant attention from the industry. The primary objective is to gain an understanding of wireless propagation through empirical methods in indoor industrial scenarios with short-range communication links. Additionally, channel statistics need to be compared with the 3GPP standard channel models. The following subsections present the *large-scale radio channel characteristics* derived from measurements in a typical industrial scenario, focusing specifically on the FR3 band: the analysis includes key metrics such as *RMS Delay Spread (DS), Path Loss,* and *Rician K-factor*, and compares the measured results with 3GPP Indoor Factory (InF) channel models (3GPP TR 38.901 2024). Results are obtained by processing 'spatial map measurements' carried out in the Industrial Production Lab at Aalborg University, Denmark, using a setup consisting of a vector network analyser (VNA), and omnidirectional antennas at FR1, FR2, and FR3 frequencies, even though the present analysis is mainly focused on the FR3 band. More details about the measurement setup are available in Vitucci (2024). During the measurements, the receiving antenna was kept fixed in three different locations (Rx1, Rx2, and Rx3), while the transmitter (Tx) unit was

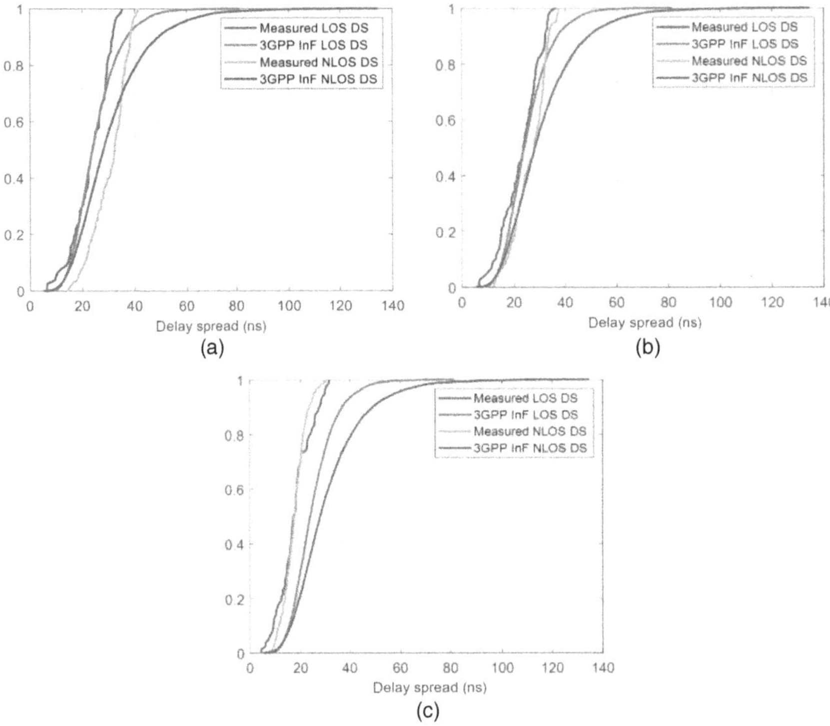

Figure 4.16 Delay spread CDF, (a) FR1, (b) FR3, and (c) FR2.

moved along several routes in the main and the side corridors. Power-Delay Profiles (PDPs) were finally extracted for each measurement location to derive the large-scale channel parameters.

4.3.1.1 Delay Spread Analysis

The root-mean-square DS reflects the time dispersion in the radio channel caused by multipath propagation. In the following analysis, the distribution of DS is plotted using empirical cumulative distribution functions (CDFs) for both line-of-sight (LoS) and non-line-of-sight (NLoS) links. Figure 4.16 illustrates the CDF of DS extracted from the measurements, for FR1, FR3 and FR2 frequency bands, respectively, compared to the 3GPP model.

From the results in Figure 4.16, it can be concluded that:

- DS increases with distance, with stronger dispersion observed in NLoS conditions.
- Comparison with 3GPP InF models shows a good agreement, particularly in the FR3 band, where the median values are well aligned with the standard.
- However, extreme values of DS observed in very large industrial settings are not reached, due to the limited size of the tested environment.

4.3.1.2 Path-Loss Analysis

Path loss quantifies the signal attenuation over distance. Measured data were fitted using a *floating-intercept log-distance model*, and the results were compared with the following 3GPP *InF* channel models: *InF-DL* for dense indoor environment and low transmitter height and *InF-SL* for sparse indoor environment and low transmitter height (3GPP TR 38.901 2024). Figure 4.17 shows

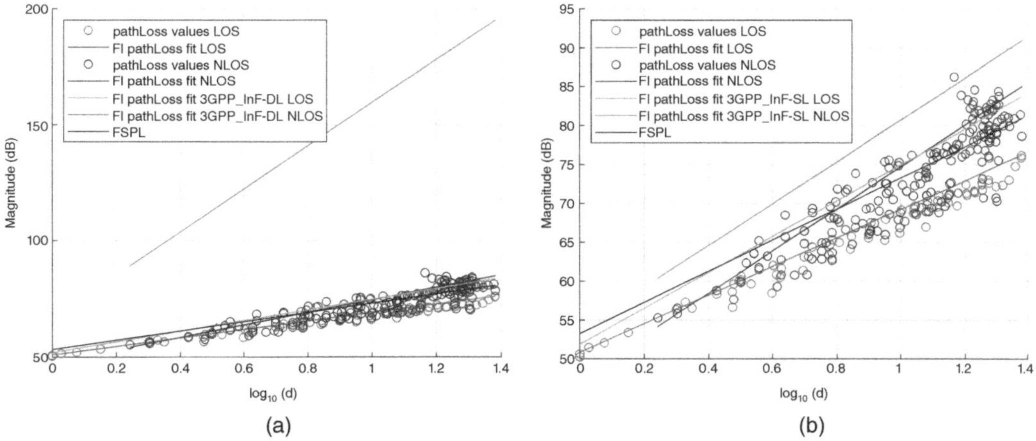

Figure 4.17 Pathloss floating intercept fit of FR3 channel and comparison with 3GPP pathloss models, (a) InF-DL and (b) InF-SL.

the path loss regression for the FR3 band and its comparison with the standard models. From the fitting lines, the *Path Loss intercept* and the *Path-Loss exponent* are also derived and compared with the 3GPP models (see Table 4.1).

From the results in Figure 4.17, it can be inferred that:

- Path loss increases predictably with distance and is strongly influenced by LoS/NLoS conditions.
- The 3GPP InF-DL model significantly overestimates the path loss for FR3 in this environment, often being completely out of range.
- The InF-SL model offers a much better fit, suggesting that industrial layouts with shorter distances and mixed reflectivity may be better represented by InF-SL than InF-DL.

Empirical histograms of shadow fading, i.e. deviation between estimated path loss sample and corresponding values of the fitted path loss regression line, are depicted in Figure 4.18, for both LoS and NLoS locations.

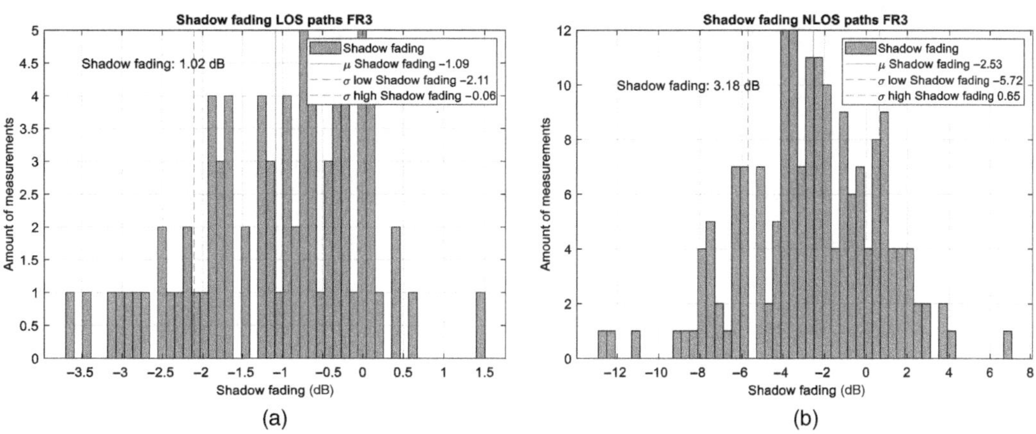

Figure 4.18 Shadow fading for FR3, (a) LoS, and (b) NLoS.

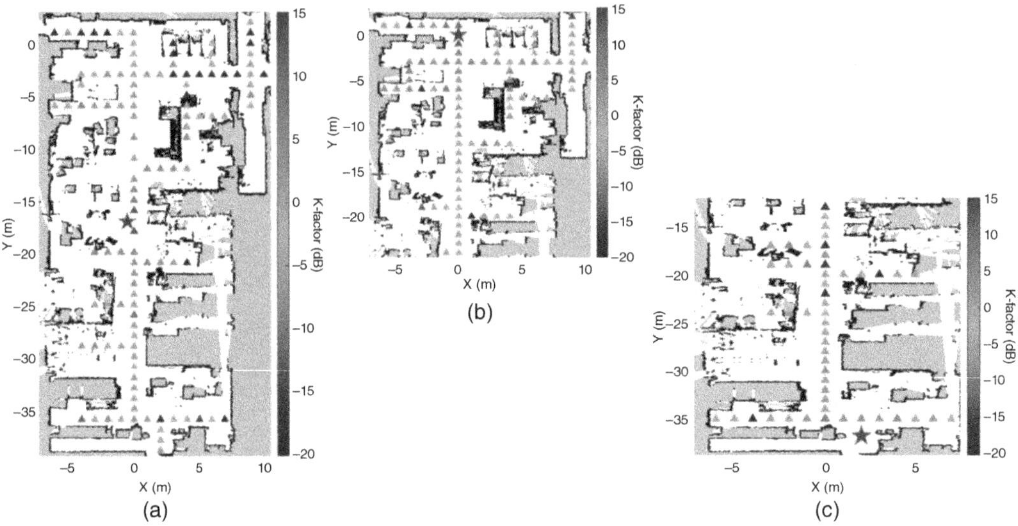

Figure 4.19 Measured FR3 channel K-factors, (a) Rx1, (b) Rx2, and (c) Rx3.

4.3.1.3 Analysis of the Rician K-Factor

The Rician K-factor characterizes the ratio of the dominant (LoS) path to the scattered paths caused by reflections on the surrounding environment. The results extracted from all the measurements' locations are reported in Figure 4.19.

Looking at the figure, it can be observed that:

- K-factor values vary significantly, even for links expected to be LoS.
- Many apparent LoS links showed negative or near-zero K-factors, likely due to strong reflections from metallic surfaces and machinery.
- Compared to 3GPP InF models, the measured K-factors are consistently lower, indicating a less pronounced Rician fading behaviour.

This deviation suggests the need to adjust the standard model for indoor industrial applications, especially in highly reflective environments.

Table 4.2 summarizes average values (μ) and standard deviation (σ) of the analysed large-scale parameters for the FR3 band. This analysis provides valuable insights into the propagation characteristics of mid-band frequencies in real-world industrial environments. While existing 3GPP InF models are mostly reliable for DS, they tend to overestimate K-factors and path loss, particularly in settings with high reflectivity and confined geometries. The InF-SL variant emerges as a more suitable reference for such conditions in FR3. These results suggest the *need to refine or re-parameterize* standard channel models for accurate use in emerging FR3-based industrial communication systems.

4.3.2 Macroscopic Channel Modelling for RIS

RISs can be realized with different technologies, as previously described in Section 4.2.3. In order to assess the overall performance of RIS-aided wireless networks, it is necessary to integrate RIS reradiation models into deterministic planning tools, such as ray tracing, and use them in coverage- or link-level simulations. Such RIS models must be accurate and general enough to be independent

Table 4.2 Statistics of estimated FR3 (10–12 GHz) large-scale parameters and comparison to corresponding 3GPP model parameters.

	InF measured		InF 3GPP	
	LoS	NLoS	LoS	NLoS
PL exponent	18.34	27.06	SL: 23.03 DL: 23.03	SL: 26.89 DL: 93.08
PL intercept (dB)	50.89	47.63	SL: 51.93 DL: 51.93	SL: 53.82 DL: 66.53
PL μ (dB)	65.84	73.07	SL: 74.90 DL: 74.90	SL: 83.27 DL: 168.43
PL σ (ns)	6.38	6.92	SL: 9.02 DL: 9.02	DL: 7.10 SL: 26.66
SF σ (dB)	1.02	3.18	SL: 0.56 DL: 0.56	SL: 0.35 DL: 2.73
DS μ (ns)	22.63	26.27	25.03	30.70
DS σ (ns)	7.30	6.50	8.89	14.11
K-factor μ (dB)	2.66	−4.03	7	—
K-factor σ (dB)	3.82	7.95	8	—

of the adopted technology and the specific RIS implementation. This is the main idea behind the 'macroscopic' RIS modelling approach, which is presented later.

4.3.2.1 Fully Ray-Based Macroscopic Modelling

Ray-based macroscopic modelling of scattering from RIS focuses on large-scale effects of wave interactions with the surface, employing an efficient physics-based formulation. This approach considers the RIS as a continuous scatterer described by a few parameters and a *spatial modulation coefficient*, which accounts for the desired phase, amplitude, and polarization modulation of impinging waves, disregarding the underlying network technique or microscopic phenomena that generate such a modulation. Such an approach can be applied to several frequency bands, ranging from sub-6GHz to mm-wave and sub-THz, and allows determination of the reradiated field in the far-field and radiative near-field region of the RIS.

As a first step, multiple *reradiation modes* – including both the desired and the parasitic reradiated waves – must be identified using Floquet's theory under the assumption of *locally periodic* RIS configuration (Díaz Rubio and Tretyakov 2021). Each mode is associated with an amplitude coefficient and a Spatial Modulation Dyadic (SMD) coefficient $\underline{\boldsymbol{\Gamma}}$ that determines the phase and amplitude modulation at the RIS surface. An overall power balance must be enforced to make sure that the amplitudes of the incident wave and of all the reradiated waves are compatible with power conservation over the RIS slab (Degli-Esposti et al. 2022). Then, for each mode, the procedure described later is followed to achieve a ray-based description of its reradiated field.

Unlike wave-based models, this method discretizes the RIS surface into small *elements* or *tiles*, treating each element as a local plane wave interaction point. Using this discretization, the method computes the anomalously reflected rays at each tile while considering the curvature of both the incident and reradiated wavefronts (Vitucci et al. 2024). The key steps involved in this approach include:

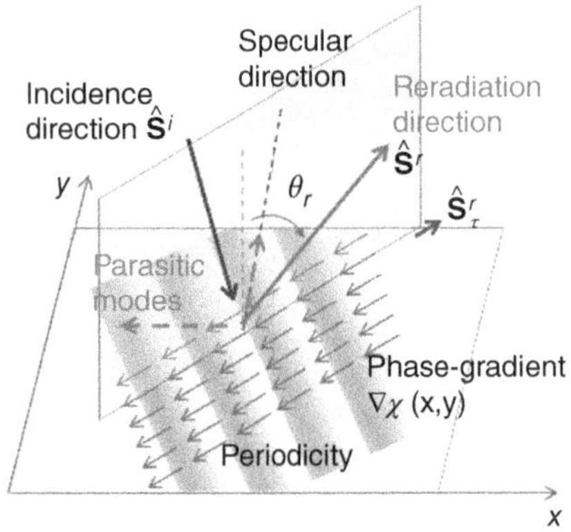

Figure 4.20 Incident and reradiated rays on a point of a periodic RIS, with representation of the total surface phase gradient.

1) **Computation of the anomalous ray direction**: Determining the reflected ray trajectory based on the phase modulation of the surface impedance (see the following Figure 4.20).
2) **Computation of the reradiated field**: Evaluating the field reradiated by each tile of the RIS.
3) **Computation of the reradiated wave spreading factor**: Calculating wave attenuation as it propagates.

The whole procedure is based on the cited SMD $\underline{\boldsymbol{\Gamma}}$ that can be expressed as:

$$\underline{\boldsymbol{\Gamma}}(\mathbf{r}') = \underline{\boldsymbol{\Gamma}}_0(\mathbf{r}')e^{j\chi^m(\mathbf{r}')} = K^m(\mathbf{r}')e^{j\chi^m(\mathbf{r}')}\cdot\underline{\mathbf{R}}^m,$$

where $\mathbf{r}'$ is the generic position on the surface; $K^m(\mathbf{r}')$ and $\chi^m(\mathbf{r}')$ are the *amplitude and phase* modulation profiles that the RIS imposes on the reradiated field, whereas matrix $\underline{\mathbf{R}}^m$ applies the polarization transformation. In step 1, the reradiated ray direction is computed on the basis of the incident ray field direction and of the phase profile gradient $\nabla\chi^m$. The reradiation direction is given by

$$\hat{\mathbf{s}}^r = (\underline{\mathbf{1}} - \hat{\mathbf{n}}\hat{\mathbf{n}})\cdot\hat{\mathbf{s}}^i - \frac{1}{k}\nabla\chi^m + \sqrt{1 - \left|(\underline{\mathbf{1}} - \hat{\mathbf{n}}\hat{\mathbf{n}})\cdot\hat{\mathbf{s}}^i - \frac{1}{k}\nabla\chi^m\right|^2}\,\hat{\mathbf{n}},$$

where the $\hat{\mathbf{s}}^i, \hat{\mathbf{s}}^r$ unit vectors represent the incidence and radiation directions on the current tile, respectively, $\hat{\mathbf{n}}$ is the normal unit vector to the RIS surface, k is the free-space wavenumber, $\underline{\mathbf{1}}$ is the identity matrix, and $\underline{\mathbf{1}} - \hat{\mathbf{n}}\hat{\mathbf{n}}$ is the *tangent projection* operator, so that when applied to a vector $\mathbf{v}$ it provides its tangent component to the RIS surface, i.e. $(\underline{\mathbf{1}} - \hat{\mathbf{n}}\hat{\mathbf{n}})\cdot\mathbf{v} = \mathbf{v} - (\hat{\mathbf{n}}\cdot\mathbf{v})\hat{\mathbf{n}} = \mathbf{v}_\tau$.

In steps 2 and 3, the reradiated field is computed using the SMD and the incident field $\boldsymbol{E}^{\mathbf{i}}(\boldsymbol{r}')$ as:

$$\boldsymbol{E}^r(\boldsymbol{r}) = \underline{\boldsymbol{\Gamma}}(\boldsymbol{r}')\boldsymbol{E}^{\mathbf{i}}(\boldsymbol{r}')A^r(s)e^{-jk|\boldsymbol{r}-\boldsymbol{r}'|} = \underline{\boldsymbol{\Gamma}}(\boldsymbol{r}')\boldsymbol{E}^{\mathbf{i}}(\boldsymbol{r}')\sqrt{\frac{\rho_1^r\rho_2^r}{(\rho_1^r + s)(\rho_2^r + s)}}e^{-jks},$$

being $\mathbf{r}$ the position vector of the observation point P, $s = |\boldsymbol{r} - \boldsymbol{r}'|$ the local coordinate along the reflected ray, and $A^r(s)$ the spreading factor depending on the curvature radii ρ_1^r, ρ_2^r of the reradiated wave, assumed a generic astigmatic wave. A more complete formulation can be found in Vitucci et al. (2024). A similar procedure can be applied to the computation of the diffracted rays on the edges of a finite-size RIS using an adaptation of the uniform geometrical theory of diffraction (UTD), as shown in Vitucci et al. (2024).

Once the reradiated field is computed for each ray, the total reradiated field by an RIS is obtained through coherent summation of all the fields in the receiving locations.

The proposed method provides results of equal accuracy compared to well-established methods in the literature, such as physical optics, but is computationally more efficient by an order of magnitude.

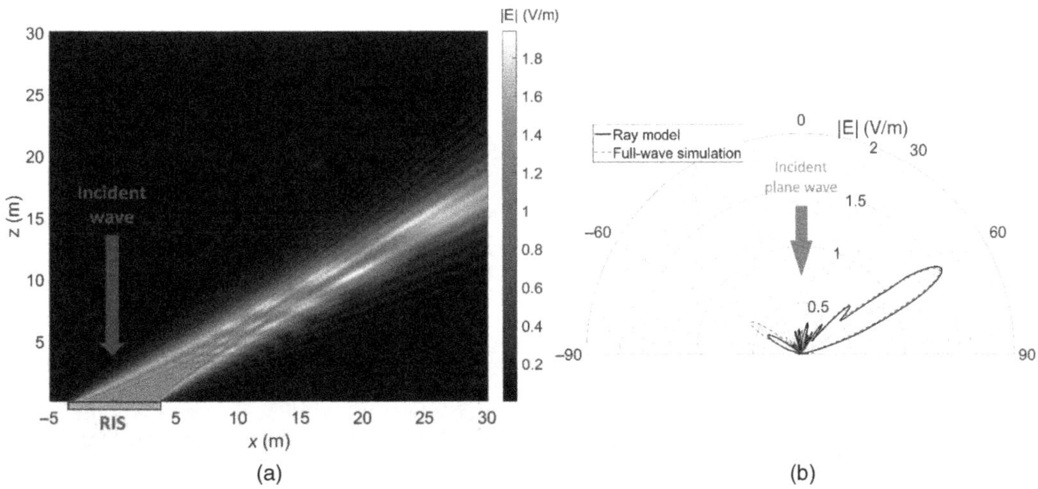

Figure 4.21 Simulation of reradiated field (V/m), (a) in the *xz* plane for an ideal anomalous reflector; (b) along a semicircle on the *xz* plane centred on the RIS, at a distance $r = 1$ m, considering parasitic Floquet's modes and comparison with full-wave simulation.

Figure 4.21(a) shows an example of reradiated field prediction using the fully-ray-based macroscopic model for an ideal anomalous reflector of size $7 \times 7\,\text{m}^2$ laying on the *xy* plane, working at $f = 3$ GHz and illuminated by a plane wave (normal incidence angle $\theta_i = 0°$, and desired reradiation angle at $\theta_r = 60°$). Figure 4.21(b) shows the performance of a real anomalous reflector with parasitic Floquet's modes at $\theta_r = 0°$ and $\theta_r = -60°$, in addition to the desired reradiation mode at $\theta_r = 60°$: in the figure, the reradiated field predicted with the model along a semicircle at a distance $r = 1$ m from the RIS is compared with a full-wave EM simulation, showing a very good agreement.

4.3.3 Modelling of Sub-THz Channel Dispersion in the Presence of Beamforming

Sub-THz frequencies are central to future 6G systems due to their capacity for ultra-high data rates. However, their directional nature and high path loss require accurate channel models that reflect how antenna patterns affect what the Rx observes. Most traditional models consider only the propagation environment, ignoring the directional reception imposed by beamforming.

To address this gap, measured channels across multiple indoor and outdoor propagation environments at sub-THz frequencies (142 GHz) are analysed in Zhang et al. (2025), reconstructing PDPs and power angular spectra (PAS) from multipath component data. Then, synthetic beam patterns are applied to PDPs and PAS at the Rx side, allowing evaluation of channel dispersion metrics under different beamforming strategies: single-beam (max gain), local maxima (LM), multibeam, and null-steered linear constrained minimum variance (LCMV).

Beamforming is shown to significantly reduce temporal dispersion. DS and maximum excess delay (MED) are both minimized when using narrow beams, particularly in the single-beam case. For example, the median DS decreases from 18.62 ns (omnidirectional) to 7.23 ns (max-beam), and MED from 154.89 to 71.03 ns when using a beam with a half-power beamwidth (HPBW) of 26.2 degrees. These results are illustrated in Figure 4.22(a), which shows how DS scales with HPBW. Figure 4.22(b) complements this by presenting the cumulative distribution functions of DS under various beam strategies, highlighting the consistent advantage of directional reception over omnidirectional, in reducing temporal dispersion.

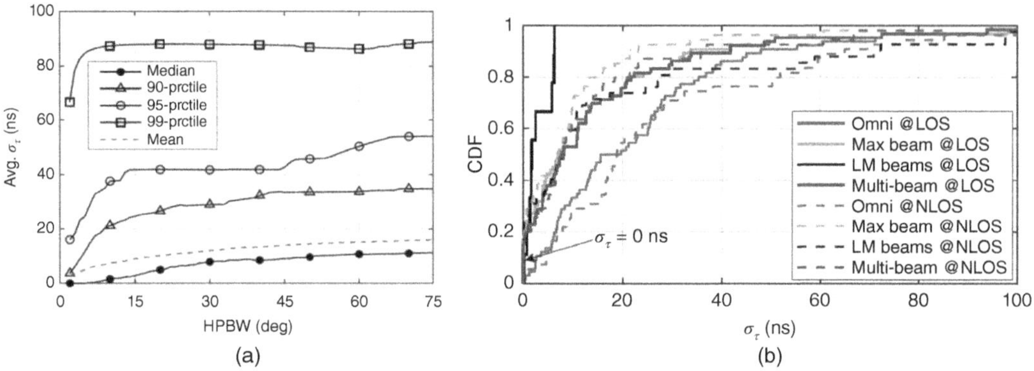

Figure 4.22 Behaviour of delay spread at 142 GHz, (a) in indoor links versus HPBW in the case of the 3GPP beam patterns; (b) CDFs of delay spread in outdoor links across multiple beamforming strategies in the case of a 4-element ULA pattern.

The study compares two distinct beam patterns: a 3GPP-defined pattern and that of a uniform linear array (ULA) of N isotropic elements. The 3GPP pattern offers high granularity in controlling HPBW but suppresses sidelobes completely. In contrast, the ULA pattern, though less flexible in HPBW tuning, includes sidelobes and better reflects realistic antenna behaviour. Despite these differences, the influence of sidelobes on time dispersion is found to be minor, suggesting that main lobe characteristics dominate dispersion behaviour.

The azimuth angle spread of arrival (ASA) is also examined. In LoS conditions, where multipath is limited, beamforming can increase ASA slightly by capturing energy from side directions. In non-NLoS cases, where scattering dominates, beam patterns have a smaller effect on ASA. The directionality of reception is more impactful in spatially sparse environments, while in rich multipath scenarios, ASA remains largely unchanged regardless of beam strategy.

In conclusion, beamforming does not merely improve link budget; it fundamentally alters the temporal and spatial structure of the observed radio channel. The dependence of delay spread and ASA on beamwidth, propagation condition, and beamforming strategy confirms that channel models for 6G systems at sub-THz frequencies must support incorporating beam patterns to predict performance accurately. This approach moves beyond propagation-only modelling and better reflects the true radio channel as experienced by a directional Rx.

4.3.4 Modelling of Hardware Non-Idealities

4.3.4.1 Sub-THz Non-Idealities Modelling

As sub-THz communication systems continue to evolve, modelling HW non-idealities becomes increasingly critical. Compared to lower-frequency systems, sub-THz transceivers introduce several unique challenges, including higher carrier frequencies, wider bandwidths, and a larger number of antennas. Those factors lead to a need for model extensions of HW impairments from PA, PN, in phase & quadrature (I/Q) imbalance, and ADC/DAC.

In sub-THz PA designs, CMOS, SiGe, GaAs, and InP technologies are widely used, each offering distinct trade-offs in terms of output power, efficiency, and integration capability. Across all these technologies, the achievable output power and energy efficiency remain constrained due to the fundamental challenges of device scaling and high-frequency losses in sub-THz operation. Concerning PA non-linearity, the memoryless Rapp model is recommended (ETSI 2025; Li 2024). The corresponding model parameters, particularly for saturation voltage V_{sat}, are updated based

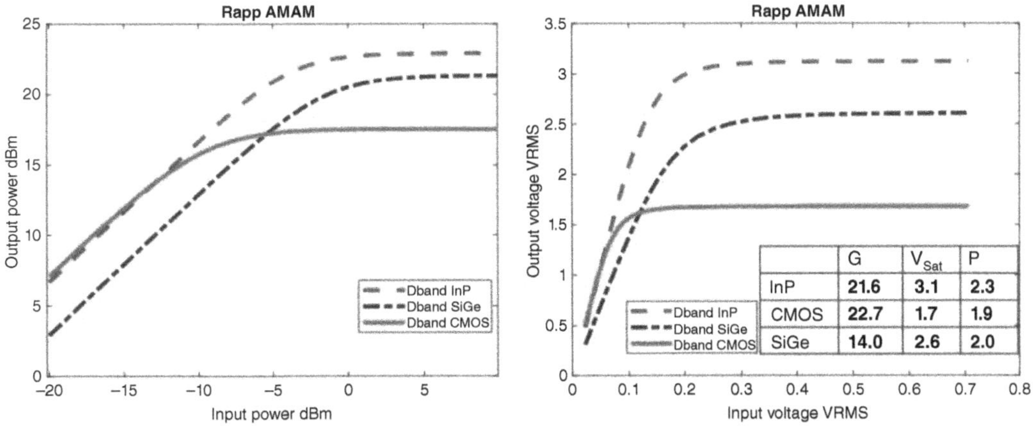

Figure 4.23 Rapp AM-AM (amplitude modulation) modelling for D-band PAs in different technologies.

on state-of-the-art PA designs in D-band as reported in Figure 4.23, together with other model parameters.

As the bandwidth increases, the short-term memory effects resulting in frequency-selective gain need to be considered. The memory-aware models, such as Wiener (Celka et al. 2001), Hammerstein (Nordsjo and Zetterberg 2001), general memory polynomial (GMP) (Morgan et al. 2006), and artificial neural networks (ANNs) (Yan et al. 2016), are essential for accurately capturing both the non-linear behaviour and the memory effects of PAs. The look-up table-based parallel Hammerstein (LUT-PH) model simplifies the approach for extracting the frequency basis vectors compared with the conventional parallel Hammerstein model. Compared with the MP model, the LUT-PH model exhibits greater robustness, particularly when considering varying operational conditions, such as different bandwidths and input power levels. However, ultra-high operating frequencies – especially between 170 and 300 GHz – present significant challenges in measurement accuracy. Moreover, there is a lack of PA models in the literature for those frequencies. The authors of Tervo et al. (2023), hence, utilize measurement data combined with transistor modelling simulations to extract PA model parameters. They recommend the Rapp model over more complex alternative models.

PN scales quadratically with increasing carrier frequency, making it more problematic in sub-THz bands. Additionally, the integrated PN level is directly influenced by the signal bandwidth. In sub-THz systems, the wider bandwidth results in a higher integrated PN level, further degrading signal quality. Finally, the distribution of the LO signal across larger antenna arrays introduces potential decorrelated PN contributions at individual antennas, affecting the overall system performance.

Frequency-selective I/Q imbalance might become pronounced for wideband systems. However, the measurements of a Tx at 140 GHz with a bandwidth of 5 to 10 GHz show only minimal frequency-selective effects with post-chip calibration (Zhang et al. 2024). With this observation, a non-frequency-selective I/Q imbalance model can still be used for system studies, assuming proper chip calibration has been performed. For the ADC/DAC models, specific effects such as clock jitter and nonlinearity can be adequately represented by modelling quantization noise through a reduced effective number of bits (ENoB). Unless the system has specific requirements, there is no need for additional high-level models beyond the classical ADC/DAC models.

4.3.4.2 FR3 Power Amplifier Modelling

For 6G, new frequency bands are expected: FR3 was proposed at the World Radio Conference 2023 to include frequencies from 7.127 to 24.25 GHz. As compared to FR2 (going up to 52.6 GHz initially and extended to 71 GHz later) and even more to proposed bands for FR4 (71 to 114.25 GHz) and FR5 (114.25 to 275 GHz), the lower carrier frequency translates into a larger wavelength, hence fewer antennas per unit of area. This is especially constraining for UEs, where antenna arrays would be too large, and the maximum number of antennas needs to be limited to one or two antennas. Consequently, the required output power per PA might be higher, and delivering such power at high efficiency will benefit from non-CMOS technologies, e.g. GaN or GaAs. GaN is the most popular choice for high-power applications at Tx, while GaAs and SiGe are common for medium to low-power applications and are commonly used in Rx systems due to their low-noise performance (Wang et al. 2023). Pushing for high power and efficiency in design leads to sacrificing linearity to some extent, which requires accurate models to capture. Digital pre-distortion (DPD) techniques are required to mitigate those non-linear effects and maintain signal quality while optimizing power efficiency (Haider et al. 2022; Farhadi et al. 2023). The long-memory effects due to trapping in GaN devices need to be considered in the DPD models and techniques (Yu et al. 2024).

RF design at a lower frequency is also less constraining. This can ease heterogeneous integration and make it possible to combine in the same technology the PA with other components close to the antenna, such as LNA or Tx/Rx switch, which is too challenging at higher sub-THz frequencies. Simultaneously, advanced PA topologies such as Doherty or polar can also be more easily realized at those frequencies where RF design is simpler, and hence further boost the PA efficiency.

4.4 MIMO Architectures and Transmission Schemes

MIMO technologies are at the core of 5G systems and are expected to play an even more significant role in the next generation. In 6G systems, the number of antennas is bound to increase substantially to serve a growing number of data-intensive UEs and to support operations in the FR3 and higher bands. Emerging multi-antenna technologies, such as RISs and D-MIMO, are anticipated to be key enablers not only of faster and more reliable communications but also of advanced applications such as ISAC. EM signal processing and low-resolution MIMO architectures are being explored as potential physical-layer solutions to enable scalable and energy-efficient wireless systems.

4.4.1 Hybrid Architectures Exploiting 'Over-the-Air' EM Signal Processing

MIMO communication has significantly improved data rates and spectrum efficiency. Additionally, the exploration of higher-frequency bands, such as mmWave and sub-THz, is expected to unlock new opportunities, including access to large bandwidths, massive device deployments, and the exploitation of the near-field propagation regime (Dang et al. 2020; Zhang et al. 2023). These advantages can be fully realized through the adoption of XL-MIMO systems, enabling unprecedented levels of communication and sensing performance, flexibility, and resolution (Wang et al. 2024; Gonzlez-Prelcic et al. 2024; Gong et al. 2024). Despite advancements in power-efficient hybrid architectures (Björnson et al. 2023), major challenges remain regarding the feasibility, scalability, and sustainability of XL-MIMO deployments for 6G and beyond, necessitating innovative solutions.

A promising approach involves shifting part of the signal processing directly to the EM level. This technique, known as *'over-the-air' EM signal processing* (Dardari et al. 2024a), is based on electromagnetic signal and information theory (ESIT) (Zhu et al. 2022; Di Renzo and Migliore 2024).

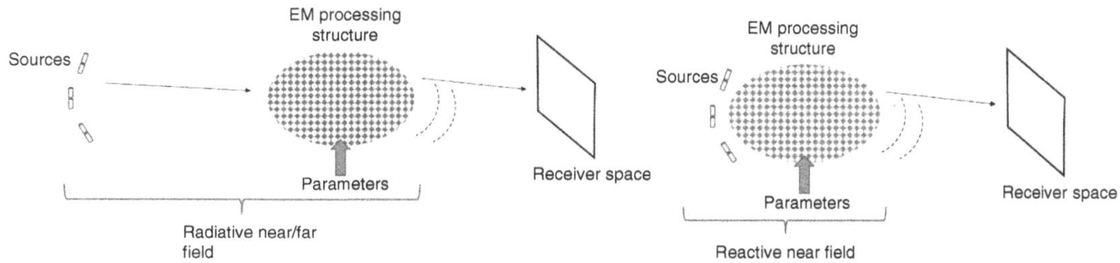

Figure 4.24 EM signal processing performed in the radiative near-/far-field region (left) or in the reactive near-field region (right) of the sources using an EM processing structure. *Source:* Dardari et al. (2024a) / arXiv / CC BY 4.0.

Recent advancements in metamaterials and EM technologies have facilitated the development of EM structures capable of realizing advanced manipulations of the EM field, such as beam forming/focusing, frequency splitting, and spatial first derivative. (Barbuto et al. 2021), which has the potential to revolutionize signal processing. These innovations offer significant advantages, including lower power consumption, reduced latency, and more compact device designs compared to traditional digital or hybrid solutions.

As illustrated in Figure 4.24, typical EM processing structures consist of multiple reconfigurable passive scattering elements whose purpose is to manipulate the EM field generated by active sources to obtain the desired response in the Rx space with minimal energy consumption and almost zero latency. The processing takes place 'over-the-air' because the final response is the result of the coupling between the scattering elements and, possibly, with the sources. In the absence of the EM processing structure, the same response would require bigger active sources equipped with more RF chains and suitable precoding performed at the digital level, with consequent additional power consumption and latency. EM processing structures can be classified into two categories: (i) scattering elements positioned in the radiative region of the signal sources (i.e. the transmitter) and (ii) scattering elements located in the reactive region of the signal sources, where inductive coupling occurs. The first category includes RISs (Di Renzo et al. 2020), metaprisms (Dardari and Massari 2021), and smart skins (Barbuto et al. 2021), which are EM structures without embedded RF sources that can be deployed within the radio environment as a cost-effective alternative to additional access points. These structures can be configured to manipulate incident EM waves, enabling functionalities such as anomalous reflection/transmission, beam splitting, and polarization control (Di Renzo et al. 2020). Recent advancements have introduced RIS variants with sensing capabilities and the ability to embed data in the backscattered signal, including backscatter communication (BackCom) systems (Liang et al. 2022) and self-conjugating metasurfaces (SCMs) (Dardari et al. 2023). The second category includes stacked intelligent metasurfaces (SIMs) (An et al. 2023), dynamic scattering arrays (DSAs) (Dardari 2024), dynamic metasurface antennas (DMAs) (Zhang et al. 2023), and fluid-based reconfigurable antennas (Wong et al. 2023). For example, DMAs feature reconfigurable subwavelength apertures and function as travelling wave antennas. They require only a single RF chain per row of the planar array, offering a practical solution that balances flexibility with a reduced number of RF chains (Zhang et al. 2023). In the following, SIMs, DSAs, and SCMs are described.

Stacked intelligent metasurfaces (SIMs): SIM technology has attracted significant interest from the research community due to its potential in EM wave manipulation (An et al. 2023). A SIM consists of a sealed vacuum structure incorporating multiple layers of metasurfaces, each composed of numerous reconfigurable meta-atoms, i.e. small, adjustable scattering elements,

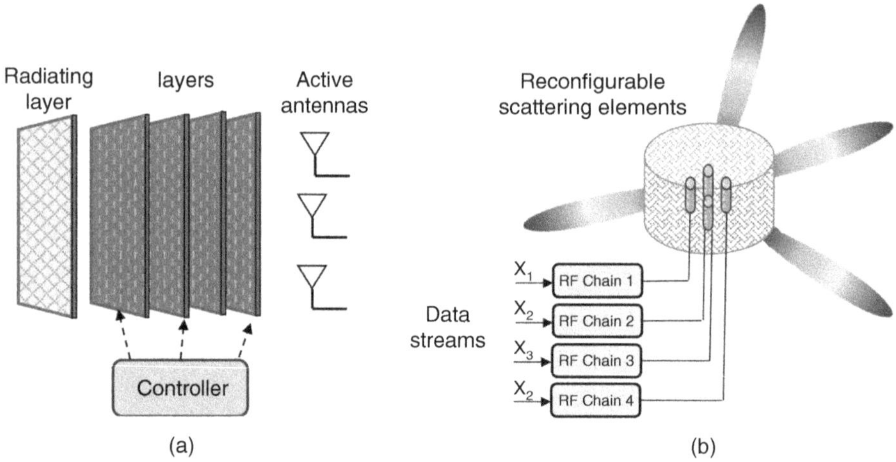

Figure 4.25 (a) Example of SIM and (b) example of EM processing at the transmitter based on a 3D DSA.

interconnected with a central controller (see Figure 4.25(a)). Based on the well-established Huygens–Fresnel principle, when an EM wave interacts with a meta-atom in any given layer, the meta-atom acts as a secondary point source, illuminating all subsequent meta-atoms in the next layer. This layered architecture, with unidirectional coupling between cells, offers greater flexibility in EM waveform manipulation and enhanced spatial-domain gain compared to conventional MIMO architectures, such as transmit arrays and hybrid MIMO antennas. Moreover, it significantly reduces the number of required RF chains, enabling highly efficient EM-level processing with minimal energy consumption. Interestingly, the unique layered structure of SIMs has inspired optimization algorithms for determining optimal transmission coefficients, drawing parallels with the well-known backpropagation algorithm used in training deep neural networks. These algorithms optimize meta-atom configurations and the phase shifts imposed on EM waves passing through the structure. Such optimizations enable the implementation of various wave-domain functionalities, including beamforming, precoding and combining, channel estimation, and signal angle-of-arrival (AoA) estimation (An et al. 2024).

Dynamic scattering arrays (DSAs): DSAs are versatile 3D structures capable of integrating wave-based computing and radiation by shifting processing from the digital domain to the EM domain. This transition minimizes the number of required RF chains, which are the primary contributors to energy consumption, system complexity, and latency. As illustrated in Figure 4.25(b), a DSA consists of a limited number of active antenna elements, each linked to an RF chain, surrounded by a dense cloud of reconfigurable passive scattering elements (e.g. dipoles). These elements interact within the reactive near field, enabling joint EM processing and radiation 'over the air' while also supporting super-directivity characteristics (Dardari 2024). A DSA can be seen as a generalization of reactively controlled arrays, originally proposed in Harrington (1978), evolved into electrically steerable passive array radiators (ESPARs), which utilize a single RF chain (Bucheli Garcia et al. 2020). Examples of EM signal processing tasks utilizing DSAs, such as MIMO EM precoding, can be found in Dardari (2024). In this case, instead of implementing the precoder in the digital domain, i.e. using N RF chains and N active antennas, with the DSA only $M \ll N$ RF chains and active antennas are needed (with M being the number of layers of the MIMO link), and the precoding and radiation tasks are performed by the N passive reconfigurable scatterers. Recent studies have also explored DSA-based designs

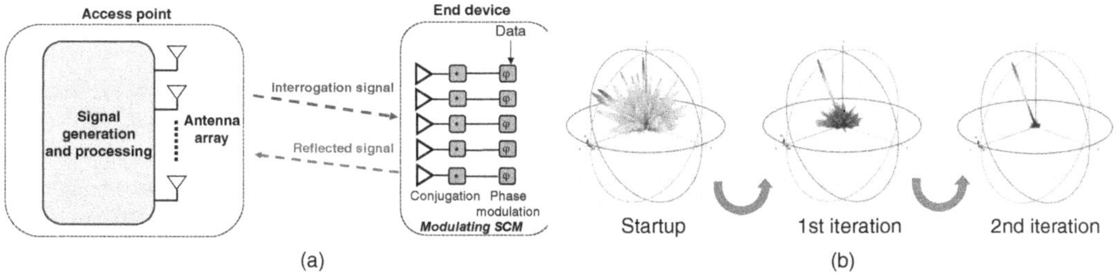

Figure 4.26 (a) Scheme of a modulating SCM-based MIMO communication and (b) beamforming time evolution.

for achieving super-directive beamforming and multi-user multiple-input single-output (MISO) communications, with extensions to frequency-selective processing introduced in Dardari (2025). DSAs offer unprecedented flexibility in managing the EM field, surpassing both conventional MIMO and SIM implementations. In fact, it is important to note that SIM performance is fundamentally constrained by the final layer surface of the structure, meaning its capabilities are inherently limited to what can be achieved by replacing the SIM with a fully digital planar or linear array of the same size and number of elements. By significantly reducing RF chains and digital processing requirements, DSAs enable the development of energy-efficient, low-latency, and cost-effective holographic MIMO systems.

Modulating self-conjugating metasurfaces (modulating SCM): In many applications, such as IoT and subnetworks (Berardinelli et al. 2023), the primary challenge is developing wireless communication and localization systems that achieve ultra-low power consumption, minimal latency, and ultra-fast beamforming simultaneously. These capabilities are especially critical when operating at high-frequency bands, such as mmWave and THz, where MIMO links are essential to compensate for the associated severe path loss. While BackCom systems based on RISs represent an interesting solution (Liang et al. 2022), they face challenges related to the need for a control link between the RIS and the access point, as well as implementing an efficient CSI estimation procedure. A recently introduced alternative is the use of modulating SCMs (Dardari et al. 2023, 2024b). As illustrated in Figure 4.26(a), an access point can establish a MIMO link with one or more low-complexity devices equipped with a modulating SCM by transmitting an interrogation signal. This signal is iteratively updated based on the conjugated backscattered response from the SCM, which also embeds data. Research has shown that, after a few iterations (see Figure 4.26(b)), the access point can effectively beamform toward the SCM-based device, which inherently exhibits retro directivity. This approach enables MIMO communication without requiring explicit CSI estimation, signalling overhead, or digital-to-analogue chains at the device side. Additionally, SCMs can play a crucial role in localization systems by leveraging AoA estimations and MIMO gains, providing fast, low-latency, and highly precise positioning solutions. This dual functionality meets the demands of next-generation wireless networks and aligns with the integrated communication and sensing paradigm essential for 6G IoT technologies.

In summary, 'over-the-air' EM signal processing is a promising approach to addressing scalability and sustainability challenges in next-generation wireless systems. It offers significant advantages over traditional digital and hybrid solutions, including reduced power consumption, lower latency, and more compact device designs. A novel architecture for XL-MIMO systems integrates low-power reconfigurable antennas with both digital and analogue precoding (see Figure 4.27(b)). This *tri-hybrid* approach overcomes key limitations of traditional and hybrid

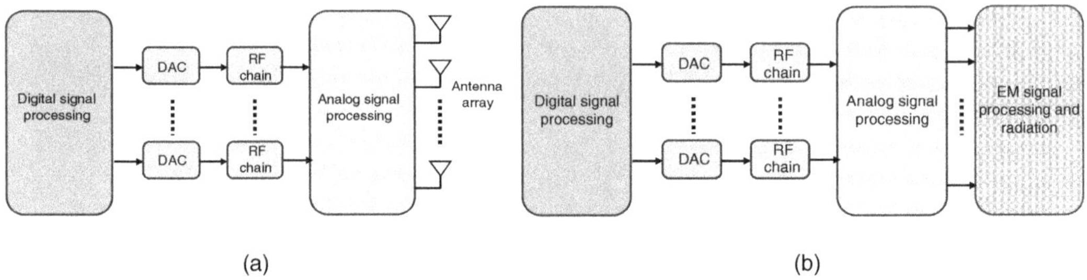

Figure 4.27 (a) Fully connected hybrid architecture and (b) fully connected tri-hybrid architecture.

MIMO systems by improving power efficiency and introducing an additional layer of signal processing (Rodrigo Castellanos et al. 2025). Looking ahead, expanding 'over-the-air' EM signal processing to wideband and time-varying systems remains a critical priority, both theoretically and practically. Promising developments include space-time modulated metasurfaces, which exploit frequency-dependent behaviours to enable tailored functionalities (Zhang et al. 2018; Mizmizi et al. 2024). To fully unlock the potential of 'over-the-air' EM signal processing, the integration of advanced technologies such as AI for the configuration of the EM structures or EM structures for AI (Liu et al. 2022a) and state-of-the-art metamaterials (Barbuto et al. 2021) will be essential. This synergy is expected to position 'over-the-air' EM signal processing as a cornerstone technology for future wireless networks.

4.4.2 Near-Field Wavefront Engineering for Integrated Sensing and Communication

To address the high requirements of future wireless connectivity, the concept of unifying sensing and communications is constantly gaining ground. With ISAC, the same equipment and resources can be used to provide high data rate communications, and high-resolution sensing, including localization, imaging and mapping, at the same time. In this landscape, the RIS offers a means for programming and reconfiguring the connectivity *beyond the transmission of bits*. This section provides an overview of RIS-aided wavefront engineering techniques for communication links and for ISAC. Moreover, it discusses near-field communications approaches and merits, which offer a promising 6G research area enabled by the use of large aperture RIS and/or higher-frequency bands.

4.4.2.1 RIS-Aided Wavefront Engineering

With the aim to meet the ever-increasing demand for bandwidth, wireless communications are nowadays shifting to higher operation frequencies, towards the mmWave and THz. A direct consequence of this shift is that the wavelength λ becomes smaller, and a radiating element of aperture size D becomes electrically larger, i.e. the ratio D/λ increases. As a result, the Fraunhofer distance $z_F = 2D^2/\lambda$ is pushed to farther distances. Hence, while wireless communications have so far operated in the far field of radiating elements, the use of larger radiating apertures and/or higher-frequency bands opens up new opportunities for operation in the near field of large radiating elements, such as RISs. The RIS is a planar surface that consists of several periodically distributed subwavelength scatterers (RIS elements) with tuneable response. By applying the proper phase shifts on the RIS elements, the RIS is able to reflect incident waves towards any desired direction, while at the same time sculpturing the wavefront into any desired beam profile.

For a relatively small RIS surface, the Fraunhofer distance is brought close to the RIS, and the reflected beam quickly enters its far field. This case corresponds to conventional beamforming.

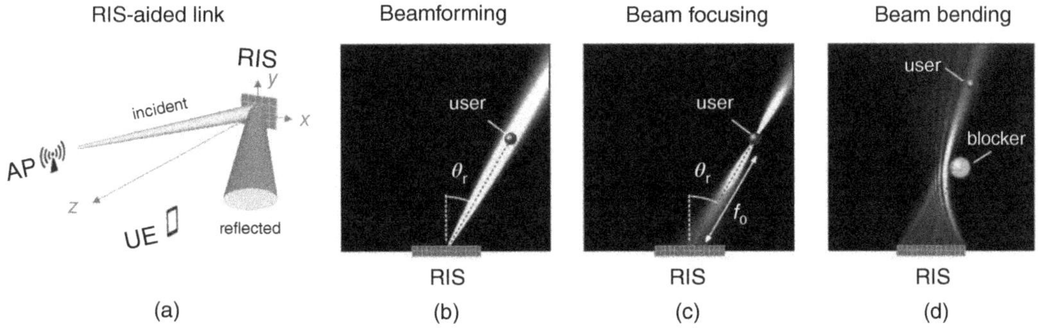

Figure 4.28 RIS-aided wavefront engineering for near-field wireless connectivity.

For larger RIS, however, the Fraunhofer distance increases and the RIS can operate in its near field, to form beams with exotic properties beyond conventional beamforming (e.g. focusing, bending, and self-healing) that can be advantageous to a number of applications, including but not limited to beam-tracking, blockage avoidance and wireless power transfer (WPT) (Stratidakis et al. 2024). To form such beams, the RIS transforms the incident beam into a reflected beam with tailored properties by properly engineering the incident wavefront, i.e. by properly phase-shifting the incident wave.

Consider a RIS that is illuminated by an access point (AP), which is equipped with a directional antenna. The RIS resides on the *xy* plane and steers beams on the *xz* plane, as schematically shown in Figure 4.28(a). The E-field of the AP beam that is incident on the RIS can be written as

$$E_i(x,y) = E_0(x,y)\exp(jk\sin\theta_i x),$$

where E_0 is the real-valued amplitude, θ_i is the angle of incidence, and $k = 2\pi/\lambda$ is the free-space wavenumber. To form a beam with the desired profile, the RIS transforms the incident beam by first correcting the tilt of the incident beam (expressed by the $\exp(jk\sin\theta_i x)$ term) and, subsequently, introducing a phase φ with tailored functional form, to shape a reflected beam with desired characteristics (Droulias and Alexiou 2022).

Mathematically, this operation is expressed as

$$E_r(x,y) = E_i(x,y)\exp(-jk\sin\theta_i x)\exp(j\varphi) = E_0(x,y)\exp(j\varphi),$$

i.e. the footprint of the incident beam on the RIS, E_0, is reshaped according to the prescribed phase φ. Depending on the functional form of φ, RIS-aided wavefront engineering results in different beams, as discussed next.

RIS-aided beamforming: Beamforming is the most fundamental of the RIS functionalities and is typically used to steer the incident beam towards a desired direction. Beam-steering is achieved by imposing linear phase shifts (with respect to the RIS element position, x) on the RIS elements (Stratidakis et al. 2021), i.e.

$$\varphi = k\sin\theta_r x,$$

where θ_r is the angle of reflection on the steering plane. An example of a numerically propagated beam that is steered at angle θ_r is shown in Figure 4.28(b). In this example, the footprint of the incident beam extends over a small section of the RIS (covering a few wavelengths) and, hence, the beam quickly evolves into its far field.

Beamforming with a RIS can be used instead of broadcasting to increase the received power at the direction of the user, establish an additional LoS link, or track the user.

RIS-aided beam focusing: Beam focusing is the ability of a RIS to focus the power of the reflected beam to a small area, similarly to how a lens focuses light onto a focal point. Beam focusing can be achieved simply by imposing a parabolic phase profile on the RIS elements (Droulias et al. 2024), i.e. of the form

$$\varphi = k \sin \theta_r x - k \frac{x^2(\cos \theta_r)^2 + y^2}{2f_0},$$

where f_0 is the focal distance. An example of a numerically propagated beam is shown in Figure 4.28(c). In this example, the footprint of the incident beam extends over the entire RIS surface (covering several wavelengths) and, hence, the beam resides entirely in its near field.

Beam focusing can be used to maximize the received power, and it is ideal for WPT applications, as well as to estimate the location of the UE.

RIS-aided beam-bending: Bending beams have the ability to propagate on bent trajectories. Beam-bending can be achieved by using a power-law phase profile of the form (Droulias et al. 2025),

$$\varphi = -kCx^\gamma,$$

where C, γ are real constants. An example of a numerically propagated beam is shown in Figure 4.28(d). In this example, the footprint of the incident beam extends over the entire RIS surface (covering several wavelengths) and, hence, the beam resides entirely in its near field.

Beams that bend can circumvent obstacles, eliminating the need for intermediate nodes and minimizing the power requirements. Importantly, in a rich blocker environment, where multiple blockers may change position and dynamically interrupt the connectivity, bending beams that can be reconfigured are ideal for dynamic blockage avoidance, minimizing the deployment complexity. The curved nature of these beams is also ideal for interference management applications, where entire areas can be enclosed by virtual boundaries, guaranteeing minimum interference. With judicious design, the power delivery to the UE can even be tailored to reach a maximum at desired locations along the UE path.

Beam-bending is also ideal for routing on a predefined trajectory that takes place at the PHY entirely, without requiring intermediate nodes. Tracking mobile UEs with conventional beamforming would require the use of codebooks and multiple beams, therefore sophisticated algorithms and overhead, depending on the UE mobility, even if the UE trajectory is known. With bending beams, these requirements are relaxed, as a single beam dedicated to the UE path guarantees uninterrupted connectivity, regardless of the UE position and speed.

4.4.2.2 Near-Field Angle-Range Localization for ISAC

The ability of the RIS to focus the power of the reflected beam to a small area is ideal for localization applications. Due to the small size of the focal area, searching the entire area of interest with beam focusing alone may require prohibitively large overhead and, therefore, searching can be assisted by hierarchical beam-tracking. The whole procedure can be split into two phases as presented in Figure 4.29(a).

In phase 1, beam-tracking is performed with beamforming through the RIS, in order to estimate the direction of the UE with respect to the RIS. In phase 2, ranging is performed with beam focusing through the RIS, in order to estimate the distance of the UE from the RIS, along the direction identified in phase 1. With information about both the direction and the distance, the location of the UE is fully determined. Both phases follow a hierarchical and binary tree structure (Stratidakis et al. 2024, 2025).

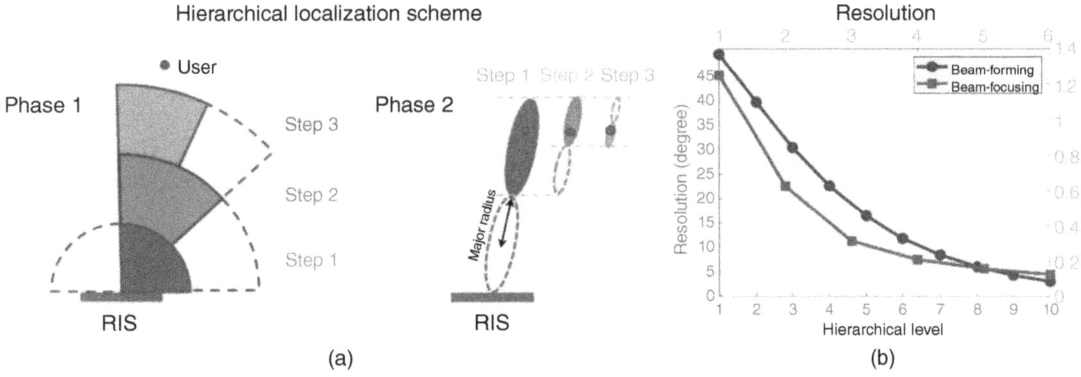

Figure 4.29 Hierarchical angle-range localization. (a) Phase 1: determination of the user's angle with respect to the RIS, via beamforming. Phase 2: determination of the user's distance from the RIS via beam-focusing. In both phases, the shaded regions mark the areas where the user location is successfully identified at each hierarchical level. (b) Example of resolution versus hierarchical level for the two phases of the hierarchical RIS-assisted localization scheme, reproduced from Stratidakis et al. (2024). In this example, localization is performed within a sector of $\pm 25°$ (phase 1) and 5 m maximum distance (phase 2).

In this scheme, the UE communicates with the AP, exchanging information about the received power levels at each step of the localization procedure. Hence, sensing (i.e. localization) is aided by communication. The angular resolution is determined by the minimum angular beam-width and beam-separation in phase 1, and the radial resolution is determined by the minimum size of the focal areas formed in phase 2 at each hierarchical level. An example of resolution at each level of phase 1 and 2 is presented in Figure 4.29(b).

4.4.3 Massive MIMO with Low-Resolution Data Converters

The combination of massive antenna arrays and broad bandwidths characterizing communications in the FR3, mmWave, and sub-THz bands calls for drastic simplifications in terms of RF architectures. In this context, 5G New Radio systems operating in the low mmWave spectrum (up to 52.6 GHz) have initially utilized hybrid analogue–digital MIMO architectures resulting from decreasing the number of RF chains compared with the number of antenna elements (Dahlman et al. 2018). This strategy, however, offers limited spatial multiplexing capabilities and, more importantly, does not scale effectively for larger antenna arrays and bandwidths due to the complexity of the extra analogue circuitry (such as phase shifters) and the resource-demanding beam management schemes required (Saxena et al. 2017).

To support flexible wideband beamforming and large-scale spatial multiplexing, adopting fully digital MIMO architectures in line with the original concept of massive MIMO (Marzetta 2010) is paramount. However, operating with large antenna arrays and bandwidths entails a compromise on the resolution of the ADCs and DACs. In fact, the power consumption of a data converter increases linearly with the bandwidth (quadratically for the ADCs with bandwidths over 330 MHz (Murmann 2024)) and exponentially with the number of resolution bits (Li et al. 2017; Jacobsson et al. 2017a, b). Despite the added quantization distortion, fully digital massive MIMO systems with 3-bit to 5-bit ADCs and DACs can perform remarkably close to full-resolution systems (Jacobsson et al. 2017a). Furthermore, fully digital MIMO systems with lower-resolution data converters (as low as 1 bit) can outperform their hybrid analogue–digital counterparts in terms of spectral and energy efficiency (Roth et al. 2018).

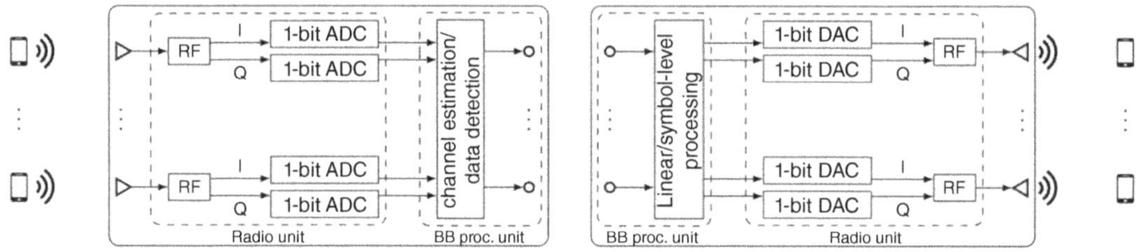

Figure 4.30 Fully digital massive MIMO BS architectures with 1-bit ADCs (left) and 1-bit DACs (right).

Reducing the resolution of the ADCs and DACs to 1 bit leads to minimal power consumption and complexity related to the data conversion (Li et al. 2017), and enables further streamlining of the RF HW. Specifically, adopting 1-bit ADCs and DACs eliminates the need for high-cost RF components such as highly linear amplifiers and automatic gain controls. Moreover, it enhances the energy efficiency at the transmitter by allowing the PAs to operate without back-off. As a result, the performance degradation due to the 1-bit quantization can be offset by adding very low-cost antenna elements and RF chains. In addition, utilizing 1-bit ADCs and DACs severely reduces the amount of raw data exchanged between the BB processing and RUs at the base station (BS) (Jacobsson et al. 2017a, b). Lastly, the use of 1-bit data converters combined with the short mmWave and sub-THz wavelengths paves the way for compact and energy-efficient massive MIMO arrays that can also be deployed at the UE (e.g. on drones and cars). Figure 4.30 illustrates fully digital massive MIMO BS architectures with 1-bit ADCs and DACs, where each antenna is connected to two 1-bit ADCs (resp. DACs) for the in-phase and quadrature components of the received (resp. transmitted) signal.

Over the past decade, massive MIMO systems with low-resolution and 1-bit ADCs and DACs have generated a significant amount of research in performance analysis (Li et al. 2017; Jacobsson et al. 2017b; Roth et al. 2018; Atzeni et al. 2021a; Ma et al. 2023), channel estimation (Choi et al. 2016; Li et al. 2017; Rao et al. 2019; Atzeni and Tölli 2022; Ding et al. 2025), data detection (Choi et al. 2016; Jeon et al. 2018; Atzeni and Tölli 2021, 2022; Radbord et al. 2024), and precoding design (Saxena et al. 2017; Jacobsson et al. 2017a; Jedda et al. 2018; Li et al. 2021). Most works in the literature generally assume coarse quantization at the BS only. However, the extreme scenario where 1-bit data converters are adopted at both the transmitter and Rx is of particular interest (Lozano 2021; Atzeni et al. 2023), as it results in fully digital systems with minimal RF power consumption, complexity, and cost.

The work in Atzeni et al. (2023) analyses a point-to-point system with fully digital massive MIMO arrays and 1-bit data converters at both the transmitter and Rx, where a simple quantized linear precoding strategy is assumed at the former. This system can be used to model, for example, a high-capacity wireless backhaul link operating at sub-THz frequencies. The analysis in Atzeni et al. (2023) explores the relationship between the transmitted data symbols (at the input of the transmitter's 1-bit DACs) and their soft estimates (at the output of the Rx's 1-bit ADCs). Through numerical results, it demonstrates that such a system, despite its simplicity, can deliver remarkable data detection performance when both the transmitter and Rx are equipped with very large antenna arrays (i.e. with 400 or more antenna elements). Interestingly, the performance of the considered system with 1-bit ADCs and DACs is similar to that of an equivalent system with 1-bit ADCs and full-resolution DACs, where the latter entails a much higher RF power consumption and complexity at the transmitter. Significant performance enhancements can be achieved by considering more sophisticated symbol-level precoding strategies at the transmitter (Saxena et al. 2017; Jacobsson et al. 2017a; Jedda et al. 2018; Li et al. 2021).

4.4.4 D-MIMO and RIS

4.4.4.1 Centralized Versus Distributed Beamforming Design in D-MIMO

D-MIMO systems consist of multiple APs distributed across a coverage area, connected to a central unit (CU) via fronthaul links for data and CSI exchange. These APs cooperate to coherently serve the UEs in the coverage area, ideally eliminating the inter-cell interference and ensuring uniform quality of service across the network. Network-wide beamforming designs based on, e.g. mean squared error minimization or zero forcing, typically require full CSI at either the APs or the CU. Network-wide beamforming designs can greatly outperform local beamforming schemes based on, e.g. maximum ratio transmission/combining, which rely on local CSI at the APs (Ngo et al. 2017; Björnson and Sanguinetti 2020). However, implementing network-wide beamforming designs at the CU requires extensive CSI exchange over fronthaul links, generating significant challenges related to latency, backhaul bandwidth, and network scalability as the number of APs and UEs increases. On the other hand, traditional distributed network-wide beamforming designs at each AP require cross-channel information from all the other APs. Thus, these designs also face similar scalability and latency issues, as the CSI exchange among the APs introduces considerable signalling overhead (Kaleva et al. 2018).

To address these challenges, a fully distributed framework for network-wide beamforming design is introduced in Atzeni et al. (2020, 2021b) considering a time-division duplex (TDD) system with channel reciprocity. This framework eliminates the need for fronthaul signalling for CSI exchange by utilizing an iterative bi-directional training procedure complemented with a novel OTA uplink signalling. Bi-directional training allows iteratively updating of the beamformers locally at each AP and UE, as shown in Figure 4.31, for a downlink (DL) transmission with a single UE and a single BS. At each iteration, each UE updates its combiner based on the preceding DL pilots simultaneously transmitted by all the APs. Similarly, each AP updates its precoders based on the preceding uplink pilots simultaneously transmitted by all the UEs, where each UE uses its combiner as a precoder. For the precoder update, the novel OTA uplink signalling resource allows each AP to obtain the cross-channel information corresponding to the other APs, which was exchanged via fronthaul links in traditional distributed beamforming designs.

The OTA signalling resources required for the distributed beamforming design can be further optimized depending on the setting. For the uplink case, an adaptive regularized Newton method is proposed in Gouda and Tölli (2021) to reduce the OTA signalling resources needed for the beamforming optimization. In a multi-group multicasting scenario, all the UEs in a multicast group can transmit the same pilots to save signalling resources without compromising the SE

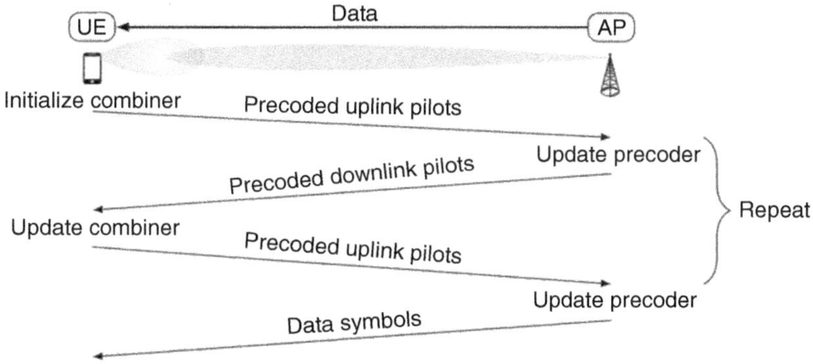

Figure 4.31 Schematic representation of iterative bi-directional training for a downlink transmission with a single UE and a single BS. To transmit the precoded uplink pilots, the UE uses its combiner as a precoder.

(Gouda et al. 2024a). For scenarios where the UEs are served in both the DL and uplink within the same resource block, the beamforming training can be further optimized by designing a single beamformer for both DL and uplink transmissions at the APs and UEs, improving the effective SE (Gouda et al. 2021, 2024b). Additionally, under time-varying channel conditions, the beamformers at both the APs and UEs can be updated periodically in a best-response fashion, enhancing the SE of the network (Gouda et al. 2024b).

4.4.4.2 ISAC D-MIMO, Scalable D-MIMO

Concept: D-MIMO systems are emerging as an important architecture in the evolution of 6G networks, particularly in combination with ISAC (Guo et al. 2025). By deploying multiple spatially distributed APs that collaboratively serve UEs, D-MIMO systems can enhance coverage, reliability, and SE (Meng et al. 2024). This distributed topology not only mitigates issues such as signal blockage but also facilitates a more uniform quality of service across the network.

ISAC in D-MIMO, multi-static sensing: In D-MIMO architectures, the integration of communication and sensing functions is achieved by sharing the same HW and spectral resources (Liu et al. 2022b). As shown in Figure 4.32, this co-design approach allows the system to perform data transmission and environmental sensing at the same time, leading to efficient spectrum utilization and reduced infrastructure costs. The distributed nature of D-MIMO enhances multi-static sensing capabilities by providing multiple observation points, thereby improving

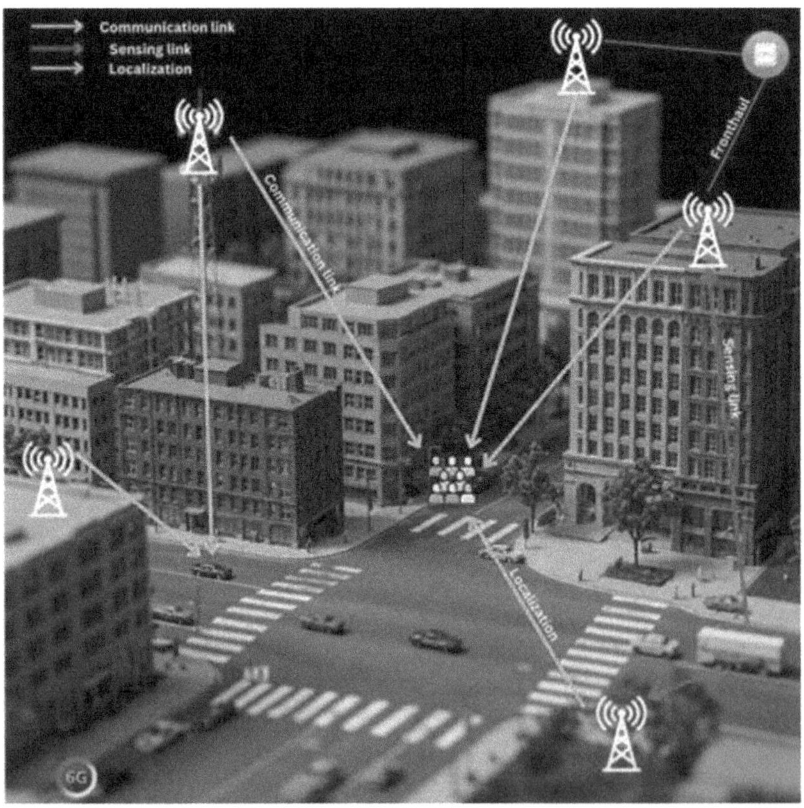

Figure 4.32 Illustration of communication, sensing, and localization functions in a D-MIMO ISAC system. *Source:* Generated with AI using Canva.

the detection and localization of objects within the network's coverage area (Guo et al. 2025). This collaborative sensing mechanism enables applications such as precise localization, object tracking, and environmental mapping.

In ISAC-enabled D-MIMO, communication prioritizes data reliability (rate, latency), while sensing focuses on environmental interpretation (resolution, detection). These differing priorities can lead to potential trade-offs and optimizations; for instance, optimal beamforming strategies for communication may not align with those for sensing (Meng et al. 2024). Therefore, ISAC systems must employ sophisticated resource management and signal processing techniques to balance and, where possible, synergize the requirements of both functions. Sensing-aided communication enhances performance by using real-time environmental data for adaptive beamforming and resource allocation, improving reliability and throughput (Guo et al. 2021). This enhances link reliability, throughput, and overall network intelligence (Guo et al. 2025).

To analyse the trade-offs between communication and sensing, we consider a DL D-MIMO ISAC model. In this setup, communication data symbols are utilized for target detection, with each AP functioning as either an ISAC transmitter or a sensing Rx. Each transmitting AP allocates a fraction of its total power to the sensing beam, while the remaining power is dedicated to communication beams for UEs. Simultaneously, the other APs receive reflections from the sensing target. We assume an isotropic target and model the sensing channel by incorporating different path delays from the distributed APs. A 100m × 100m area is considered, where APs, UEs, and the sensing target are randomly deployed within the coverage region. Phase synchronization is maintained across the distributed transmitting and receiving APs. The sensing and communication performance are quantified using the position error bound (PEB) and the UEs' sum rate metrics, evaluating different power allocation strategies for sensing and communication.

Numerical results in Figure 4.33 show that sensing accuracy is significantly affected by (i) the amount of power allocated to the sensing beam and (ii) the number of APs acting as Rxs. The sensing accuracy improves when more APs function as Rxs, as they can capture target-reflected signals

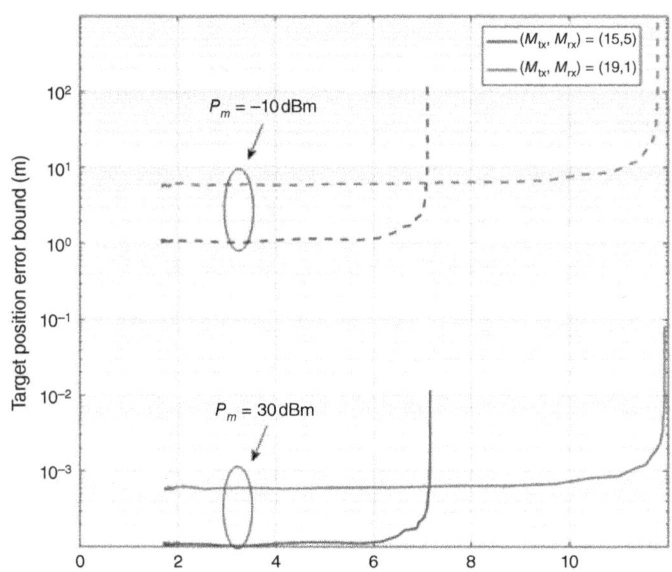

Figure 4.33 Comparison of communication SE and sensing PEB performance in a D-MIMO system for different transmit and receiver AP pairs: (i) (15,5) and (ii) (19,1). *Source:* Generated with AI using Canva.

from different locations. Similarly, communication performance is influenced by the number of APs serving as transmitters. The results indicate that if a larger number of APs act as transmitters, communication performance improves, whereas sensing performance deteriorates significantly. This highlights the need to carefully select the number of transmitting and receiving APs based on the specific sensing and communication requirements.

Achieving phase synchronization among the distributed APs is challenging in D-MIMO systems, and the phase mismatch affects both sensing accuracy and communication performance. There is a need for developing advanced synchronization techniques to mitigate phase mismatches and enhance overall system performance. Additionally, exploring dynamic AP role selection based on real-time network conditions will further improve ISAC efficiency.

4.4.4.3 RIS-Assisted D-MIMO, RIS-Assisted IAB

RISs are emerging as a transformative technology in both access and backhaul networks, enabling more energy-efficient, flexible, and intelligent wireless communication. In access networks, RISs can enhance coverage and signal strength by dynamically reflecting and steering signals, mitigating blockages, and improving performance in dense urban environments or indoor scenarios. This is particularly beneficial for mmWave and THz communications, where direct LoS links can often be obstructed. Furthermore, in wireless backhauling, RISs offer a cost-effective solution to improve the connectivity between BS, especially in locations where fibre deployment is challenging or expensive. Using intelligent signal steering over needed paths, RISs help maintain high capacity and low latency in the backhaul links.

RIS-assisted D-MIMO access: Unlike traditional MIMO systems, D-MIMO systems consist of multiple APs distributed across the deployment area to leverage spatial diversity and improve service coverage. However, the scalability of D-MIMO systems in a practically feasible way is challenging in terms of energy efficiency and cost efficiency, especially for larger networks. In Parambath et al. (2024), an indoor D-MIMO use case with a larger UE density is considered, where the service coverage is scarce due to high obstacle density. The benefits of integrating RIS to this system are explored here, mainly focusing on KPIs such as service coverage at the UEs and overall energy efficiency improvement. Here, RIS is modelled as a planar meshed surface with passive reflective elements attached to each grid, which coordinates their phase shifts using a controller to perform relative beamforming.

To improve service coverage, RISs can be strategically deployed at locations that provide higher average service coverage to the UEs in the system, typically on the ceilings or walls, where they can easily redirect the incident signal from serving APs toward the target UE. A UE-centric AP and RIS clustering strategy (Björnson and Sanguinetti 2020) ensures that a UE is served by an optimal set of APs and RIS, enhancing overall network reliability, as illustrated in Figure 4.35. To improve scheduling, a queue-based scheduling scheme (QSS) (Ye et al. 2016) can be used to dynamically allocate AP and RIS resources, ensuring an efficient and fair serving of UEs. For DL transmission, the system uses an alternating optimization algorithm for joint precoding and RIS phase-shift configuration (Zappone et al. 2021; Parambath et al. 2024) to perform codebook-based beamforming.

By addressing RF dead spots in the D-MIMO system, RIS plays a crucial role in improving the KPIs, including signal to interference and noise ratio (SINR) at UE and the average service coverage of UEs (Parambath et al. 2024). However, the average energy efficiency improvement depends on the type of RIS used. PIN-diode-based RIS suffers from notable static power dissipation at RIS elements, resulting in less significant energy efficiency improvement for a larger RIS size. In contrast,

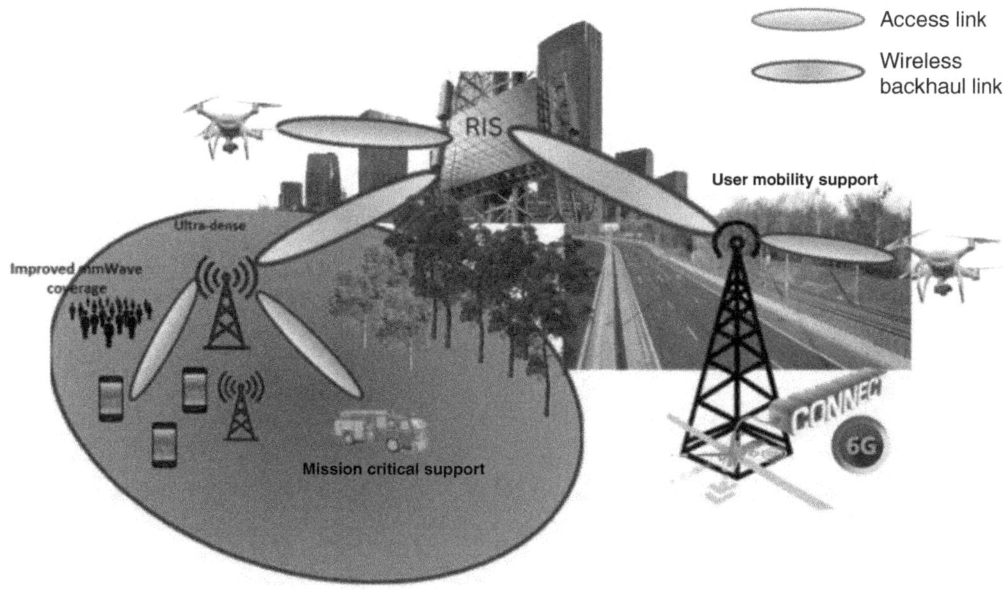

Figure 4.34 RIS-assisted access and backhaul.

RF switch-based RIS is a preferable energy-efficient alternative due to the negligible static power dissipation at the RIS elements.

Overall, by addressing service coverage and energy efficiency, RIS emerges as a key enabler in improving the access network performance of D-MIMO systems for 6G.

RIS-assisted integrated access and backhaul (IAB): The integration of RISs into integrated access and integrated access and backhaul (IAB) networks has emerged as a promising approach to enhance network resilience, particularly in suburban environments where tree foliage obstructs backhaul connectivity, as shown in Figure 4.34. In Madapatha et al. (2025), the impact of RIS deployment on service coverage probability and backhaul rates, focusing on seasonal variations that degrade the direct link reliability for backhauling, is studied. Concentrating on a representative use case of a sub-urban environment, in an area of 1 km^2 with a wide-area IAB setup where UEs are randomly distributed, it strategically places RISs to support the backhaul links to mitigate foliage-induced blockages. Simulation results demonstrate that RIS-assisted IAB networks significantly improve service coverage probability, particularly at higher frequencies.

As shown in Figure 4.36 (Madapatha et al. 2025), the RIS-assisted links provide considerable improvement in service coverage probability by bypassing the obstructions caused by dense foliage.

The mmWave frequencies are prone to foliage loss, making robust backhaul solutions important. With pre-planned RIS deployment, the propagation environment can be controlled, maintaining seamless connectivity between the IAB donor and child nodes. Compared to conventional relay-based solutions, RISs provide an energy-efficient alternative that reduces energy consumption while improving service coverage probability. While network-controlled repeaters (NCRs) also offer an alternative, they introduce high energy and financial footprint, making RISs a sustainable long-term solution. However, careful planning of terrain elevation, tree distributions, and multi-operator coexistence is needed as unintended reflections can introduce interference challenges.

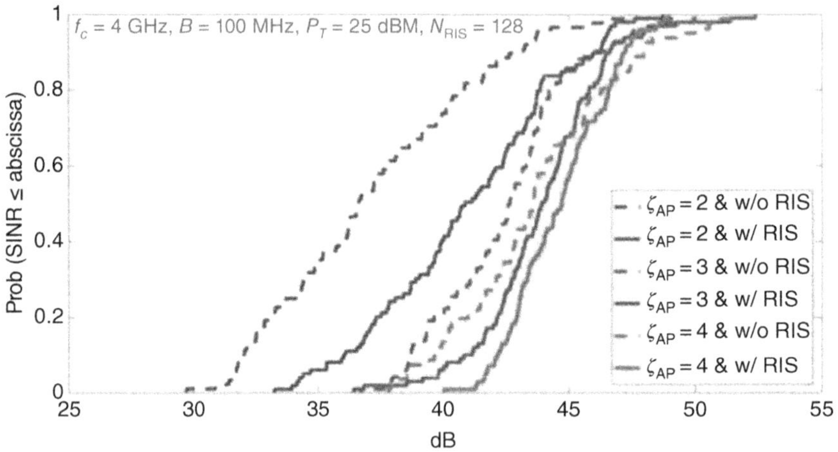

Figure 4.35 SINR cumulative distribution function (CDF) comparison between D-MIMO system with and without RIS assistance for different AP cluster sizes (ζ_{AP}) (i) 2 AP, (ii) 3 AP, and (iii) 4 AP.

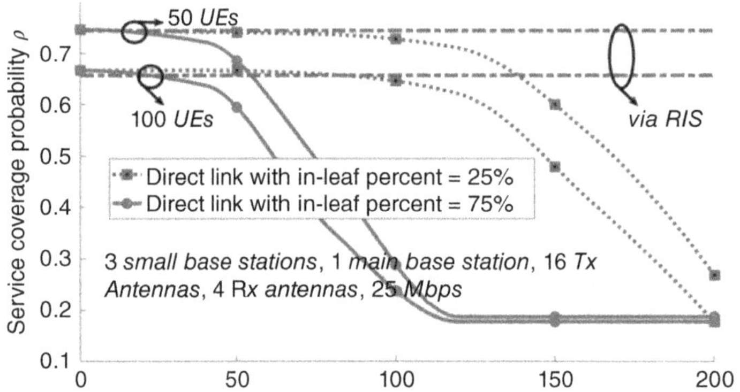

Figure 4.36 Service coverage probability as a function of total tree line depth for different in-leaf percentages.

RIS-assisted IAB networks could redefine wide-area connectivity by enabling cost-effective, adaptable backhaul solutions that respond dynamically to environmental changes. As 6G research advances, further investigations on intelligent reflective surface design, ML-based beamforming, and large-scale deployment strategies will be key to unlocking the full potential of such technologies.

4.5 6G Devices and Infrastructure

This section provides an overview of technology innovations for 6G devices, focusing on three aspects. First, a broad view is taken, looking at future directions for low-power IoT devices in general. Second, the design of a secure SoC architecture is described. It is a HW–software co-designed platform that is built on a scalable tiled HW architecture, with specific support for HW accelerators (for both AI and signal processing). The design of the AI and DSP accelerators is then further detailed. Third, focus is shifted to a new type of device, namely energy neutral (EN) devices. Such devices are powered by EH, either ambiently harvested or via WPT. The section covers a variety of relevant aspects, including EH, low-power protocol design, and passive device design.

4.5.1 Future Directions for IoT Devices

The rapid evolution of IoT devices is being shaped by advancements in energy efficiency, connectivity, and integration with intelligent systems. Future IoT ecosystems will revolve around sustainable energy solutions, enhanced connectivity for massive deployments, and AI-enabled, autonomous operations. The following key directions are expected to define the future landscape of IoT devices.

4.5.1.1 Energy Neutral Devices

EN devices represent a significant leap towards sustainability in IoT, operating relying on harvested energy from environmental sources and eliminating the need for batteries. This will enable maintenance-free deployments with ultra-low energy consumption across the device's entire life-cycle, making them ideal for applications in remote monitoring, asset tracking, and smart infrastructure. EN devices are expected to leverage advanced energy-harvesting mechanisms, efficient energy storage solutions, and optimized EH materials and manufacturing with sustainable and circular practices, as well as backscattering and low complexity Rxs. Protocol and infrastructure enablers for the seamless support and application expansion of EN devices will include lightweight and energy-aware design, WPT (especially RF-based) for controllable/predictable energy supply management, as well as adaptive duty cycling and wake-up protocols (López et al. 2025).

4.5.1.2 Enhanced LPWA

Enhanced low-power wide area (eLPWA) devices, focusing on mMTC, are poised to redefine large-scale, long-range IoT connectivity. Improvements on eLPWA devices will be designed around massive deployment scenarios, long battery life, cost efficiency, extended coverage and addressing pre-request requirements on reliability and joint communication and sensing. Future eLPWA devices will integrate power-saving techniques, adaptive low-power sleep modes, cost-effective design and operation, and enhanced coverage mechanisms. NTNs will be a key enabler to allow ubiquitous coverage for eLPWA devices. They will coexist seamlessly from day one with emerging 6G networks, supporting dense IoT ecosystems such as smart cities, industrial automation, and large-scale environmental monitoring (Balakrishnan 2023).

4.5.1.3 Intelligence with TinyML

The integration of AI at the edge, particularly through tiny machine learning (TinyML), is set to revolutionize IoT devices. By enabling real-time data processing on the device itself, TinyML minimizes latency and energy consumption while enhancing operational efficiency. This is crucial for energy-constrained devices, facilitating applications like predictive maintenance, anomaly detection, and smart agriculture. Future IoT devices will increasingly employ lightweight AI models for intelligent decision-making and context-aware functionalities.

4.5.1.4 Security and Privacy Enhancements

With the growing ubiquity of IoT devices, robust security and privacy mechanisms are critical. Future devices will incorporate lightweight cryptographic protocols, secure boot processes, and end-to-end encryption. Location-aware security measures will ensure devices operate within authorized zones, protecting data integrity and preventing unauthorized access. As IoT ecosystems scale, enhanced security solutions will be indispensable for safeguarding user data and ensuring trustworthy communications.

4.5.2 Secure Integration of SoC Accelerators

4.5.2.1 Secure SoC Architecture with Accelerator Integration Support

The next generations of mobile networks open up more and more features in sectors such as smart systems, communication devices, and autonomous driving. Besides increasing data rates and higher energy efficiencies, new platforms must address the growing security and privacy requirements as well. This poses two main challenges concerning the digital SoC, which have to be considered at the same time: First, there is the need for a holistic approach to trustworthiness including all layers from HW, operating system (OS), towards application software. Not only must the software be secure, but HW must be the basis to provide security to upper layers. Second, new 6G applications will demand a wide range of requirements related to performance and energy consumption. This requires a scalable solution ranging from small application-specific HW accelerators to highly programmable processing cores.

One approach to tackle the mentioned challenges is a HW/OS co-designed SoC platform, called M^3, that is built on a scalable tiled HW architecture with a network-on-chip (NoC) and supports a microkernel-based OS (Pauls et al. 2022) (cf., Figure 4.37). The M^3 HW architecture is designed to securely integrate untrusted heterogeneous processing cores and accelerators. Applications and software components with various performance requirements are mapped to individual tiles. Security on the HW level is achieved by a dedicated HW component called the trusted communication unit (TCU) (Haas and Asmussen 2022), which enables isolating all tiles, including untrusted components, so that no communication is possible by default. Allowed communication between tiles is facilitated through capabilities that enable access to shared memory or message passing features between applications. These capabilities are managed by the M^3 kernel and enforced by TCUs in HW. The general M^3 system architecture is available as open source (M3 2023), and a subset has successfully been implemented in a silicon research chip (Haas et al. 2024).

HW accelerators are primarily deployed to achieve very high throughputs for algorithms such as DSP in 6G modems. However, due to its efficiency, a HW accelerator has no or only limited programmability. Integrating an accelerator into the M^3 platform becomes a challenge since some degree of programmability is required to interface the TCU within a tile and, thus, to access shared memory or communicate with other tiles. The authors in Pauls et al. (2024) propose an approach to enable inter-tile communication of a HW accelerator by adding another component to the tile, called the accelerator support module (ASM). The ASM is a small software-programmable processing core and is therefore able to access the TCU and to support the accelerator for inter-tile

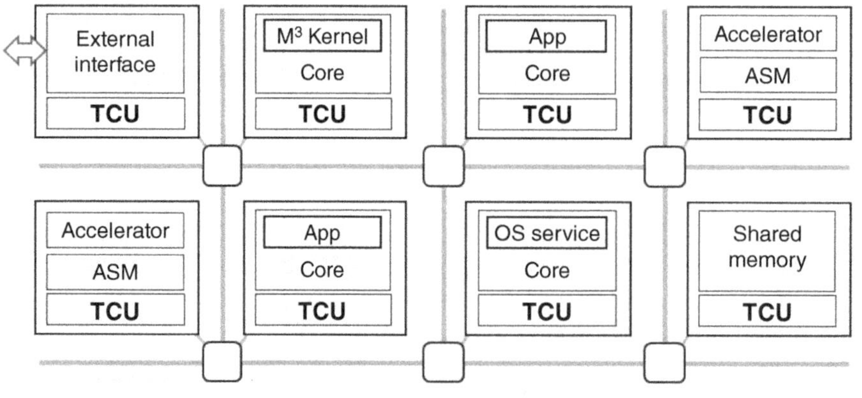

Figure 4.37 Example tiled architecture of the secure M^3 SoC platform.

communication. The ASM runs a software multiplexer, which is responsible for providing other tiles access to the accelerator and securely sharing the accelerator between different applications. The multiplexer supports arbitrary protocols such as stream processing.

Further research is necessary to investigate weaknesses of the M^3 approach, such as the impact of man-in-the-middle and side-channel attacks, as well as HW trojans (Das et al. 2024). Security features like trusted execution environments (Weinhold et al. 2023) and remote attestation will help to increase the isolation on the software level and to securely establish communication to remote components, which may span across single SoCs and devices.

4.5.2.2 AI and DSP Accelerator Capabilities

The development of efficient 6G devices requires flexible and scalable SoC architectures that integrate dedicated AI and DSP accelerators to balance performance, energy efficiency, and adaptability to evolving AI and signal processing workloads. The integration of AI and DSP accelerators into scalable SoC architectures is essential for meeting the diverse computational demands of 6G applications. However, beyond performance scalability, the impact of AI and DSP acceleration on the energy efficiency of PHY tasks must also be investigated.

AI accelerators, commonly referred to as neural processing units (NPUs), are designed to execute AI/ML algorithms efficiently by leveraging massive parallelism and optimized memory architectures. Their key design principles include data locality to reduce memory access overhead, efficient bandwidth utilization, and techniques to minimize memory conflicts and redundant computations in sparse data. To accommodate varying precision levels required for processing network layers in inference tasks, one approach is to exploit flexible processing elements (PEs) based on variable-precision bit-serial arithmetic. This approach enables dynamic precision adaptation to balance speed, power efficiency, and computational accuracy according to network and application requirements. The reference NPU design (Friedrich et al. 2024) incorporates a RISC-V core, a tightly coupled memory subsystem, and a 3D PE array optimized for parallel computations of AI workload. This architecture supports mixed-precision operations, layer fusion, and zero-skipping mechanisms to enhance efficiency. Additionally, its modular design allows for scalability across different performance requirements, ensuring adaptability for both edge and high-performance AI workloads. An AI accelerator prototype developed using GlobalFoundries' 22 nm technology achieves 6.144 TOPS (INT8 @1 GHz) with an efficiency of 0.65 TOPS/W at 8-bit precision, which would scale to approximately 2.37 TOPS/W with 7 nm technology.

On the DSP side, the computational demands of BB signal processing for real-time 6G applications must be evaluated. By modelling key DSP computation, storage, and interconnect SoC components targeting 22 nm technology as a use case, the impact of scalable signal processing on performance and energy consumption in future 6G devices is investigated. Energy consumption is analysed per MAC operation, enabling optimization across different clock frequencies and bit precisions. In addition, the study investigates the energy efficiency and performance of a fully SDR SoC architecture for 6G BB signal processing. Unlike HW-accelerated approaches, the proposed SoC relies entirely on RISC-V processors with vector extensions, providing maximum adaptability at the cost of increased energy consumption. A multi-tile architecture is considered, where tiles containing RISC-V vector cores communicate via a mesh NoC. The energy demands are analysed using the PhyBench BB benchmark, representing a 4×4 MIMO OFDM (orthogonal frequency division multiplexing) system across different bandwidths (10 MHz, 100 MHz, and 1 GHz) and a 1 ms slot duration. The study evaluates power consumption across cores, memory, and interconnects, revealing that increasing the number of cores per tile reduces inter-tile energy use while increasing intra-tile data transfer. The optimal configurations achieve minimum power consumption of

0.785 W at 10 MHz (2 tiles, 16 cores per tile), 7.929 W at 100 MHz (10 tiles, 32 cores per tile), and 89.328 W at 1 GHz (15 tiles, 64 cores per tile).

Vector processing is crucial for efficient signal processing tasks. However, the RISC-V architectures with vector extensions assumed in this study present limitations due to restricted memory and register file access, which do not fully support typical signal processing patterns or complex data types. Consequently, optimizing the microarchitecture for signal processing is essential. Strategies to enhance memory bandwidth utilization have been proposed in Razilov et al. (2023), while innovative methods for mitigating conflicts in the vector register file have been introduced in Razilov et al. (2024).

4.5.3 Energy Neutral Device Design

An energy neutral (EN) device is a category of IoT devices designed to operate autonomously, without the need for human intervention (Lopez 2025). EN devices harvest the energy from their surrounding environment, e.g. using solar, thermal, RF, and vibration, thus eliminating the need for a wired power supply or battery replacements, and in some cases, the need for batteries altogether. The design principle of EN devices focuses on the ultra-low power consumption throughout the life cycle, allowing them to operate effectively in scenarios with limited or intermittent power sources. The reduction in power consumption is achieved by using optimized HW and communication protocols involving sleep modes with minimalistic power requirements. The remainder of this section provides an overview of various HW and SW components of EN devices.

4.5.3.1 Energy Harvesting

Multi-source Energy Harvesting Circuit Ambient energy sources are often intermittent and unpredictable, which leads to reliability concerns in ambient IoT. In many cases, ambient IoT devices rely on a single energy source with low energy density, resulting in a limited energy budget. The concept of the internet of everything (IoE), where every ‘thing’ is connected, intelligent and EN, faces challenges due to these inherent limitations of ambient energy sources. However, it may be possible to overcome these limitations by introducing multisource harvesting architectures, where the system can harvest energy from many sources in its surroundings. Multisource harvesters allow the device to expand its energy budget as well as to ensure system reliability by providing a backup source when the other source is weak. For instance, solar radiation and soil-air thermal energy can be used in outdoor deployment – during sunny days, both sources are available, whereas at night or on cloudy days, the soil thermal energy can provide a low-energy backup to keep the system running (Puluckul and Weyn 2024).

Although harvesting energy from multiple sources seems ideal, it is not a trivial task. Multisource harvesting requires efficient power management integrated circuits (PMICs), which can seamlessly work with multiple sources and charge an energy buffer. A PMIC regulates the unstable energy output from transducers, such as solar cells, in EH systems. Typically, a PMIC is tailored for a specific transducer to match its I-V characteristics. This means that integrating multiple transducers (energy sources) into a design may require multiple PMICs, which is highly inefficient, increasing both cost and form factor. Additionally, when multiple PMICs are included, it would require separate energy storage for each PMIC. In order to use single energy storage, the PMICs need to be connected in parallel, through a diode OR circuit (Puluckul and Weyn 2024). However, this method is inefficient as it leads to uneven utilization of the sources, with the strongest source being harvested all the time. An alternative to this approach is to use a specialized circuit called an energy combiner (EC), which can multiplex energy from many sources into a single energy storage. These

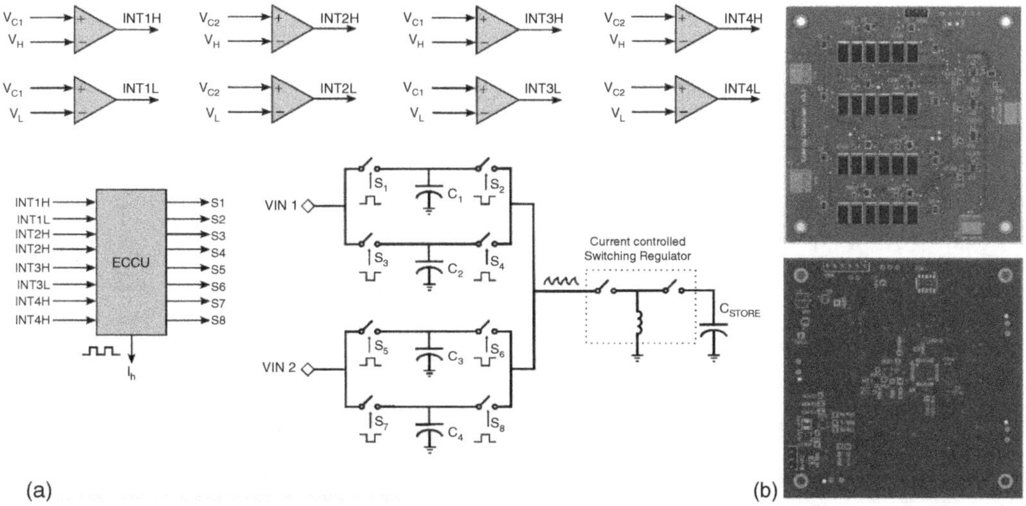

Figure 4.38 Architecture of the EC: (a) high-level schematic and (b) 3D render of the PCB showing both top and bottom sides. *Source:* Adapted from Puluckul and Weyn (2023); Lopez (2025).

ECs often use inductor-sharing architectures with multiple temporary buffers, enabling them to provide each source exclusive access to the storage without causing reverse flow or source underutilization (Puluckul and Weyn 2023). Figure 4.38(a) illustrates an EC design that uses multiple buffers of regular capacitors to store energy from each harvester temporarily and transfer this energy to a supercapacitor using a switching regulator (Lopez 2025). The proposed design was implemented in a two-layer PCB (cf., Figure 4.38(b)) for validation.

The ideal solution for multisource EH would be a single PMIC capable of accommodating various sources at the same time. Nevertheless, this can considerably increase the self-power consumption of the PMIC as it must include maximum power point tracking for all the sources. Moreover, seamlessly integrating multiple harvesters requires each PMIC input to be compatible with a wide range of voltages and currents, increasing HW complexity and energy consumption (Abuellil et al. 2019). Despite these challenges, a single PMIC solution remains ideal for many low-form-factor applications such as smart labels, hearables, and wearables.

Radio Frequency Wireless Power Transfer (RF-WPT) RF-WPT may be pivotal in supporting the widespread deployment and operation of energy-neutral IoT devices by providing a controllable and predictable energy supply (López et al. 2025). Unlike other WPT technologies, RF-WPT allows broadcasting energy over relatively long distances and charging multiple devices simultaneously, even in NLoS conditions. Moreover, RF EH circuit form factors and manufacturing costs allow seamless integration of this technology in existing devices and enable dual EH from both dedicated and ambient energy sources.

6G BS may be designed and configured to support WPT with predefined performance guarantees in addition to legacy and enhanced data transmission services. However, this might be only possible in highly dense network deployments since the system efficiency decreases exponentially with the charging distance. In other scenarios, the deployment of dedicated WPT nodes, usually referred to as power beacons (PBs), is needed. Although this entails higher upfront costs, overall network costs can be reduced compared with traditional IoT setups, especially as the number of IoT devices increases and considering that inaccurate HW power profiling, battery imperfections, and/or operating conditions can significantly and unexpectedly shorten battery lifespans (Rosabal et al. 2023). Qualitative results in Rosabal et al. (2023) indicate that traditional solutions relying on batteries are cost-effective when powering a few devices, while deploying multiple PBs to provide the wireless charging support becomes more appealing as the network grows.

The adoption of low-power/cost multi-antenna architectures at the PBs and lightweight and efficient dedicated protocols is still needed to increase charging efficiency and promote economic feasibility (López et al. 2023). Recent advancements in hybrid beamforming and reconfigurable architectures, such as DMA (cf., Section 4.4.1), RIS, and movable antennas, have demonstrated significant potential for scalable and energy-efficient WPT (López et al. 2024; Rosabal et al. 2025). Regarding RIS-assisted WPT, beyond-diagonal surfaces provide significant gains over traditional diagonal structures for NLoS channels (Azarbahram et al. 2025), increasing with the charging waveform's degrees of freedom, i.e. number of subcarriers. Finally, much more work is needed on efficient and cautious strategies and protocols to boost the end-to-end conversion efficiency, especially considering the complex interactions among WPT system components, as well as to transparently address safety constraints that ultimately hold the adoption of the technology (López et al. 2024).

4.5.3.2 Protocols for Active Energy Neutral Devices

Lightweight Protocols One significant simplification could be targeted at higher layers to make the communication link less strenuous by removing states, as opposed to the current protocol design, where the user is defined with various RRC states. Resultant, the devices do not require to maintain continuous RAN connectivity, especially in a dedicated manner, resulting in fewer handshakes between devices and the RAN node. This should be done in synchronization with improved lightweight non-access stratum (NAS) security between device and 6G Core Network, and improved lightweight user plane (UP) security between device and 6G RAN node or 6G UP function (UPF) tailored for devices' requirements.

The lightweight protocols (López et al. 2025) are aimed at devices requiring no stringent dedicated handling or connection, stringent quality-of-service (QoS) flow, i.e. no dedicated bearer provisioning, and device context in the network will be short-lived if applicable. This reduces handshakes, monitoring load, and signalling or channel overhead. At the RAN side, the transmission payload can be reduced by defining shorter identifiers, and since dedicated handling is disregarded, the transmissions can be made self-contained, inclusive of identifier and optionally data in the uplink (e.g. over contention resources) and DL (e.g. over paging resources). Additionally, once the transmission is performed and the device is identified, the network can allocate resources for optional subsequent data transmissions if needed and/or optional short-lived context (some signature or AS identifier) if the device is expected to be followed up later in time. Such non-dedicated handling of transmissions can induce collisions, and, thus, various transmission designs can be considered that assist in collision detection, optimizing re-access subject to the application use case/traffic scenario, cell load, device capability, and harvesting infrastructure. The design options can consider equipping a self-contained transmission payload with or without preamble and/or control information. The preamble utilization can offer advantages related to synchronization, assistance in collision detection, especially if preambles are orthogonal or near orthogonal. On the other hand, the control information provides flexibility by allowing selection of transmission encoding parameters subject to reliability needs. However, the downside with every bit of information included is increased transmission size, and thereby increasing complexity for low-power devices, e.g. driving up processing resources, EH needs, and storage requirements.

Further, up in the higher layer, an optimized UP solution (Ericsson 2025) can be targeted to leverage additional gains, which can be compounded with optimized RAN solution catering towards EN devices. The possible optimization can be done by enhancing the PDU session UP (i.e. individual N3 tunnel per UE) to a common UP (i.e. a common N3 tunnel shared by all devices) between RAN and UPF. Unlike UEs, there are no data radio bearers (DRBs) in RAN for the devices, and thus, it does not require per PDU session UPs, which map to the DRBs over the air. And considering the massive number of devices, UPF and RAN may run out of resources if an individual UP needs to be kept for each device. The advantage is that based on device feedback, the network will trigger UP information exchange over the pre-established common UP between RAN and UPF, which ensures the proper message routing for UL and DL data. The protocol can be built by modifying legacy CP and UP stacks. The stacks differ from legacy stacks, where lower layers are simplified according to the requirements of the device and network type. For instance, the lower layers may comprise of only PHY and media access control (MAC) layers. Herein, the functionalities related to RLC and PDCP are either compressed to MAC or removed. To enable optimized UP solution, a modified CP procedure can be adopted, which enables mapping of device ID to a common tunnel specific to RAN and UPF communication link for UP data transportation. Pertinent to this, an ID identifying a device as part of the general packet radio system tunnelling protocol user plane (GTP-U) extension header between RAN and UPF is introduced. If a payload comes from N6 point-to-point tunnel with user datagram protocol (UDP)/internet protocol (IP) encapsulation, then UPF decapsulates it, appends a GTP header with the device ID, and delivers it to the proper RAN. The RAN gets the device ID from the GTP header, delivers the payload to the device, and if the device successfully decodes the transmission containing the payload, then the device may send a response to the RAN, subject to the feedback policy. In the case of uplink, the RAN appends the GTP header with device ID and delivers it to the proper UPF. The UPF gets the device ID from the GTP header, encapsulates the payload, and delivers it to the proper N6 point-to-point tunnel. With modified stacks, the air interface is simplified, e.g. only those scheduling features are implemented which are energy efficient, e.g. scheduling policies in harmony with transmission design options based on self-contained transmission.

Energy-Aware Protocols Energy-aware protocols are crucial in a communication system where the devices operate using harvested energy. This is because energy resources may not be available or sufficient during an entire transmission or reception, particularly if they are ambient or natural resources. To allow for a reliable communication system, it is important that the communication protocol is designed such that both the UL transmission and the DL reception are designed to fulfil the target service requirements, but also to be adapted to the device's EH capability and energy storage size.

During a communication, the stored energy level of the EN device may drop below a certain level. Under such conditions, the device may need to stay in sleep mode or an EH state for a certain time before it can reach an energy level where it can perform transmission or reception. Therefore, when scheduling UL transmission operations, certain gaps may need to be inserted between UL transmission blocks based on the device harvesting capability. The transmission time may also need to be adjusted if the device has the capability to simultaneously perform data transmission and EH. For DL operations, duty-cycled low-power wake-up radios are typically used to save power at the Rx. The on duration of a duty-cycling is selected based on the device's energy availability to operate a low-power wake-up Rx listening for a potential wake-up signal. If a wake-up signal is detected, similar to UL scheduling, there may be a need to insert time gaps between DL reception blocks or extend the reception period for devices, which have the capability of simultaneous reception and EH.

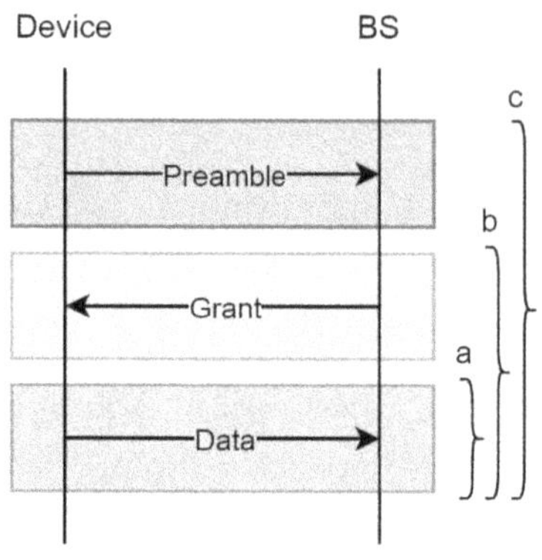

Figure 4.39 Signalling steps for the (a) grant-free, (b) fast uplink grant, and (c) random access schemes.

Low-Power Channel Access Next-generation cellular communication technologies will have to be aligned with the trend of massive deployment and increasing energy efficiency of the IoT, preferably using existing network infrastructure. Narrowband IoT (NB-IoT) stands out as a promising candidate due to its low cost and easy installation, because of its long-term evolution (LTE)-based architecture. However, the conventional uplink access procedure of NB-IoT limits its energy efficiency and predictability (El-Tanab and Hamouda 2021). To this end, the achievable connection density (i.e. the number of connected devices supported by the network with a certain quality of service) and energy efficiency of the conventional random-access scheme have been compared with novel grant-free and fast-uplink access techniques.

In the traditional random-access scheme, as shown in Figure 4.39, a device begins its uplink connection by sending a preamble to the BB. Once the BB receives the preamble, it schedules a time–frequency slot for the device and responds with a grant containing this information. The device then transmits its data within the allocated slot. In contrast, the grant-free access scheme aims to save energy by eliminating the preamble and grant steps, allowing the device to transmit data directly in a randomly chosen time–frequency slot. Meanwhile, the fast-uplink access technique enables the BB to predict which devices need to transmit data and when, allowing it to proactively send grants before transmission occurs (Mitsiou et al. 2023).

Through analytical methods, link-level simulation, and system-level simulations, the findings reveal that optimal random-access schemes can support three times as many devices as grant-free schemes and therefore perform the best in terms of connection density (Moons et al. 2024). However, when it comes to energy efficiency, random access performs the worst, with energy consumption up to 12 times higher than grant-free schemes. Fast-uplink access fulfils its expectations as the most promising technique for next-generation IoT communication, supporting a connection density twice as high as grant-free access, at a similar power consumption. It is, however, important to note that the performance of the fast-uplink access technique is highly dependent on the quality of the prediction mechanism.

4.5.3.3 Passive Energy Neutral Devices

Ambient backscatter communication (AmBC) is emerging as a promising solution for the next generation of IoT. This type of low-power wide area (LPWA) networks and devices was defined and studied by 3GPP (2022, 2024). Unlike conventional radios that generate and process their own carrier waves, backscatter devices (BDs) simply reflect and modulate existing ambient signals, such as those from TV broadcasts or cellular BBs (Liao et al. 2025; Phan-Huy et al. 2022) to convey information. This strategy significantly reduces HW complexity and power consumption, making

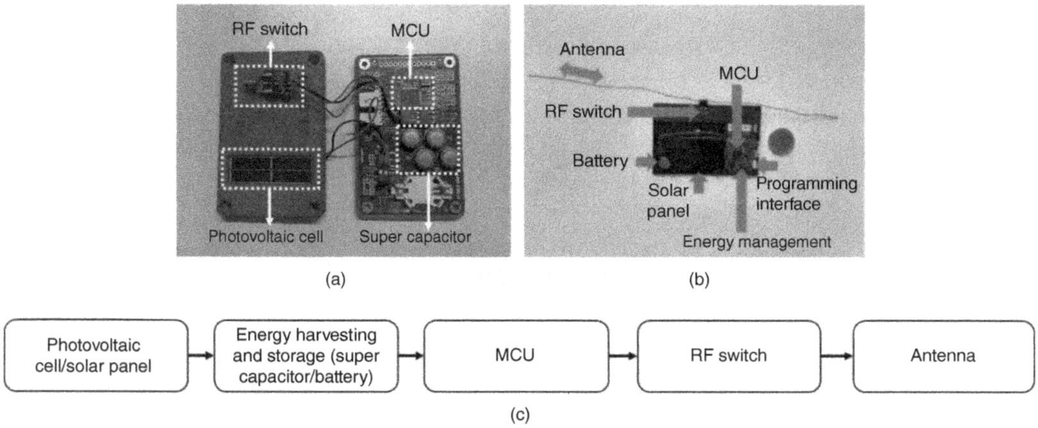

Figure 4.40 First (a) (*source:* Aalto University) and second (b) tested ambient IoT devices (*source:* Orange), and (c) block diagram view (*source:* Aalto University and Orange) with the below.

it an ideal candidate for EN devices that can operate by harvesting ambient energy. Unlike conventional backscatter (e.g. used by RFID), AmBC does not require a dedicated reader to transmit the carrier signal, but instead makes use of existing (ambient) RF waves for backscatter purposes.

Two prototypes that illustrate the practical implementation of this technology have been developed (cf., Figure 4.40). The first prototype (Liao et al. 2025) is designed around an ultra-low-power microcontroller unit (MCU) STM32L562, and an EH system. It utilizes a photovoltaic cell to harvest light energy, which is then managed by an ADP5091 energy harvester and stored in supercapacitors. The MCU operates primarily in a low-power sleep mode and only wakes to set timers for symbol generation. Binary frequency shift keying (BFSK) is implemented using pulse width modulation (PWM), and an RF switch ADG919, connected to the antenna modulates the backscattered signal. This design achieves extremely low average power consumption on the order of tens of microwatts, enabling the device to function using energy harvested even under normal indoor lighting conditions. The second prototype (Phan-Huy et al. 2022) further refines this approach using an MCU MSP430 to control all system components, including an RF switch BGS12WN6 connected to a dipole antenna. Here, energy is harvested from ambient light with a solar panel, stored in a 3V coin-cell battery, and managed by a dedicated energy management component BQ25570. Initial implementations employed FM modulation for backscatter, but later enhancements reprogrammed the device to support square-wave FSK modulation. Field studies have demonstrated that, under favourable conditions, this prototype can sustain periodic transmissions (e.g. 96-bit messages every 10 seconds) continuously day and night, highlighting the potential for indefinite energy autonomy.

The seamless integration of ambient BDs with cellular networks is one of the key advantages of this technology. The integration relies on leveraging existing signals in both the DL and UL channels, allowing coexistence with primary communications. In 4G LTE, the DL contains cell-specific reference signals (CRS) that are transmitted independently of user data (Liao et al. 2023; Phan-Huy et al. 2022). These CRS can serve a dual purpose: while they help UE in channel estimation for primary data, they also provide a reliable carrier for backscatter modulation. When a BD modulates the incoming carrier, its secondary data is superimposed onto the primary LTE waveform. Similarly, the uplink uses periodic sounding reference signals (SRS), which can be exploited for the reception of backscatter messages (Liao et al. 2025; Elsanhoury et al. 2025). 5G NR introduces more flexibility with configurable signals. Positioning reference signals (PRS) can be used to fulfil a similar role as

CRS for DL backscatter (3GPP 2020). PRS are broadcasted in a cell-wide manner, enabling them to serve as a robust carrier for backscatter modulation. On the UL, 5G NR offers enhanced configuration options for SRS, which can be tailored to optimize the reception of backscatter signals. For the future 6G network, the advanced waveform design and network architecture are expected to offer more versatile reference signal configurations, increased bandwidth, and tighter ISAC, potentially allowing ambient BDs to seamlessly integrate into a hyper-connected, energy-efficient IoT ecosystem.

4.6 Conclusions

The design of 6G radio systems reflects a shift toward a value-driven and holistic engineering approach, balancing technical excellence with sustainability and scalability. The transceiver, as the heart of the radio system, must support diverse application requirements across use cases ranging from high-capacity wireless access to ultra-low-power IoT. To enable this, 6G radio design integrates multiple enabling technologies, frequency bands, and architectural innovations into a modular and flexible framework.

Spectrum expansion, particularly into FR3 and sub-THz bands, introduces vast opportunities but also significant challenges. While FR3 provides a practical step forward for early 6G deployments with its favourable balance between coverage and capacity, sub-THz research remains in scope for extreme data rate and sensing applications in future dense scenarios. Progress in transceiver FE technologies, including photonic solutions, RTDs, and RIS, demonstrates promising paths to overcome the physical limitations of current electronics. Notable findings show that RTD-based FEs can sustain reliable D-band communication over 30 m using QPSK, with even greater throughput over shorter distances using higher-order modulation. Photonic-based transceivers have also demonstrated high gain and bandwidth, offering viable solutions for overcoming traditional RF limitations in sub-THz regimes.

Modelling remains a cornerstone of 6G radio design for reliable simulation, validation, and optimization. Sub-THz systems require refined propagation models that incorporate beamforming effects, directionality, and frequency-dependent behaviour. Measurements at 142 GHz confirm that narrow beam patterns significantly reduce delay spread – a factor critical for accurate system-level performance evaluation. At FR3, existing models such as 3GPP InF were found to overestimate path loss and Rician K-factors, whereas the InF-SL variant showed better agreement with real measurements, reinforcing the need for model refinement. Furthermore, RIS-specific macroscopic models were proposed to predict reradiated fields efficiently while accounting for parasitic effects.

Transceiver design is also evolving to accommodate new multi-antenna architectures. Advanced MIMO systems, including mMIMO, XL-MIMO, and D-MIMO, are essential for achieving spatial precision and capacity goals. Low-resolution (1–5 bit) ADC/DAC architectures were shown to provide near-optimal performance with significant reductions in energy consumption and HW complexity. Innovations such as OTA EM signal processing, RIS-assisted wavefront shaping, and tri-hybrid beamforming architectures open new pathways for compact, energy-efficient, and intelligent radio FEs.

Device-centric radio design is another key dimension, where the radio interface must be tailored to the characteristics and constraints of end devices. Secure, AI-enabled SoC platforms, such as the M^3 architecture, demonstrate how modular HW with embedded accelerators can support both high-performance computing and trusted operation. Energy-neutral IoT devices powered by multi-source EH or RF WPT were explored alongside passive communication solutions like ambient backscatter. Lightweight and energy-aware communication protocols – such as grant-free or predictive uplink access – were also shown to improve scalability and efficiency under strict energy constraints.

In summary, 6G radio design is not a monolithic blueprint but a flexible and integrated framework that brings together diverse technologies to meet a wide range of application and infrastructure demands. A unified design approach – encompassing simulation, prototyping, and adaptable functional splits across HW, PHY, MAC, and higher-layer protocols – enables scalable deployment and optimization across heterogeneous environments. This chapter has explored a selected set of innovations in 6G transceivers and radio design that highlight the direction of ongoing research, while representing only a portion of the broader landscape. Achieving the ambitious value-based goals of 6G will ultimately depend on the integration of diverse technologies, disciplines, and design principles. A truly holistic radio system must bring together expertise in HW, software, propagation modelling, signal processing, AI, and energy management. The convergence of these domains is essential to realize flexible, efficient, and resilient wireless platforms that meet the diverse and evolving demands of next-generation applications.

Acronyms and Abbreviations

I-V	I–V (current–voltage)
3D	Three-dimensional
3GPP	3rd generation partnership project
6G	Sixth generation
ADC	Analogue-to-digital converter
AM	Amplitude modulation
ANN	Artificial neural network
AOA	Angle-of-arrival
AP	Access point
ASM	Accelerator support module
BackCom	Backscatter communication
BB	Baseband
BER	Bit error rate
BFSK	Binary frequency shift keying
BS	Base station
BW	Bandwidth
CMOS	Complementary metal-oxide-semiconductor
CRS	Cell-specific reference signal
CSI	Channel state information
CU	Central unit
DAC	Digital-to-analogue converter
DC	Direct current
DL	Downlink
DMA	Dynamic metasurface antenna
D-MIMO	Distributed MIMO
DS	Delay spread
DSP	Digital signal processing
DTPA	Domino-tiled-phased array
DU	Distributed unit
EH	Energy harvesting

eLPWA	Enhanced low-power wide area
EM	Electromagnetic
ENoB	Effective number of bits
ESIT	Electromagnetic signal and information theory
ESPAR	Electrically steerable passive array radiator
EVM	Error vector magnitude
FR2	Frequency range 2
GMP	General memory polynomial
GTP-U	General packet radio system tunnelling protocol user plane
HW	Hardware
IAB	Integrated access and backhaul
IC	Integrated circuit
ID	Identifier
InF	Indoor factory to the list
IoE	Internet of everything
IoT	Internet of Things
IP	Internet protocol
ISAC	Integrated sensing and communication
I/Q	In-phase / quadrature
JCAS	Joint communication and sensing
KPI	Key performance indicator
LNA	Low-noise amplifier
LO	Local oscillator
LoS	Line-of-sight
LTE	Long-term evolution
LUT-PH	Look-up table-based parallel Hammerstein
MCU	Microcontroller unit
MIMO	Multiple input multiple output
MISO	Multiple input single output
mMTC	massive machine-type communication
mmWave	millimetre-wave
NAS	Non-Access Stratum
NB-IoT	Narrowband IoT
NCR	Network-controlled repeater
NLoS	Non-LoS
NoC	Network-on-chip
NPU	Neural processing unit
OFDM	Orthogonal frequency-division multiplexing
OS	Operating system
OTA	Over-the-air
PA	Power amplifier
PB	Power Beacon
PCA	Photoconductive antennas
PCB	Printed circuit board
PD	Photodiode
PDCP	Packet data convergence protocol
PE	Processing element
PEB	Position error bound
PHY	Physical layer
PIN	Positive-intrinsic-negative
PMIC	Power management integrated circuits

PN	Phase noise
PRS	Positioning reference signal
PTRS	Phase-tracking reference signal
PWM	Pulse width modulation
QoS	Quality-of-service
QPSK	Quadrature phase shift keying
RAN	Radio access node || Radio access network
RF	Radio frequency
RIS	Reconfigurable intelligent surfaces
RISC	Reduced instruction set computer
RLC	Radio link control
RRC	Radio resource control
RTD	Resonant tunnelling diode
RU	Radio unit
SCM	Self-conjugating metasurface
SINR	Signal to interference and noise ratio
SLL	Side lobe level
SoC	System-on-chip
SRS	Sounding reference signal
TCU	Trusted communication unit
TDD	Time-division duplex
THz	Terahertz
TinyML	Tiny machine learning
TOPS	Tera operations per second
TTPA	Tetromino-tiled-phased array
Tx	Transmitter
UDP	User datagram protocol
UE	User equipment
UP	User plane
UPF	UP function
UTC	Uni-travelling-carrier
WPT	Wireless power transfer

References

3GPP TR 38.901 V18.0.0 (2024-03) (2024). Study on channel model for frequencies from 0.5 to 100 GHz (Release 18).

3rd Generation Partnership Project (3GPP) (2020). NR; physical channels and modulation. 3GPP TS 38.211 V16.3.0, June.

3rd Generation Partnership Project (3GPP) (2022). Study on ambient IoT (internet of things) in RAN. 3GPP TR 38.848 V0.3.0, December.

3rd Generation Partnership Project (3GPP) (2024). Study on solutions for ambient IoT (internet of things). 3GPP TR 38.769 V0.1.0, August.

Abuellil, A., Estrada-López, J.J., Bommireddipalli, A. et al. (2019). Multiple-input harvesting power management unit with enhanced boosting scheme for IoT applications. *IEEE Transactions on Industrial Electronics* 67 (5): 3662–3672.

Al-Khalidi, A., Wang, J., and Wasige, E. (2018). Compact J-band oscillators with lm RF output power and over 110 GHz modulation bandwidth. *2018 43rd International Conference on Infrared, Millimeter, and Terahertz Waves (IRMMW-THz)*, Nagoya, pp. 1–2.

Aminu, M.U., Lehtomäki, J., and Juntti, M. (2019). Beamforming and transceiver optimization with phase noise for mmWave and THz bands. *2019 16th International Symposium on Wireless Communication Systems (ISWCS)*, Oulu, Finland, pp. 692–696. doi: https://doi.org/10.1109/ISWCS.2019.8877230.

An, J., Xu, C., Ng, D.W.K. et al. (2023). Stacked intelligent metasurfaces for efficient holographic MIMO communications in 6G. *IEEE Journal on Selected Areas in Communications* 41 (8): 2380–2396.

An, J., Yuen, C., Guan, Y.L. et al. (2024). Two-dimensional direction-of-arrival estimation using stacked intelligent metasurfaces. *IEEE Journal on Selected Areas in Communications* 42 (10): 2786–2802.

Arora, A., Tsinos, C.G., Rao, B.S.M. et al. (2021). Analogue beamforming with antenna selection for large-scale antenna arrays. *ICASSP 2021 – IEEE International Conference on Acoustics, Speech and Signal Processing (ICASSP)*, Toronto, ON, Canada, pp. 4795–4799. doi: https://doi.org/10.1109/ICASSP39728.2021.9414673.

Atzeni, I. and Tölli, A. (2021). Uplink data detection analysis of 1-bit quantized massive MIMO. *Proceedings of the IEEE International Workshop on Signal Processing Advances in Wireless Communications (SPAWC)*.

Atzeni, I. and Tölli, A. (2022). Channel estimation and data detection analysis of massive MIMO with 1-bit ADCs. *IEEE Transactions on Wireless Communications* 21 (6): 3850–3867.

Atzeni, I., Gouda, B., and Tölli, A. (2020). Distributed joint receiver design for uplink cell-free massive MIMO. *Proc. IEEE International Conference on Communications (ICC)*.

Atzeni, I., Tölli, A., and Durisi, G. (2021a). Low-resolution massive MIMO under hardware power consumption constraints. *Proc. Asilomar Conference on Signals, Systems, and Computers (ASILOMAR)*.

Atzeni, I., Gouda, B., and Tölli, A. (2021b). Distributed precoding design via over-the-air signaling for cell-free massive MIMO. *IEEE Transactions on Wireless Communications* 20 (2): 1201–1216.

Atzeni, I., Tölli, A., Nguyen, D.H.N., and Swindlehurst, A.L. (2023). Doubly 1-bit quantized massive MIMO. *Proc. Asilomar Conference on Signals, Systems, and Computers (ASILOMAR)*.

Avramopoulos, H (2024). TERAWAY project final report. Available at: https://cordis.europa.eu/project/id/871668/results (accessed January 2024).

Azarbahram, A., López, O., Clerckx, B., Di Renzo, M., and Latva-aho, M. (2025). Beyond diagonal reconfigurable intelligent surfaces for multi-carrier RF wireless power transfer. *Proceedings of IEEE Wireless Communications and Networking Conference (WCNC)*.

Balakrishnan, B. (ed.) (2023). Characteristics and classification of 6G device classes. HEXA-X-II deliverable D5.2.

Barbuto, M., Hamzavi-Zarghani, Z., Longhi, M. et al. (2021). Metasurfaces 3.0: a new paradigm for enabling smart electromagnetic environments. *IEEE Transactions on Antennas and Propagation* 70 (10): 8883–8897.

Basar, E., Di Renzo, M., De Rosny, J. et al. (2019). Wireless communications through reconfigurable intelligent surfaces. *IEEE Access* 7: 116753–116773.

Berardinelli, G., Adeogun, R., Coll-Perales, B. et al. (2023). Boosting short-range wireless communications in entities: the 6G-SHINE vision. *2023 IEEE Future Networks World Forum (FNWF)*, Baltimore, MD, USA, pp. 1–6. doi: https://doi.org/10.1109/FNWF58287.2023.10520553.

Björnson, E. and Sanguinetti, L. (2020). Making cell-free massive MIMO competitive with MMSE processing and centralized implementation. *IEEE Transactions on Wireless Communications* 19 (1): 77–90.

Björnson, E., Eldar, Y.C., Larsson, E.G. et al. (2023). Twenty-five years of signal processing advances for multiantenna communications: from theory to mainstream technology. *IEEE Signal Processing Magazine* 40 (4): 107–117.

Bozorgi, F., Sen, P., Barreto, A.N. et al. (2021). RF front-end challenges for joint communication and radar sensing. *Proceedings of the IEEE International Online Symposium on Joint Communications & Sensing (JC&S)*, Dresden, Germany, pp. 1–6. doi: https://doi.org/10.1109/JCS52304.2021.9376387.

Bucheli Garcia, J.C., Kamoun, M., and Sibille, A. (2020). Low-complexity adaptive spatial processing of ESPAR antenna systems. *IEEE Transactions on Wireless Communications* 19 (6): 3700–3711.

Carpintero, G. Headland, D., Schwanke, G. et al. (2024). Photonic-enabled terahertz phase arrays using dielectric rod waveguides for 6G wireless communications. *IEEE MTT-S International Microwave Symposium Digest*, pp. 74–77. doi: https://doi.org/10.1109/IMS40175.2024.10600234.

Celka, P., Bershad, N.J., and Vesin, J.-M. (2001). Stochastic gradient identification of polynomial Wiener systems: analysis and application. *IEEE Transactions on Signal Processing* 49 (2): 301–313.

Chen, J. et al. (2017). Does LO noise floor limit performance in multi-gigabit millimeter-wave communication? *IEEE Microwave and Wireless Components Letters* 27 (8): 769–771. https://doi.org/10.1109/LMWC.2017.2724853.

Choi, J., Mo, J., and Heath, R.W. (2016). Near maximum-likelihood detector and channel estimator for uplink multiuser massive MIMO systems with one-bit ADCs. *IEEE Transactions on Communications* 64 (5): 2005–2018.

R. D'Errico (Ed.) (2023). *RISE-6G Deliverable 3.4: Optimised RIS Prototypes for PoCs and Model Assessment Test.* https://rise-6g.eu/Documents/LIVRABLES/RISE-6G_WP3_D3.4_Rev.pdf.

Dahlman, E., Parkvall, S., and Skold, J. (2018). *5G NR: The Next Generation Wireless Access Technology*. Academic Press.

Dahlman, E., Parkvall, S., and Sköld, J. (2020). mmW LO generation and phase noise aspects. In: *5G NR – The Next Generation Wireless Access Technology*, 2e, Chapter 26, §26.2. ISBN: 9780128223208.

Dang, S., Amin, O., Shihada, B., and Alouini, M.-S. (2020). What should 6G be? *Nature Electronics* 3: 20–29.

Dardari, D. (2024). 3D electromagnetic signal processing. *Proc. Asilomar Conf. on Signals, Systems and Computers.*

Dardari, D. (2025). Frequency-selective dynamic scattering arrays for over-the-air EM processing. Submitted to URSI-B EMTS 2025, Bologna, arXiv, https://arxiv.org/abs/2502.07336 (accessed February 2025).

Dardari, D. and Massari, D. (2021). Using MetaPrisms for performance improvement in wireless communications. *IEEE Transactions on Wireless Communications* 20 (5): 3295–3307.

Dardari, D., Lotti, M., Decarli, N., and Pasolini, G. (2023). Establishing multi-user MIMO communications automatically using retrodirective arrays. *IEEE Open Journal of the Communications Society* 4: 1396–1416. https://doi.org/10.1109/OJCOMS.2023.3289326.

Dardari, D., Torcolacci, G., Pasolini, G., and Decarli, N. (2024a). An overview on over-the-air electromagnetic signal processing. [Online]. Available at: https://arxiv.org/abs/2412.14968 (accessed December 2024).

Dardari, D., Lotti, M., Decarli, N., and Pasolini, G. (2024b). Grant-free random access with backscattering self-conjugating metasurfaces. *IEEE Transactions on Cognitive Communications and Networking* 10 (5): 1620–1634.

Das, N., Pauls, F., Hasler, M. et al. (2024). Hardware attack models in tiled chip multi-core processors: a survey. *2024 IEEE 17th International Symposium on Embedded Multicore/Many-Core Systems-on-Chip (MCSoC)*, pp. 215–222. IEEE.

Degli-Esposti, V., Vitucci, E.M., Renzo, M.D., and Tretyakov, S.A. (2022). Reradiation and scattering from a reconfigurable intelligent surface: a general macroscopic model. *IEEE Transactions on Antennas and Propagation* 70 (10): 8691–8706.

Desset, C., Wambacq, P., Zhang, Y. et al. (2020). A flexible power model for mm-wave and THz high-throughput communication systems. *Proceedings of the IEEE International Symposium on Personal, Indoor and Mobile Radio Communications (PIMRC).*

Desset, C., Collaert, N., Sinha, S., and Gramegna, G. (2021). InP/CMOS co-integration for energy efficient sub-THz communication systems. *Proceedings of the IEEE Globecom Workshop on Emerging Topics in 6G Communications.*

Desset, C. et al. (2024). HEXA-X-II Deliverable 5.3 initial design and validation of technologies and architecture of 6G devices and infrastructure.

Deumer, M., Stiewe, O. et al. (2024a). Optoelectronic heterodyne THz receiver for 100-300 GHz communication links. *IEEE Access* 12 (January): 27158–27166. https://doi.org/10.1109/ACCESS.2024.3366448.

Deumer, M. et al. (2024b). Ultra-wideband PIN-PD THz emitter with > 5.5 THz bandwidth. *Journal of Infrared, Millimeter, and Terahertz Waves* 45 (9–10): 831–840. https://doi.org/10.1007/s10762-024-01001-z.

Deumer, M., Nellen, S., Berrios, S. et al. (2025). Advancing terahertz photomixers through the integration of photoconductive antennas with optical waveguides. *APL Photonics* 10 (3): https://doi.org/10.1063/5.0246801.

Di Renzo, M. and Migliore, M.D. (2024). Electromagnetic signal and information theory. *IEEE BITS the Information Theory Magazine* 1–13.

Di Renzo, M., Zappone, A., Debbah, M. et al. (2020). Smart radio environments empowered by reconfigurable intelligent surfaces: how it works, state of research, and the road ahead. *IEEE Journal on Selected Areas in Communications* 38 (11): 2450–2525.

Díaz Rubio, A. and Tretyakov, S.A. (2021). Macroscopic modeling of anomalously reflecting metasurfaces: angular response and far-field scattering. *IEEE Transactions on Antennas and Propagation* 69 (10): 6560–6571.

Ding, M., Atzeni, I., Tölli, A., and Swindlehurst, A.L. (2025). On optimal MMSE channel estimation for one-bit quantized MIMO systems. *IEEE Transactions on Signal Processing* 73: 617–632.

Droulias, S. and Alexiou, A. (2022). Reconfigurable intelligent surface: an angular spectrum representation approach. *56th Asilomar Conference on Signals, Systems, and Computers*, pp. 413–418.

Droulias, S., Stratidakis, G., and Alexiou, A. (2024). Near-field engineering in RIS-aided links: beamfocusing analytical performance assessment. *IEEE Access* 12: 29536–29546.

Droulias, S., Stratidakis, G., and Alexiou, A. (2025). Bending beams for 6G near-field communications. *IEEE Transactions on Wireless Communications* 24 (2): 1467–1480.

Elsanhoury, A., Galal, I., Alkady, K. et al. (2025). First real-time detection of ambient backscatters using uplink sounding reference signals of a commercial 4G smartphone. Submitted to IEEE Journal of Radio Frequency Identification, January.

El-Tanab, M. and Hamouda, W. (2021). An overview of uplink access techniques in machine-type communications. *IEEE Network* 35 (3): 246–251. https://doi.org/10.1109/MNET.011.2000513.

Ericsson (2025). KI#1, #2, #3, New Solution: UP Solution using AIOTF for Topology 1. 3GPP S2-2406133.

Ertuğrul, Y., Kapusuz, K.Y., Desset, C., and Pollin, S. (2024a). Domino-tiled phased arrays for beyond 100 GHz multi-user MIMO communication. *ICC 2024-IEEE International Conference on Communications*, June, pp. 3177–3182. IEEE.

Ertuğrul, Y., Kapusuz, K.Y., Desset, C., and Pollin, S. (2024b). Performance assessment of irregular array configurations for beyond 100-GHz multi-user MIMO systems. *2024 Joint European Conference on Networks and Communications & 6G Summit (EuCNC/6G Summit)*, pp. 564–569. IEEE.

ETSI GR (2025). ETSI GR THz 004: TeraHertz modeling (THz); RF hardware modeling.

Farhadi, H. and Sundberg, M. (2020). Machine learning empowered context-aware receiver for high-band transmission. *IEEE Globecom Workshops.*

Farhadi, H., Haraldson, J., and Sundberg, M. (2023). A deep learning receiver for non-linear transmitter. *IEEE Access* 11: 2796–2803.

Friedrich, S., Wittig, R., Matúš, E. et al. (2024). A 22 nm 10 TOPS mixed-precision neural network SoC for image processing with energy-efficient dilated convolution support. *Proceedings of IEEE Symposium on Low-Power and High-Speed Chips (COOLCHIPS 2024)*, Tokyo, Japan.

Gong, T., Gavriilidis, P., Ji, R. et al. (2024). Holographic MIMO communications: theoretical foundations, enabling technologies, and future directions. *IEEE Communications Surveys & Tutorials* 26 (1): 196–257.

Gonzlez-Prelcic, N., Keskin, M.F., Kaltiokallio, O. et al. (2024). The integrated sensing and communication revolution for 6G: vision, techniques, and applications. *Proceedings of the IEEE* 112 (7): 676–723.

Gouda, B. and Tölli, A. (2021). Distributed joint receiver design for cell-free massive MIMO with fast convergence. *Proc. IEEE International Conference on Communications (ICC).*

Gouda, B., Atzeni, I., and Tölli, A. (2021). Joint minimum DL-UL rate maximization for cell-free massive MIMO. *Proc. International ITG Workshop on Smart Antennas (WSA).*

Gouda, B., Atzeni, I., and Tölli, A. (2024a). Pilot-aided distributed multi-group multicast precoding design for cell-free massive MIMO. *IEEE Transactions on Wireless Communications* 23 (8): 9282–9298.

Gouda, B., Arvola, A., Atzeni, I., and Tölli, A. (2024b). Combined DL-UL distributed beamforming design for cell-free massive MIMO. *IEEE Wireless Communications Letters* 13 (6): 1621–1625.

Guo, H., Makki, B., Phan-Huy, D.T. et al. (2021). Predictor antenna: a technique to boost the performance of moving relays. *IEEE Communications Magazine* 59 (7): 80–86.

Guo, H. et al. (2025). Integrated communication, localization, and sensing in 6G D-MIMO networks. *IEEE Wireless Communications* 32 (2): 214–221. https://doi.org/10.1109/MWC.007.2400117.

Gupta, Y.D., Binet, G., Diels, W. et al. (2023). Implementation, modelling and verification of high-speed Mach-Zehnder phase modulators in an open access InP Foundry Platform. *Journal of Lightwave Technology* 41 (11): 3498–3504. https://doi.org/10.1109/JLT.2023.3244129.

Haas, S. and Asmussen, N. (2022). A trusted communication unit for secure tiled hardware architectures. *2022 29th IEEE International Conference on Electronics, Circuits and Systems (ICECS)*, pp. 1–4. IEEE.

Haas, S., Dunkel, C., Pauls, F. et al. (2024). Trustworthy silicon: an MPSoC for a secure operating system. *2024 IEEE Nordic Circuits and Systems Conference (NorCAS)*, pp. 1–7. IEEE.

Haider, M.F., You, F., He, S. et al. (2022). Predistortion-based linearization for 5G and beyond millimeter-wave transceiver systems: a comprehensive survey. *IEEE Communications Surveys & Tutorials* 24 (4): 2029–2072.

Harrington, R. (1978). Reactively controlled directive arrays. *IEEE Transactions on Antennas and Propagation* 26 (3): 390–395.

ITU-R (2023). Recommendation ITU-R M.2160-0: framework and overall objectives of the future development of IMT for 2030 and beyond.

Izumi, R., Suzuki, S., and Asada, M. (2017). 1.98 THz resonant-tunneling-diode oscillator with reduced conduction loss by thick antenna electrode. *42nd International Conference on Infrared, Millimeter, and Terahertz Waves (IRMMW-THz).*

Jacobsson, S., Durisi, G., Coldrey, M. et al. (2017a). Quantized precoding for massive MU-MIMO. *IEEE Transactions on Communications* 65 (11): 4670–4684.

Jacobsson, S., Durisi, G., Coldrey, M. et al. (2017b). Throughput analysis of massive MIMO uplink with low-resolution ADCs. *IEEE Transactions on Wireless Communications* 16 (6): 1304–1309.

Jedda, H., Mezghani, A., Swindlehurst, A.L., and Nossek, J. (2018). Quantized constant envelope precoding with PSK and QAM signaling. *IEEE Transactions on Wireless Communications* 17 (12): 8022–8034.

Jeon, Y.-S., Lee, N., Hong, S., and Heath, R.W. (2018). One-bit sphere decoding for uplink massive MIMO systems with one-bit ADCs. *IEEE Transactions on Wireless Communications* 17 (7): 4509–4521.

Kaleva, J., Tölli, A., Juntti, M. et al. (2018). Decentralized joint precoding with pilot-aided beamformer estimation. *IEEE Transactions on Signal Processing* 66 (9): 2330–2341.

D. E. Knuth (2000) Dancing links. arXiv preprint cs/0011047.

Leinonen, M.E. (ed.) (2021). HEXA-X-II Deliverable 2.2: initial radio models and analysis towards ultra-high data rate links in 6G.

Li, M. (2024). ETSI THz(24)000158: Sub-THz Transmitter_Power_Amplifier_Modeling.

Li, Y., Tao, C., Seco-Granados, G. et al. (2017). Channel estimation and performance analysis of one-bit massive MIMO systems. *IEEE Transactions on Signal Processing* 65 (15): 4075–4089.

Li, A., Masouros, C., Swindlehurst, A.L., and Yu, W. (2021). 1-bit massive MIMO transmission: embracing interference with symbol-level precoding. *IEEE Communications Magazine* 59 (5): 121–127. https://doi.org/10.1109/MCOM.001.2000601.

Liang, Y.-C., Zhang, Q., Wang, J. et al. (2022). Backscatter communication assisted by reconfigurable intelligent surfaces. *Proceedings of the IEEE* 110 (9): 1339–1357.

Liao, J., Wang, X., Ruttik, K. et al. (2023). In-band ambient FSK backscatter communications leveraging LTE cell-specific reference signals. *IEEE Journal of Radio Frequency Identification* 7: 267–277.

Liao, J., Zhang, T., Ruttik, K., Jäntti, R. & Phan-Huy, D.-T., (2025). Ambient backscatter communication in LTE uplink sounding reference signal. arXiv preprint, arXiv:2501.10952.

Liu, C., Ma, Q., Luo, Z.J. et al. (2022a). A programmable diffractive deep neural network based on a digital-coding metasurface array. *Nature Electronics* 5 (2): 113–122.

Liu, F., Cui, Y., Masouros, C. et al. (2022b). Integrated sensing and communications: toward dual-functional wireless networks for 6G and beyond. *IEEE Journal on Selected Areas in Communications* 40 (6): 1728–1767.

Lopez, O. (ed.) (2025). HEXA-X-II Deliverable D5.5: final design of enabling technologies for 6G devices and infrastructure.

López, O., Rosabal, O., Ruiz-Guirola, D. et al. (2023). Energy-sustainable IoT connectivity: vision, technological enablers, challenges, and future directions. *IEEE Open Journal of the Communications Society* 4: 2609–2666.

López, O., Rosabal, O., Azarbahram, A. et al. (2024). High-power and safe RF wireless charging: cautious deployment and operation. *IEEE Wireless Communications* 31 (6): 118–125. https://doi.org/10.1109/MWC.017.2300462.

López, O., Singh, R.K., Phan-Huy, D.T. et al. (2025). Zero-energy devices for 6G: technical enablers at a glance. *IEEE Internet of Things Magazine*.

Lozano, A. (2021). 1-bit MIMO for terahertz channels. *Proc. International ITG Workshop on Smart Antennas (WSA).*

M3 (2023). Open-source software and hardware. Available at: https://github.com/Barkhausen-Institut/M3, https://github.com/Barkhausen-Institut/M3-hardware (accessed March 2025).

Ma, M., Nguyen, N.T., Atzeni, I., and Juntti, M. (2023). Analysis of oversampling in uplink massive MIMO-OFDM with low-resolution ADCs. *Proc. IEEE International Workshop on Signal Processing Advances in Wireless Communications (SPAWC).*

Madapatha, C., Makki, B., Guo, H. and Svensson, T. (2025). Reconfigurable intelligent surfaces-assisted integrated access and backhaul. arXiv preprint arXiv:2502.12011

Mahmood, N.H., Atzeni, I, and Nimr, A. (eds.) (2025). Hexa-X-II deliverable D4.5 final results of 6G radio key enablers [Online]. Available at: https://zenodo.org/records/15773214 (accessed 29 Aug 2025).

Marzetta, T.L. (2010). Noncooperative cellular wireless with unlimited numbers of base station antennas. *IEEE Transactions on Wireless Communications* 9 (11): 3590–3600.

Meng, C., Wei, Z., Ma, D. et al. (2024). Multi-objective optimization-based transmit beamforming for multi-target and multi-user MIMO-ISAC systems. *IEEE Internet of Things Journal* 11 (18): 29260–29274. https://doi.org/10.1109/JIOT.2024.3413687.

Mitsiou, N.A., Tegos, S.A., Diamantoulakis, P.D. et al. (2023). *Proactive scheduling for zero-energy device networks with fast uplink grant. IEEE Wireless Communications Letters* 12 (7): 1184–1188. https://doi.org/10.1109/LWC.2023.3265993.

Mizmizi, M., Tagliaferri, D., and Spagnolini, U. (2024). Wireless communications with space-time modulated metasurfaces. *IEEE Journal on Selected Areas in Communications* 42 (6): 1534–1548.

Moons, L., Nasser, S., Sabovic, A., et al. (2024). Evaluating fast and grant-free uplink access in next-generation cellular IoT networks. *3rd International Conference on 6G Networking (6GNet).*

Morgan, D.R., Ma, Z., Kim, J. et al. (2006). A generalized memory polynomial model for digital predistortion of RF power amplifiers. *IEEE Transactions on Signal Processing* 54 (10): 3852–3860.

Murmann, B. (2024). ADC performance survey 1997–2024. [Online]. Available at: https://github.com/bmurmann/ADC-survey (accessed April 2024).

Nellen, S., Ishibashi, T., Deninger, A. et al. (2020). Experimental comparison of UTC- and PIN-photodiodes for continuous-wave terahertz generation. *Journal of Infrared, Millimeter, and Terahertz Waves* 41 (4): 343–354. https://doi.org/10.1007/s10762-019-00638-5.

Nellen, S., Qian, T., Schwanke, G. et al. (2022a). Photonic-enabled beam steering at 300 GHz using a photodiode-based antenna array and a polymer-based optical phased array. *Optics Express* 30 (25): 44701. https://doi.org/10.1364/oe.472507.

Nellen, S., Lauck, S., Peytavit, E. et al. (2022b). Coherent wireless link at 300 GHz with 160 Gbit/s enabled by a photonic transmitter. *Journal of Lightwave Technology* 40 (13): 4178–4185. https://doi.org/10.1109/JLT.2022.3160096.

Ngo, H.Q., Ashikhmin, A., Yang, H. et al. (2017). Cell-free massive MIMO versus small cells. *IEEE Transactions on Wireless Communications* 16 (3): 1834–1850.

Nimr, A. (ed.) (2023). Hexa-X Deliverable D2.3 Radio Models and Enabling Techniques Towards Ultra-High Data Rate Links and Capacity in 6G.

Nishigami, N., Nishida, Y., Diebold, S. et al. (2018). Resonant tunneling diode receiver for coherent terahertz wireless communication. *2018 Asia-Pacific Microwave Conference (APMC)*, Kyoto, pp. 726–728.

Nordsjo, A.E. and Zetterberg, L.H. (2001). Identification of certain time-varying nonlinear Wiener and Hammerstein systems. *IEEE Transactions on Signal Processing* 49 (3): 577–592.

Ohara, T., Ishibashi, T., Kawamoto, Y. et al. (2023). 2-mW-output power uni-traveling-carrier photodiodes in 300-GHz-band. *Asia-Pacific Microwave Conference Proceedings*, APMC, pp. 676–678. doi: https://doi.org/10.1109/APMC57107.2023.10439767.

Oshima, N., Hashimoto, K., Suzuki, S., and Asada, M. (2016). Wireless data transmission of 34 Gbit/s at a 500-GHz range using resonant tunnelling diode terahertz oscillator. *Electronics Letters* 52 (22): 1897–1898.

Oshima, N., Hashimoto, K., Suzuki, S., and Asada, M. (2017). Terahertz wireless data transmission with frequency and polarization division multiplexing using resonant-tunneling-diode oscillators. *IEEE Transactions on Terahertz Science and Technology* 7 (5): 593–598.

Parambath, A.V., Flordelis, J., Madapatha, C. et al. (2024). Integrating reconfigurable intelligent surfaces (RISs) into indoor D-MIMO networks for 6G. *2024 IEEE 100th Vehicular Technology Conference (VTC2024-Fall)*, Washington, DC, USA, pp. 1–6.

Pauls, F., Haas, S., Köpsell, S. et al. (2022). On trustworthy scalable hardware/software platform design. *Smart Systems Integration Conference and Exhibition (SSI).*

Pauls, F., Haas, S., and Verma, Y. (2024). Integration of IP-cores for the M3 architecture with low area overhead: accelerator support module. *2024 21st International SoC Design Conference (ISOCC)*, pp. 340–341. IEEE.

Phan-Huy, D.-T., Barthel, D., Ratajczak, P. et al. (2022). Ambient backscatter communications in mobile networks: crowd-detectable zero-energy-devices. *IEEE Journal of Radio Frequency Identification* 6: 660–670.

Puluckul, P.P. and Weyn, M. (2023, October). Infiniteen: a multi-source energy harvesting system with load monitoring module for batteryless internet of things. *2023 IEEE 9th World Forum on Internet of Things (WF-IoT)*, pp. 1–8. IEEE.

Puluckul, P.P. and Weyn, M. (2024). Harvesting energy from soil-air temperature differences for batteryless IoT devices: a case study. *IEEE Access* 12: 85306–85323.

Qi, Z., Hunukumbure, M., Nam, H., and Yoo H. (2018). On the phase tracking reference signal (PTRS) design for 5G new radio (NR). *2018 IEEE 88th Vehicular Technology Conference (VTC-Fall)*, pp. 1–5. doi: https://doi.org/10.1109/VTC-Fall.2018.8690852.

Radbord, A., Atzeni, I., and Tölli, A. (2024). Enhanced uplink data detection for massive MIMO with 1-bit ADCs: analysis and joint detection. [Online]. Available at: https://arxiv.org/pdf/2312.04183v2 (accessed September 2024).

Rao, S., Mezghani, A., and Swindlehurst, A.L. (2019). Channel estimation in one-bit massive MIMO systems: angular versus unstructured models. *IEEE Journal of Selected Topics in Signal Processing* 13 (5): 1017–1031.

Rasekh, M.E., Abdelghany, M., Madhow, U., and Rodwell, M. (2021). Phase noise in modular millimeter wave massive MIMO. *IEEE Transactions on Wireless Communications* 20 (10): 6522–6535. https://doi.org/10.1109/TWC.2021.3074911.

Ratajczak, P., Bennadji, R., Séguenot, E., and Phan-Huy, D.-T. (2024). Experimental demonstration of reflected beamforming and interference nulling at sub6GHz thanks to varactor-based reconfigurable intelligent surface. *Annales des Telecommunications* 79: 633–648. https://doi.org/10.1007/s12243-024-01037-1.

Razilov, V., Zhong, J., Matúš, E., and Fettweis, G. (2023). Dual vector load for improved pipelining in vector processors. *Proceedings of IEEE Symposium on Low-Power and High-Speed Chips (COOLCHIPS 2023)*, Tokyo, Japan.

Razilov, V., Gecin, I., Matúš, E., and Fettweis, G. (2024). Conflict management in vector register files. *ACM Transactions on Architecture and Code Optimization (TACO)* 1–19.

Rivera-Lavado, A., García-Muñoz, L.E., Generalov, A. et al. (2017). Design of a dielectric rod waveguide antenna array for millimeter waves. *Journal of Infrared, Millimeter, and Terahertz Waves* 38 (1): 33–46. https://doi.org/10.1007/s10762-016-0310-9.

Robins, W.P. (1984). *Phase Noise in Signal Sources: Theory and Applications*, vol. vol. 9. Edison, NJ, USA: IET.

M. Rodrigo Castellanos, S. Yang, C-B Chae, and R. W. Heath Jr, (2025) Embracing reconfigurable antennas in the tri-hybrid MIMO architecture for 6G, arXiv:2501.16610v1.

Roeloffzen, C.G.H., Hoekman, M., Klein, E.J. et al. (2018). Low-loss Si3N4 triplex optical waveguides: technology and applications overview. *IEEE Journal of Selected Topics in Quantum Electronics* 24 (4): https://doi.org/10.1109/JSTQE.2018.2793945.

Rosabal, O., López, O., Alves, H., and Latva-Aho, M. (2023). Sustainable RF wireless energy transfer for massive IoT: enablers and challenges. *IEEE Access* 11: 133979–133992.

Rosabal, O., López, O., di Renzo, M., Souza, R., and Alves, H. (2025). Movable antennas-enabled wireless energy transfer for the Internet of Things (submitted).

Roth, K., Pirzadeh, H., Swindlehurst, A.L., and Nossek, J. (2018). A comparison of hybrid beamforming and digital beamforming with low-resolution ADCs for multiple users and imperfect CSI. *IEEE Journal of Selected Topics in Signal Processing* 12 (3): 484–498.

Sadhu, B., Gu, X., and Valdes-Garcia, A. (2019). The more (antennas), the merrier: a survey of silicon-based mm-wave phased arrays using multi-IC scaling. *IEEE Microwave Magazine* 20 (12): 32–50.

Sarajlić, M., Tervo, N., Pärssinen, A. et al. (2023). Waveforms for sub-THz 6G: design guidelines. *2023 Joint European Conference on Networks and Communications & 6G Summit (EuCNC/6G Summit),* Gothenburg, Sweden, 2023, pp. 168–173. doi: https://doi.org/10.1109/EuCNC/6GSummit58263.2023.10188254.

Saxena, A.K., Fijalkow, I., and Swindlehurst, A.L. (2017). Analysis of one-bit quantized precoding for the multiuser massive MIMO downlink. *IEEE Transactions on Signal Processing* 65 (17): 4624–4634.

Stratidakis, G., Droulias, S., and Alexiou, A. (2021). Analytical performance assessment of beamforming efficiency in reconfigurable intelligent surface-aided links. *IEEE Access* 9: 115922–115931.

Stratidakis, G., Droulias, S., and Alexiou, A. (2024). Perceptive, resilient, and efficient networks: programming the wireless environment with reconfigurable intelligent surfaces. *IEEE Vehicular Technology Magazine* 19 (4): 36–43. https://doi.org/10.1109/MVT.2024.3432252.

Stratidakis, G., Droulias, S., and Alexiou, A. (2025). Near-field hierarchical localization for integrated sensing and communication. *IEEE Access* 13: 17766–17776.

Tervo, N., Leinonen, M.E., Singh, S.P. et al. (2023). Parametrization of simplified memoryless amplifier models at 300 GHz. *2023 IEEE 34th Annual International Symposium on Personal, Indoor and Mobile Radio Communications (PIMRC)*, Toronto, ON, Canada, pp. 1–6.

Vitucci, E. (ed.) (2024). 6G-SHINE Deliverable 2.3: radio propagation characteristics for in-X subnetworks.

Vitucci, E.M., Albani, M., Kodra, S. et al. (2024). An efficient ray-based modeling approach for scattering from reconfigurable intelligent surfaces. *IEEE Transactions on Antennas and Propagation* 72 (3): 2673–2685.

Wang, J., Al-Khalidi, A., Wang, L. et al. (2018). 15-Gb/s 50-cm wireless link using a high-power compact III–V 84-GHz transmitter. *IEEE Transactions on Microwave Theory and Techniques* 66 (11): 4698–4705.

Wang, H., Wang, F., Nguyen, H.T. et al. (2023). Power amplifiers performance survey 2000-present. [Online]. Available at: https://gems.ece.gatech.edu/PA_survey.html (accessed April 2023).

Wang, Z., Zhang, J., Du, H. et al. (2024). Extremely large-scale MIMO: fundamentals, challenges, solutions, and future directions. *IEEE Wireless Communications* 31 (3): 117–124. https://doi.org/10.1109/MWC.132.2200443.

Weigel, M., Kleinert, M., Kresse, M. et al. (2024). Design and fabrication of crossing-free waveguide routing networks using a multi-layer polymer-based photonic integration platform. *Journal of Lightwave Technology* 42 (5): 1511–1517. https://doi.org/10.1109/JLT.2023.3320908.

Weinhold, C., Asmussen, N., Göhringer, D., & Roitzsch, M. (2023). Towards modular trusted execution environments. *Proceedings of the 6th Workshop on System Software for Trusted Execution* (pp. 10–16).

Wong, K.-K., New, W.K., Hao, X. et al. (2023). Fluid antenna system – part I: preliminaries. *IEEE Communications Letters* 27 (8): 1919–1923.

Yan, S., Shi, W., and Wen, J. (2016). Review of neural network technique for modeling PA memory effect. *IEEE MTT-S International Conference on Numerical Electromagnetic and Multiphysics Modeling and Optimization (NEMO)*, 2016.

Ye, Q., Bursalioglu, O.Y., Papadopoulos, H.C. et al. (2016). User association and interference management in massive MIMO HetNets. *IEEE Transactions on Communications* 64 (5): 2049–2065.

Yu, Z., Zhu, Y., and Gajadharsing, J. (2024). DPD algorithm with long-term memory effects compensation for AlGaN/GaN HEMTs. *2024 IEEE Topical Conference on RF/Microwave Power Amplifiers for Radio and Wireless Applications (PAWR)*, San Antonio, TX, USA, pp. 51–54.

Zappone, A., Di Renzo, M., Shams, F. et al. (2021). Overhead-aware design of reconfigurable intelligent surfaces in smart radio environments. *IEEE Transactions on Wireless Communications* 20 (1): 126–141.

Zhang, L. et al. (2018). Space-time-coding digital metasurfaces. *Nature Communications* 9 (1): 4334.

Zhang, H., Shlezinger, N., Guidi, F. et al. (2023). 6G wireless communications: from far-field beam steering to near-field beam focusing. *IEEE Communications Magazine* 61 (4): 72–77.

Zhang, Y., Vaesen, K., Mangraviti, G. et al. (2024). A 56Gb/s zero-IF D-band transmitter for a beamformer in 22nm FD-SOI. *Proceedings of the IEEE Radio Frequency Integrated Circuits Symposium (RFIC)*, pp. 347–350.

Zhang, P., de Guzman, M.F., Lyu, Y. et al. (2025). Evaluating the beamforming impact on channel dispersion characterization using multiscenario sub-THz channel measurements. *IEEE Transactions on Antennas and Propagation* 73 (1): 469–481.

J. Zhu, Z. Wan, L. Dai, M. Debbah, and H. V. Poor (2022). Electromagnetic information theory: fundamentals, modeling. Applications, and open problems, arXiv e-prints, p. arXiv:2212.02882.

5

Architecture Enablers for 6G

Mårten Ericson[16], Ozgur Umut Akgul[1], Panagiotis Botsinis[4], Sameh Eldessoki[4], Nicolas Chuberre[2], Dorin Panaitopol[3], Joachim Sachs[5], Hasanin Harkous[6], Iman Hmedoush[7], Luis G. Uzeda Garcia[8], Pere Garau Burguera[9], Antonio de la Oliva[10], János Harmatos[11], Halina Tarasiuk[12,13], Marcin Ziółkowski[12,13], Karol Kuczyński[12,13], Alperen Gundogan[4], Milan Zivkovic[4], Diamanti Maria[14], Pilar Andres Maldonado[15], Stefan Wänstedt[16], Vignesh Raman[17], Bassem Arar[17], Riccardo Bassoli[17], Frank H.P. Fitzek[17], Vasileios Tsekenis[18], Sokratis Barmpounakis[18], and Antonio Varvara[19]

[1] *Nokia, Espoo, Finland*
[2] *Thales Alenia Space, Cannes, France*
[3] *Thales SIX GTS France, Gennevilliers, France*
[4] *Apple, Munich, Germany*
[5] *Ericsson, Bremen, Germany*
[6] *Nokia, Germany*
[7] *Nokia, Massy, France*
[8] *Nokia Networks, Massy, France*
[9] *Aalto University, Espoo, Finland*
[10] *University Carlos III of Madrid, Leganes, Madrid, Spain*
[11] *Ericsson Research, Budapest, Hungary*
[12] *Orange Innovation Poland, Warsaw, Poland*
[13] *Warsaw University of Technology, Warsaw, Poland*
[14] *Institute of Communication and Computer Systems (ICCS), Zografou, Athens, Greece*
[15] *Nokia, Aalborg, Denmark*
[16] *Ericsson Research, Luleå, Sweden*
[17] *Technische Universität Dresden, Dresden, Germany*
[18] *WINGS ICT Solutions, Athens, Greece*
[19] *TIM, Torino TO, Italy*

The design of the 6G architecture faces a multifaceted challenge. It must not only accommodate the ever-growing demand for existing mobile services (MS) like mobile broadband (MBB), machine-type communication (MTC) and ultra-reliable low-latency communications (URLLC) but also pave the way for entirely new 6G services and 6G use cases that were further delved into in Chapter 2. This expansion must be achieved sustainably, considering social, economic and environmental factors as explained in Chapter 3. Furthermore, 6G must play a crucial role in bridging the digital divide, ensuring equitable access to technology for all. Beyond traditional communication, 6G aims to build a data-driven architecture capable of supporting advanced applications (Apps) like artificial intelligence (AI), sensing, and compute offloading. This necessitates a flexible system that can integrate these new services and access mechanisms, while remaining economically viable. This flexibility requires modularity, simplicity, and cloud nativeness in the 6G systems.

6G to Build a Sustainable Future, First Edition. Edited by Mikko A. Uusitalo, Patrik Rugeland, Mauro R. Boldi, and Ahmad Nimr.

Traditionally, the mobile communications services have been user-to-user or user-to-server oriented, and therefore, the architecture is also developed for that. The typical example of this is a voice call, where one user initiates a call to another user within the network, leading to the setting up of bearers and/or IP addresses for this call. However, recently there has been a strong push to extend the mobile communication services to other types of services. One prominent example is the sensing service, which can be initiated by a user demand, such as the desire to 'see' around a corner in a city. However, unlike traditional services, the receiver (Rx) of a sensing request is more complex, involving multiple entities like base stations (BS) and devices. This necessitates a re-evaluation of the network architecture, as outlined in Section 5.1. For a more in-depth explanation of sensing, please refer to Chapter 7. Another significant service type is compute offloading, which also demands innovative architectural solutions involving the introduction of novel functional entities and procedures. Furthermore, the integration of AI, both within the network and AI as a service (AIaaS) for users and Apps, will have a profound impact on the 6G architecture. Similar to sensing, Section 5.1 deals with the architecture impact; for a more in-depth knowledge of AI, see Chapter 6.

To accommodate the dynamic and evolving demands of 6G, the network architecture must be highly flexible. This necessitates a modular design, as outlined in Section 5.2, where the network is composed of easily deployable modules that can adapt to changing needs. Modularization should be approached with an end-to-end (E2E) perspective, considering not only the granularity of the network functions (NFs) but also the required interfaces and deployment options. This allows for seamless integration of existing and new use cases and services. Emerging solutions combine the flexibility of cloud technology with distributed processing nodes, creating self-contained modules with minimal dependencies. This modular approach enables stepwise network expansion and scaling. Furthermore, simplifying the architecture is crucial, especially as mobile systems face increasing service and requirement complexity. This simplification can be achieved through streamlining operations and procedures, reusing existing infrastructure, and developing new services that deliver tangible value.

Achieving increased network flexibility in 6G also necessitates several architectural improvements, including the introduction of subnetwork solutions and simplified but improved multi-connectivity capabilities (see Section 5.3). Crucial to achieving digital inclusion and expanding network coverage is the seamless integration of non-terrestrial networks (NTN) with terrestrial networks (TNs), requiring a unified architecture that efficiently supports both (see Section 5.4). While flexibility is paramount, it must be balanced with network predictability. Thus, the 6G architecture must support the emerging use cases such as industrial automation, collaborative robotics, and connected vehicles, which require predictable service performance metrics such as bounded latency, low jitter, and ultra-high reliability (see Section 5.5).

To effectively meet the demands of 6G, a re-evaluation and optimization of the 6G radio interface and protocols is essential, see Section 5.6. This presents an opportunity to address shortcomings in the 5G air-interface and protocols, aiming for a simplified protocol design, enhanced cross-layer awareness, improved security, reduced signalling overhead, and streamlined mobility procedures. These improvements are crucial for ensuring the efficiency, reliability, and performance of the 6G network. Last, this chapter also includes the possible use of quantum to enhance the functionality of the 6G architecture, see Section 5.7.

5.1 Novel Services

Since mobile networks have universal coverage worldwide, there are several ideas to expand the use of mobile networks beyond communication and introduce new types of services, such as

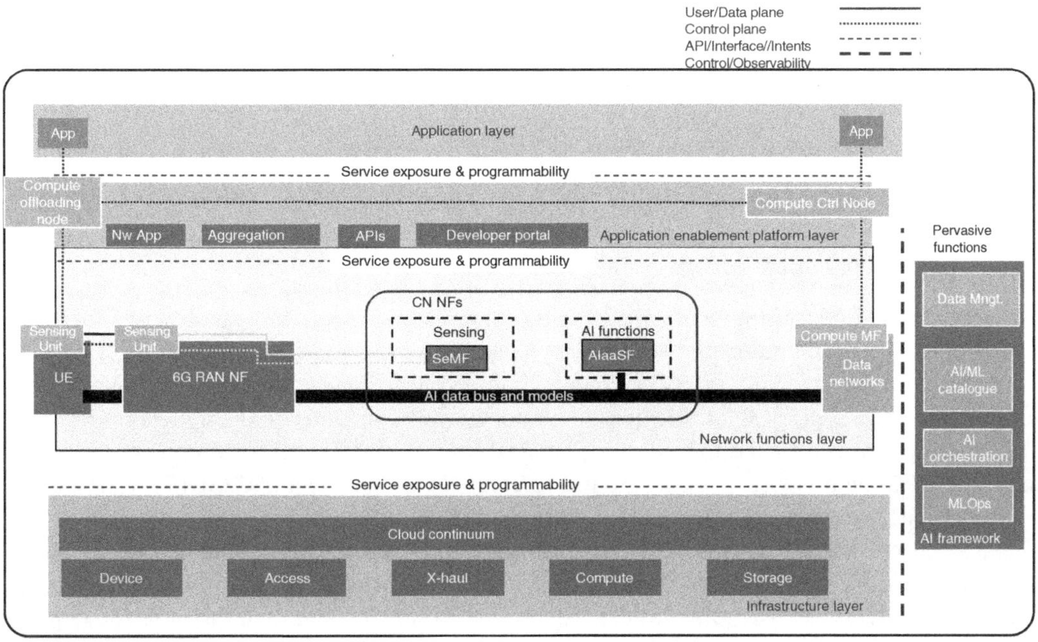

Figure 5.1 Novel services on the 6G E2E system blueprint.

sensing, compute offloading, and AIaaS. In addition to introducing the new services, this section also presents methods for how to place functionalities (where to execute the functions) within the network to meet the demands of the new services. Figure 5.1 shows an overview of the 6G E2E system blueprint from Chapter 3, including an overview of how the necessary sensing architecture functionalities can be added. With the already deployed radio, sensing can be used to detect and track any objects within the network coverage. Sensing will use the same hardware and radio resources as communication, allowing both the user equipment (UE) and the 6G radio access network (RAN) to function as sensing units (SUs), see Figure 5.1, which in turn is controlled by a sensing management function (SeMF). More architectural details of this can be found in Section 5.1.1, but the full details are given in Chapter 7.

Compute offloading is another novel service in 6G, where services in the UE are allowed to use the hardware in the network to perform calculations, either due to complexity or to save UE power. The compute offloading framework is depicted in Figure 5.1. At the UE, there is a compute intelligence that determines when to offload a task to the compute resources, which is done by communicating via the compute control node and the compute management function. Note that the compute management function can be located outside the network (i.e. in the data network in Figure 5.1), see Section 5.1.2 for more details. For most use cases, the idea is to keep the most private data at the UE; however, as more computational power is needed, using the hardware in the network closest to the device is the second best option.

The third novel service depicted in Figure 5.1 is the AI framework. Note that the AI functions are placed as pervasive functions in Figure 5.1 and are mainly handled by management and orchestration (see Chapter 6 for more details). In short, the management and operations (M&O) can, based on the needs from in-network or external use, initiate AIaaS functions, together with necessary machine learning (ML) functions and data collection. A short overview of the architecture needs is given in Section 5.1.3, while the full details are given in Chapter 6.

Finally, not explicitly shown in Figure 5.1, there is a service that is helping the new services and other services too, called App function (AF) placement, which is essential to achieve optimal performance in terms of, e.g. energy efficiency, latency, and security, see Section 5.1.4 for more details.

In the following sections, functions needed in the architecture to support such new services are introduced and described.

5.1.1 Sensing Functional Architecture

The main idea behind sensing in 6G networks is that radio resources can be repurposed for Apps beyond traditional communication services, i.e. integrated sensing and communication (ISAC). The network leverages the inherent resources and properties to detect, locate, and identify objects within its coverage area, even if those objects are not actively connected to the network itself.

The sensing architecture has to be designed taking into account a set of requirements, e.g. those from an 3GPP System aspects working group 1 (SA1) study on sensing for 5G (3GPP, 22.837 2024), and the various use cases presented in Almodóvar et al. (2023).

Figure 5.2 depicts a functional architecture for sensing. A sensing procedure is initiated by an App sending a sensing request over the control plane (CP) to a sensing request function. This function receives requests, makes sure that the requester is authorized and then forwards the request to the correct sensing control function (SCF) within the SeMF. The SeMF is responsible for efficient coordination of sensing resources and procedures, considering various aspects such as sensing requirements, sensing capabilities, and sensing constraints. The SeMF can be designed as a dedicated core network (CN) NF, since it is enabling new functionality for next-generation networks. The SeMF consists of the SCF and the sensing process function (SPF). The SCF indirectly configures the measurement nodes, both BSs and UEs. This process involves translating the sensing request to the required measurement nodes for the area of interest. Specifically, it determines which BS and UEs that should be involved to effectively cover the desired area. Additionally, it provides configurations to these identified nodes to enable the actual measurement. Note that when the UE is involved in sensing, the UE needs to be connected to the network via a serving BS. Even though the latter is drawn as a separate node in Figure 5.2, the serving BS could be one of the involved sensing nodes performing measurements, i.e. being either a transmitter (Tx) or a Rx BS as depicted in Figure 5.2, see Chapter 7 for more details. The system shall be able to authorize and configure a UE for sensing operation (e.g. based on location, time, device capability, and user consent) and for

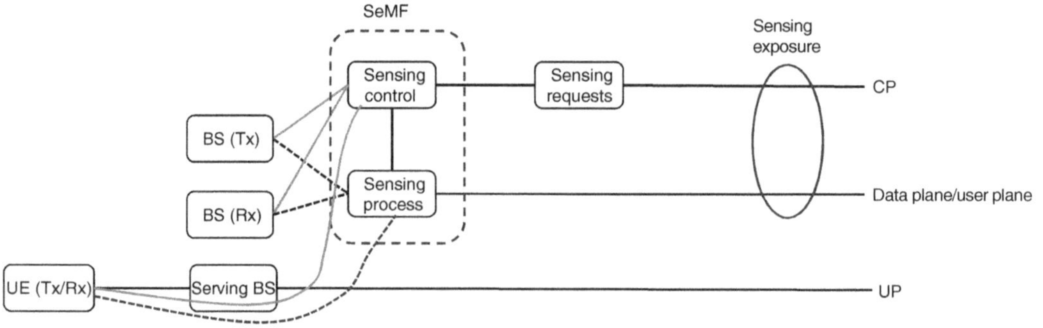

Figure 5.2 Functional architecture for sensing with UE involvement, i.e. UEs are involved either as transmitters or receivers of sensing signals. CP is the control plane; UP is the user plane. *Source:* Akgul et al. (2025) / Hexa-X-II.

establishing the communication connection needed to assist the sensing service. This means that the control function asks the scheduler in the BS for resources; however, the scheduler decides what resources are made available for sensing. The SCF also configures the SPF for the type of processing necessary.

The selected nodes transmit sensing signals according to respective configurations and scheduler, and the affected signal is received by configured Rx nodes. Next, the system shall be able to support means to enable entities of the RAN and UEs to transfer sensing measurement data to the SPF, which is responsible for processing and aggregating the data. After the processing of the sensing data has been performed, the results need to be exposed to the requested App or user. This can be done in a number of different ways. The user plane (UP) can be suitable for continuous streaming or large results directly to the App or user. Alternatively, for small sensing results, the data can be transferred via the CP using the network exposure function (NEF). An alternative to the UP is to utilize a new data plane, dedicated to exposing a large amount of data from several sources.

5.1.2 Compute Offloading

The emergence of new Apps, heterogeneous and computationally heavy use cases, and potentially new devices requires the tight integration of computing and communication in the 6G system. Power-constrained devices, or devices with limited computational resources, should be able to offload some computations to an external and more capable node. In order to make computational offloading appealing for the 6G user device, some requirements on latency, power consumption, offloaded data accuracy and privacy will be imposed; hence, the integration of computing and communication should take them into consideration.

Figure 5.3 shows a proposal for a possible compute offloading architecture. To minimize the overhead and also allow the possibility to offload to remote execution environments, the proposed framework is 'loosely' integrated into the cellular network. This approach is similar to that of 3GPP System aspects working group 6 (SA6) EDGEAPP (3GPP, 23.558 2024), as it requires no modifications to the RAN and non-access stratum (NAS) protocols. Consequently, it is less disruptive to existing cellular standards compared to other solutions.

The offloading node (ON) is responsible for handling the interaction with the UE App on which compute task(s) are to be offloaded to one or more compute nodes (CompNs). The ON is also responsible for interacting with the Compute Control Node (CCN) on behalf of the UE App, including discovery and registration procedures. The CCN is the corresponding node for the ON in the offload cluster. It receives the registration, authentication, and offloading requests from the ONs, and coordinates with the resource manager (i.e. compute resource management service (CRMS) and compute process management service (CPMS)) to find suitable CompNs, typically based on the App requirements expressed in the request by the ON. Note that the ON may coordinate directly with CompNs, bypassing CCN coordination, for the offloading of the computation tasks. The CompNs are wireless network nodes with certain processing capabilities to perform offloaded compute task(s) and produce compute result(s). The resource manager is responsible to keep track of the capabilities and availabilities of the allocated resources (i.e. CompNs) assigned to the offloading cluster. The proposed loose integration has the advantages that it requires no changes to RAN or NAS protocols and also allows access to remote execution environments. As a result, such loose integration is less disruptive with respect to existing standards and has no dependencies on UE chipset designs, opening up for faster time to market realization. On the other hand, it still requires richer interfaces and new procedures to interact with the network, for instance, through AF via NEF. In order to reduce the power consumption of the ON, certain

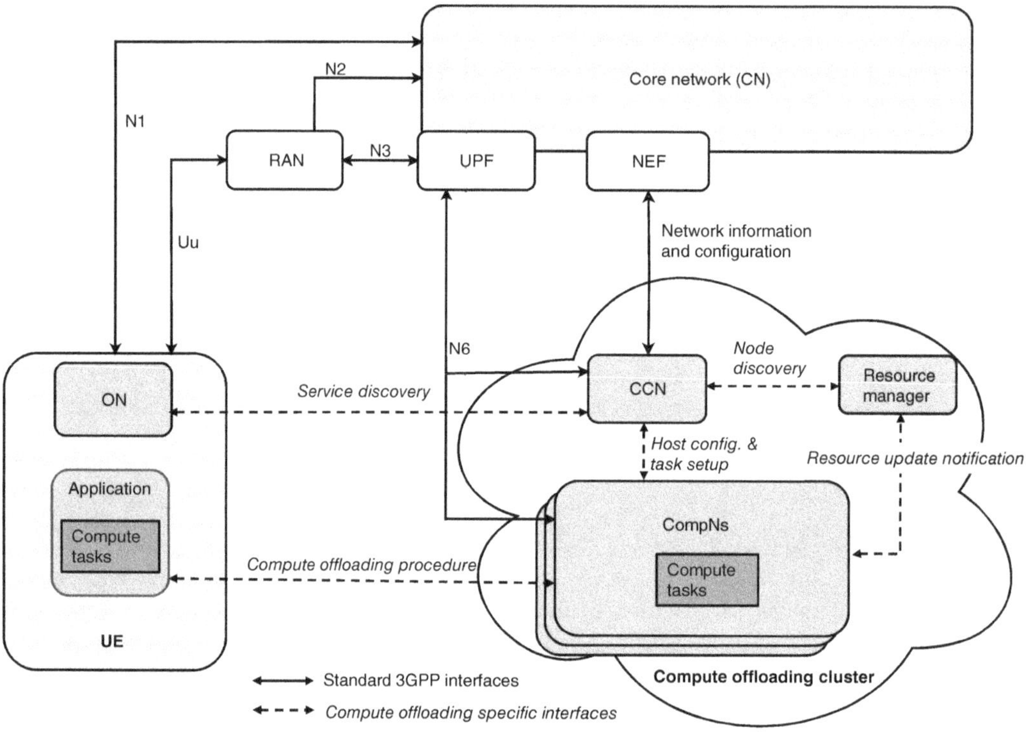

Figure 5.3 Compute offloading architecture framework.

compute offload deployments could also adopt novel RAN procedures, e.g. planned/configured downlink (DL) scheduling of compute results (Akgul et al. 2025).

Before the device, acting as an ON, offloads a computation, it would need to discover and select the candidate CompNs (Ericson et al. 2024) that are capable of performing the requested computation while satisfying the associated key performance indicators (KPIs). Then, each CompN should estimate the task(s) execution complexity and required resources (i.e. computation and storage) based on a common characterization of the offloaded compute workloads (i.e. compute tasks) determined by various requirements, such as:

- Traffic class (i.e. one-time versus multi-iteration, one-node versus multi-collaborative-nodes).
- Computation complexity (e.g. number of FLOPS and memory).
- Communication requirements (e.g. size of compute payload to be transferred).
- Precision requirement (e.g. quantization level of the compute data and operations).
- Quality of compute service classes (QoCS), such as latency-sensitive and precision-sensitive.

This characterization serves as the basis for the discovery, capability exchange and compute offloading coordination among the different nodes.

Furthermore, the computation offloading procedure can be foreseen as a staged procedure, as illustrated in Figure 5.4. After the 'Stage#0: Initial Registration Procedure', the CompN capabilities are identified in 'Stage#1: Node Discovery Phase 1', followed by the request for computation offloading in 'Stage#2: Node Discovery Phase 2', and, finally, the exchange of computing tasks and compute results in 'Stage#3: Computational Offload Procedure'. The different stages of computation offloading are associated with service layers that exist amongst the nodes to establish a process

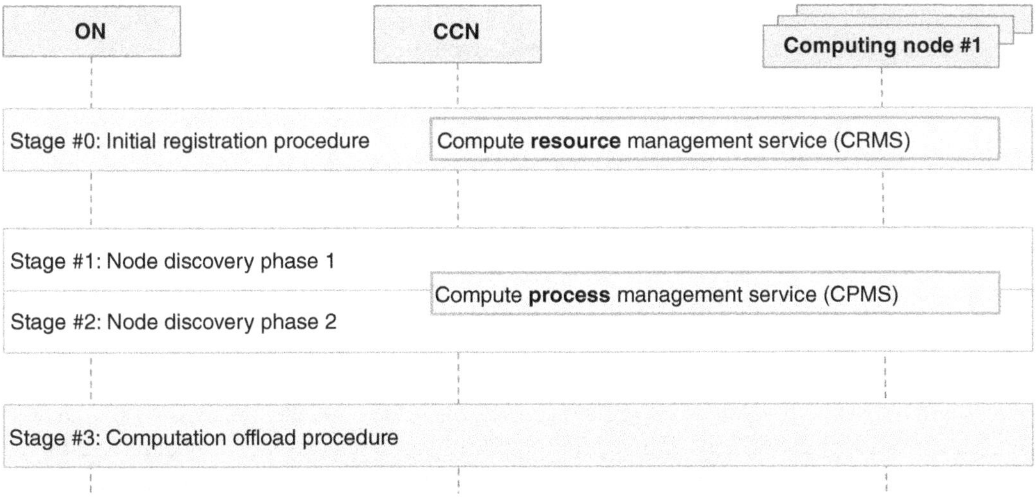

Figure 5.4 Distribute compute: general functional architecture and high-level flow procedure.

and provide functions to maintain synchronization amongst nodes. Two service layers for computational offloading are introduced. The CRMS layer is responsible for registration, assignment, and request of compute services and discovery of nodes. The CPMS layer regulates the process management (i.e. the discovery and allocation of compute resources) and the event management (i.e. maintaining the status of the requests).

5.1.3 AI as a Service

AIaaS is a key component of AI-driven 6G networks, helping automate tasks in Apps, network management, and resource management. By leveraging machine learning operations (MLOps), AIaaS provides managed AI/ML services – including pre-built models, on-demand training, inference, and lifecycle management – through standardized exposure application programming interfaces (APIs) such as Third-Generation Partnership Project (3GPP) NEF (3GPP, 29.522 2024), service enabler architecture layer (SEAL) (3GPP, 23.434 2024), and CAMARA (The Linux Foundation 2024). This allows Apps to consume AI functionalities without requiring dedicated infrastructure, enabling seamless AI integration across diverse network domains.

AIaaS includes key capabilities such as digital twins, transfer learning, and federated learning, which enhance AI/ML lifecycle management and enable intelligent network orchestration. These capabilities optimize model performance, facilitate distributed learning, and support adaptive AI-driven decision-making across network environments. Please see details in Chapter 6. Additionally, AIaaS introduces secure AI exposure mechanisms, allowing AI-driven Apps to interact with real-time network insights, improving automation, predictive analytics, and decision-making.

In an AI-native 6G architecture, AIaaS plays a crucial role by embedding intelligence across distributed data infrastructures, enabling zero-touch network management and real-time adaptation to dynamic service demands. By interacting with NWDAF, AI/ML Enablement Service (AIMLE), and Common API Framework (CAPIF), AIaaS facilitates efficient AI model storage, retrieval, training, deployment, and inference, supporting real-time data-driven decision-making across heterogeneous edge-cloud continuum environments. The AIaaS framework ensures that AI/ML models are efficiently managed and continuously optimized, while AI exposure APIs offer developers access to advanced AI capabilities for innovative App development.

A key innovation in AIaaS is the enablement of MLOps-as-a-service (MLOps-aaS), federated learning as a service (FLaaS), and federation as a service (FEDaaS), allowing collaborative AI model training and lifecycle management across multiple edge, cloud, and telecom environments. AIaaS integrates open-source tools such as Prefect (2025) (workflow orchestration), MLFlow (2025) (model management), KServe (2025) and Seldon (2025) (deployment), and OpenFL (2025) (federated learning) to support distributed AI operations at scale. Through AIaaS software prototypes, AI models can be dynamically trained, deployed, and exposed via secure and programmable APIs, ensuring AI-driven network optimization and service assurance.

Furthermore, AIaaS aligns with 3GPP SA6 (AIMLE) (3GPP, 23.482 2024) and 3GPP System aspects working group 5 (SA5) AI/ML management frameworks (3GPP, TR128908 2024), introducing standardized AI lifecycle management procedures. AIaaS includes standardized AI model management functionalities, such as discovery, registration, metadata-based exposure, and inference optimization. These capabilities ensure interoperability across CSPs and third-party AI Apps, enabling scalable and efficient AI deployments in 6G networks. Please see details in Chapter 6. AIaaS also integrates security and privacy-preserving AI mechanisms, addressing concerns related to data access, federated model training, and trustworthy AI deployment in 6G. With AI-native design principles, AIaaS will transform telecommunications networks into intelligent, adaptive platforms, optimizing performance, enhancing user experiences, and enabling new AI-powered Apps across industries.

5.1.4 Consumer Application Function Placement Optimization

The evolution of 6G networks extends beyond enhanced communication capabilities to encompass a wide array of novel 3GPP AFs. Effective placement of Apps within the network infrastructure is essential to achieve optimal performance, security, and energy efficiency. This section explores the mechanisms and strategies employed to strategically position AFs, leveraging the App enablement platform (AEP) described in Chapter 3 (Le Sauze and Kerboeuf 2025) and advanced placement algorithms to identify the most suitable nodes within the network.

At the core of the AF placement strategy lies the AEP, which serves as a vital bridge between NFs and App services. The AEP is responsible for exposing a comprehensive set of APIs that allow Apps to access vital network information and services. These APIs expose novel service capabilities, such as sensing data, compute resource availability, offloading capabilities, and AI models and services.

As explained in Chapter 3 (see Figure 3.4), the AEP includes several key functions. Below is a concise recap of those functions most relevant to App placement:

- **API exposure:** The AEP provides real-time access to essential network metrics, enabling Apps to retrieve up-to-date information on latency, bandwidth, and resource availability. This facilitates informed decision-making for App placement and resource allocation.
- **Data aggregation:** By collecting and processing data from the network functions layer (NFL), including services like compute-as-a-service and AIaaS, the AEP ensures that Apps receive a consolidated view of network performance and capabilities. This aggregation is crucial for maintaining a coherent and efficient data flow between the network and Apps.
- **Aggregation layer:** The AEP acts as an aggregation layer, exposing endpoints from the infrastructure to Apps, enabling them to consume tailored network data for operational needs.

Placing AFs optimally within a 6G network involves leveraging intelligent placement algorithms that consider various performance and resource metrics. These algorithms ensure that Apps are deployed on the most suitable nodes to meet their operational requirements. Key factors influencing placement decisions include:

- **Latency minimization:** Deploying Apps closer to data sources reduces E2E delay, which is crucial for real-time and latency-sensitive services.
- **Energy efficiency:** Selecting nodes that consume less power contributes to the overall sustainability of the network.
- **Data exposure minimization:** Strategically placing Apps to limit the distance data travels across the network enhances privacy and security.
- **Resource utilization optimization:** Efficiently leveraging existing computational and storage resources prevents bottlenecks and maintains balanced network loads.

To illustrate the placement process, consider an obstacle detection App that processes data from various sensors to identify and alert about potential obstacles in real time. Utilizing the AEP, the App accesses APIs that provide data from the NFL, including sensor inputs, current network latency, available compute resources, and energy consumption metrics from different nodes, among others. Then, the intelligent placement algorithm evaluates potential nodes – ranging from extreme edge to core nodes – based on their ability to handle the App's computational load, maintain low latency, and ensure minimal data exposure. Deploying the obstacle detection App at an edge node significantly reduces latency due to its proximity to the data sources, ensuring rapid processing and timely alerts. This placement is ideal for latency-sensitive tasks that require immediate responses. On the other hand, placing the App at a core node offers greater computational resources for more complex processing tasks, but with slightly higher latency. This option suits Apps that require intensive data processing but can tolerate a bit more delay.

Implementing strategic placement of AFs within a 6G network architecture involves addressing several key challenges to ensure successful deployment and operation. One major challenge is managing dynamic network conditions. 6G networks experience rapidly changing traffic patterns and fluctuating resource availability, requiring the placement algorithm to continuously monitor these variations and adapt App placements in real time to maintain optimal performance and reliability. Scalability is another critical factor, especially as the number of Apps increases. The placement strategy must efficiently handle growing demand without causing network congestion or degrading performance, necessitating robust algorithms capable of seamless scaling and infrastructure that supports expanding App needs.

Interoperability poses significant challenges when Apps interact with multiple NFs and services from different providers. Ensuring seamless communication and integration across diverse Apps and network components requires standardized protocols and interfaces, fostering a cohesive network ecosystem where various Apps can operate harmoniously. Additionally, security enhancements are very important in the placement strategy. While minimizing data exposure inherently boosts security by reducing potential attack surfaces, the placement strategy must also incorporate comprehensive security measures to protect against unauthorized access, data breaches, and cyberattacks. Implementing robust authentication, encryption, and continuous monitoring mechanisms is essential to safeguard sensitive information and maintain user trust. Addressing these challenges requires a holistic approach that combines advanced algorithms, real-time monitoring, and stringent security protocols, enabling network operators to optimize App placement effectively while ensuring resilience and security within the 6G network.

5.2 6G Cloud-Native Architecture

The service-based architecture (SBA) of 5G CN aimed to be cloud-friendly and flexible, enabling service reuse. While the SBA achieved extensibility with many standardized NFs, actual NF reuse has been lower than anticipated. This has led to increased integration costs and signalling overhead due to the need to deploy and operate numerous NFs. As discussed in Chapter 3, a key design principle of 6G is the 'separation of concerns for network functions', which entails a transition toward more self-contained modules and a reduction in interdependencies among various functions. This may involve bundling NFs, reducing the need for new NFs for every new functionality. This approach aims to decrease signalling overhead, improve performance, and simplify management by introducing a single point of contact for new procedures, while maintaining backwards compatibility with 5G. Key considerations for this bundling approach include the design of the modules and the optimal granularity of network functionality allocated to each module, as discussed in Section 5.1.2. Furthermore, the interactions between these modules, depending on their deployed location (e.g. core, RAN, or edge), and their required interfaces are crucial (Section 5.1.3). Finally, user behaviour, including location and traffic generation, varies throughout the day, with events like sporting matches potentially causing significant local traffic surges. Optimizing module placement is crucial for minimizing network latency, signalling traffic, and costs. Novel cloud functions and multi-cloud federation support module placement in cloud continuum, as detailed in Section 5.1.4.

5.2.1 Modular Network Architecture for 6G

Although SBA has many properties for modular design, there is always room for improvement. The following principles are identified to be important aspects to consider when 6G CN NFs are defined:

Hide internal procedures: Hide functions not needed for external communication (i.e. make it more self-contained). This simplifies the standard and allows room for vendor innovation.

Separation of concerns: This is achieved by avoiding splitting responsibility for a certain objective of a procedure across different modules (or services), e.g. so that large amounts of information can be contained in the module rather than being signalled several times to different NFs. Consequently, when possible, merge functions that have tight connections. The motivation for this principle is to reduce signalling between NFs, thus improving manageability and possible evolution.

Information only by request: This principle states that the NF interested in information requests the information. With this principle, the network avoids providing unsolicited information that is not needed. Note that there may be cases where this principle might impact performance, in which case it has to be weighed whether to give more importance to this principle or to performance.

Avoid unnecessary functional proxies: The intention of this principle is that all modules can communicate directly with other modules, without a proxy between them. With this principle, long signalling chains can be avoided. Further, proxying in App logic is avoided, which leads to less coupling between NFs. However, there is (at least) one case where the benefits of using a proxy outweigh this principle, namely, the access and mobility management function (AMF) between the RAN/CN and the UE and other NFs, e.g. AMF guarantees that there is a single NF in charge of handling UE mobility and UE reachability (instead of duplicating such functionalities to many CN NFs).

Smaller transactions: If a procedure requires that more than two modules need to communicate, aim to redesign the modules to reduce the need for inter-module communication.

A coarse granular design of 6G NFs or modules reduces the number of modules involved in executing certain system procedures. This results in reducing the volume of signalling needed between NFs to execute the procedures, shortening the time needed to execute the procedures because fewer NFs are involved in the execution path, as well as simplifying the management of the consistency of the state of NFs involved in procedure execution. However, this coarse granular design of modules or NFs reduces the flexibility in deploying and placing these coarser-grained NFs in different physical locations, increases the overhead in terms of isolating faults when they occur, and limits the efficacy of scaling operations compared to more fine-grained modules that can be scaled independently to provide a granular level of control over resource utilization.

For example, the CP of the 5G core (5GC) system is designed as a group of NFs that interact with each other based on defined procedures to deliver a certain functionality, such as registering the UE to the system or establishing a protocol data unit (PDU) session. A new design for the core system CP is proposed in Goshi et al. (2023), where each system procedure is consolidated into a single NF. For example, the UE registration procedure is designed and implemented as a single procedure-based NF where the execution logic, used to span multiple NFs (AMF, Authentication Server Function, Policy Control Function, Network Repository Function, Unified Data Management, and Unified Data Repository) in 5G, is consolidated into a single UE Registration NF. This approach leverages the advantages of coarse-granular design of modules, such as shorter execution time and less signalling overhead, while also paying the cost in terms of reduced flexibility in module placement and limited independence of scaling operations as described earlier.

6G systems will require new networking and air interface solutions to support the increasingly stringent requirements in terms of providing ubiquitous, reliable connectivity (Jiang et al. 2021). Combining joint processing and macro diversity from area-distributed access points, user-centric distributed multiple-input multiple-output (D-MIMO) has been proposed as a radio interface technology with the aim of providing ubiquitous high data rates, solving one of the main drawbacks of current cellular systems. This solution requires a certain level of centralization of the baseband processing, which makes monolithic architectures unscalable and incapable of offering the promising performance gains of D-MIMO. In a modularized RAN, functionalities are disaggregated into different entities. Considering a low-layer split (LLS), where most baseband functions are performed by a central processing unit, whereas the low physical layer functions lie on the radio units (RUs), is crucial for deploying a scalable and cost-effective user-centric D-MIMO network. Open radio access network (O-RAN) has evaluated the different functional LSS options between RAN entities, consisting of RUs, distributed units, and centralized units (called O-RU, O-DU, and O-CU in O-RAN specifications, respectively). Focusing on the Radio Unit (RU) and the Distributed Unit (DU) split, O-RAN proposes an LLS, referred to as 7.2x (O-RAN 2025). Such a split comes as a balance between RU simplicity and the latency and data rate requirements on the RU-DU interface (or RU to RAN as in Figure 5.8). The RUs perform radio frequency (RF) operations, as well as Fast Fourier transform (FFT) and cyclic prefix addition/removal, while the DUs implement the rest of the physical layer and the medium access control (MAC) and radio link control (RLC) layers, while the CU manages the packet data convergence protocol (PDCP) and higher layers.

Modular design also brings challenges not only in the module design but the feasibility of their App in the 6G system itself. As alternative deployment models of core NFs are considered, both via designing and creating new NFs (such as data function (DataF) and AIaaSF), and via evaluating a disaggregated CN (both logically and geographically, such as Cloud Continuum), it becomes more apparent that these enablers should not occur at the cost of reliability of the network. Moreover, with the emergence of new use cases of 6G networks as described in Chapter 2 (such as from robots to cobots, massive twinning, immersive telepresence for enhanced interactions), reliability and high uptime remain paramount. The main challenges for ensuring a high level of reliability in cloud-native environments that allow for the implementation of modular system design are:

- definition of a set of metrics that will allow for the prediction of failures in such systems,
- selection of appropriate monitoring/observability tools that will not negatively affect the system performance,
- adequately short data processing time that will allow for an early response to failures, and
- selection of appropriate prediction algorithms.

For example, in Akgul et al. (2025), 5GC failure prediction experimental results are presented. Based on the obtained results, key takeaways for the implementation of the failure prediction use case in the 5GC were formulated:

- DataOps enabler offers insights into the state of the core NFs, supplying ML models with data necessary for failure prediction.
- To minimize the time for data processing and the ML algorithms execution with preprocessing, the critical attributes for failure prediction should be selected. Then, only for these attributes should failure prediction be performed.

5.2.2 Inter-module Interactions and Interfaces

The distributed nature of the modules brings out the question of how to interact between different modules at different locations, e.g. modules in CN interacting with modules in RAN or modules between edge nodes. The interactions between RAN and CN continue to be a rather critical aspect of 6G systems. In 5G, the RAN and CN communicate through the N2 interface, with all interactions routed through the AMF. This means any CN function accessing the RAN, or vice versa, must pass through the AMF. The N2 interface supports a range of procedures, including PDU session management, UE context management, UE mobility management, paging, transport of NAS messages, and interface management. The N2 interface utilizes the NG application protocol (NGAP), where NGAP packets are encapsulated by internet protocol (IP) and stream control transmission protocol (SCTP), general-purpose protocols found in other specifications beyond 3GPP. The key requirements for RAN-CN interactions and interfaces are as follows. RAN-CN CP interaction for 6G shall adhere to the 6G system architecture design principles (see Chapter 3) and enable the support of envisioned new services and their requirements. The variety of deployment options brings the need for these interactions and interfaces to support the various RAN deployments and architecture while being efficient in terms of latency, payload size, reliability, availability, and resilience. Moreover, from the angle of economic sustainability, these interactions should neither hinder the deployment and uptake of 6G SA nor break the support of the baseline connectivity services and their requirements. Based on these requirements, Akgul et al. (2025) analyse and conclude that the RAN-CN interactions/interfaces presented in Figure 5.5 are the most feasible one.

Although some limitations have been identified for this RAN-CN interaction (Ericson et al. 2024), there are several key advantages. One important aspect is that since the communication between RAN and CN goes via a proxy (the AMF), RAN and CN are isolated from each other. Staying with the point-to-point RAN–CN interface in option A does not prevent possible protocol enhancements on that interface, e.g. NGAP could be enhanced and/or the underlying SCTP could be enhanced or replaced, thereby addressing some of the 'limitations' and improving the overall performance on that interface. One additional benefit is that during a change of gNB, the active/idle state change needs to be communicated to and known/managed by only a single NF, i.e. the AMF. In addition,

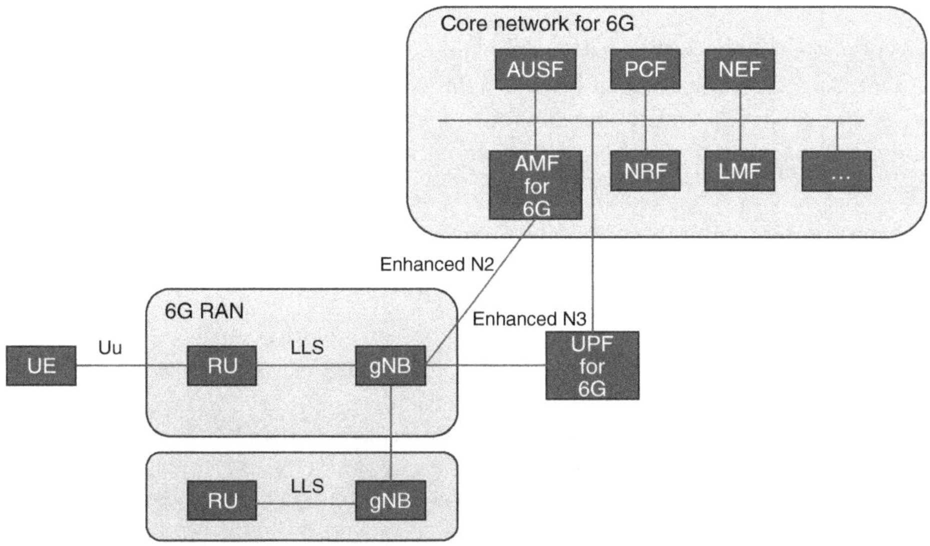

Figure 5.5 RAN-CN CP option for 6G, i.e. via enhancing the NGAP.

the UE only needs to establish and maintain one NAS association (including security). Furthermore, the current solution inherently supports (both 5G and) 6G. Overall, despite its issues, the current N2 solutions have many positive characteristics and are a well-proven multi-vendor interface between RAN and CN.

In addition, as 5G networks evolve beyond the centralized cloud-native architecture of 5G SBA, they are expected to become fully distributed across the entire compute continuum. This shift requires new service routing capabilities for optimal resource utilization. In Ericson et al. (2024), to address this challenge, two key concepts are introduced: data-centric networking (DCN) and data flow programming (DFP). DCN shifts away from traditional IP-style addressing and focuses on routing based on App-specific data, making it more suited for 6G's edge-native network requirements. When combined with DFP, which models Apps as directed graphs with operators and First-in First-out data streams, it enables the creation of serverless, dynamic, and highly granular NFs. This approach supports a flexible and scalable architecture, improving efficiency, resiliency, performance, and security across distributed 6G systems (Baldoni et al. 2024).

5.2.3 Integration of Extreme Edge

Traditional multi-access edge computing (MEC) architectures (Multi-access Edge Computing 2024), designed to offload computation from user devices to edge servers, face challenges in supporting ultra-low latency Apps, maintaining service continuity in mobile environments, and efficiently utilizing constrained resources at the extreme edge. To address these limitations, constrained MEC (cMEC) extends the MEC paradigm by enabling a lightweight, decentralized computing layer within user devices and resource-constrained nodes, such as mobile terminals, drones, or in-vehicle systems (Rojas and Guimarães 2024). This approach integrates the extreme edge into the cloud continuum, distributing computing and storage across multiple infrastructure layers dynamically.

The cMEC architecture introduces a four-layer model that expands the traditional MEC framework. While standard MEC (referred to as telco MEC (tMEC)) resides within telecom infrastructure, cMEC operates within constrained end-user devices or nearby mobile computing units, executing a subset of MEC functionalities. Unlike tMEC, which depends on centralized orchestration for resource allocation, cMEC enables local processing, reducing reliance on remote infrastructure and ensuring service availability even in disconnected environments.

To facilitate seamless interaction between cMEC and tMEC, the architecture introduces new interconnection mechanisms and workflow modifications to standard MEC operations. These include:

- **Lightweight functionalities:** cMEC nodes only support a subset of MEC functions due to their limited computational capacity. For instance, high-level orchestration remains in tMEC, while cMEC manages local execution and data processing.
- **Inter-layer coordination:** Instead of full federation models requiring business agreements, cMEC relies on direct interconnection with tMEC through newly defined interfaces for service discovery, App lifecycle management, and workload distribution.
- **Dynamic task offloading:** Apps can execute tasks locally on cMEC, offload them to tMEC when additional resources are needed, or distribute workloads across multiple cMEC nodes.
- **Resilient service continuity:** If a device loses connectivity to tMEC, cMEC ensures uninterrupted operation by handling processing locally, later synchronizing with the broader MEC system when reconnected.
- **Security and privacy enhancements:** By allowing sensitive data to be processed within end-user devices, cMEC reduces exposure to external infrastructure while maintaining compatibility with MEC security frameworks.

This architectural extension requires modifications to the standard MEC framework, particularly in orchestration, service availability, and inter-layer communication. MEC orchestrators (traditionally centralized in tMEC) must be adapted to recognize and interact with distributed cMEC nodes. New cross-system interfaces are introduced to allow App provisioning across both layers, enabling tMEC to dynamically allocate resources based on available cMEC capacity. Additionally, enhanced service discovery mechanisms ensure Apps can locate and utilize cMEC-based resources without disrupting existing MEC workflows.

By integrating constrained devices into the broader MEC ecosystem, this approach optimizes latency-sensitive Apps, enhances bandwidth efficiency, and increases system resilience, particularly in mobility-intensive or connectivity-challenged environments. The transition towards a fully distributed, multi-layer MEC architecture aligns with the evolution of cloud-to-edge computing, enabling a more scalable, flexible, and adaptive framework for next-generation network services.

5.3 Flexible Networks

Flexible networks represent networks that can efficiently be adapted to varying and dynamic requirements, which can utilize, e.g. flexible topologies, multi-connectivity, and spectrum sharing techniques to achieve this. Figure 5.6 depicts a specific view of the system blueprint of Figure 3.1, focusing on the flexible networks. It should be noted that even though NTN also belongs to the flexible networks category, they are separately analysed in Section 5.4. Subnetworks, described in Section 5.3.1, involve a UE that manages a set of trusted devices while being in coordination with the overlay 6G network. Subnetworks may enable new types of devices and lower the power consumption of the system (Akgul et al. 2025). A summary of the relevant procedures and their impact on the system is illustrated in Figure 5.6. Additionally, Section 5.3.2 introduces a single architectural solution as the design goal for multi-connectivity in 6G, in order to reduce

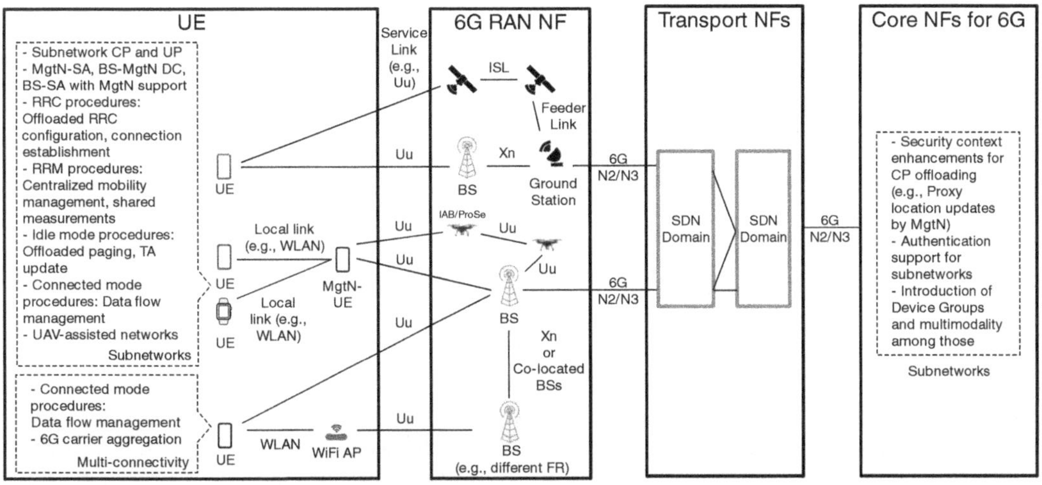

Figure 5.6 Flexible networks' specific view of the system blueprint. *Source:* Le Sauze and Kerboeuf (2025) / Hexa-X-II; iconsgraph / Adobe Stock Photos; Sharmin / Adobe Stock Photos.

the implementation and deployment cost. That solution should combine the benefits of carrier aggregation (CA) and dual connectivity (DC) while addressing the complexities and drawbacks associated with them. A type of multi-connectivity, integrating wireless local area network (WLAN) in the RAN, is shown in Figure 5.6. Finally, in Section 5.3.3, the importance of having an efficient spectrum co-existence solution from Day-1 in 6G is highlighted.

5.3.1 Subnetworks

Traditional networks in 6G may face challenges due to the rapid growth in the number and diversity of connected devices. Additionally, 6G will introduce requirements for higher data rates, increased service coverage, lower power consumption, increased resilience, and user privacy, compared to 5G. Subnetworks may aid in achieving these KPIs by offloading functionalities from one node to another and by providing connectivity to devices that are out of network coverage. Additionally, subnetworks are designed to allow for more flexible and efficient communication between trusted devices, which would further enable the operation of devices of varying capabilities (e.g. compute power, battery, and cellular). Their architecture can differ based on various criteria, such as device type, App type or geographical location of the nodes. The adoption of a distributed architecture would allow the BS to offload certain functionalities to the subnetworks. The specific set of functions that user and network nodes currently have may be insufficient and/or inefficient in the context of subnetworks. Therefore, new node roles and responsibilities should be defined, targeting a lower complexity for low-capability nodes. Trusted nodes may need to securely exchange information about the devices' status and their Apps to aid with the subnetwork management. Without coordination, communication may be inefficient, leading to increased latency, suboptimal resource management, as well as link failures.

A subnetwork is formed voluntarily by UEs based on their mutual trust. A UE may assume the new role of a management node (MgtN) in a subnetwork and become the subnetwork's primary node, which can communicate with the BS and other UEs. The MgtN would take over the coordination within a subnetwork, while still being in coordination with the global 6G network. Even though the subnetwork is formed among UEs with limited network configuration and awareness, the subnetwork itself may be associated with the 6G network, along with all local UEs, as depicted in Figure 5.7. Moreover, the subnetwork shall be integrated into the 6G network, so that local UEs are able to seamlessly move between the subnetwork and the 6G network.

Various subnetwork architectures have been considered based on the connections between the BS and the UEs, including MgtN standalone (MgtN-SA), BS-MgtN DC, and BS standalone with MgtN support. The CP entities of local devices can be flexibly deployed on the MgtN, and the global network does not have to be aware of this CP offloading. Figure 5.7 depicts an example of an MgtN-SA architecture, where the whole or part of UE3's CP is deployed at the MgtN. The subnetwork may then use a new lightweight subnetwork CP (snCP) between the MgtN and the UE. The snCP is transparent to the overlay 6G network, since it includes the offloaded UE CP information and the procedures that run only within a subnetwork, as explained in more detail in Ericson et al. (2024) and Akgul et al. (2025). With the concept of subnetworks, certain procedures can be offloaded from lower-capability UEs to higher-capability UEs acting as the MgtN to save power. This includes radio resource control (RRC) procedures (e.g. RRC configuration), radio resource management procedures (e.g. mobility management and shared measurements), idle mode procedures (e.g. offloaded paging and tracking are updated), and connected mode procedures (e.g. data flow management and relaying). Still referring to Figure 5.7, the MgtN-BS interface would be the 6G Uu-equivalent interface, while the MgtN-UE3 interface may be any access network (e.g. WiFi, Sidelink, and Uu). As far as the higher UP layers are concerned (e.g. PDU session and IP addresses), they could terminate at the local device, while the lower UP layers may terminate at the MgtN, making the MgtN-UE interface subnetwork-specific by using the subnetwork UP (snUP).

5.3.2 Multi-connectivity

5G has two solutions for multi-connectivity, namely, DC and CA. In DC, the master cell group (MCG) and potentially also the secondary cell group (SCG) are connected to the CN. The MCG is a group of serving cells associated with the master node (MN), comprising the PCell and optionally one or more SCells, while the SCG is a group of serving cells associated with the secondary

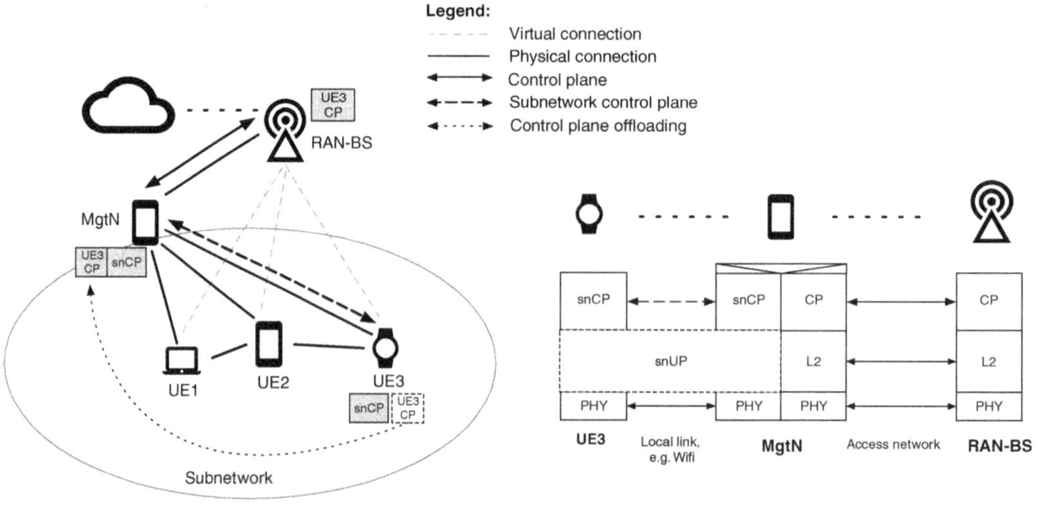

Figure 5.7 Subnetwork with a management node (MgtN) connected to three UEs and to the 6G base station. UE3's control plane (CP) is deployed at the MgtN, and both the subnetwork CP (snCP) and the subnetwork User Plane (snUP) are used within the subnetwork.

node (SN), comprising the PSCell and optionally one or more SCells (3GPP, 37.340 2024). DC supported multi-vendor operations (i.e. MCG from one vendor and SCG from another vendor). The two cell groups communicate with each other with a standardized set of procedures via the Xn interface to exchange UP and CP data. DC can improve throughput or reliability, especially in networks where both frequency range 1 (FR1) and FR2 frequencies are deployed. For example, the MCG may use low frequencies (i.e. FR1) to provide reliability, while the SCG may use high frequencies (i.e. FR2) to increase the user throughput. Within a cell group, CA relies on the configuration of multiple component carriers (CCs), which can increase the UE's throughput when activated. The CCs may be contiguous in the same frequency band and non-contiguous in the same or different bands. Typically, CA is used as an intra-site deployment due to the high demands on the synchronicity between the different CCs.

One of the main drawbacks of DC during the evolution from 4G to 5G was the number of options specified in 3GPP (3GPP, 37.340 2024). These options increased the complexity of the specifications and necessitated a greater number of test cases to prevent multi-vendor interoperability issues. In addition, flow control, i.e. determining whether a particular data packet should be sent via the MCG or the SCG, can cause 'stalling' of the packet delivery due to the latency over the Xn interface. The CP messages between the MCG and the SCG (e.g. UE capability coordination) further increase the implementation complexity. An advantage of DC is that when the connection to a cell group is degraded, the UE may report that event to the other cell group in order to recover the connection. In addition, in DC, the same packet can be sent via both the MCG and the SCG to create redundant paths for reliability. This can also be done in CA for different CCs, but they are typically not transmitted from different physical sites. At the same time, UL control is coupled to the cell group in DC, which is not an issue in CA. Finally, the process to activate the SCG may be deemed slow in 5G, since the session may often be completed before the SCG has completed its activation. CA only employs one scheduler, which controls the cells in the CA; hence, there is no need for flow control. CA can thereby avoid the stalling problems of DC. UL control scheduling can be either on both the PCell and an SCell or only on the PCell, which means that DL and UL are not coupled and therefore provide better UL coverage compared to DC. The main disadvantage of CA is that it cannot handle inter-vendor operability since there is no multi-vendor protocol specified in 3GPP. Possible 6G improvements for CA can be a faster method to include an SN, e.g. with a pre-configuration of adjacent cells (similar to conditional handover [CHO]).

For simplicity, only one architectural option (i.e. CA or DC) should be standardized, while ensuring that the selected solution has the benefits of both architectural options. An improved CA architecture is proposed to be the sole solution in 6G (Akgul et al. 2025). A proposed way forward for 6G MC is depicted in Figure 5.8. There, the UE can connect via several non-co-located RUs and using LLS, to the network (see Section 5.2.1 for more details on LLS), aggregating the resources from each RU using CA like 6G MC. Multiple improvements for CA have been introduced in detail in Ericson et al. (2024) [HEX225-D35].

Potential CA improvements in 6G may include faster PCell change similar to existing layer 1 and layer 2 handover-based solutions, and the use of a set of inactive connections for fallback. UE mobility may lead to radio link failure (RLF), especially at the cell edge. When RLF occurs in single cell group connectivity, the UE initiates the RRC-re-establishment procedure and tries to find a suitable cell to continue connectivity, which leads to a non-negligible service interruption time. However, in the case of DC, upon MCG failure, there are mechanisms to recover connectivity using the SCG, without the need to initiate the re-establishment procedure. Hence, mechanisms for solving the same issue are required in CA to reduce the service interruption time. It is proposed that upon RLF on PCell, the UE may use the SCell to recover the connectivity when CA is enabled. Regarding the

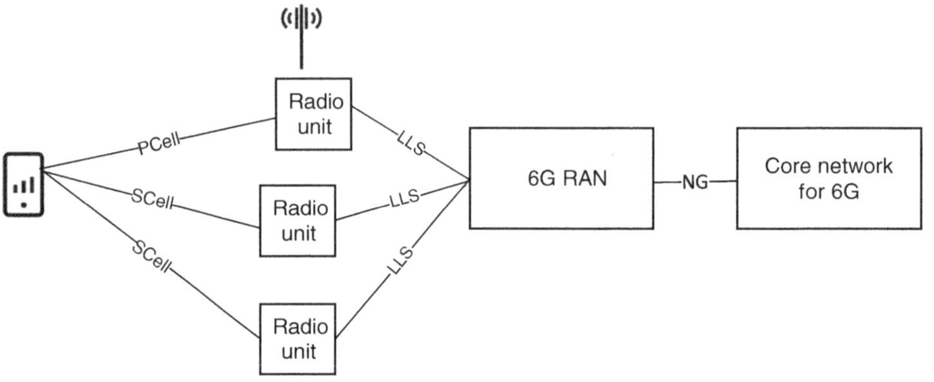

Figure 5.8 6G carrier aggregation proposal.

need for a faster addition of cells, the early measurements report (EMR) feature was introduced in 5G by 3GPP to allow for proactive UE measurements on network pre-configured frequency layers, which would be ready for potential reporting once it transitions from idle/inactive to connected mode. However, this would require the UE to keep measuring the configured frequency layers in idle/inactive mode for 10–300 seconds, deeming the EMR feature too power hungry. An enhanced mechanism for SCell addition when transitioning from idle/inactive to connected mode could be introduced, where in idle/inactive mode, the UE will have internal logic to determine if it will perform measurements. If it does, it will only perform measurements of a subset of the pre-configured cells and not on all configured frequency layers.

As another type of multi-connectivity, different access networks could be integrated with 6G cellular networks in the RAN by using adaptation protocol layers. A RAN-integrated WLAN and 6G solution leverages a trusted WLAN terminal (WT), which is within cellular coverage, to support UEs. Both the UE and the WT have established a connection with the BS and with each other. The proposed WLAN-cellular aggregation (WCA) framework aggregates the direct UE-BS and UE-WT-BS paths on the RAN level at the BS where the radio bearer originated from. An advantage of using a WT as an intermediate hop between the UE and the BS would be that the WT-BS link may use a higher FR than the FR of the UE-BS link, due to the expectation that the WT is more stationary than the UE. This would not only result in different (i.e. non-competing and non-interfering) cellular resources being assigned to the UE and to the WT but also to higher WT-BS throughput. Another use case is when a remote UE is within WLAN range of a WT within cellular coverage, acting as a relay. The current and predicted usage of quality of service (QoS) flows and radio bearers necessitates exploring how this framework integrates into a WT. The BS can configure the UE to transmit the same or different packets over the UE-BS and UE-WT-BS paths. Packet duplication increases reliability, while packet splitting may increase throughput. Aggregation occurs below the IP layer, ensuring privacy as the WT does not have an IP connection with the cellular network for packet relaying. Finally, this feature could be enabled by the UE, in coordination with the cellular network, based on its internal logic and on the availability of trusted WTs in its vicinity. Where applicable, the aggregation of the UE-BS and UE-WT-BS paths can be done on the PDCP layer. When a packet is transmitted over the WT (i.e. via the UE-WT-BS path), the WLAN relay adaptation protocol (WRAP) is used (Akgul et al. 2025). The WRAP header includes the information required by the WT and by the endpoint (i.e. BS in UL or UE in DL) to determine which UE-BS entities and parameters (e.g. cellular UE ID, radio bearer ID) the transmission corresponds to.

The WT shall read the content of the WRAP header to determine to which logical channel (i.e. WT-UE in DL or WT-BS in UL) it should map the packet.

5.3.3 5G–6G Spectrum Co-existence: Multi-RAT Spectrum Sharing

While 6G probably will have new allocated spectrum bands, the existing 5G frequency bands (especially in FR1) are still very important for 6G coverage and capacity. One straightforward solution for 5G–6G coexistence in FR1 is static refarming, which implies a hard partitioning of the existing spectrum between two radio access technologies (RATs). The technique has been largely utilized during previous migration rounds, such as from 3G to 4G. As a result, operators have a deep understanding of the pros and cons of refarming, and it can be utilized as a benchmark when evaluating the performance of 5G-6G coexistence techniques. Despite its conceptual simplicity, the main drawback of refarming is its inherent inability to adapt the allocation of spectral resources to instantaneous traffic fluctuations.

One important enabler for 5G–6G spectrum migration is multi-RAT spectrum sharing (MRSS). With MRSS, a 6G RAT can be employed in 5G spectrum dynamically and allowing the 6G UEs to use the 5G spectrum, meaning that the time–frequency resources in a cell are dynamically assigned to either 5G or 6G according to traffic demands. Hence, MRSS aims to be a more spectrally efficient sharing approach between 5G and 6G compared to dynamic spectrum sharing (DSS) between 4G long term evolution (LTE) and 5G new radio (NR).

DSS implementation faced limitations due to legacy requirements – specifically, the need to continuously transmit multiple control and reference signals for both technologies within a shared DSS cell. This added signalling overhead due to 4G design and DSS limiting implementations has impacted the capacity of the DSS cell. Building on insights gained from DSS, the lean and flexible design for the physical layer is a key design for MRSS. Notably, 5G NR has already moved in this direction by eliminating persistent signals such as the LTE cell-specific reference signals (CRS), which improves spectral efficiency. Another key design to learn from DSS is the necessity for intelligent, coordinated scheduling mechanisms to dynamically allocate spectrum across different generations of cellular technology. DSS was designed to provide a flexible allocation of spectral resources between 4G LTE and 5G NR. However, DSS faced several legacy and practical restrictions that negatively impacted the overall spectral efficiency of the system. In a DSS cell, several 4G and 5G control and reference signals need to be transmitted regularly. For example, one of the major limitations of DSS design was the constraint imposed by the LTE CRS that are typically transmitted more than once in every LTE subframe (1 ms). These signals represented a non-negotiable overhead and had a direct and negative impact on the available resources for 5G NR data transmission. This will not be an issue for MRSS, since 5G does not have as frequent 'always on' signals as CRS, and the plan is not to introduce such signals in 6G either. MRSS is a key feature for the migration to 6G as it will allow NR and 6G cells to share the same carrier(s) dynamically and efficiently while adapting to real-time traffic requirements. In addition, some NR control signals could be reused by 6G to minimize overhead and save energy. Akin to the 6G physical layer design, the exact architecture options remain undefined. One can assume, however, that 6G will likely be deployed in the existing FR1 and FR2 as well as new candidate bands including the essential 7–15 GHz range, as depicted in Figure 5.9.

The recommended approach for moving forward is to adopt a single-step migration based on 6G SA. While this implies a simplified migration and avoids coexistence issues between definitive and interim architectural solutions, it also means that not only is MRSS needed to provide 6G coverage via FR1, but also that Day-1 support for UL and DL CA is essential to aggregate FR1 carriers with

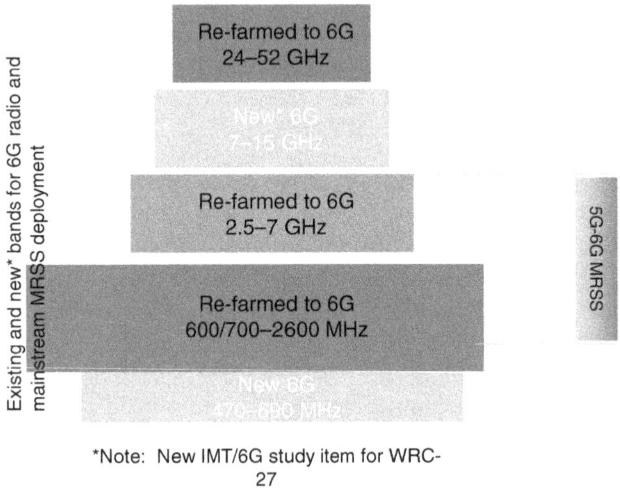

Figure 5.9 6G frequency bands, including those that will be shared between 5G and 6G.

mid- and high-bands for enhanced capacity. In other words, MRSS and 6G CA are key technology components for migration. Moreover, MRSS shall not require any changes to 5G UEs, and 6G UEs shall support MRSS through 6G radio design and basic 6G radio functionalities.

Ultimately, the MRSS framework must effectively utilize the available flexibility to minimize the fixed overhead and exploit different resource partitioning solutions such as time division multiplexing (TDM), frequency division multiplexing (FDM), or spatial division multiplexing (SDM) in isolation or combined. The ideal domain(s) and the optimal granularity depend on several factors. For instance, data channels such as the physical downlink shared channel (PDSCH) are expected to require more flexibility and faster adaptation than the physical downlink control channel (PDCCH).

5.4 Non-terrestrial Networks

NTN refers to access networks based on satellites at any orbit or high-altitude platforms (HAPS). Support for NTN in 3GPP specifications was introduced late in the definition of the 5G system, creating some limitations on the features and performance that can be provided when using NTN. With 6G, there is an opportunity to define a 6G system which natively supports both TN and NTN components, as shown in Figure 5.6, and hence to provide capabilities taking the most out of NTN or from the combination of NTN and TN.

In this section, the specific deployment scenarios of NTNs and what constraints they may have on the definition of the 6G system architecture are reviewed. The capabilities enabled by the integration of NTN and TN components, as well as on-board edge computing, will also be discussed.

5.4.1 Rationale for NTN in 6G

Adding an NTN component to the 6G system would allow connecting users anywhere in the world, including in the most remote and sparsely populated areas, to enable the targeted ubiquitous connectivity and improve on the digital inclusion. Major disasters around the world have shown how critical mobile communications are for both governmental organizations and the population as part of the relief. Given that NTN (Satellite and HAPS) infrastructure is not subject to the same disruptions as terrestrial cellular networks, NTN is considered critical for disaster relief. NTN in 6G is expected to play a crucial role in enhancing the overall resilience of the 6G system by supplementing other access technologies.

5.4.2 NTN Deployment Scenarios

The objective of the section is mainly to understand how the characteristics of an NTN deployment scenario have an impact on the service capabilities that can be provided by the NTN and the types of terminals that can be addressed. It will also set constraints on the 6G system architecture. Each NTN deployment scenario refers to a network involving specific space or airborne segment characterized by the orbit type around the Earth, the altitude with respect to the Earth, the satellite payload type (transparent or regenerative), the frequency band of the service link, the types of radio beams generated (Earth fixed or Earth moving) on the service link. The selection of a given NTN deployment depends on the targeted terminals and service capabilities for the satellite communications, namely, the orbit type around the Earth, the satellite payload characteristics, the frequency band of the service link, and the type of radio cells.

The following reference orbit types are generally considered: Geostationary Earth Orbit (GEO), Low Earth Orbit (LEO), and Medium Earth Orbit (MEO). Note that a highly eccentric orbit (HEO) is considered to provide services over polar regions, but they will not be addressed here. Table 5.1 summarizes the main characteristics impacting the service capabilities (see e.g. 3GPP, TR 38.821 2023; 3GPP, TR 38.811 2020):

Table 5.1 NTN orbit types.

	Geosynchronous orbit	**Non-geosynchronous orbit**[a]		**HAPS**
Orbit types	***Including geostationary orbit***	***Medium Earth orbit***	***Low Earth orbit***	***HAPS***
Altitude	35 786 km	From 7000 up to 25 000 km	From 300 up to 1500 km	Typically around 20 km
UE to serving satellite propagation delay[b]	Min: 120 ms (at 90° elevation angle) Max: 136 ms (at 10° elevation angle)	Min: 24 ms (at 90° elevation angle) Max: 99 ms (at 10° elevation angle)	Min: 1 ms (at 90° elevation angle) Max: 13 ms (at 10° elevation angle)	Min: <1 ms (at 90° elevation angle) Max: <1 ms (at 10° elevation angle)
Revisit time[c]	0 ms (permanent coverage of the UE)	Min: ∼100 minutes for terminal in polar region and satellite on a polar orbit Max: ∼12 hours for terminal at equator and satellite on a polar orbit	0 ms (in case of quasi-stationary HAPS)	

a) The inclination angle of the orbital planes followed by the satellite with respect to the equator may take different values: 0° means that the orbital plane is the equatorial plane, which enables providing coverage around the equator. ∼90° refers to a polar orbit to provide coverage in polar regions. Depending on the targeted latitude, the angle may be adjusted between 0° and 90°.

b) Propagation delay between a user equipment and a serving satellite (see Table 7.4.1-1 in 3GPP, 22.261 2025).

c) The revisit time applies to constellation not able to provide service continuity over a given area. In such case, the revisit time, is the duration, between two successive satellites able to provide service over a given area.

The 6G RAT is expected to support all these orbit types around the Earth.

The satellite payload refers to those elements of the spacecraft tasked with transmitting and receiving the communication signals. The characteristics of the communication payload are mainly constrained by the power provided by the spacecraft, the sizing under the rocket's payload fairing and the targeted life duration. Great care is taken to maximize the throughput per satellite, which leads to a highly energy-efficient design.

In the context of 6G, two types of communication payload can be distinguished:

- **Transparent:** The NTN payload transparently forwards (repeating) the 6G signals received from the UE (via the service link) to the NTN gateway (NTN-GW) (via the feeder link) and vice versa.
- **Regenerative:** The NTN payload processes the radio protocol received from the UE (via the service link) and creates another radio protocol towards the NTN-GW (via the feeder link) and vice versa.

The mapping of 3GPP NTN architecture onto satellite and gateways in the case of a transparent payload architecture is illustrated in Figure 5.10 and in the case of a regenerative payload architecture, it is illustrated in Figure 5.11, respectively.

In the case of a transparent architecture, satellite access node (SAN) is part of the NG-RAN. SAN interfaces with the 5GC network via the NG interface and serves the UE via the NR-Uu interface. The 5G NR satellite system architecture is composed of NTN infrastructure and non-NTN infrastructure with gNB functionalities on the ground in the case of transparent architecture. This architecture further considers:

- The NTN service link provisioning system, which maps the NR-Uu radio protocol over radio resources of the NTN infrastructure (e.g. beams, channels, and Tx power).
- The non-NTN infrastructure gNB functions which implement the gNB logical function, possibly with different split options (e.g. gNB-DU/CU).

The NTN service link provisioning system is formed by: (i) NTN payload(s) on board satellite(s); (ii) (NTN-GWs); and (iii) feeder link(s), which are a radio interface between the satellite and the NTN-GW. The NTN control function controls the radio resources of the SAN via operation and maintenance (O&M) and provides key information about the space segment, such as the satellite Ephemeris information.

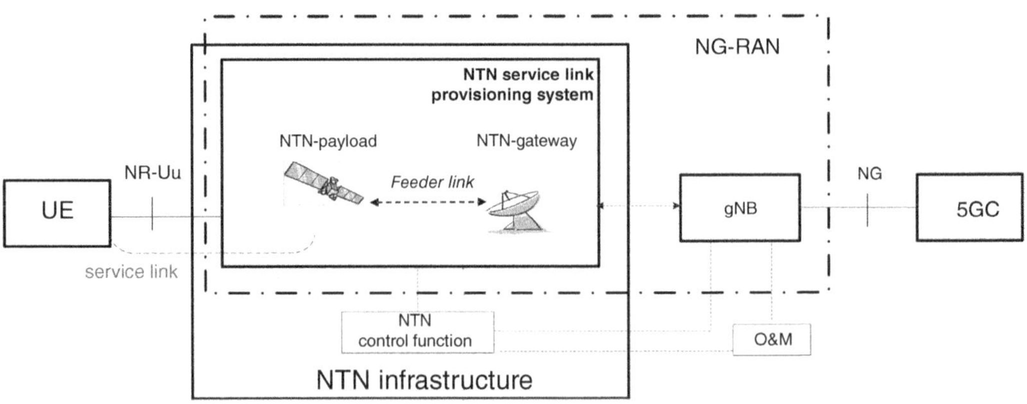

Figure 5.10 Transparent payload with NG-RAN using ground gNB.

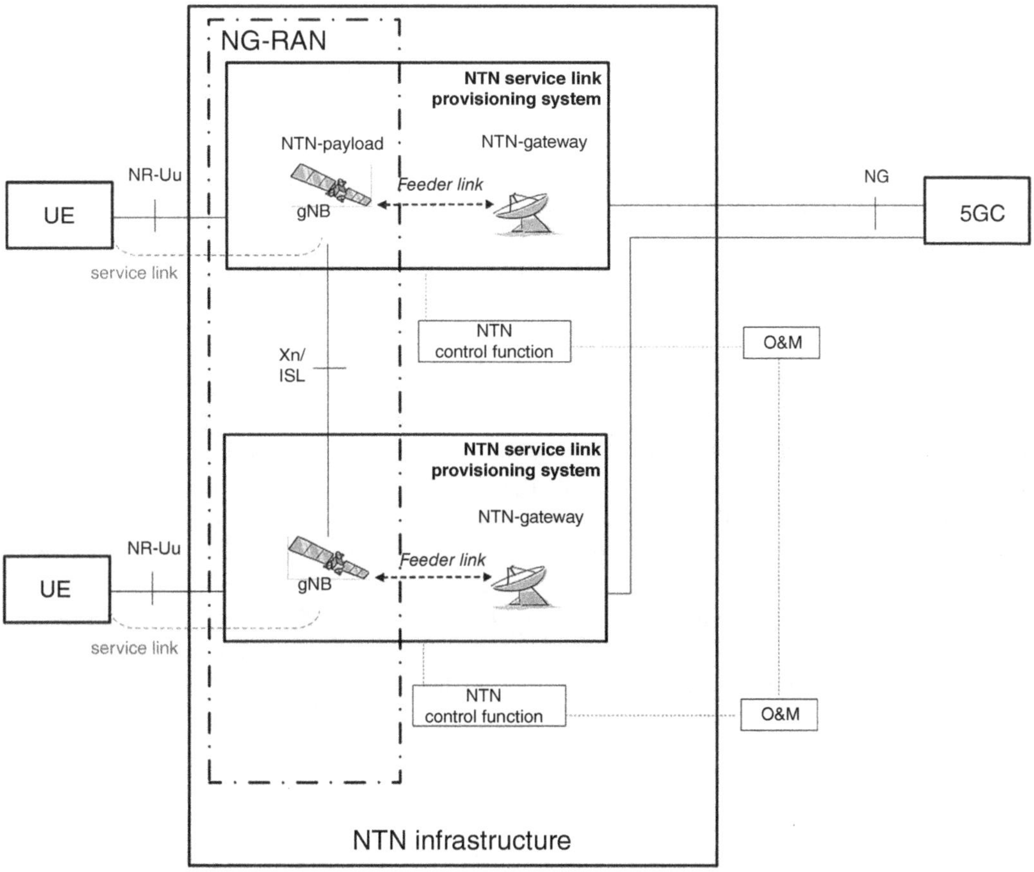

Figure 5.11 Regenerative payload with gNB (NG-RAN) on board and Xn over ISL.

Regenerative payload requires processing on board and is typically used to provide services in areas where a feeder link cannot be deployed (e.g. NTN-GW not available) or when the feeder link bandwidth is drastically constrained. A typical regenerative payload mission considered for a 6G system may provide:

- Store-and-forward capabilities to support non-real-time services;
- Inter-satellite-links (ISL), including in-plane/inter-plane, inter-orbit ISL to support real-time services;
- Mesh connectivity via satellite (i.e. UE-satellite-UE communication);
- Positioning, navigation, and timing (PNT) capability; and/or
- ISAC capability for large-scale monitoring, also known as joint (or integrated) communication and sensing (JCAS/ICAS).

In the case of a regenerative payload, gNB (NG-RAN) can be embarked on board and Xn transported over ISL between two adjacent satellites.

5.4.2.1 Frequency Band of the Service Link

For 5G NTN, 3GPP has defined several bands for NTNs. Requirements take into account the protection of TNs in adjacent bands. The frequency bands currently defined as part of Release 17, 18 are

Table 5.2 Operating bands for 5G NR-NTN (3GPP, 38.108 2025; 3GPP, 38.101-5 2025).

5G NR satellite *operating band*	**UpLink *operating band* (SAN receive/UE transmit) $F_{UL,low}-F_{UL,high}$**	**DownLink *operating band* (SAN transmit/UE receive) $F_{DL,low}-F_{DL,high}$**	**Duplex mode**	**Targeted terminals**
n256	1980–2010 MHz	2170–2200 MHz	FDD	Smartphone
n255	1626.5–1660.5 MHz	1525–1559 MHz	FDD	Smartphone
n254	1610–1626.5 MHz	2483.5–2500 MHz	FDD	Smartphone
n512	27 500–30 000 MHz	17 300–20 200 MHz	FDD	VSAT UE
n511	28 350–30 000 MHz	17 300–20 200 MHz	FDD	VSAT UE
n510	27 500–28 350 MHz	17 300–20 200 MHz	FDD	VSAT UE

allocated to satellite services and are described in Table 5.2 (see 3GPP, 38.108 2025; 3GPP, 38.101-5 2025) and Table 5.3 (see 3GPP, 36.108 2025; 3GPP, 36.102 2025).

In addition, as part of the Rel-19, some other bands are being considered for NR-NTN and eMTC/NB-IoT NTN, respectively, such as the satellite Ku-bands (10.7–14.5 GHz) and other potential combinations of satellite S-bands (around 2 GHz) and satellite L-bands (around 1.6 GHz). For direct satellite connectivity to smartphones or IoT devices, an NTN-capable UE follows the same RRM and RF requirements as a TN-capable UE. Some of the frequency bands described in Table 5.2 and Table 5.3, as well as additional satellite service allocated bands below 8 GHz (e.g. C-band) and above 10 GHz (e.g. Q/V-band), may be considered for NTN in 6G.

5.4.2.2 Radio Cells

The satellite radio cells correspond to the footprint on Earth of the beams generated by the satellites on the service link. Several categories of radio cells can be distinguished, as the 6G system is expected to support all these radio cell types:

- Earth-fixed: provisioned by beam(s) continuously covering the same geographical areas all the time (e.g. the case of GSO satellites).
- Quasi-Earth-fixed: provisioned by beam(s) covering one geographic area for a limited period and a different geographic area during another period (e.g. the case of Non-GSO satellites generating steerable beams). This is typically considered for a narrow beam footprint to prevent excessive handover rate, e.g. for broadband services.

Table 5.3 Operating bands for eMTC/NB-IoT NTN [36.108] [36.102].

E-UTRA operating band	**UpLink operating band (SAN receive/UE transmit) $F_{UL,low}-F_{UL,high}$**	**DownLink operating band (SAN transmit/UE receive) $F_{DL,low}-F_{DL,high}$**	**Duplex mode**	**Targeted terminals**
256	1980–2010 MHz	2170–2200 MHz	FDD	Smartphone, IoT devices
255	1626.5–1660.5 MHz	1525–1559 MHz	FDD	Smartphone, IoT devices
254	1610–1626.5 MHz	2483.5–2500 MHz	FDD	Smartphone, IoT devices
253	1668–1675 MHz	1518–1525 MHz	FDD	Smartphone, IoT devices

- Earth-moving: provisioned by beam(s) whose coverage area slides over the Earth's surface (e.g. the case of Non-GSO satellites generating fixed or non-steerable beams). This is typically considered for a wide beam footprint, e.g. for narrowband services.

5.4.3 Impact on 6G System Architecture

In the case of regenerative payloads, great care shall be taken when designing the payload and overall system to ensure that the available spacecraft power is appropriately allocated to respectively the antenna and RF front-end part (mostly responsible for the satellite performances in terms of data rate and throughput) and to the digital processing part (which may implement some of the 6G functions as well as edge computing resources). Furthermore, the feeder link may be subject to varying propagation impairments/Doppler/latency and constrained in bandwidth. Therefore, the 6G system and RAN architecture should support various functional splits able to accommodate the feeder link characteristics and allow to selection of the minimum set of 6G/RAN functions to be embarked on board (in view of the energy limitation) for the mission purpose.

5.4.4 Support of NTN-TN Integration

Unification of TN and NTN components in 6G is expected to facilitate the ubiquitous connectivity, the resiliency and sustainability of the network and the overall spectrum usage efficiency.

5.4.4.1 Ubiquitous Connectivity

While in 5G, NTN/TN mobility schemes have been defined to ensure global service including the most remote and low densely populated areas, 6G will allow to support a quasi-seamless service continuity with minimal interruption for real-time services especially with novel NTN/TN mobility scheme in connected mode based, e.g. on simultaneous multi connectivity in parallel with handover procedure and enhanced triggers.

Two options are foreseen for TN-NTN faster switching, using the 5G DC framework [TS 37.340] as baseline, depending on which node (TN or NTN) will be the MN and which one will be the SN. The option where the TN node is the MN is termed as 'Option 1'. UE performs initial context setup with TN network, and TN is configured as MN while NTN is configured as SN. NTN SN is selected either based on UE measurements or UE location. NTN circuitry is switched off. User data transfer takes place between the UE and the CN via the TN network. When UE slips towards the end of TN coverage, it provides an indication to TN. UE may internally activate the NTN circuitry, and the TN node sends an activation message to the NTN node. Alternatively, TN sends a message to UE for activating NTN, and UE performs timing synchronisation with NTN. User data transfer should now take place between UE and CN via the NTN. UE may perform TN measurements, and as soon as it comes back to good TN coverage, it sends an indication to the NTN. NTN may either directly perform the switch if the UE is time aligned with TN, or the UE performs time alignment before the switch.

The option where the NTN node is the MN is termed as 'Option 2'. In this option, NTN acts as MN and TN as SN. UE connects to TN and performs user data transfer. UE indicates loss of TN coverage to NTN, and NTN performs the switch immediately. UE NTN circuitry is always activated because NTN acts as a MN and is already synchronized and able to perform measurement. In this case, UE may keep the TN circuitry also switched on because the interruption may be temporary. So, this option does not have the benefit of UE power saving as compared to Option 1.

The DC framework is used to allow pre-configuration for handover and faster switching. This is different when compared to the traditional DC, which is mainly used for throughput enhancements. This faster switching procedure could alternatively be called layer 3 handover, including the benefits of layer 2 triggered mobility (LTM) and conditional handover + dual active protocol stack (DAPS). The edge of TN coverage could be based on fine-tuning of existing measurement events or defining new events so that this end-of-coverage message is not lost, a similar motivation as CHO. The other leg is active before the radio connection for the main leg is broken, similar motivation as DAPS. Timing synchronisation will be performed with NTN prior to TN coverage is lost and vice versa, incorporating the main benefit of LTM.

5.4.4.2 Resiliency

The NTN will be designed to intrinsically be able to operate independently from third-party systems such as global navigation satellite systems (GNSS). NTN is subject to different disruptions compared to TN. Hence, the 6G system architecture will allow the NTN to provide alternative communication paths to maintain service provisioning where the terrestrial cellular network may be subject to local and temporary network failures.

5.4.4.3 Network Energy Efficiency/Sustainability

6G will allow the introduction of energy-based criteria for the selection of RATs for a given UE, depending on the traffic load as well as on the device's connection state.

Table 5.4 provides a high-level comparison between NTN and TN radio access technologies in terms of energy consumption profile and, consequently, of environmental footprint.

5.4.4.4 Spectrum Usage Efficiency

Currently, 3GPP has defined NTN bands considering that mobile and satellite services are allocated to different but possibly adjacent frequency bands at the national regulatory level. On the basis that MS allocated bands may not be used in remote areas, there is an interest in reusing the unused MS spectrum in these areas for satellite services. This creates interference issues at the edge of the TN coverage, but also across borders.

In the frame of the 6G system, some novel radio resource management and signal processing techniques may allow to support a mix of separated and partially overlapped bands for mobile and satellite services with the aim to maximize the overall spectrum usage in time and space.

5.4.5 On-Board Edge Capabilities

The 6G system architecture shall support the deployment of edge capabilities on board satellites as well as the transfer of context between successive serving satellites. Integrating satellite edge nodes

Table 5.4 NTN and TN comparison in terms of environmental footprint.

	Terrestrial network	Non-terrestrial network
Deployment phase	Transport and installation of the base stations (on new or existing sites)	Transport of satellites to the launch pads Satellite launches
Operational phase	Electrical power consumption of the base stations (millions around the world), depending on the traffic load	Solar-powered satellite payloads Electrical power consumption of Gateways (few tens for a global NGSO constellation)

with the current terrestrial computing infrastructure will enable computation task offloading from terrestrial or aerial UEs to the network edge and in-orbit task processing among satellites, facilitating the high processing demands of Apps originating from the terrestrial and NTN segments.

Similar to conventional MEC supported by terrestrial nodes such as gNBs, air/space-enabled edge computing alleviates the load on centralized cloud platforms, reduces transmission delays, and minimizes backhaul resource consumption, thereby enhancing overall network efficiency. Beyond this, air/space-enabled edge extends computing capabilities ubiquitously, allowing UEs to access computing resources anywhere and anytime, ultimately fostering the development of innovative Apps. A promising example of this is computing for remote Internet of Things (IoT) Apps, where IoT devices operate in areas or challenging environments where cellular infrastructure is unavailable or impractical. Such devices are usually characterized by constrained battery and computing capacity, making offloading to NTN the only viable solution. In this context, effective task offloading strategies should strike a balance between transmission and propagation overhead with local computation overhead, focusing on the energy efficiency and time-critical nature of the App, which in turn imposes constraints on task completion.

At the same time, satellites involved in data collection, sensing, and communications also generate a substantial amount of computational tasks that demand significant computing resources. For instance, in Earth observation scenarios, satellites capturing images to monitor environmental changes or detect natural phenomena require extensive data processing to extract meaningful insights in real time. However, the computing capabilities of LEO satellites, such as CubeSats, are typically limited by their restricted size, weight, and onboard energy. Therefore, offloading tasks through inter-satellite links to other satellites within the constellation can reduce processing delays and enhance resource utilization among them.

5.5 Dependable Networking

Dependable networking over converged networks addresses a new direction in communication technologies, aiming to deliver predictable, reliable, and low-latency data transfer across diverse and complex environments. In the context of next-generation networks, such as 6G, dependable networking emerges as a response to the demands of emerging use cases, including industrial automation, collaborative and cloud robotics, connected vehicles, real-time digital twins and immersive experience including high-fidelity virtual reality, which require guaranteed service performance metrics such as bounded latency, low jitter, and ultra-high reliability (see e.g. Chapter 2, 5G-ACIA 2021). The stochastic nature of wireless communication, but also the variability in computational resources perceived by virtualized Apps, and the need for real-time adaptation present significant challenges, particularly for safety-critical and business-critical operations. To this end, a paradigm shift from deterministic to dependable networking is needed, which integrates deterministic and adaptive capabilities, ensuring dependable operation while accommodating inherent variabilities in communication and computation (Sharma et al. 2023). Dependable service provisioning plays a central role in achieving these objectives by enabling consistent performance and reliability even under dynamic conditions.

The objective of dependable networking is to create an E2E framework that ensures consistent performance across multiple domains and heterogeneous technologies. Unlike conventional best-effort networks, dependable networks provide precise guarantees on packet delivery timelines and data integrity, irrespective of the complexity of the underlying infrastructure. This involves understanding any delay variations that different system components may incur, orchestrating and managing a network architecture that integrates capabilities across wired and wireless domains, spanning multiple administrative and technological boundaries. Such integration is essential to cater to real-time, mission-critical Apps, where delays or packet loss can lead to operational failures. From an E2E perspective, dependable networking extends beyond isolated segments to encompass the entire communication path, ensuring seamless and predictable service quality across diverse environments. This includes addressing challenges such as varying link-layer capabilities, inter-domain resource management, and cross-technology orchestration. Multi-domain operation is particularly critical as it enables the unification of disparate network segments, each potentially governed by different standards, capabilities, and administrative controls, into a coherent system that can deliver dependable services.

The motivation for advancing dependable networking research is rooted in its potential to redefine digital infrastructure by enabling new Apps and driving industrial transformation. For wired networks, current deterministic solutions, such as those found in time-sensitive networking (IEEE 802.1 TSN 2025) and deterministic networking (IETF DetNet 2025), have shown promise within specific domains but remain limited in geographic range, scalability, and interoperability. These limitations highlight the need for a holistic approach that integrates advanced data-plane and control-plane mechanisms, leverages AI for network orchestration, and employs digital twin technologies for predictive management.

Furthermore, as networks evolve to support mixed traffic types, including time-critical and best-effort flows, the complexity of ensuring consistent service quality increases. Dependable networking seeks to address this by providing robust mechanisms for traffic prioritization, resource allocation, and fault tolerance, while also ensuring compliance with stringent QoS requirements. The integration of AI-driven frameworks and multi-stakeholder CPs can enable proactive resource management, congestion avoidance, and real-time adaptability, ensuring the resilience and scalability of dependable networks.

In short, a **dependable network** can be defined as a network that can quantitatively ascertain that it will deliver the required service performance for the network services according to the performance level(s) that were agreed upon (Höftberger 2025).

5.5.1 Enablers for Dependable Networking

Several technical building blocks are the foundation for providing a dependable network, which will be highlighted in the following subsections.

5.5.1.1 Performance Observability and Predictability

In order to be able to commit to providing a dependable network service, the network requires the capability to assess its own network performance at run-time. A critical App is typically defined by one or more KPIs for the required QoS, where at least one KPI defines a minimum performance level. Often, this is a maximum tolerable packet delay for delivering messages from the Tx to the Rx. The App depends on this minimum performance being met for the App, in order to achieve its operation. This means best-effort service delivery is insufficient, and high communication service availability needs to be provided for achieving at least the minimum performance.

A dependable network must be able to quantitatively judge the performance that it can deliver; this requires that the network is capable of estimating or measuring the delay that it introduces for the transfer of App data (see Jardim 2025). In a mobile network, this performance observability needs to be achieved in both uplink (UL) and DL directions. From system state evolution as well as observable past performance, a prediction can be obtained by the network to judge if the minimum

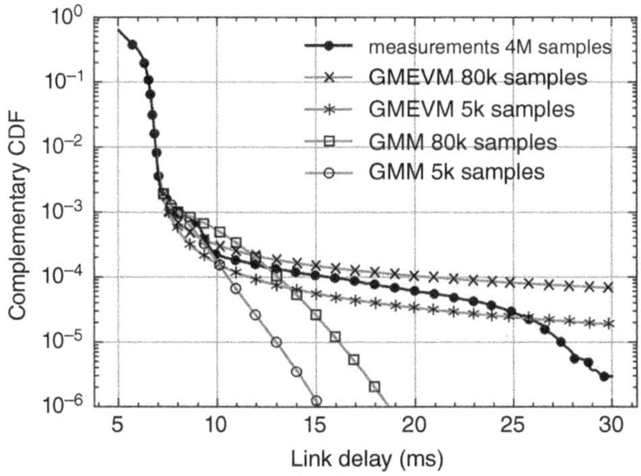

Figure 5.12 Example of the uplink latency prediction, based on different training sample sizes for different training models. *Source:* Adapted from Gross and Sharma (2023); Mostafavi et al. (2023).

service performance can be delivered with sufficient probability (i.e. to meet the communication service availability). Data-driven approaches that apply ML for predicting the achievable performance at the current network operation state show promising capabilities (Sharma et al. 2023; Mostafavi et al. 2023; Gross and Sharma 2023; Jardim 2025). In particular, the ability to predict tail probabilities of the performance enables us to judge which level of commitment or guarantee the network can provide to the App on achieving the minimum performance required (Gross and Sharma 2023; Mostafavi et al. 2023), see e.g. Figure 5.12.

5.5.1.2 Dependable Edge Cloud Integration

Edge computing allows for placing an App workload in proximity to a device and can thereby provide reduced latency between the client and the server App, enabling the support of real-time Apps. However, the edge computing paradigm cannot circumvent the uncertainties of the compute domain (e.g. latency variation of the execution time). Consequently, dependable computing is not inherently ensured. In order to offer dependable computing, the harmful uncertainties of the computing components must be mitigated by leveraging the resource allocation, Central processing unit (CPU) scheduling, and isolation toolset of cloud computing in a coordinated way to provide time-bounded execution of Apps. Furthermore, the timely and coordinated delivery of messages of time-critical Apps towards the dependable network domain must also be addressed. Ultimately, the goal is to achieve real-time performance of virtualized Apps in the edge cloud that is similar to a dedicated physical machine. From an App perspective, the E2E latency requirements need to estimate a compute delay budget, in addition to a respective delay budget planned for the communication.

First, service requirement decomposition is needed between the communication and the compute domains to specify the requirements applicable to the compute domain. Then, a service description in the form of a service-level agreement (SLA) is required between the App owner and the compute service provider by covering all the required metrics and aspects of dependable compute. To facilitate this, a proper description of predictability properties and assurance mechanisms in the cloud stack must be explored towards the client App by the cloud management.

For the deployment of a dependable compute service, the cloud management must ensure proper isolation and scheduling for workloads with real-time requirements. However, the trade-off between the level of workload isolation and the overhead of the virtualization technique must be considered; higher isolation results in larger virtualization overhead, which negatively impacts the real-time capabilities. Regarding the CPU scheduling, the cloud management configures the scheduling parameters (e.g. time budget and period) based on the requirements of the Apps. The usage of a real-time operating system (RTOS) can ensure strict timing guarantees, while the flexibility of cloudification is still kept. RTOS typically employs deterministic and priority-driven scheduling (e.g. earliest deadline first) to manage task execution within defined time boundaries, which fits perfectly for ensuring dependable compute services (Hadziaganovic 2024). It is important to note that virtualization introduces the complexity of multiple schedulers influencing App behaviour, making bare metal a preferred option over a virtual machine or container-based deployment, due to its ability to minimize virtualization overhead.

The above-mentioned facilitators for compute resources and the continuously evolving traffic handling features in the virtualized networking stack enable the tight harmonization of the traffic handling in the compute and communication network domains. Indeed, this makes it possible to ensure seamless, dependable operation in an E2E scope (Hadziaganovic 2024).

5.5.1.3 Packet Delay Correction for Deterministic Delay Performance

A dependable network needs to ensure that the network performance stays within the range that is required by the App, as explained in Section 5.4.1. This means that for a time-critical App, the delay performance, i.e. the envelope of the performance distribution, must remain within the delay tolerance of the App. To this end, it is required to observe and predict the delay performance as shown in Section 5.4.2.1. However, there are also Apps that may require deterministic delay performance, which means that any packet delay variation should be negligible (see e.g. 5G-ACIA 2021). Wireless transmission is inherently subject to stochastic variations, e.g. caused by the time-varying nature of a radio channel. Reliable transmission can be efficiently provided by fast *hybrid automatic repeat request* (HARQ) retransmissions. While this achieves high reliability, it comes at the cost of packet delay variations. Typical wireless communication systems experience a non-negligible packet delay variation, as discussed and analysed in Gross and Sharma (2023). To drastically reduce packet delay variation of the wireless transmission may be difficult or come at a high resource cost (Sharma et al. 2023). However, if the packet delay variation can be observed in the system, it is possible to apply *packet delay correction* (PDC), which compensates delay variations by means of hold-and-forward buffering (Gross and Sharma 2023; Jardim 2025), as shown in Figure 5.13. Transmission with very low packet delay variation may be desirable, e.g. in TSN networks that apply time-aware scheduling (i.e. IEEE802.1Qbv) (Egger et al. 2025). It is shown in Egger et al. (2025) that PDC can significantly increase the efficiency and capacity in a TSN network with wireless nodes.

5.5.1.4 Network Programmability and Communication–Control–Compute Co-design

For dependable communication, the network has to ascertain that it fulfils the quantitative performance requirements. To this end, the App needs to be able to provide to the network a service request, which specifies the performance levels needed by the App (Höftberger 2025). Such a service request may be provided in combination with an SLA that defines the commercial agreement between the network communication service provider and the App owner. Such programmability of the network is enabled via network exposure interfaces, including an API.

In many cases, Apps can support multiple different levels of operation, which provides some adaptivity of the App to operate at different meaningful performance levels provided by the network (Höftberger 2025). This opens the path for co-designing the usage of the E2E system, comprising the

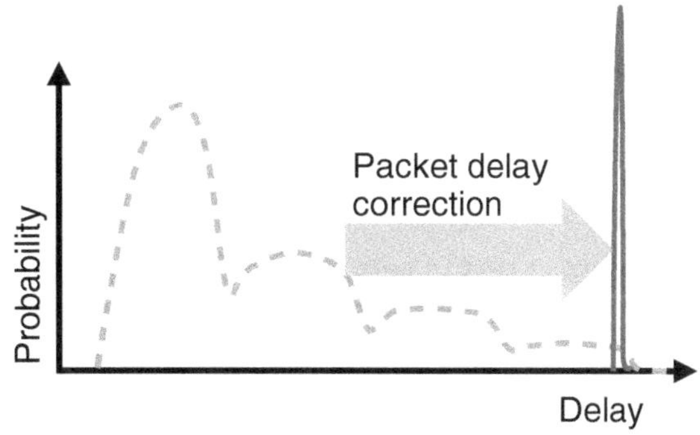

Figure 5.13 Packet delay correction to compensate for delay variations introduced during transmission (dotted line), resulting in a transmission behaviour with deterministic latency characteristics (bold line). (*Source:* Gross and Sharma 2023).

different domains of the communication (provided by the network), the compute domain for running the AF (provided as edge cloud by the network or by a compute infrastructure provider) and the App itself, as provided by an App developer (Höftberger 2025; Grosjean et al. 2025). The related work on co-design has mostly focused on control Apps. As shown in Figure 5.14, by sharing insights from the operation states in the communication and compute domains and the resulting expected delay performance, an efficient configuration and operation in these domains can be enabled to provide dependable service performance to the App. In challenging situations, it may happen that it is at risk if sufficient latency performance can be provided to the App, e.g. due to high communication and/or compute load and poor radio conditions. In such cases, it can be exploited if multiple levels of operations are feasible. By making the App adapt the level of operation to achievable performance levels, an efficient and scalable realization becomes possible.

5.5.1.5 Bringing Dependability to the Multi-domain Multi-technology Data Plane

Modern Apps require networking infrastructures capable of providing predictable, time-sensitive, and highly reliable communications across diverse environments. These characteristics of dependable networks (sometimes also referred to as deterministic networking) are defined as a network featuring:

- **Predictability:** Ensures that network behaviour remains consistent, allowing Apps to function with predefined performance guarantees.
- **Reliability:** Guarantees minimal packet loss and mechanisms for redundancy and recovery (e.g. packet replication and elimination).

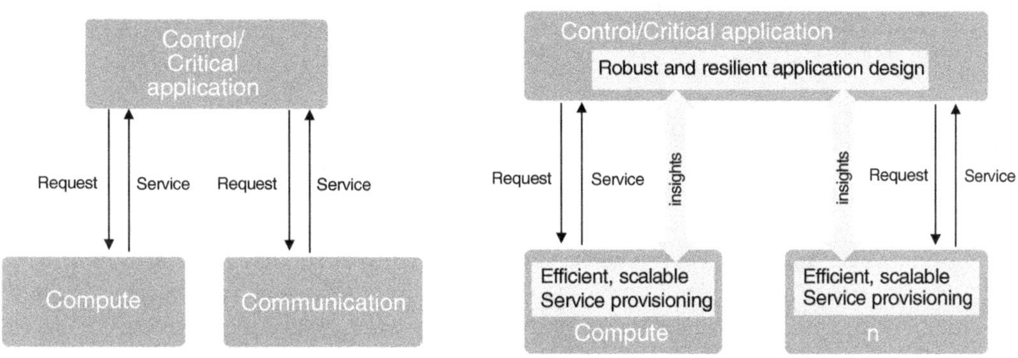

Figure 5.14 Communication–compute–control co-design (right) in contrast to a traditional design (left). *Source:* Adapted from Höftberger (2025); Grosjean et al. (2025).

- **Time-sensitivity:** Maintains strict latency and jitter constraints, ensuring that real-time Apps receive their data within well-defined time windows.

Traditional best-effort networks cannot guarantee the performance required by Apps such as industrial automation, remote-controlled robotics, and critical IoT services. As a result, a multi-domain, multi-technology dependable network is needed to:

- **Support critical applications:** Industrial automation, cloud robotics, and real-time control systems require low-latency, highly reliable networks to function effectively.
- **Unify disparate technologies:** Wi-Fi, Ethernet, and 5G offer different advantages, but no single technology alone can provide complete E2E multi-domain dependable guarantees.
- **Enable E2E dependability:** Most dependable solutions operate within a single domain (e.g. TSN for Ethernet, URLLC for 5G), but service requirements span across multiple domains, necessitating a cross-technology, Layer-3 approach.
- **Future-proof networking:** The shift towards 6G and beyond will further emphasize the need for dependable networking across diverse infrastructures.

Approaches such as the one described in Contreras et al. (2023) have as objective to enable dependable communications over heterogeneous networking technologies, including Institute of Electrical and Electronics Engineers (IEEE) 802.11 Wi-Fi future enhancements, IEEE 802.1 TSN (wired Ethernet), and 3GPP 5G, using an Internet Engineering Task Force (IETF) Deterministic Networking (DetNet) overlay data plane (Ferrari et al. 2024). This integration ensures that the network provides predictability, reliability, and time-sensitive communication, which are essential characteristics for Apps such as industrial automation, digital twins, and remote control robotics.

The data plane (Rico-Menendez et al. 2025) of this architecture serves as a bridge between these different technologies by implementing DetNet's Layer-3 overlay, which manages deterministic traffic across the domains. This integration addresses the challenge of ensuring bounded latency, low jitter, and high reliability while traversing multiple network segments with different transmission characteristics. The DetNet data plane is responsible for (i) encapsulation and forwarding, as it enables the seamless transition of packets across different domains using a multiprotocol label switching (MPLS) over UDP/IP mechanism; (ii) traffic engineering and QoS as each domain applies traffic shaping, scheduling, and congestion control to maintain the deterministic nature of flows; and (iii) service protection and reliability, as DetNet replicates and eliminates duplicate packets to ensure reliability in environments prone to transmission errors.

In Szilagyi et al. (2024), the authors evaluate the performance of each domain and the overall E2E performance of the multi-domain data plane. Experiments show that deterministic traffic maintains stable latency and zero packet loss, even under heavy background traffic, due to the traffic isolation and prioritization mechanisms implemented in each domain.

5.5.2 Architecture Support for Dependable End-to-End Communication with 6G

In order to provide dependable networking, certain functionality needs to be integrated into the network architecture (Sachs 2024), as shown in Figure 5.15. The Apps (indicated as 1) use the network exposure APIs (indicated as 2) to request a dependable network service and provide its service specification with required performance levels. The exposure interface is also used for possible App adaptations according to the communication–compute–control co-design approach. Providing dependable service performance for time-critical Apps calls for time-awareness (e.g. supporting time synchronization, indicated as 4) and reliable low-latency transmission capabilities. In addition, it is needed that the network can observe the latency performance that it delivers. Even more,

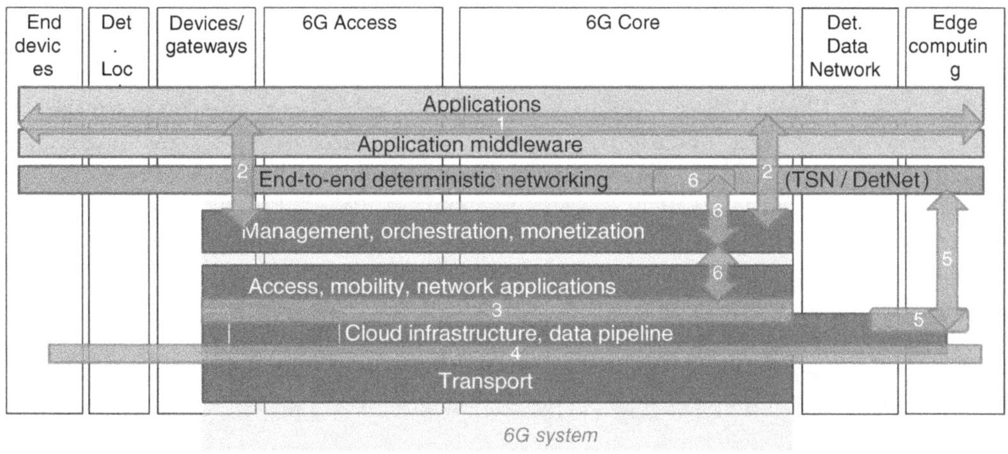

Figure 5.15 6G mobile network architecture enhancements for dependable networking (see Sachs 2024).

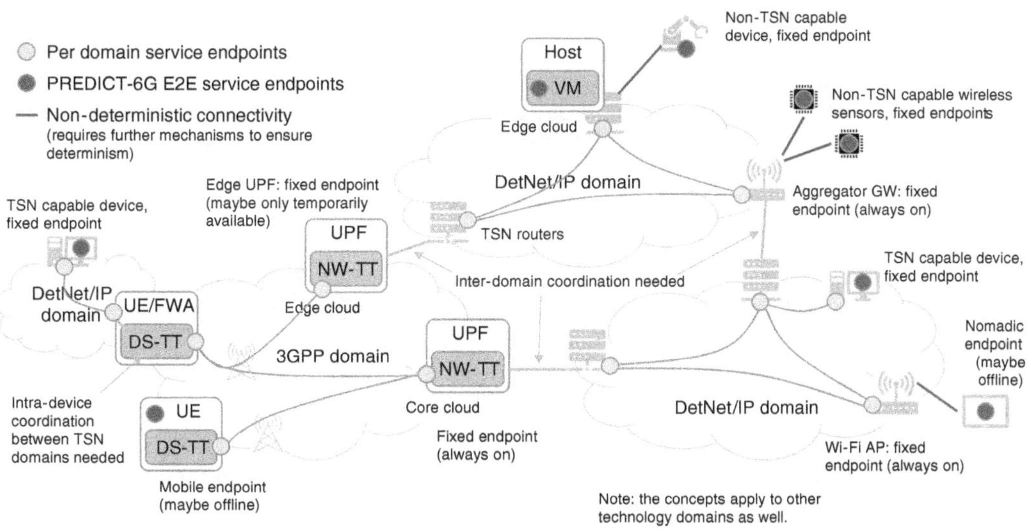

Figure 5.16 Multi-domain multi-technology implementation. *Source:* Contreras et al. (2023) / PREDICT-6G / CC BY 4.0.

based on its observed service performance, the network must be able to predict if it can deliver on a service request with the required level of guarantee. This is depicted in Figure 5.15 as with the numeral 3. The integration of dependable edge computing as described in Section 5.4.2.2 is shown as 5. In order to support E2E time-sensitive communication functionality, as provided by TSN for Ethernet LAN networks or DetNet for IP networks, the integration of the mobile network domain with the E2E domain is needed, as indicated in 6. This can include optimizations for multi-domain operation, where, e.g. the predictable stochastic latency variation of the mobile network can be considered in the multi-domain configuration (see Muzaffar et al. 2023).

The architecture depicted in Figure 5.15 can be implemented considering a multi-domain, multi-technology approach as defined in Figure 5.16. The overarching design goal of this architecture is to enable dependable E2E services over multiple technologies through a fully automated life-cycle management and quality assurance. In this context, Figure 5.16 illustrates the multi-domain approach of the architecture for dependable networks, showcasing the interconnectivity and interaction of various domain-specific management services and entities (as depicted by PREDICT-6G E2E service endpoints, points of ingestion/consumption of dependable traffic) (Contreras et al. 2023). This visual representation underscores the seamless integration and coordination across different network technologies, which is critical for the realization of smart manufacturing, critical communications, and multi-domain dependable communication. Multiple technologies are meant to be different networking stacks, protocols, and control- and user-plane mechanisms, which are defined and governed by their own standards. Prominent technologies within the scope are 3GPP, IETF DetNet[2], IEEE 802.1 TSN[1] and Wi-Fi. The system enables devices connected to network segments implemented through these various technologies to connect to each other via dependable services, to enable use cases such as smart manufacturing, critical communications, or multi-domain dependable communication.

5.6 Radio Protocols

In cellular networks, the RAN connects users (devices) to the CN. The RAN manages wireless connectivity and adapts its radio resources to changing radio conditions. This operation requires close collaboration between the CN and RAN to understand what resources are needed to provision and what is achievable by the air interface. One of the factors significantly impacting the RAN operation is the access network topology, as presented in Section 5.3. This section goes one step further from the research on flexible networks of Section 5.3, to focus on the 6G radio interface and protocols between the UE and the RAN node. Research on 6G radio interface and protocols aims to optimize RAN performance and enable new services to meet a KPI/key value indicator (KVI) requirement. This research leverages the learnings and challenges from 5G deployments and explores support for NR capabilities. Challenges identified in 5G deployments have fuelled new directions for innovation in the RAN protocols. These 5G challenges include complex protocol design structures, complexity arising from 'one solution fits all' configurability, lack of cross-layer awareness for functionalities targeting the same KPI, lower-layer security risks, signalling overhead, and inefficiencies in mobility-related procedures.

To address those challenges, seven principles have been identified for designing future 6G radio interface and protocols (Porambage 2023):

- Focus on fundamental 6G requirements needed from 'day one'.
- One flexible protocol stack that is easily extensible, building on top of modular components for future enhancements.
- Simplicity in comparison to previous generations in implementing and deploying.
- Practical optimization for the needs identified in the use cases, not only extreme cases.
- Fast and reliable protocol operation.
- Separation of protocol layer concerns within the multi-layer protocol stack.
- Design with spectrum sharing in mind.

These principles focus on the RAN domain and are aligned with the key 6G architecture principles introduced in Hexa-X (Ericson et al. 2021; Khorsandi et al. 2022) and later extended in Hexa-X-II (Porambage and Kerboeuf 2023), which emphasize flexibility, scalability, and network simplification. These principles are further delved into in Section 3.1.1.

To illustrate the envisioned 6G radio interface and protocols, Figure 5.17 captures the main innovation areas within the RAN. This RAN protocol close-up diagram focuses on the App, UE, and 6G

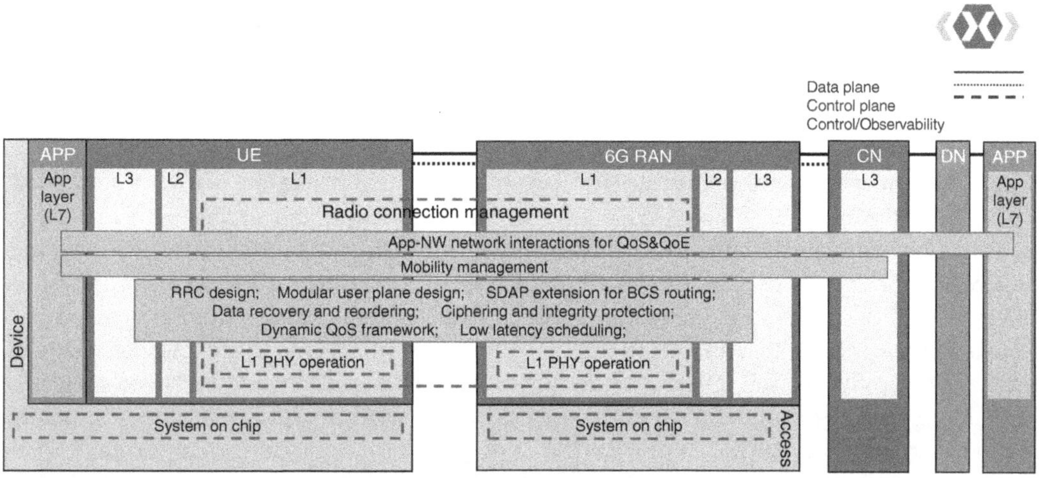

Figure 5.17 Simplified 6G radio protocol close-up diagram from radio protocol. *Source:* Le Sauze and Kerboeuf (2025) / Hexa-X-II.

RAN NF blocks of the 6G E2E system, highlighting the key layers where different functionalities reside. The radio protocol stack is synthesized into four different components. These components influence the management of the RAN interface directly or indirectly and are App layer (Layer 7 on the OSI protocol stack model), Layer 3 (network layer), Layer 2 (data link layer), and Layer 1 (physical layer). Each component may have one or multiple protocol layers within, e.g. Layer 2 in the figure consists of a structure of protocols for CP and/or UP operation in the radio interface (3GPP, 38.300 2025) (i.e. service data adaptation protocol [SDAP], PDCP, RLC, and MAC), while Layer 3 in the figure consists of NAS or RRC protocols. The horizontal boxes in the figure represent a subset of the RAN-related enablers developed in Hexa-X-II. These enablers serve as building blocks for managing radio connections within the 6G RAN node.

5.6.1 Radio Control Plane

The CP in the RAN domain manages the operation of the air interface between the device and the RAN node and directs how the communication between both entities should happen in order to access the cellular network and ensure connectivity. To direct the RAN operation, the RRC protocol layer plays a crucial role in managing radio resources, as it is responsible for key management for security, connection establishment between the device and the RAN node, configuration setup for radio interface operation, measurement configuration, etc. RRC protocol has been used since 3G and has evolved with each new cellular generation. In 5G, the RRC protocol became quite complex, with many different configuration options designed to handle a wide range of configuration scenarios. However, for 6G, the fundamentals of RRC protocol design should be re-evaluated. 6G RRC design should aim for a more scalable, flexible, and resilient design, while ensuring fast and efficient signalling exchanges between the device and the RAN node. Innovations in this research area have envisioned the redesign of the RRC protocol to target a simpler RRC design in 6G. Two main design directions are proposed (Zou and Rugeland 2024):

- **Defining a flat RRC structure:** The target is to simplify RRC encoding structure and optimize RRC signalling required to update or release a configuration, compared to 5G. A combination of multiple recommendations can be followed to achieve this, including: an RRC structure that is tied to one single architecture option in 6G, eliminating hierarchical scope levels of configurability in RRC messages, avoiding including restrictions in the messages encoding and using the procedure description to establish a restriction, and optimizing RRC signalling based on updating RRC configurations only requiring to signal the part of the configuration that has changed (i.e. delta configuration).

- **Defining a modular RRC protocol:** The target is to use a modular design of RRC protocol that relies on RRC profiles to enhance scalability, resilience, and faster transitions between distinct operational modes the device supports. RRC profiles would be configurations stored by both the device and the RAN node. The RAN node and the device negotiate the RRC profiles and tailor them to the device operation and performance profile. This way, the RRC profiles can ensure that both device capabilities and network feature implementation support are considered as part of the RRC profile determination. As a result, each defined profile would be linked to device capabilities and the requested service categories the device will use with the network (e.g. power saving, voice call, and MBB).

Both design directions are complementary and can be applied independently or together to achieve a simplified and agile configuration of the air interface with RRC in 6G. While 5G supports multiple architecture options, both design directions assume RRC is built based on a single architecture option in 6G.

5.6.2 Radio User Plane

The radio UP is responsible for the transmission of users traffic from a network node to a device, and vice versa. To enable the data transmission, the radio UP protocol stack is divided in different layers that prepare the data packet to go through the radio interface and are responsible of different functionalities, e.g. security, resource distribution, rate control, data recovery, discard operation, and reassembly. In Porambage (2023), Zou and Rugeland (2024), and Le Sauze and Kerboeuf (2025), several proposals for enhancements to UP design and parts of the UP-protocol stack (e.g. SDAP, MAC, RLC, and PDCP) were introduced to offer an overall better performance and/or reduced power consumption and a simplified system complexity.

UP design proposals target the simplification of the UP protocol stack using a modular design (Zou and Rugeland 2024; Le Sauze and Kerboeuf 2025). The modularization relies on two main components: anchor module and additional modules. Those modules can be configured to support essential control/UP functions or functionality related to higher data rates support. Therefore, different device types optimized for different purposes can natively be supported with a combination of modules.

Focusing on the UP protocol stack, another proposal explores modifications to SDAP layer for handling the data generated by beyond communication services (BCS) not destined for an external IP address (Zou and Rugeland 2024). To avoid superfluous processing in the user plane function (UPF) and time to transmit data to and from the UPF, the proposal modifies the SDAP header to replace QoS flow identifier (QFI) with a generic service flow identifier (SFI). SFI would allow the BS to identify which NF the data should be forwarded to.

Going further down in the UP protocol stack and its functionalities, other proposals are concerned with data recovery and reordering mechanisms. These mechanisms ensure reliability of transmission, namely the enhancement of the automatic repeat request (ARQ) and HARQ retransmission procedures of legacy cellular systems. Proposals in this area introduce cross-layer interaction-based mechanisms between RLC and PDCP layers to avoid RLC window

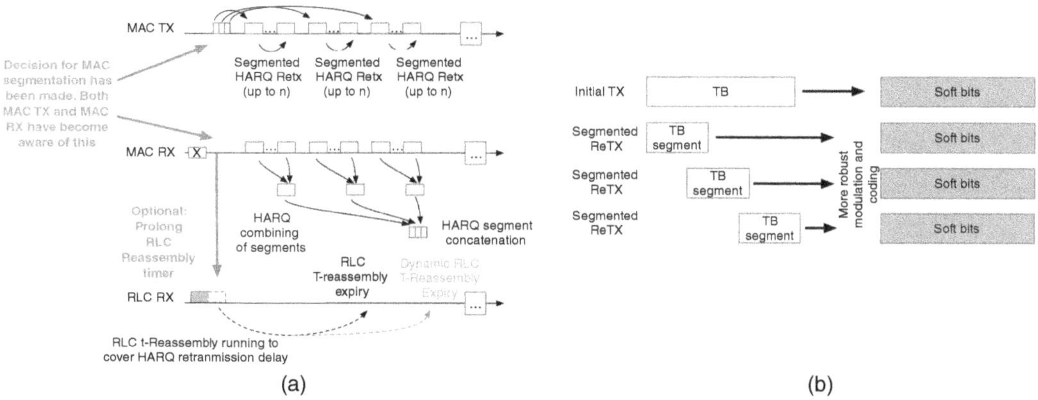

Figure 5.18 MAC segmentation for HARQ retransmissions: (a) example flow highlighting MAC segmentation and HARQ combining along with RLC timer; (b) visualisation of MAC segmentation and soft bits generation.

blocking or unnecessary delays by allowing window moving operations, or between RLC and MAC layers to early terminate RLC timers associated with retransmission processes (Porambage 2023; Zou and Rugeland 2024). Another proposal tackles the issue of the unnecessary HARQ retransmissions in legacy cellular systems (e.g. in case of highly degraded channel conditions during the reception of the initial transport block [TB]) that leads to wasting resources and increasing the latency of other transmissions. To combat this, a new mechanism is introduced, where the MAC layer shall determine conditions where (further) HARQ retransmissions would not help the decoder and instead reset the transmission of the TB to avoid wasting resources (Le Sauze and Kerboeuf 2025). The MAC layer shall have a new function 'MAC segmentation for HARQ retransmissions' to aid with such retransmissions. This function shall be applicable in both DL and UL and is independent of the MAC PDU's content (i.e., whether it includes full or partial RLC PDUs). Figure 5.18 provides an example of the MAC segmentation for HARQ retransmissions.

A final proposal for data recovery and reordering mechanisms tackles the RLC layer, more specifically the need for RLC acknowledged mode (AM) in UL direction (Le Sauze and Kerboeuf 2025). The BS that controls the HARQ procedure knows when a TB has been correctly or erroneously received, and hence there is no need for the UE to implement RLC AM in the UL direction. Therefore, a new scheme for RLC is proposed, namely the Asymmetric RLC protocol, where the UE would only require RLC unacknowledged mode (UM) functionality for both UL and DL with some RLC AM-related functionality only for the DL path, or vice versa, where the RLC AM-related functionality is only for the UL path.

Moreover, proposals targeting ciphering and integrity protection in legacy cellular systems were introduced to offer an overall better protection, as well as to allow for reduced user complexity. In Porambage (2023) and Zou and Rugeland (2024), proposals in this area explore how to extend MAC operation for 6G access stratum security. One proposal moves the security functionality from PDCP to MAC to ensure the protection of the layer 2 headers that were kept unprotected in legacy 3GPP cellular systems (e.g. 5G) and to avoid the issues that arise from this. Another proposal introduces the option to include a 'protected zone' per TB (MAC PDU) that is encrypted and integrity protected by the MAC layer before the transport block is passed to the physical layer.

5.6.3 Mobility Procedures

Mobility procedures are a crucial part of wireless communication, which ensures service continuity. One of the crucial mobility procedures is handover, where a serving cell hands a device over to a neighbouring cell if the signal quality of the serving cell degrades. In 5G, handover relies on the signal measurement reports based on the pre-configured events, such as the A3 event, which indicates that a neighbouring cell is stronger than the serving cell. In other words, the UE currently reacts when the pre-determined reporting condition is satisfied and either waits for a handover command in the baseline case or performs handover to a neighbouring cell in the case of CHO or lower layer triggered mobility. Successful handovers can cause up to around 40 ms service interruption time (3GPP, 38.744 2023), whereas handover failures can cause up to couple of hundreds milliseconds service interruption time, due to cell selection and re-establishment procedures.

Multiple proposals have been explored in Porambage (2023), Zou and Rugeland (2024), and Le Sauze and Kerboeuf (2025) to optimize mobility decision, mobility procedure and related operations, as well as mobility awareness, etc. One proposal in this area explores the harmonization of mobility procedures, unifying their control commands. Other proposals target a leaner system mobility decoupling operations and information needed for IDLE and CONNECTED mode. To reduce service interruption time, one proposal introduces proactive decisions with the help of AI/ML models that leverage the mobility experiences of the UEs to predict the signal degradations or the potential target cells over time. Another proposal aiming to reduce service interruption time is to skip the random access channel (RACH) procedure for baseline handover using an early time advance acquisition mechanism.

Other proposals target the avoidance of the costly handover preparation procedure, in terms of time, signalling, and resources. One proposal introduces the concept of unprepared handover. In the unprepared handover, a (candidate) target node prepares a default (cell-specific) configuration. Then when a source node has a UE that needs to be handed over to a new cell and satisfies the associated criteria corresponding to the default target configuration, the handover command can be generated and sent to the UE. Resulting in the avoidance of the costly handover preparation phase between neighbouring nodes. At the same time, the UE information is transmitted to the target node to initiate preparation of the UE-specific configuration at target node and is delivered to the UE when it arrives. Another proposal relies on delegating more autonomy to lower layers (PHY and MAC) to control the mobility management. This approach will further improve the benefits of the LTM mechanism, already introduced in 5G, for 6G.

Additionally, the mobility procedures should support computation offload use cases. Existing procedures, such as RRC connection establishment and mobility, are agnostic to the computation capability of the hardware and only aim to improve the communication quality (e.g. signal quality and throughput). The UE should evaluate its own computation requirements as well as the computation capabilities of the BS(s) before initiating mobility procedures. To support resilient distributed computing, true convergence of communication and computing is required. The computation-aware mobility (Zou and Rugeland 2024) enables the different computation offload use cases by taking into consideration both the computation requirements and the computation capabilities of the cellular system in order to make mobility-related decision.

5.6.4 App-Network Interactions for Service Differentiation and QoS/QoE Management

Service differentiation in cellular networks enables operators to offer distinct services to their customers. The network capabilities and resources can be tailored, e.g. QoS, to deliver different performance levels to the customers, following the customer needs and preferences. There are multiple approaches for utilizing App-network interactions for 6G service differentiation and QoS/quality of experience (QoE) management enhancement that were introduced in Porambage (2023), Zou and Rugeland (2024), and Le Sauze and Kerboeuf (2025). These approaches can be divided into two categories, ones that target radio resource allocation optimization for low latency, and others that target enhancing the QoS framework.

In the first category on radio resource allocation, there are proposals that explore how to enhance the UL grant configuration and minimize latency from UL packet arrival to actual transmission. One proposal (Porambage 2023; Zou and Rugeland 2024) presents enhancements for UL scheduling, namely low-latency scheduling, exploring two schemes. In one scheme, the UE indicates what traffic characteristics (e.g. number of upcoming streams, packet sizes, cadence, jitter constrains, and expected packets arrival time) to expect via a new reporting mechanism, called periodic cadence report (PCR), sent to the BS. In the second scheme, a service-specific scheduling request (SR) as a further enhancement to the regular SR used in legacy dynamic grant methods is introduced. The idea is that the network may configure multiple SR configurations for different Apps, services, QoS flows, or traffic profiles.

Another proposal in the radio resource allocation (Zou and Rugeland 2024; Le Sauze and Kerboeuf 2025) introduces a new mechanism using MAC CEs that would allow for the dynamic setup, reconfiguration, and release of data radio bearers (DRBs) based on App or service requirements in comparison to legacy cellular systems, enabling the anticipated new services and QoS requirements in 6G.

In the second category of the QoS model, proposals explore a more dynamic QoS framework that can rapidly adjust the operation between the UE and the network. One proposal introduces a new two-stage procedure (Le Sauze and Kerboeuf 2025). In the first stage, the UE agrees with the network on multiple quality levels that can then be used as a reference for the second stage. In the second stage, the UE is informed on a regular basis about the quality level the network may serve to this UE currently and/or in the future steering the UE to known and sustainable levels. Such a procedure would allow for more flexibility and more fairness in scheduling by the network as well as better planning and more efficiency on the RAN level. Another proposal in this category introduces an additional resource type for 6G QoS framework that can complement the current 5G resource types available. This new resource type is optimized for high bandwidth adaptive Apps that tolerate adaptation of their QoS requirements in a controlled and responsive manner within RAN congestion timescales (e.g. tens of milliseconds), and with the criterion for adaptation being to optimize the achievable rate while fulfilling a maximum packet delay budget. The objective of this additional resource type is to enable the RAN to use soft upper bounds to autonomously upgrade/downgrade QoS metrics (i.e. bit rate, packet delay budget, and packet error budget) within the given range configured for the QoS flow. Per QoS metric, the tolerable range can be defined by the App.

5.7 Quantum-Enhanced Network Functionalities

The emergence of quantum information science (Dowling and Milburn 2003) has been preceded by significant contributions made by numerous distinguished scientists. Quantum technology can be used for manipulation, transmission, storage, and measurement of quantum information, enabling the construction of quantum computers and quantum networks that are capable of providing a lot of computational and communicational advantages (Ferrara et al. 2021) in the context of 6G networks. Quantum technology can play a role in mobile communication to meet specific KPIs without the typical compromises associated with classical technology solutions. The modular design of 6G is a key enabler to allow quantum services to operate within a hybrid regime that supports seamless interaction between classical and quantum components. This hybrid quantum-classical integration builds on current network functionality and does not require architectural changes to the 6G design. In this section, the emphasis is on the use of quantum technology and IoT or its forthcoming iteration known as the tactile internet.

Quantum networks will allow the transmission of quantum bits (qubits) between quantum information processors through quantum channels. The effectiveness of a quantum network in accomplishing specific tasks can lead to significant advantages over classical communication, particularly in areas such as latency reduction, provably secure cryptography, and steganography. Furthermore, utilizing quantum networks in conjunction with classical networks can provide numerous benefits for transmitting both classical and quantum information within 6G communication networks, tactile internet networks, and similar frameworks. By viewing quantum and classical networks not as separate, static layers but as interconnected systems, it becomes feasible to employ concepts like shared entanglement to balance workloads and potentially decrease communication complexity.

As discussed in Section 5.4, 6G will seamlessly integrate terrestrial cellular communications with satellite-based non-terrestrial communication systems, potentially incorporating quantum technology. In a dynamic classical-quantum hybrid network, satellite nodes can participate in tasks such as the distribution of quantum entanglement or facilitate space-based quantum key distribution (QKD) to enhance security for both classical and quantum computations at various nodes. Consequently, quantum communication technologies supported by satellites and space-based quantum networks significantly bolster the resilience of information against malicious security threats. A notable example of this resilience, which is demonstrably secure and operates optimally when the physical health of the node is maintained, is the use of satellite QKD to establish secure keys between RAN components. These components must be deployed in a manner that ensures secure coupling to mitigate network security vulnerabilities, such as man-in-the-middle attacks or eavesdropping, particularly in networks that adhere to stringent security requirements as outlined by current ultra-reliable low-latency communication (URLLC) and future xURLLC technology standards.

In the realm of IoT integration within 6G and tactile internet networks, the App of quantum communication technology presents significant benefits in meeting various KPIs established by the industry. For example, quantum technology may enhance the resilience of the solutions developed against noise, cyberattacks, energy leaks, and other vulnerabilities. Many industries face challenges in fulfilling the latency and computational demands of their industrial and consumer technology solutions. This often results in substantial costs associated with suboptimal energy efficiency or inadequate security measures. Therefore, quantum technology may play a role in bolstering the resilience of existing classical solutions, ensuring compliance with evolving standards in terms of security and energy efficiency.

One notable instance is the cloud RAN based on 'quantum entanglement' (Khalif et al. 2019), which demonstrated a reduction in time delay during the handover process (X2AP) compared to classical C-RAN by approximately 0.3 ms. This reduction is significant, particularly in the realm of tactile internet networks that demand high time sensitivity and low latency. Such networks necessitate specialized and proficient IoT controllers to ensure classical resilience in traffic scheduling, queuing, and other functions. Additionally, these IoT controllers must maintain security while transmitting data to the associated IoT devices and haptic robotic systems, all while striving to achieve energy efficiency to meet sustainability requirements essential for successful implementation. To enhance resilience through improved security, recent advancements have employed QKD between the controllers and connected devices to thwart potential malicious attacks on the wireless links established between them. Conversely, another initiative has focused on achieving energy savings for IoT devices by integrating QKD with software-defined networking for their fibre-optic connected controllers.

The potential for quantum technology to leverage a different form of QKD is evident in the continuous variable QKD (CV-QKD). In contrast to the previously mentioned discrete variable QKD (DV-QKD), which necessitates the use of coherent sources and single photon sources that are challenging to integrate with current telecommunication systems, CV-QKD employs technologies such as homodyne and heterodyne detectors that are compatible with classical optical communication. This compatibility not only facilitates near-term deployment at reduced costs but also allows for the App of existing resilience studies relevant to classical technologies. However, it is important to highlight that in the context of free-space and atmospheric quantum communication, the use of continuous variable quantum entanglement with squeezed light states is generally discouraged due to significant channel losses that can compromise the integrity of the quantum information. Although initial findings regarding CV-QKD were met with scepticism from the telecommunications sector, recent evaluations (Kundu et al. 2023) of secret key rates (SKR) in multiple-input multiple-output (MIMO) CV-QKD systems under limited eavesdropping conditions – reflecting more realistic scenarios where an eavesdropper lacks complete control over the environment – have revealed substantial improvements in SKR compared to unrestricted conditions. Furthermore, it has been observed that increased squeezing can lead to a degradation of the SKR. These findings underscore the potential for non-squeezed coherent state-based CV-QKD protocols and represent a significant advancement in near-term implementable quantum technology.

There are ongoing investigations into the use of quantum network simulators in conjunction with classical network emulators. These studies aim to illustrate how classical information can be transmitted over quantum channels at enhanced data rates. Future research should focus on employing simulation tools to explore the benefits of quantum networks and quantum communication channels for both classical and quantum data transmission. It is essential to utilize more sophisticated platform-specific quantum simulators that can manage multi-qubit interactions while addressing issues such as crosstalk and fluctuations in connectivity caused by quantum interference. This approach will enable a more effective demonstration of the trade-offs in the resiliency curve when optimizing specific NFs, such as latency and throughput.

5.8 Conclusions

As mentioned in the introduction of this chapter, the aim of the 6G architecture is to sustainably expand existing MSs while enabling new Apps and bridging the digital divide through a flexible, modular, and cloud-native system that considers social, economic, and environmental factors. These **novel services** are sensing, computing, and AI. Sensing in 6G opens up a wide range of new services and Apps, allowing the network to find objects which are not normally seen. To support this, the architecture in 6G needs functionality for handling the sensing requests, controlling and configuring relevant sensing units (e.g. UEs or gNBs), processing the sensing data, and exposing it to the end users. Another novel service is the Compute offloading, which allows power- or CPU-constrained devices to offload tasks to a compute offloading cluster. A loosely integrated framework is proposed for this, as it requires no modifications to the RAN and NAS protocols and also allows the compute offloading cluster to use a remote execution environment. Consequently, it is less disruptive to existing cellular standards compared to other solutions. Several procedures to handle the compute offloading, such as discovery, registration, and actual offloading of compute tasks, need to be defined. The third novel service is AI and AIaaS. AIaaS ultimately facilitates access to advanced AI services that can be utilized both within the network and by end users. AIaaS needs to include standardized methods, such as discovery and registration, to handle the training and updates of the AI and ML models (MLOps). An essential pillar for the novel services is the optimal placement of AFs. By using real-time network insights and advanced algorithms via the AEP's interfaces, 6G networks can provide services where and when needed, achieving lower latency, better energy efficiency, and enhanced reliability, even in dynamic and diverse scenarios.

The **6G cloud native** architecture in Section 5.2 introduces a modular 6G network design capable of supporting innovative cloud functions and facilitating the integration of novel services, and offers flexibility for dynamic network configurations. Designing a modular 6G architecture involves a trade-off between performance and flexibility. Although virtualization and cloud-based NFs enhance network flexibility, they also increase complexity and pose reliability challenges due to the dynamic nature of modular components. Here, we introduce a novel approach on how to designing modular NFs for the CN CP is proposed here, which involves grouping NFs to streamline interactions and improve performance. This design leads to faster procedure execution times and reduced message counts. However, it also limits the flexibility in deploying larger, more coarse-grained NFs.

Looking ahead to 6G, the analysis presented here suggests maintaining the AMF (or its 6G equivalent) as a central point for RAN-CN communication, primarily for security and 5G-to-6G migration. In 6G, the focus shifts to building upon existing principles while prioritizing security and seamless migration from 5G. Data-centric SBA for the CN emerges as a promising approach, offering significant advantages in automation, service composability, and performance. Furthermore, network federation will enable seamless cross-border services, enhancing user experience and fostering global connectivity. As part of this evolution, the hybrid integration of quantum technologies within the modular 6G architecture can further enhance network services like synchronization, security, and distributed computing without structural changes. To summarize, the adoption of cloud-native architectures promises enhanced resource availability, flexibility, and efficiency, paving the way for a more agile and responsive network infrastructure.

Flexible networks, introduced in Section 5.3, represent a key differentiator for 6G compared to legacy cellular technologies. They address challenges like rapid growth in connected devices, higher data rates, wider service coverage, lower power consumption, enhanced resilience, and improved user privacy. One way to achieve this is through the introduction of subnetworks that offload functionalities across nodes, extend coverage to out-of-network devices, and enable efficient communication between devices of varying capabilities.

Another significant aspect of flexible networks is multi-connectivity, where the goal for 6G is to implement a single architecture, avoiding all architectural DC options that were part of 5G (e.g. E-UTRAN New Radio dual connectivity (EN-DC), NR E-UTRAN DC (NE-DC), and NR-DC 3GPP, 37.340 2024). The proposal is to have an enhanced CA solution as the single architecture, which would include valuable features that only existed in 5G DC. For instance, the enhanced CA will offer faster PCell changes, improved SCell additions via enhanced EMR, and reduced service interruptions with the introduction of SCell recovery like DC in 5G. Additionally, a RAN-integrated WLAN and 6G solution is proposed, which leverages a trusted WT that is within cellular coverage to potentially support UEs with either increased reliability or throughput.

Finally, spectrum co-existence is addressed with the introduction of MRSS for 5G-6G spectrum sharing, allowing for a more dynamic sharing approach between 5G and 6G compared to the spectrum sharing between LTE and NR (DSS). The recommended path forward includes a single-step migration based on 6G SA, where MRSS and 6G CA are critical for enabling such a transition.

In the context of 6G, the **NTN** component is anticipated to extend the coverage of the cellular TN while also enhancing the resiliency of the infrastructure and potentially improving sustainability in specific scenarios. In this chapter, different reference NTN deployment scenarios have been presented with different characteristics, such as transparent or regenerative satellite payload design, different radio beams strategies for the service link, such as Earth fixed or Earth moving beams. Regenerative payloads are likely to be increasingly deployed since they can provide advanced capabilities such as store-and-forward capabilities, inter-satellite links, UE-satellite-UE communications, position navigation and timing capabilities, and/or ISAC.

In order to support time-critical Apps with strict performance requirements and a need for high reliability and availability, 6G needs to provide **dependable connectivity**. The 6G network needs to be able to observe its latency and data rate performance and predict which performance commitments it can provide reliably to an App. 6G needs to support dependable networking E2E, which is often provided over multiple network domains, e.g. using deterministic networking technologies like Ethernet TSN or IP DetNet.

Radio protocols, introduced in Section 5.6, explore innovations for 6G radio protocols compared to legacy 5G design. These innovations are focused on current 5G challenges in four different areas: radio UP design, radio CP design, mobility procedures, and App-network interactions. Innovations in 6G radio protocols have aimed to simplify radio protocol design, enable tighter coordination between protocol layers to optimize their operation, enhance mobility management, and allow more adaptiveness in the radio interface operation. To achieve these goals, the 6G radio protocol innovations presented utilize the existing ecosystem of devices and infrastructure, positioning 6G radio design as an evolution of 5G.

Acronyms and Abbreviations

3GPP	3rd generation partnership project
AEP	Application enablement platform
AF	Application function
AI	Artificial intelligence
AIaaS	AI as a service
AIMLE	AI/ML enablement service
AMF	Access and mobility management function
AP	Action point
API	Application programming interface
BCS	Beyond communication services
BS	Base station
CA	Carrier aggregation
CAMARA	A project by The Linux Foundation
CAPIF	Common API framework
CC	Component carrier
CCN	Compute control node
CHO	Conditional handover

CN	Core network
CP	Control plane
CPU	Central processing unit
C-RAN	Cloud radio access network
CRMS	Compute resource management service
CU	Centralized unit
CV-QKD	Continuous variable quantum key distribution
DAPS	Dual active protocol stack
DC	Dual connectivity
DCN	Data-centric networking
DetNet	Deterministic networking
DFP	Data flow programming
DL	Downlink
D-MIMO	Distributed multiple-input multiple-output
DP	Data plane
DSS	Dynamic spectrum sharing
NR-Uu	NR-Uu interface
E2E	End to end
EDGEAPP	Edge applications
EMR	Early measurements reports
EPC	Evolved packet core
ETSI	European Telecommunications Standards Institute
FDD	Frequency division duplex
FLOPS	Floating-point operations per second
FR	Frequency range
FTT	Fast Fourier transform
GEO	Geostationary Earth Orbit
GNSS	Global navigation satellite systems
GSO	Geostationary orbit
HAPS	High-altitude platforms
HEO	Highly eccentric orbit
IEEE	Institute of electrical and electronics engineers
IETF	Internet Engineering Task Force
IP	Internet protocol
ISAC	Integrated sensing and communication
ISL	Inter-satellite links
JCAS/ICAS	Joint/integrated communication and sensing
KPI	Key performance indicator
KVI	Key value indicator
LEO	Low Earth Orbit
LLS	Low-layer split
LTE	Long term evolution
LTM	Lower layer triggered mobility
MAC	Medium access control
MBB	Mobile broadband
MC	Multi-connectivity
MCG	Master cell group

MEC	Multi-access edge computing
MIMO	Multiple-input multiple-output
M&O	Management and operations
MRSS	Multi-RAT spectrum sharing
MTC	Machine-type communication
NAS	Non-access stratum
NEF	Network exposure function
NFs	Network functions
NG	Next generation
NGAP	Next generation application protocol
NG-RAN	Next-generation radio access network
NGSO	Non-geostationary orbit
NR	New radio
NTN	Non-terrestrial networks
NWDAF	Network data analytics function
OA	Open access
ON	Offloading node
O-RAN	Open radio access network
PCell	Primary cell of the master cell group
PSCell	Primary cell of the secondary cell group
PDCCH	Physical downlink control channel
PDCP	Packet data convergence protocol
PDSCH	Physical downlink shared channel
PDU	Protocol data unit
PNT	Positioning navigation and timing
QKD	Quantum key distribution
QoCS	Quality of compute service
QoE	Quality of experience
QoS	Quality of service
RAN	Radio access network
RLC	Radio link control
RLF	Radio link failure
RRC	Radio resource control
RTOS	Real-time operating system
RU	Radio unit
RX	Receiver
SA	Standalone
SA1	3GPP System aspects working group 1
SA5	3GPP System aspects working group 5
SA6	3GPP System aspects working group 6
SAN	Satellite access node
SBA	Service-based architecture
SCell	Secondary cell
SCF	Sensing control function
SCG	Secondary cell group
SCTP	Stream control transmission protocol
SDAP	Service data adaptation protocol

SDM	Spatial division multiplexing
SEAL	Service enabler architecture layer
SeMF	Sensing management function
SFI	Service flow identifier
SLA	Service-level agreement
SN	Secondary node
SPF	Sensing processing function
SR	Scheduling request
TDM	Time division multiplexing
TN	Terrestrial networks
TSN	Time-sensitive networking
TX	Transmitter
UE	User equipment
UL	Uplink
UPF	User plane function
URLLC	Ultra-reliable low-latency communications
WLAN	Wireless local area network
WT	WLAN terminal
xURLLC	Extreme ultra-reliable low-latency communications

References

3GPP, 22.261 (2025). 5G; Service requirements for the 5G system. 3GPP TS 22.261, V18.17.0, April 2025.

3GPP, 22.837 (2024). Feasibility study on integrated sensing and communication. 3GPP TR 22.837, V19.4.0, June 2024.

3GPP, 23.434 (2024). Technical Specification Group Services and System Aspects; Service enabler architecture layer for verticals (SEAL); Functional architecture and information flows. TS 23.434, V19.2.0, June 2024.

3GPP, 23.482 (2024). Technical Specification Group Services and System Aspects; Functional architecture and information flows for AIML Enablement Service. 3GPP TS 23.482, V19.0.0, December 2024.

3GPP, 23.558 (2024). Architecture for enabling edge applications. TS 23.558, V19.3.0, September 2024.

3GPP, 29.522 (2024). Technical specification group core network and terminals; 5G System; Network Exposure Function Northbound APIs. TS 29.522, V19.0.0, September 2024.

3GPP, 36.102 (2025). LTE; Evolved Universal Terrestrial Radio Access (E-UTRA); User Equipment (UE) radio transmission and reception for satellite access. 3GPP TS 36.102, V18.8.0, April 2025.

3GPP, 36.108 (2025). LTE; Evolved Universal Terrestrial Radio Access (E-UTRA); Satellite Access Node radio transmission and reception. 3GPP TS 38.108, V18.8.0, April 2025.

3GPP, 37.340 (2024). Evolved Universal Terrestrial Radio Access (E-UTRA) and NR; Multi-connectivity; Overall Description; Stage 2. TS 37.340, v18.4.0, December 2024.

3GPP, 38.101-5 (2025). NR; User Equipment (UE) radio transmission and reception; Part 5: Satellite access Radio Frequency (RF) and performance requirements. 3GPP TS 38.101-5, V18.9.0, April 2025.

3GPP, 38.108 (2025). NR; Satellite Access Node radio transmission and reception. 3GPP TS 38.108, V18.6.0, April 2025.

3GPP, 38.300 (2025). NR; NR and NG-RAN Overall description; Stage-2. 3GPP TS 38.300, V18.5.0, March 2025.

3GPP, 38.744 (2023). Study on Artificial Intelligence (AI)/Machine Learning (ML) for mobility in NR; (Release 19) and NR; Multi-connectivity; Overall Description; Stage 2. TR 38.744, v18.4.0, December 2024.

3GPP, TR 38.811 (2020). Study on New Radio (NR) to support non-terrestrial networks. 3GPP TR 38.811, V15.4.0, September 2020.

3GPP, TR 38.821 (2023). Solutions for NR to support non-terrestrial networks (NTN). 3GPP TR 38.821, V16.2.0, March 2023.

3GPP, TR128.908 (2024). Study on Artificial Intelligence/Machine Learning (AI/ML) management. 3GPP TR 128.908, V18.1.0, October 2024.

5G-ACIA (2021). Integration of 5G with time-sensitive networking for industrial communications. February 2021.

Akgul, O., Ericson, M., Botsinis, P. et al. (2025). Hexa-X-II Deliverable D3.5 "Final architectural framework and analysis," Hexa-X-II project, 2025. [Online]. Available at: https://zenodo.org/records/15774984 (accessed August 2025).

Almodóvar, J. and Vinagre, C., et al. (2023). Deliverable D1.2 – 6G use cases and requirements. [Online]. Available at https://zenodo.org/records/16779960 (accessed August 2025).

Baldoni, G., Quevedo, J., Guimarães, C. et al. (2024). Data-centric service-based architecture for edge-native 6G network. *IEEE Communications Magazine* 62 (4): 32–38. https://doi.org/10.1109/MCOM.001.2300178.

Contreras, L.M., de la Oliva, A., Vázquez, J. et al. (2023). D1.2, PREDICT-6G framework architecture and initial specification, December 2023. PREDICT-6G Project. [Online]. Available at: https://zenodo.org/records/12167871 (accessed August 2025).

Dowling, J.P. and Milburn, G.J. (2003). Quantum technology: the second quantum revolution. *Philosophical Transactions of the Royal Society A* 3611655–1674.

S. Egger, F. Dürr, B. Varga, et al. (2025). "Wireless-aware TSN engineering: implications for 5G and upcoming 6G networks," Submitted for Publication, 2025.

Ericson, M., Flinck, H., Vlacheas, P. et al. (2021). Hexa-X Deliverable D5.1: initial 6G architectural components and enablers, December 2021. [Online]. Available at: https://hexa-x.eu/wp-content/uploads/2022/03/Hexa-X_D5.1_full_version_v1.1.pdf (accessed August 2025).

Ericson, M., Botsinis, P., Saimler, M. et al. (2024). Hexa-X-II Deliverable D3.3 "Initial analysis of architectural enablers and framework," Hexa-X-II project, 2024, [Online]. Available at: https://zenodo.org/records/16995133 (accessed August 2025).

Ferrara, R., Bassoli, R., Deppe, C. et al. (2021). The computational and latency advantage of quantum communication networks. *IEEE Communications Magazine* 59 (6): 132–137. Article 9475166. https://doi.org/10.1109/MCOM.011.2000863.

Ferrari, E., Rosales, R., Szilagyi, P. et al. (2024). Release 2 of PREDICT-6G MDP innovations. PREDICT-6G project. [Online]. Available at: https://zenodo.org/records/14865183 (accessed August 2025).

Goshi, E., Stahl, R., Harkous, H. et al. (2023). PP5GS—an efficient procedure-based and stateless architecture for next-generation core networks. *IEEE Transactions on Network and Service Management* 20 (3): 3318–3333. https://doi.org/10.1109/TNSM.2022.3230206.

Grosjean, L., Sachs, J., Ansari, J. et al. (2025). A framework for communication-compute-control co-design in cyber-physical systems. *Electronics* 14 (5): https://doi.org/10.3390/electronics14050864.

Gross, J. and Sharma, G.P. (2023). DETERMINISTIC6G Deliverable D2.1 "First Report on 6G-centric enablers," DETERMINISTIC6G project, 2023. [Online]. Available at: https://deterministic6g.eu/index.php/library-m/deliverables (accessed August 2025).

Hadziaganovic, A. (2024). DETERMINISTIC6G Deliverable D3.3 "Edge computing and situational awareness via digital twining," 2024. [Online]. Available at: https://deterministic6g.eu/index.php/library-m/deliverables (accessed August 2025).

Höftberger, O. (2025). DETERMINISTIC6G Deliverable D1.3, "Report on Dependable Service Design," DETERMINISTIC6G project, 2025. [Online]. Available at: https://deterministic6g.eu/index.php/library-m/deliverables (accessed August 2025).

IEEE 802.1 TSN (2025). Time-Sensitive Networking (TSN) task group. [Online]. Available at: https://1.ieee802.org/tsn/ (accessed August 2025).

IETF DetNet (2025). IETF Deterministic Networking (DetNet). [Online]. Available at: https://datatracker.ietf.org/group/detnet/about/ (accessed August 2025).

Jardim, M.A. (2025). DETERMINISTIC6G Deliverable D2.3 "Second report on 6G-centric enablers," DETERMINISTIC6G project, 2025. [Online]. Available at: https://deterministic6g.eu/index.php/library-m/deliverables (accessed August 2025).

Jiang, W., Han, B., Habibi, M. A. et al. (2021). "The road towards 6G: a comprehensive survey," *IEEE Open Journal of the Communications Society*, vol. 2, pp. 334–366, 2021.

Khalif, B.N.A., Hasan, J.A.K., and Alhumaima, R.S. (2019). Performance analysis of quantum based cloud radio access networks. *IEEE Access* 8: 18123–18133.

Khorsandi, B.M., Hoffmann, M., Uusitalo, M. et al. (2022). Hexa-X Deliverable D1.3 "Targets and requirements for 6G – initial E2E architecture," March 2022. [Online]. Available at: https://hexa-x.eu/wp-content/uploads/2022/03/Hexa-X_D1.3.pdf (accessed August 2025).

KServe (2025). KServe documentation. Available at: https://kserve.github.io/website/latest/ (accessed April 2025).

Kundu, N.K., McKay, M.R., Conti, A. et al. (2023). MIMO terahertz quantum key distribution under restricted eavesdropping. *IEEE Transactions on Quantum Engineering* 4: 1–15.

Le Sauze, N. and Kerboeuf, S. (2025). Hexa-X-II Deliverable D2.5 "Final overall 6G system design," Hexa-X-II project, 2025 [Online]. Available at: https://zenodo.org/records/15796361 (accessed August 2025).

MLflow (2025). MLflow. Available at: https://mlflow.org/ (accessed April 2025).

Mostafavi, S., Sharma, G.P., and Gross, J. (2023). Latency probability prediction for wireless networks: focusing on tail probabilities. *IEEE Globecom* 2023: 2023.

Multi-access Edge Computing (MEC) (2024). Framework and reference architecture, V3.2.1, 2024-04.

Muzaffar, R., Hadziaganovic, A., and Hamidovic, D. (2023). DETERMINISTIC6G Deliverable D3.1 "Report on 6G convergence enablers towards deterministic communication standards," DETERMINISTIC6G project, 2023. [Online]. Available at: https://deterministic6g.eu/index.php/library-m/deliverables (accessed August 2025).

OpenFL (2025). OpenFL. Available at: https://openfl.io/ (accessed April 2025).

O-RAN (2025). O-RAN Alliance, "Control, user and synchronization plane specification", O-RAN.WG4.TS.CUS.0-R004-v17.01, 2025.

Porambage, P. (2023). Hexa-X-II Deliverable D2.2 "Foundation of overall 6G system design and preliminary evaluation results" December 2023. [Online]. Available at: https://zenodo.org/records/16995047 (accessed August 2025).

Porambage, P and Kerboeuf, S. (2023). Hexa-X-II Deliverable D2.1 "Draft foundation for 6G system design" June 2023. [Online]. Available at: https://zenodo.org/records/16995021 (accessed August 2025).

Prefect (2025). Prefect. Available at: https://www.prefect.io/ (accessed April 2025).

Rico-Menendez, D., Picazo-Martínez, P., Barroso-Fernández, C. et al. (2025). Experimentation on a wireless multi-domain deterministic network: the SLICES-Madrid approach. *IEEE Communications Magazine* 63 (2): 54–60. https://doi.org/10.1109/MCOM.001.2400320.

Rojas, E., Guimarães, C., and de la Oliva, A. (2024). Beyond multi-access edge computing: essentials to realize a mobile, constrained edge. *IEEE Communications Magazine* 62 (1): 156–162. https://doi.org/10.1109/MCOM.017.2300056.

Sachs, J. (2024). DETERMNISTIC6G Deliverable D1.2 "First report on DETERMINISTIC6G architecture," DETERMINISTIC6G project, 2024. [Online]. Available at: https://deterministic6g.eu/index.php/library-m/deliverables (accessed August 2025).

Seldon (2025). Seldon core documentation. Available at: https://docs.seldon.io/projects/seldon-core/en/v2/index.html (accessed April 2025).

Sharma, G.P., Patel, D., Sachs, J. et al. (2023). Toward deterministic communications in 6G networks: state of the art, open challenges and the way forward. *IEEE Access* 11: 106898–106923. https://doi.org/10.1109/ACCESS.2023.3316605.

Szilagyi, P., Contreras-Murillo, L.M., Menéndez, D.R. et al. (2024). 6G architecture for enabling predictable, reliable and deterministic networks: the PREDICT6G case. *2024 IEEE Wireless Communications and Networking Conference (WCNC)*, Dubai, United Arab Emirates, 2024, pp. 1–6, doi: https://doi.org/10.1109/WCNC57260.2024.10571097.

The Linux Foundation (2024). Camara project. Available at: https://camaraproject.org (accessed 23 September 2024).

Zou, Z. and Rugeland, P. (2024). Hexa-X-II "Deliverable D2.3: Interim overall 6G system design," June 2024. [Online]. Available at: https://zenodo.org/records/16995059 (accessed August 2025).

6

6G Intelligence

Hamed Farhadi[1], *Dani Korpi*[2], *Nabeel Nisar Bhat*[3], *Pawani Porambage*[4], *Halina Tarasiuk*[5], *Marcin Ziółkowski*[5], *Karol Kuczyński*[5], *José Miguel Mateos Ramos*[6], *Christian Häger*[6], *Henk Wymeersch*[6], *Ricard Vilalta*[7], *Merve Saimler*[8], *Leyli Karacay*[8], *and Milan Zivkovic*[9]

[1] *Ericsson Research, Stockholm, Sweden*
[2] *Nokia Bell Labs, Espoo, Finland*
[3] *University of Antwerp-IMEC, Antwerp, Belgium*
[4] *VTT Technical Research Centre of Finland, Oulu, Finland*
[5] *Warsaw University of Technology, Warsaw, Poland*
[6] *Chalmers University of Technology, Gothenburg, Sweden*
[7] *Centre Tecnològic de Telecomunicacions de Catalunya (CTTC-CERCA), Castelldefels (Barcelona), Spain*
[8] *Ericsson Research, Istanbul, Turkey*
[9] *Apple Munich, Germany*

6.1 The Motivations for AI/ML in 6G

6G is envisioned to widely benefit from intelligence within the networks and to have a crucial role for large-scale deployments of intelligence in the wider society (Uusitalo et al. 2025). This is expected to be realized by the integration of various AI-driven or ML-driven methods within the networks to enhance various performance metrics, and enablers for realizing AI/ML-driven services across different verticals. 6G is the first generation that has been designed right from the beginning with ML/AI in focus through all the layers of the technology, including, e.g. the physical layer as well (Merluzzi et al. 2023). This section presents the motivations for data-driven architecture and AI-driven methods for cellular communication networks, including AI-driven radio air interface and AI-driven M&O. The need for trustworthy AI/ML methods and AI/ML for 6G trustworthiness is also discussed in this section.

6.1.1 The Needs for Data-Driven Architecture

The integration of AI into 6G networks marks a significant shift in the evolution of communication technologies (Ericson, M. (ed.) (2023)). By incorporating AI-driven capabilities, these networks can optimize resources in real time, predict maintenance needs, and adapt seamlessly to varying network conditions. This advancement goes beyond the capabilities of previous generations, where network management was often reactive rather than adaptive. However, to fully leverage AI in 6G, it is imperative to establish a robust, data-driven architecture that can accommodate

6G to Build a Sustainable Future, First Edition. Edited by Mikko A. Uusitalo, Patrik Rugeland, Mauro R. Boldi, and Ahmad Nimr.

the complexities of advanced technologies, manage large-scale data, and address the operational challenges associated with ML operations (MLOps).

A data-driven architecture is fundamental to AI integration, yet the realization of this vision presents several challenges. These include the inherent complexity of the network, data availability constraints, and resource limitations, which complicate the effective deployment of AI technologies. Interoperability issues, where different network components fail to communicate efficiently, further exacerbate these challenges. Standardized frameworks, such as the 3GPP network exposure function (NEF) and service enabler architecture layer (SEAL), provide mechanisms for ensuring secure and efficient data exchange across various network entities (Boneza 2023). These frameworks allow for a cohesive approach to data sharing while maintaining privacy and security standards.

To facilitate the seamless integration of AI into 6G, it is essential to establish comprehensive architectural frameworks and communication protocols. These protocols should enable data and model exchanges between network components while safeguarding privacy. Privacy-preserving learning methods, such as federated and split learning (SL), are crucial for ensuring that data privacy is maintained while supporting decentralized decision-making across the network (European Commission 2021). However, balancing computational overhead, energy consumption, and model accuracy remains a challenge, as these methods introduce trade-offs that must be carefully managed.

Efficient AI model training and inference are also contingent upon the development of robust MLOps frameworks. In scenarios where data mobility is restricted due to privacy or regulatory concerns, distributed ML approaches, which enable local data processing while exchanging model parameters, offer potential solutions. However, these approaches present new challenges in maintaining model consistency and performance across diverse cellular nodes. Therefore, ensuring real-time data synchronization and the use of privacy-preserving aggregation functions are essential to maintaining model integrity and performance (Ickin et al. 2023).

Interoperability is another critical challenge for AI-driven 6G networks. As these networks are expected to support a wide range of applications, from healthcare to autonomous systems, it is essential that they can integrate seamlessly with various systems and platforms. Standardized application programming interfaces (APIs) will play a vital role in enabling different stakeholders to exchange data and AI models, ensuring compatibility across heterogeneous systems. In addition, the establishment of trust mechanisms will be necessary to ensure the secure exchange of AI services between parties (Charatsaris et al. 2024). Moreover, the implementation of explainable AI (XAI) solutions is essential to enhance the interpretability of AI predictions, thereby fostering greater confidence among stakeholders.

The allocation of resources within AI-native networks further complicates operations. The real-time optimization of computing, storage, and communication resources is critical to ensuring the smooth functioning of the network while meeting new quality of service (QoS) requirements. These include considerations such as data quality, training latency, and inference accuracy, all of which must be addressed to ensure optimal performance across the network. Efficient resource management will require the dynamic allocation of resources based on the evolving demands of AI-driven applications.

Finally, intent-based management (IbM) presents an important area for development. This approach aims to automate network management by translating high-level business objectives into actionable network configurations. However, achieving this objective is complex, particularly when managing the integration of policies across heterogeneous networks. Sophisticated algorithms are necessary to optimize performance while balancing energy consumption and cost efficiency. Additionally, real-time monitoring and reconfiguration mechanisms must be implemented to detect deviations from intended network behaviour and correct them swiftly, ensuring alignment with overarching business objectives.

In conclusion, realizing AI-native 6G networks will require addressing several challenges, including the development of privacy-preserving frameworks, ensuring efficient interoperability, and optimizing resource management strategies. A data-driven architecture will serve as the foundation for AI-driven automation, efficiency, and security within these networks. As 6G continues to evolve, future advancements must focus on creating scalable, secure, and efficient solutions capable of supporting the diverse and dynamic demands of next-generation networks.

6.1.2 The Needs for AI/ML for Physical Layer Signal Processing

The *AI-driven air interface* is a radio air interface for communication and/or sensing, where AI-based methods replace one or more functionalities in the lower layers across the transmitter and/or receiver (Farhadi et al. 2025a,b). There are several key motivations for the development of this innovative technology for both communication and integrated sensing and communication (ISAC). In the context of communications, the following motivations are outlined (Figure 6.1).

Enhanced efficiency: Legacy air interface methods often face limitations due to model deficiencies, particularly with respect to radio propagation channel models and/or radio frequency (RF) hardware models. AI-driven air interfaces can address these limitations by learning and adapting in real time, resulting in improved performance under a variety of conditions. For instance, AI-based models can predict complex wireless propagation environments and adapt to the changing environments, e.g. via predictive beamforming (Farhadi et al. 2025a,b), more effectively compared to the classical non-AI approaches (Nimr et al. 2025). As another example,

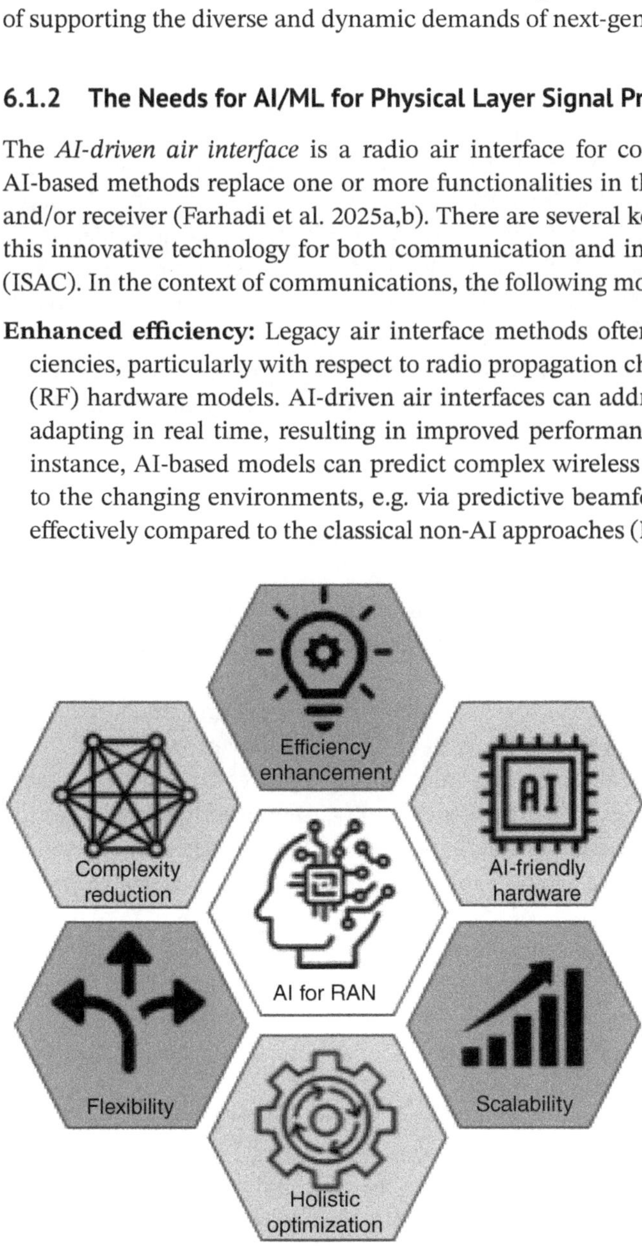

Figure 6.1 The motivations for AI/ML in RAN.

AI-driven methods have been shown to be more effective in mitigating hardware imperfections, e.g. oscillator phase noise in high-frequency bands (Farhadi et al. 2023), and power amplifier (PA) nonlinearities (Farhadi and Sundberg 2020) compared to the legacy non-AI methods.

Improved flexibility: AI-driven air interfaces provide enhanced flexibility, allowing autonomous operation under changing conditions. For example, these systems can dynamically adapt to fluctuations in the radio propagation environment, user equipment (UE) mobility, RF hardware variations, and traffic patterns (Rydén et al. 2023). This adaptability ensures nearly optimal performance across a broad spectrum of scenarios without manual intervention, making the system more resilient and responsive to changes.

Driving scalability: AI-driven solutions offer scalability through re-training capabilities. By leveraging AI's ability to learn from new collected data, air interfaces can be retrained to cope with different scenarios and environments. As a result, these systems are better suited for diverse deployment scenarios, from dense urban networks to rural coverage, supporting a wide range of communication needs. For AI-driven air interfaces to be effectively deployed, standard support for model life cycle management (LCM) is necessary. This involves developing standardized approaches for managing the AI model's life cycle, including training, deployment, updating, and retraining processes (3GPP TR 38.843 2023). Such standards are critical for ensuring smooth integration and operation in real-world 6G networks.

Leveraging on AI-friendly hardware: To support the computational demands of AI-driven air interfaces, specialized hardware is essential. AI inference-dedicated chipsets optimized for real-time and low-power processing with low precision are needed to perform AI computations efficiently during inference operation (Deng et al. 2020). On the training side, chipsets with high-memory bandwidth and high precision are required to handle the intensive data processing required for AI model training (Moore 2024). These hardware advancements will be pivotal in realizing the full potential of AI in air interfaces.

These motivations reflect the growing need for AI-driven air interfaces to address the limitations of traditional systems and unlock new capabilities, making them a critical component of future communication networks such as 6G.

In addition to the motivations mentioned for AI-driven air interface for communication, the need for ISAC arises from the unique capabilities AI offers in optimizing complex tasks beyond the limits of traditional signal processing methods. Traditionally, signal processing relies on established models that often struggle to account for real-world impairments and uncertainties in the environment. However, AI has the ability to adapt and address these impairments dynamically, enhancing both communication and sensing functionalities in ways that classical methods cannot, as outlined in the following:

AI for complex task solving: AI excels in solving intricate tasks such as object classification and human pose estimation, which are crucial for advanced sensing capabilities (Farhadi et al. 2025a,b). By integrating these AI-driven solutions into the air interface, networks can enhance their ability to sense and interpret environmental data, leading to improved situational awareness and intelligent decision-making in real time.

Holistic optimization: Another significant motivation for the integration of AI/ML in the physical layer is the potential for joint optimization across the system. AI-driven air interfaces enable end-to-end or cross-layer optimization, where in the former case, multiple functionalities in the transmitter and/or receiver can be optimized to improve the overall system performance, while in the latter case, multiple functionalities across different system layers can be adjusted simultaneously to maximize overall system performance. The holistic optimization across different system components enables moving away from the classical block-level design and optimizations. In the context of ISAC systems, this implies that AI can jointly optimize communication and sensing operations, leading to performance improvements in scenarios where both functions are required simultaneously. For example, AI can facilitate semantic communications, where the meaning or context of transmitted data is prioritized, alongside sensing functions, enabling networks to better interpret and respond to their environment. Moreover, AI can handle multi-objective optimization tasks, balancing the demands of both communication and sensing without compromising either function.

Improved resolution at lower frequencies: AI has the potential to enhance resolution in sensing applications, even at lower frequencies where classical methods may struggle. Techniques such as fingerprinting, where AI models learn to recognize and distinguish between various patterns in the signal environment, allow for more precise sensing. This is particularly important in environments with limited bandwidth or in scenarios in which energy efficiency is a priority.

In summary, the integration of an AI-driven air interface for both communication and ISAC is essential for meeting the advanced requirements of 6G and beyond. AI's ability to dynamically adapt, solve complex tasks, and perform holistic optimizations can transform how communication systems interact with their surroundings, leading to more intelligent, efficient, and responsive networks. The AI-driven air interface is integrated in the 6G blueprint as outlined in Section 6.2, and the training and monitoring of the AI/ML models for the AI-driven air interface need to be supported by the standard LCM and the enablers for the AI-native architecture as outlined in Section 6.3.

6.1.3 The Needs for AI-Driven Management and Orchestration

The integration of AI-driven mechanisms for 6G M&O enables 6G networks to support a wide range of verticals and applications such as immersive augmented reality (AR), virtual reality (VR), autonomous vehicles, and smart cities. Indeed, it is challenging to support this wide range of verticals and applications with the classical approaches: on the one hand, those applications have strict requirements including ultra-high data rates, extensive coverage, ultra-low latency, and enhanced scalability and reliability. On the other hand, 6G networks are expected to address sustainability goals by minimizing energy consumption and reducing carbon footprints.

To satisfy those requirements, M&O mechanisms for networks and services ought to optimize multiple, sometimes conflicting objectives over complex and dynamic virtualized environments in real time. In this context, AI/ML methods can support the 6G requirements and use cases by automating and optimizing the orchestration processes for complex systems and providing decisions for function placement and resource allocation as well as post-deployment LCM operations such as the migration and scaling of services. Indeed, AI-driven solutions enable real-time analysis of the environment state and adjustments, allowing the network to dynamically respond to changing conditions and user demands without human intervention. This leads to improved overall network performance, user experience and enhanced energy efficiency. Additionally, AI/ML methods can be used for predicting the future state of the network to detect environment changes that can affect network and service performance, which can enable proactive actions that can help to prevent or minimize service disruptions.

To fulfil these advancements, AI/ML model training requires high-quality datasets with sufficient representative data points as input in order to ensure satisfactory model performance. Further, in the case of reinforcement learning (RL) approach, the model interacts with the environment and explores the solution space by providing actions and adapting the decision policy based on the received feedback. However, it is not feasible to apply possibly sub-optimal exploration actions on the network infrastructure due to their risk of negatively impacting the networks and services. In this context, network digital twins (NDTs) are a useful tool that provides a near real-time virtual representation of the network infrastructure state, properties and behaviour. Using the NDT, the impact of network management actions can be safely checked before being applied to the real network, which can prevent undesirable effects. Additionally, NDTs can be used to generate synthetic data to augment or replace datasets for training AI/ML models.

6.1.4 The Needs for Trustworthy AI/ML and AI/ML for 6G Trustworthiness

Trustworthiness and AI/ML have a two-way relationship in the context of 6G research. On one hand, it has been used as a tool to enhance security and privacy in 6G by automating security operations with the improved efficiency and accuracy of threat and anomaly detection. On the other hand, the trustworthiness of AI/ML should also be improved and guaranteed as it spans almost all the operations in the RAN, core network, management layer and exposure layer, of the data-driven 6G end-to-end system. The wide adoption of AI/ML in 6G systems will also introduce new forms of security vulnerabilities, threat vectors, data breaches, and dynamic risks associated with adversarial learning methods and intelligent agents.

AI/ML techniques, including supervised, unsupervised, or RL, can be widely adopted in anomaly and intrusion detection in 6G networks by developing both reactive and proactive countermeasures to safeguard the networks. The integration of large language models (LLMs) and generative AI (GenAI) into the 6G security automation framework will enhance the capabilities of cyber threat intelligence, including security monitoring, root-cause analysis, and automating remediation actions. Nevertheless, it is crucial to address the challenges, such as high computational requirements, privacy risks, adversarial ML attacks, and latency constraints, while incorporating such AI/ML techniques in 6G networks.

NDTs provide the possibilities for safe integration of security algorithms with ML models or AI agents into the network M&O framework, as they provide a way to train the utilized models with up-to-date data from the physical networks, as well as to test the consequences of the resulting network management decisions outside the actual production network environment. This enables the identification of potential network security vulnerabilities and anomalies, leveraging data collected to train ML models for anomaly detection, which can be utilized in the 6G end-to-end system. For instance, the integration of intelligent security techniques into more complex 6G systems may require efficient deployment strategies, distributed learning methods, and robust defence mechanisms against AI-driven threats.

6.2 6G System Blueprint: AI/ML-Specific View

The 6G system blueprint is presented in Chapter 3 of this book. The presented architecture in Figure 6.2 represents an AI/ML-specific view of the 6G system blueprint and presents a structured flow of operations and data exchanges across multiple layers, ensuring the seamless integration of AI/ML processes into network operations. At the application layer, where end-user applications or enterprise services request specific resources and capabilities, extending beyond conventional network resources to include services like compute-as-a-service (CaaS) and AI-as-a-service (AIaaS). These applications span various domains, including communication services, media streaming, industrial automation, and AI-driven analytics. At this stage, the service exposure interface plays a

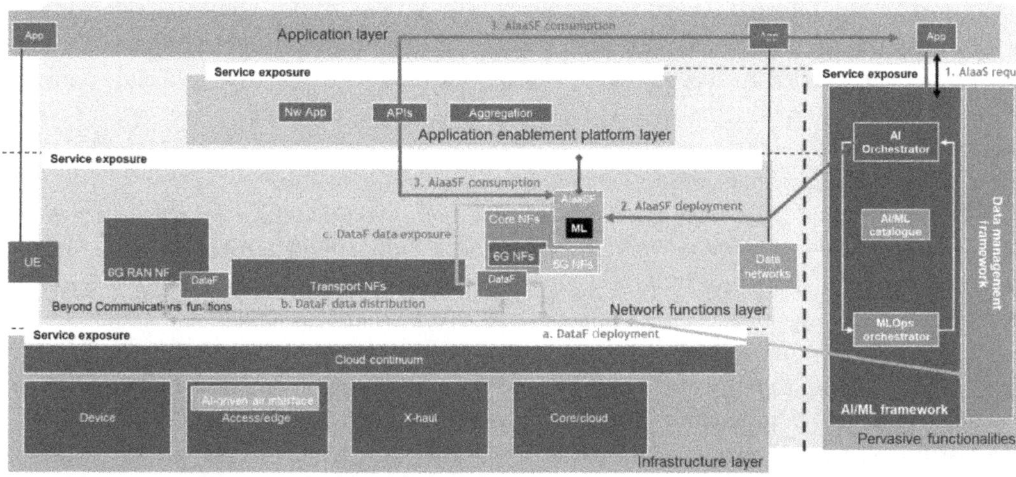

Figure 6.2 AI/ML specific view of the 6G system blueprint.

critical role by offering standardized APIs and aggregation services tailored for AI functionalities, enabling applications to leverage AI-driven features – such as model inference, training orchestration, and data processing – without requiring deep integration into the network core.

Once a service request is initiated, it is processed by the *application enablement platform layer*, which acts as an intermediary between applications and network functions (NFs). This layer ensures that requests are contextually enriched, properly formatted, authenticated, and directed to the relevant network components. It also leverages AI to enhance service orchestration, translating high-level application demands into network-specific commands while optimizing for dynamic conditions and resource availability.

At the *NFs layer*, service requests are handled by 6G RAN, transport, and 5G/6G core NFs. AI-powered RAN components manage wireless access, optimize connectivity, and allocate resources proactively to maintain seamless communication between UEs and the network. AI-driven transport NFs improve data routing, transmission efficiency, and fault prediction, while core NFs incorporate AI for mobility management, session control, and intelligent data routing, enabling adaptive and context-aware service delivery. The AI-driven air interface (Farhadi et al. 2025a,b), part of the *access infrastructure layer*, enhances communication and sensing by replacing classical physical signal processing and medium access control (MAC) layer functions with AI-based methods, including AI-assisted waveform design, modulation, coding, channel state information (CSI) acquisition, multi-input multi-output (MIMO) transmission, and hardware impairment compensation. Integrating AI into radio access networks necessitates new protocols, procedures, and signalling mechanisms.

AI/ML integration is central to enabling intelligent, adaptive, and efficient network operations. The AI/ML framework, a comprehensive suite of tools, ensures the implementation of advanced AI-driven workflows across the network. It consists of three core components: the *AI orchestrator*, the *AI/ML catalogue*, and the *MLOps orchestrator*. Clearly defining AI capabilities as services is crucial for positioning the AI/ML Framework as a fundamental element of 6G networks. The adoption of an AIaaS approach enables the delivery of tailored AI solutions across application and 6G management layers. Unlike generic AI services, these are designed for deep integration within the 6G ecosystem, leveraging network-native features as AI-driven RAN, edge AI execution, and cross-layer AI insights for performance optimization, scalability, and reliability. Standardized exposure APIs simplify access for vertical industries and application providers, even those lacking AI expertise, by offering telecom-specific AI capabilities, such as AI model management, training, validation, testing, deployment, inference, and monitoring. These three core components are elaborated further in the following.

The AI orchestrator is responsible for deploying, executing, and managing AI services, ensuring smooth network operations for use cases such as real-time traffic optimization, video streaming enhancement, low-latency autonomous vehicle communication, fault detection, predictive maintenance, and resource allocation. It dynamically selects and deploys AI models based on specific network conditions and application requirements, optimizing both AI and network performance. The AI orchestrator also delivers AI/ML services through the AIaaS function (AIaaSF), integrating them with NFs and exposing them via interfaces such as the common API framework (CAPIF). Rather than functioning as a standalone network entity, AIaaSF standardizes AI service access within existing NFs. It collaborates with the M&O system, complementing its role in AI model integration rather than replacing it.

The AI/ML catalogue serves as a repository for pre-trained AI models, algorithms, and metadata, supporting a wide range of network use cases, including anomaly detection, QoS prediction, and energy efficiency optimization. This repository offers flexibility, hosting models developed in-house or sourced externally. The catalogue provides AI services through components such as AIaaSF, the data function (DataF), and the MLOps function (MLOpsF). AIaaSF facilitates standardized AI model exposure, DataF handles data preprocessing and sharing while maintaining privacy, and MLOpsF oversees AI model LCM, ensuring continuous improvements. Model versioning within the catalogue enables networks to adopt the latest AI advancements while ensuring interoperability across diverse network environments.

The MLOps Orchestrator is vital for end-to-end AI/ML LCM, overseeing model deployment, performance monitoring, retraining, and adaptation. Unlike the AI orchestrator, which deploys models to NFs, the MLOps Orchestrator ensures their long-term relevance and accuracy through automated retraining and updates based on real-time performance metrics. This capability is critical for maintaining AI model effectiveness in evolving network conditions. Operating within a cloud-native or virtualized environment, it collaborates with AIaaSF to manage both network and application-centric AI services, orchestrating MLOps workflows without directly exposing capabilities via APIs.

AI/ML integration extends beyond individual network components to enable cross-layer optimization. By analysing data across RAN, transport, and core networks, the AI framework generates insights that enhance network-wide performance. AI models can predict congestion by examining traffic patterns, enabling dynamic resource allocation to prevent bottlenecks. Similarly, predictive maintenance models assess equipment health to identify faults before failures occur, reducing downtime and operational costs. These capabilities foster a proactive approach to network management.

The *data management framework* system ensures the effective collection, processing, and governance of network data, which is essential for AI/ML operations. It gathers performance, traffic, user behaviour, and environmental data from UEs, RAN, transport, and core NFs while enforcing robust privacy measures such as encryption, anonymization, and differential privacy (DP). The system aggregates and normalizes data within centralized or distributed storage solutions, enabling AI models to make informed, context-aware decisions. Additionally, it supports real-time data streaming to facilitate AI-driven optimizations. The Data Framework executes all DataF, managing data distribution and storage across network layers to provide AIaaSF and MLOpsF with high-quality data for inference, training, and performance monitoring.

A crucial aspect of the data framework is regulatory compliance. Given the increasing focus on data privacy, especially in telecommunications, the framework aligns with legal requirements such as general data protection regulation (GDPR) by incorporating access controls, audit trails, and user-consent-based processing. These mechanisms ensure that AI/ML operations are both effective and ethically sound, while also maintaining data quality throughout their lifecycle.

Once AI/ML-driven processes have been executed, their results are fed back into the NFs layer and then propagated to the Application Layer. This feedback loop ensures that service requests are met with nearly optimal performance, delivering enhancements such as reduced latency, improved throughput, and dynamic resource allocation. Applications benefit from near real-time insights and network optimizations, enhancing user experience and service reliability.

6.3 AI-Native Architecture

This section provides a comprehensive overview of the key architectural enablers that support an AI-native network design, namely the AIaaS, MLOps, and DataOps. These components respectively correspond to the *AI orchestrator*, *MLOps orchestrator*, and *data management framework functional* domains and play a central role in enabling intelligent automation, scalable lifecycle management of AI models, and efficient data handling. Their roles, interactions, and placement within the architecture are illustrated and summarized in Figure 6.2 and are discussed in further detail in the following.

6.3.1 DataOps

DataOps facilitates the consistent preparation and provisioning of datasets for applications such as AI-driven network management, digital twins (DTs), and predictive analytics. While the overall requirements for AI/ML trustworthiness in 6G are introduced in Section 6.6.2, the DataOps-related mechanisms for privacy-preserving data and model aggregation are further discussed here. Combining DP with secure aggregation techniques enables us to reliably learn statistical information from aggregated UE data, without revealing personally identifiable information, neither to the network nor to application providers.

Privacy-preserving data aggregation: The emergence of distributed data applications and heterogeneous networks in 6G requires distributed data-driven decision-making approaches, such as distributed ML and federated learning (FL). Enabling distributed ML among cellular nodes requires data and model exchange between the cellular network and UEs. One of the key challenges in both collecting and learning from data is to enable privacy preservation when sharing privacy-sensitive UE data with the network and other UEs. Therefore, there is a need to introduce privacy-preserving techniques for data collection and learning that should provide a trade-off between the resulting computation overhead, energy consumption, and model/learning efficiency.

Private FL and private data aggregation enable privacy-preserving learning on user data. Privacy is preserved by combining the local noise addition with secure aggregation techniques in such a way that correct statistical information of data can still be learnt without revealing personally identifiable information. The aggregated data originating from several UEs is collected from a defined sampling area and over a defined sampling time window. The data shared with the network does not capture individual private data, but it could consist of statistical (e.g. mean, variance, and distribution) and/or analytical insights of the users in the network. The size of aggregated data samples should be sufficiently large not to reveal personal information, while providing adequate privacy-utility trade-offs.

The privacy-preserving architecture, proposed in Porambage (2023) and Botsinis (2024), introduces new architectural elements, such as the UE aggregation and data-driven network control units. The UE aggregation unit performs privacy-preserving data aggregation by leveraging secure aggregation techniques and/or data anonymization to collect statistical information of UE data at the collector, which is an entity that produces the statistics of the aggregated UE data. At the same time, the data-driven network control unit is an entity at the network side that configures the base station by performing automation, optimization, and intelligence services, based on both network data and statistics of the aggregated UE data shares located at the Collector unit.

This architecture is based on privacy-preserving cryptographic protocols like Prio (Corrigan-Gibbs and Boneh 2017), depicted in Figure 6.3, currently being standardized in the Internet Engineering Task Force (IETF) Privacy-Preserving Measurement Workgroup (Geoghegan et al. 2024). It consists of a collector, and aggregators (i.e. leader and helper), which are responsible for UE data collection. The *leader* (i.e. main aggregator) is responsible for coordinating the data aggregation by receiving the encrypted UE data and orchestrating the process of computing the aggregate result as requested by the *collector*. The *helper* is an aggregator assisting the leader with the computation. Privacy is preserved under the assumption that the leader and the helper do not collide, i.e. they do not contain the same data shares. The leader and the helper send aggregated shares to the collector that computes the final aggregation results.

A possible realization of the Prio protocol in future cellular network architecture can be based on a network data analytics function (NWDAF) service framework (3GPP TS 23.288 2024). NWDAF provides analytics to 5G Core NFs to enable autonomous network operations and service management, allowing for the integration of various data-driven AI/ML technologies into the 5G networks. However, the NWDAF framework introduces additional privacy-related requirements because some analytics may need to use privacy-sensitive UE data.

To address these concerns, the physical realization of the different Prio entities in the cellular network architecture can have different variants. The leader and helper can be realized either as an NWDAF service or reside in the data network (DN), while the collector can reside in the DN. An example of a privacy-preserving architecture realization is shown in Figure 6.4. The leader is implemented in NWDAF, thus leveraging mostly core network resources for aggregating encrypted

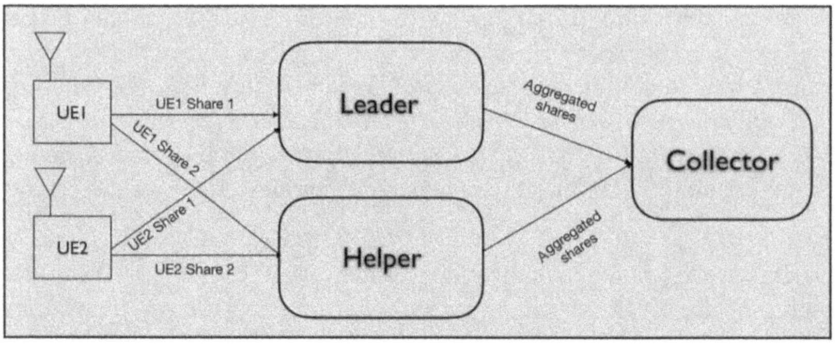

Figure 6.3 Prio-based aggregation system architecture.

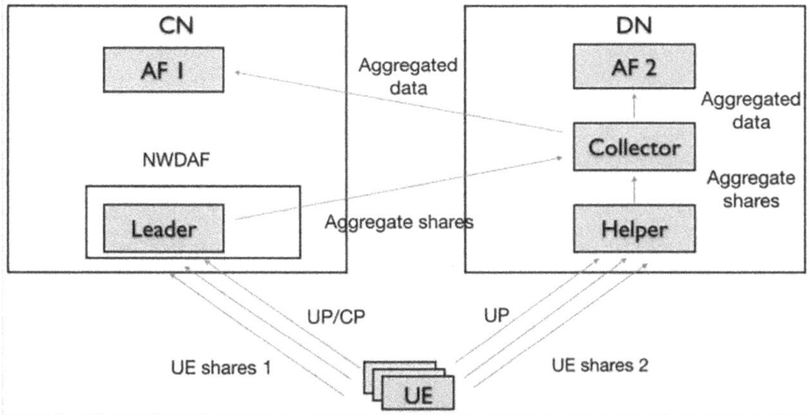

Figure 6.4 An example of privacy-preserving architecture in 6G network.

UE data shares, while the helper resides in DN, assisting the leader. The application function can be in the core network (AF1), e.g. for network optimization/O&M, or in the DN (AF2), e.g. for analytics. Here, the privacy is preserved with aggregators residing in different trust domains (i.e. leader in CN and helper in DN). In addition, the collector has access to the statistics of the collected data without having direct access to the privacy-sensitive data of the UE. Another alternative realization of the Prio protocol in 6G cellular network architecture would assume the deployment of the aggregators and the collector in the RAN, therefore having an impact on RAN procedures (Nimr et al. 2025).

6.3.2 MLOps

MLOps operationalizes ML workflows and enhances lifecycle management of AI models by offering tools and methodologies for deploying and maintaining distributed AI functions. This includes support for cooperative learning, including FL and SL for scalable model training and inference.

Cooperative learning: Cooperative learning is an ML-based cooperative intelligence technique, where network nodes collaboratively share data and/or models towards achieving a common task, by taking advantage of each other's knowledge and experience. The learning task requires the participation of multiple nodes at the same time and needs multiple successive steps. The potential scenarios may include cooperative rendering, where each UE shares a partial result with the aggregator, which in turn combines these partial results and shares the full result back to the UEs. The different types of data can be characterized by their privacy sensitivity level that determines the data and model cooperative sharing variant (Botsinis 2024):

- **Privacy level 0**: Individual user private data can contain precise UE location, exact used application, exact behaviour (e.g. sleeping), and personal information (e.g. age and sex). This information is not shared with the network. The network shares with the UE the inference model, trained in a privacy-preserving manner, so that no individual private UE data can be revealed by inference. The UE then performs inference on the model using its private data and sends the inferred result back to the network, as shown in Figure 6.5 (UE 1).
- **Privacy level 1**: Individual telemetry data capturing statistical information about non-private connectivity activity patterns (e.g. QoS categories and traffic prediction) of the UE. The UE

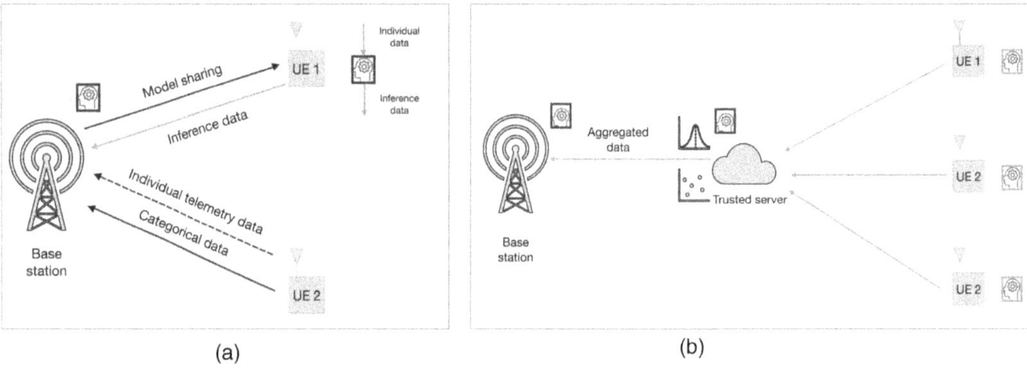

Figure 6.5 Cooperative data and model sharing: (a) inferring the individual private data on network-shared model (UE1) and sharing of individual non-private data (UE2); (b) sharing of aggregated multi-user data and models.

might conditionally share this data with the network for inference, as depicted in Figure 6.5(a) (UE 2).

- **Privacy level 2**: Categorical data containing contextual information of the UE (i.e. high/low/no traffic) that can be shared as a class/category instead of exact UE information, as shown in Figure 6.5(a) (UE2).
- **Privacy level 3**: Aggregated multi-user data. As illustrated in Figure 6.5(a), by using privacy-preserving methods, individual private data can be aggregated across multiple-UEs in such a way that it is a statistical and analytical information about the UEs without revealing individual private information.

Based on this privacy-aware data classification and trust levels between the UEs and the network, UE data will be shared either partially, totally, or owned locally. Depending on the level of privacy of data, privacy-preserving measures will be used for data and model sharing, while privacy-utility trade-offs need to be characterized. In Akgul (2025), the privacy-aware cooperative learning in 6G architecture is proposed by introducing the cooperative session concept, to enable:

- Training over a distributed network of UEs, grouped in cooperative groups, e.g. geographically in subnetworks, that are collaboratively learning a model in a privacy-preserving manner.
- Cooperative inference over multiple (groups of) UEs given a pre-trained model.

To enable a privacy-preserving model training, UEs may adopt an FL framework in which UE data remains locally, while in each training iteration, the UEs share only the model parameters (gradients) with the model aggregator at the network side. However, the model parameters may still expose sensitive information of training data to the outside adversaries, e.g. by means of gradient-based inversion attacks. Therefore, instead of a classical centralized FL, a privacy-preserving hierarchical FL can be utilized. Here, the UEs create cooperative groups based on different criteria such as geographical distribution, task requirements, and privacy levels. Each cooperative group has a group head (GH), which aggregates the intra-group results and shares the cooperative group's aggregated results (e.g. model parameters) with other cooperative groups through the network. The cooperative groups are not necessarily formed based on distances using proximity-based services (e.g. in subnetworks), but they can be formed based on privacy levels or the type of environment, e.g. in a collaborative remote gaming session, creating collaboration groups. Here, the network can help with the creation of collaboration groups based on the information shared by the UEs, such as UE communication and computation capabilities and their privacy budget.

If the trust level among UEs within a cooperative group excludes the possibility of potential adversarial attacks, the model can be trained using FL among UEs in that cooperative group, aggregating the results/model at GH. GH(s) can share this aggregated data with the NW, by using privacy-preserving aggregation, with the aid of the Prio-protocol, where the aggregated data is split into shares, so that the NW only obtains the statistics of the aggregated data shares.

AI/ML model layer offloading: Distributed learning techniques such as SL enables flexibility via model layer offloading between distributed computation nodes (e.g. between UE and network) seamlessly without interruption in training and inference. Conventionally, via full AI/ML model offloading, both the ML models and data are moved from one place to another. This requires adding/removing workloads in the cloud. Moreover, sensitive data and models in the UEs may not be well preserved. The techniques discussed in the collaborative learning section above are one solution, and in addition, via SL as demonstrated within a component proof of concept (Akgul 2025), both the ML models and the datasets are kept where they are originally located. Only the intermediate layers, i.e. only part of the neural network (NN) model, are offloaded. For example, in the case of devices offloading AI/ML workloads caused by the computation in the intermediate ML layers to the network cloud, this helps reduce the storage, compute, and energy requirements on limited endpoints such as devices running the network applications. In the scope of MLOps, model layer offloading needs to be managed by an ML model coordinator to decide on when and how to offload the AI/ML model layers, as this decision depends on parameters including energy consumption, privacy, and model efficacy.

6.3.3 AI as a Service

With the emergence of 6G networks, AIaaS is poised to become a key enabler of automation, intelligent orchestration, and AI-driven service delivery. Unlike traditional AI implementations that require dedicated infrastructure, AIaaS provides managed AI/ML capabilities through standardized exposure APIs such as 3GPP NEF, SEAL, and CAMARA (Camara Project 2024). This allows applications to consume AI functionalities on-demand, streamlining AI model training, deployment, inference, and lifecycle management (LCM) without requiring direct ownership of AI resources such as compute infrastructure, storage, datasets, AI models, MLOps tools, and security frameworks (Figure 6.6).

6G networks introduce a data-driven, AI-native architecture, where AI is integrated intrinsically within network operations rather than being an external add-on. AIaaS leverages MLOps (Tamburri 2020) frameworks to support a broad range of AI/ML services, including pre-built models, real-time inference, FL, and AI/ML model LCM. By enabling AI-driven automation in network resource management, DTs, and service optimization, AIaaS plays a foundational role in enhancing efficiency, reliability, and security across telecom ecosystems (Saimler et al. 2024).

AIaaS provides a wide range of AI/ML LCM capabilities that are exposed to applications, network management systems, and third-party AI consumers via standardized APIs. AIaaS enables AI/ML model LCM, which includes AI/ML model training, validation, performance monitoring, model deployment, and inference. Additionally, it supports FL and Transfer Learning, allowing distributed AI training across multiple edge/cloud nodes. The AI exposure APIs facilitate secure access to AI functionalities through 3GPP NEF, SEAL, and the CAMARA framework, ensuring seamless interaction between AI models and NFs. AIaaS also enhances intelligent network orchestration, supporting intent-based networking, dynamic resource allocation, and self-optimizing networks (SONs), ultimately driving zero-touch automation.

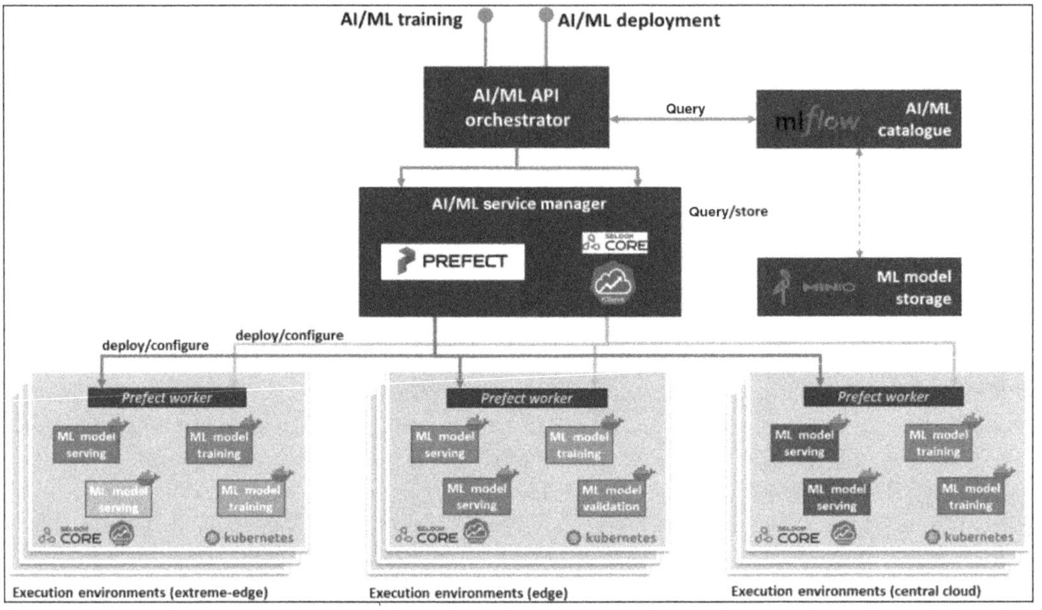

Figure 6.6 AIaaS architecture and functional split.

AIaaS is central to the realization of an AI-native 6G ecosystem, embedding intelligence at every layer of the network. This architecture is characterized by intelligence everywhere, distributed data infrastructure, zero-touch management, and scalable AI service delivery. AIaaS interacts with key 6G network components, including NWDAF, which provides real-time analytics for AI/ML-based network optimization, AI/ML enablement (AIMLE) layer, which defines AI model exposure and training functions within 3GPP, and CAPIF (3GPP TS 23.222 2024), which ensures secure API management for AI-driven services.

AIaaS extends beyond traditional cloud-based AI deployments by enabling federated learning as a service (FLaaS) and federation as a service (FEDaaS). These approaches facilitate distributed AI model training across edge-cloud environments, enhancing privacy-preserving AI, real-time AI inference, and multi-party AI collaboration. The AIaaS platform integrates open-source tools such as prefect (2025) for workflow orchestration, MLFlow (2025) for model management, KServe (2025) and Seldon (2025) for model deployment, and OpenFL (2025) for FL, ensuring robust AI capabilities in a scalable environment.

AIaaS aligns with 3GPP SA6 (AIMLE) (3GPP TR 23.700-82 2024) and SA5 AI/ML management frameworks (3GPP TS 28.105 2024; 3GPP TR 28.858 2024), ensuring standardized AI LCM that enables model discovery, registration, and inference optimization. It promotes interoperability across communication service providers (CSPs) and third-party AI applications, facilitating AI-driven innovation across telecom ecosystems. AIaaS also integrates security and trust mechanisms, implementing AI governance, federated model validation, and privacy-aware AI training. Furthermore, AIaaS is aligned with non-real-time RAN intelligent controller (O-RAN Non-RT RIC) (O-RAN Alliance 2021) to support AI/ML model registration, storage, discovery, and deployment, ensuring predictable QoS, low latency, and enhanced security in AI-powered services.

As AIaaS continues to evolve, its role in 6G networks extends beyond traditional AI service models, incorporating advanced lifecycle management and intelligent orchestration mechanisms.

A critical component of this evolution is the integration of AI-native frameworks that ensure seamless AI/ML model exposure, closed-loop automation, and dynamic model updates.

To enable AIaaS within 6G systems, an AI-native framework was developed for managing AI/ML models and providing AI/ML exposure to external services (Toumi and Dimitrovski 2024). The architecture supports closed-loops AI/ML model management by triggering automated data collection and DT-based model training and updates. Additionally, the LCM framework considers model dependencies, ensuring that performance degradation in a DT prompts necessary AI/ML model updates. For example, if a DT loses accuracy, the AI models trained on that Twin must also be retrained to maintain reliability.

By leveraging this AI-native framework, 6G networks can dynamically detect model performance degradation, identify root causes, and implement mitigation strategies. This ensures that AIaaS remains robust, adaptive, and capable of delivering intelligent automation across the telecom ecosystem.

Figure 6.7 illustrates the proposed framework, which includes the following set of functions to support the AI/ML model LCM:

- **AI/ML model orchestration:** This function is the main decision component of the AI-native framework. It manages the lifecycle of the AI/ML models and DTs to maintain overall performance by triggering different orchestration operations, such as new data collection or model updates.
- **AI/ML model training function:** This function is responsible for AI/ML model training and update. It retrieves the required training data from the repository function and/or the data collection function to create the input dataset, then performs AI/ML model training and validation. The model training operation may be distributed over multiple training functions in the case of, e.g. FL. This function may use a sandbox environment including DTs for safe AI/ML model training and validation, where the sandbox is a controlled environment that allows experimentation for AI/ML model training without impact on the network. The function can also use base AI/ML

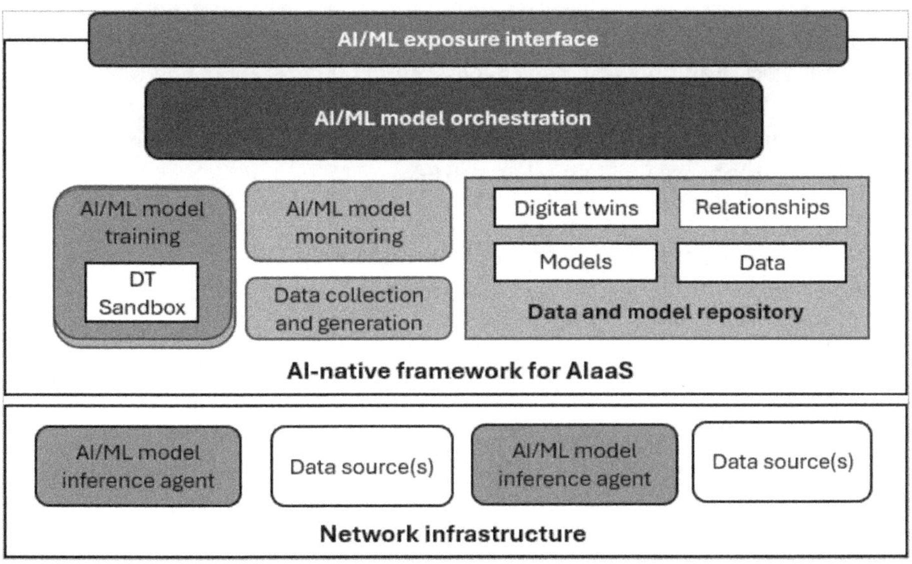

Figure 6.7 AI-native architecture for AIaaS.

models for transfer learning, which is when a model that has been trained on a certain task is used as a basis to train another model for a similar task.

- **AI/ML model inference agent:** This function performs inference using the trained AI/ML model that is stored in the repository. To ensure model performance, the inference agent sends regular performance reports to the model monitoring function.
- **AI/ML model monitoring:** This function subscribes to and collects performance reports from the AI/ML model inference agents to assess the model's performance across agents. The function notifies the AI/ML orchestration function when model performance drops below a certain threshold.
- **Data collection function:** This function collects data from different sources and performs pre-processing prior to storage in the repository. The data collected can be used to train models or to update DTs.
- **Data and model repository:** This repository serves to store the system's AI/ML models, DTs, and datasets together with their related metadata, such as accuracy and dependency relationships.

6.4 AI-Driven Radio Air Interface

The AI-driven air interface explores the convergence of wireless technology evolution and AI/ML tools, aiming to meet the increasing demands posed by 6G networks. It is part of the radio access in the infrastructure layer, as shown in the 6G system blueprint in Figure 6.2. It benefits from a systematic exploration of AI-driven solutions across four pivotal aspects of radio interface design: modulation and coding, CSI acquisition and compression, MIMO transmissions, and compensation for hardware impairments (Farhadi et al. 2025a,b). The application of AI to waveform modulation and coding not only optimizes established techniques but also introduces novel signal design approaches, promising significant enhancements in spectral efficiency and reliability. The advancements in AI-based CSI acquisition highlight the role of intelligent compression and prediction techniques in minimizing overhead in feedback transmissions while maximizing spectral utilization. Additionally, AI-enhanced MIMO transmissions present innovative solutions for specific challenges, such as imperfect or limited CSI, power consumption optimization, and pilot assignment. Furthermore, it highlights the potential of AI to mitigate hardware impairments, including PA non-linearities, to enhance spectral efficiency, improve energy efficiency, and extend coverage and oscillator phase noise to enhance reliability (Figure 6.8).

6.4.1 AI-Driven Methods for Hardware Impairment Compensation for Communication

An AI-driven method is proposed in Farhadi and Sundberg (2020) with the legacy demapper replaced with an AI/ML-based demapper to compensate for the impact of RF hardware impairments. The evaluations in the mmWave band have shown the effectiveness of the method to compensate for the impact of oscillator phase noise and improve the reliability of the communication. Another AI-driven method is proposed in Farhadi et al. (2023) for compensating the impact of PA nonlinearity. This technique enhances various performance metrics within the networks depending on how it is utilized. The performance evaluations have shown improvements in uplink throughput for a given level of PA nonlinearity, increasing coverage by enabling the UE to transmit at higher power, and enhancing UE energy efficiency by enabling the UE to operate in a

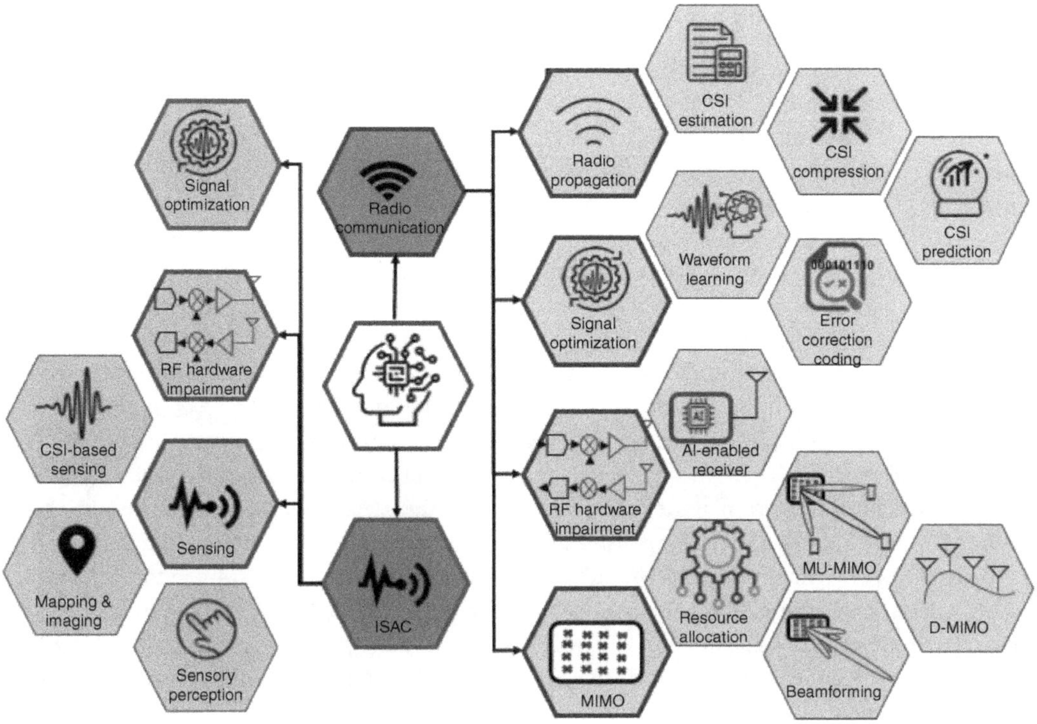

Figure 6.8 An overview of the use cases for AI-driven air interface.

more nonlinear and hence more energy-efficient operating point. These all need to be handled by setting the operating points in such a way that the requirements on the out-of-band emissions are fulfilled.

6.4.2 End-to-End Optimized Physical Layer Using AI/ML Algorithms

When using AI/ML-based transmitter and/or receiver components, further performance benefits can be obtained by optimizing the whole system end-to-end. These benefits stem from two different factors. Firstly, all existing radio systems have been designed for conventional algorithms, which in some cases have different requirements than native AI/ML-based algorithms. One example is the necessary amount of channel estimation pilots, which is typically larger for conventional receiver solutions (Honkala et al. 2021). Therefore, when relying on AI/ML-based algorithms, such aspects should be re-engineered to be better-suited for their needs.

Secondly, if both the transmitter and receiver rely on ML-based processing, they can be trained jointly end-to-end to maximize the target key performance indicator (KPI) (typically throughput). In practice, such joint training is performed in a simulator, where the whole communication link is implemented as a differentiable model. The elements representing the non-learned parts, such as the wireless channel, are included as fixed layers. Such an approach can result in higher performance than could be otherwise achieved (Korpi et al. 2023). The evaluation results based on simulations have shown 15% to 20% increase in spectral efficiency compared to a conventional MIMO system. Also, over-the-air measurements using a hardware proof of concept have been performed, where similar performance gains were observed (Nimr et al. 2025).

6.4.3 Model-Based Learning for Hardware Impairment Compensation in ISAC

ISAC is expected to operate at higher carrier frequencies than previous generations, given the benefits of larger bandwidths and carrier frequencies for sensing. The design of hardware for high frequencies is challenging, and systems are more likely to be affected by hardware impairments. Under these conditions, conventional approaches degrade in performance, which can be compensated for by AI. Among AI methods, model-based learning (MBL) parameterizes conventional methods, learning the components subject to performance degradation through data. MBL requires fewer parameters to learn and less labelled data than deep learning and offers explainability and performance guarantees. MBL has been used to compensate for hardware impairments in an ISAC scenario. An ISAC transceiver was considered equipped with a uniform linear array (ULA) and affected by antenna displacement impairments, to sense several targets in the environment and a single-antenna user for communications, as represented in Figure 6.9.

The MBL method proposed in Mateos-Ramos et al. (2025) performs end-to-end learning of the transmitter and receiver. It parameterizes existing target and communication message estimation algorithms to optimize the transmitter and receiver algorithms via gradient descent. Results show that the MBL method effectively compensates for hardware impairments as shown in Figure 6.10. Moreover, MBL also compensates for the suboptimality of the conventional transmitter algorithm, yielding a better communication performance than the baseline under normal operation.

In addition, this work was extended in Mateos-Ramos et al. (2024) to include an unsupervised loss function that reduces the amount of labelled data to collect, which is a time-consuming task in sensing applications. The MBL framework described was also used to compensate for gain-phase errors requiring no labelled data (Nimr et al. 2025).

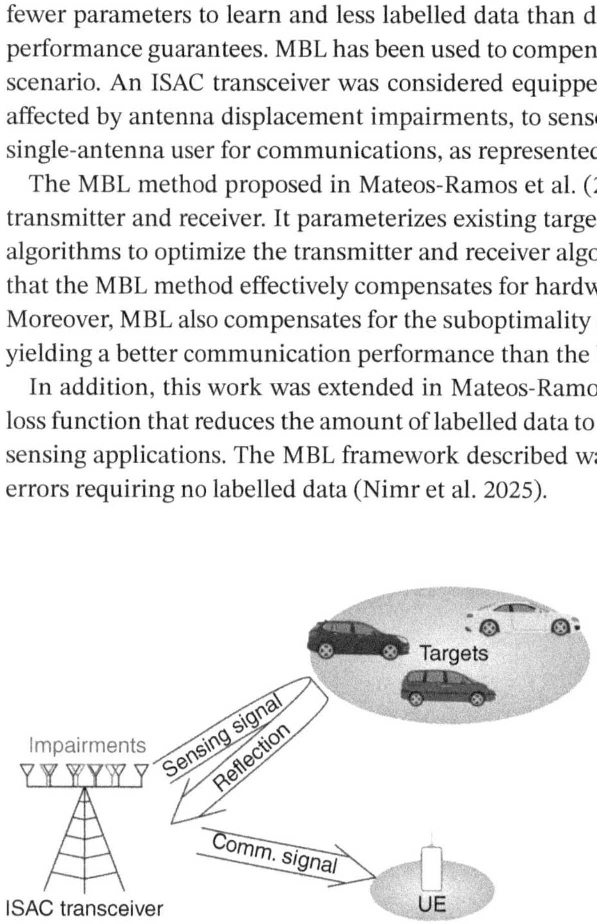

Figure 6.9 An ISAC scenario in the presence of hardware impairments.

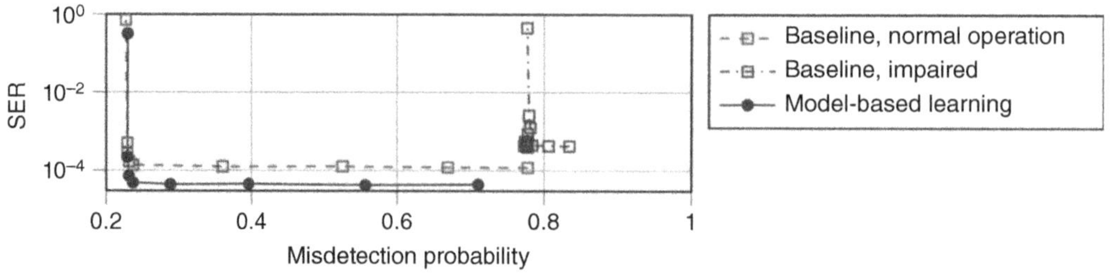

Figure 6.10 Hardware impairment compensation for ISAC.

6.4.4 Data-Driven Sensing with Wireless Signals

Wireless sensing has gained significant traction in recent years, primarily because it allows the use of existing communication infrastructure for sensing applications such as gesture recognition and pose estimation. Deep learning has achieved state-of-the-art performance on many such tasks. Particularly at mmWave, large antenna arrays and high bandwidth offer improved sensing resolution (Bhat et al. 2023a) extracted beam signal-to-noise ratio (SNR) from commercial off-the-shelf Wi-Fi access points (APs) and leveraged it for gesture recognition. Beam SNR is extracted from the sector-sweep operation employed by Institute of Electrical and Electronics Engineers (IEEE) 802.11ad APs. A convolutional neural network (CNN) was used to match changes in SNR to a set of 10 gestures, including head rotations. The system performed well, achieving 96% accuracy in a controlled environment. However, when more users and environments were included, accuracy dropped to 87%, and the model had to be retrained for each new situation. In recent work (Bhat et al. 2023b), the authors leveraged mmWave CSI extracted from commercial-off-the-shelf (COTS) Wi-Fi APs for pose estimation, predicting the 3D position of 25 individual joints of the human body. Since wireless signals don't provide clear visual details, they combined mmWave data with skeleton data from a Kinect camera to train the CNN. Once trained, the system could predict poses using only mmWave data, with an average error of 6.9 to 8.2 cm. However, a major challenge with data-driven sensing is dataset collection and generalization to unseen people or environments. Generative adversarial networks have shown promising results for augmenting the wireless sensing dataset and generalization (Xiao et al. 2019).

6.5 Smart Network Management

This section provides an overview of the representative examples for the integration of AI/ML in the network M&O.

6.5.1 AI-Based Solutions for Resource Allocation

Edge computing enables the deployment of services closer to the end users, which reduces communication latency and the computational load on the core. It also allows more efficient resource management: services that are usually deployed on centralized servers can then be placed over multiple distributed Edge servers depending on user location, and their resources can be adapted on demand based on local requirements. To further optimize computation resource usage, the physical resources of a node can be shared between the services and functions based on the current demand without a strict allocation scheme, which may lead to resource under-utilization. However, this leads to a competition of the deployed functions and services for the available resources, and possible service level agreement (SLA) violations in case of increased user demand that cannot be accommodated by the node. Therefore, an efficient resource allocation process should reach a trade-off between user admission and SLA satisfaction, where user demands are accepted only if the QoS can be satisfied. The solution described below uses a deep reinforcement learning (DRL) (Arulkumaran et al. 2017) model for dynamic and intelligent network resource allocation and user demand admission and routing in edge environments with shared resources, where the objective is to maximize user demand admission and SLA satisfaction per user, while minimizing resource and energy consumption by performing service consolidation.

As shown in Figure 6.11, the developed solution consists of an architecture comprising a DRL model which has been trained using the deep deterministic policy gradient (DDPG) algorithm

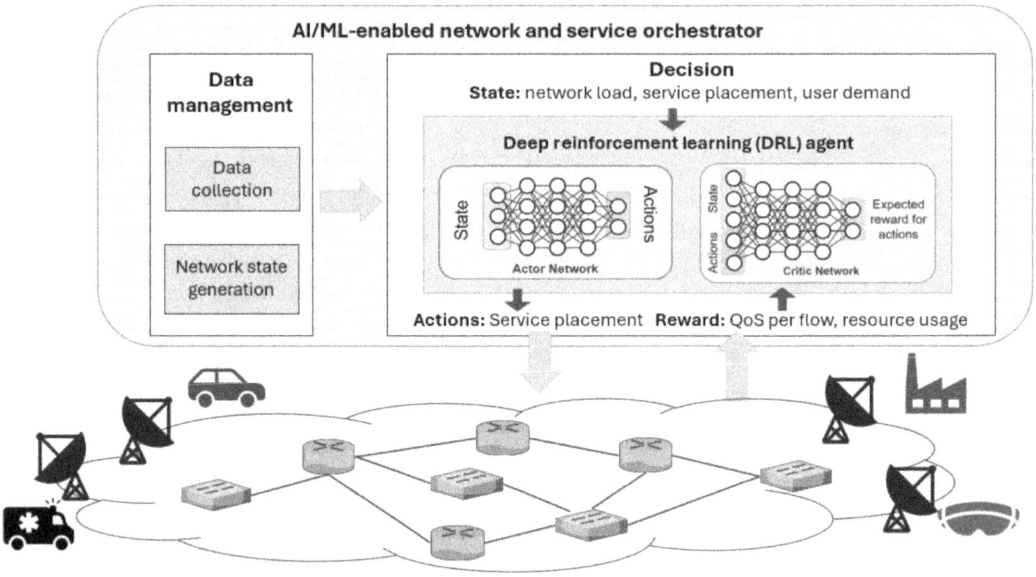

Figure 6.11 AI/ML-based resource orchestration solution.

(Lillicrap et al. 2016), which comprises an actor for providing actions and a critic for evaluating the decisions provided by the actor. The DDPG agent is trained using a network simulator to learn a decision policy for energy-efficient resource allocation. The input for the agent includes information on the current user demands for services and their location, as well as the state of the network with information on the current load on computing and network resources, and the placement of the currently deployed services. Based on the state information provided, the agent provides as output a new placement configuration for the services. The output can be translated into the deployment or termination of service instances, or their migration to a different server. Additionally, the agent performs load balancing by providing a user distribution matrix for accepting and directing the service requests towards the deployed instances or rejecting the requests if they cannot be satisfied. The rewards for the model are calculated based on the trade-off between performance and energy efficiency. The achieved performance is evaluated based on the collected QoS metrics per served flow, where the packet delay and packet drop are compared to the maximum values allowed by the SLA requirements of the service. Additionally, to achieve function consolidation and energy efficiency, a penalty is added for each utilized server. Finally, a penalty is added for each rejected user flow.

6.5.2 Network Digital Twins

As previously discussed, introducing AI/ML into network and service M&O allows further automation and optimization. However, applying the decisions provided by AI/ML models on the real network infrastructure directly carries the risk of negatively impacting the network and the deployed services if the model is not properly trained on all relevant scenarios, or if its performance degrades after a certain point. To support AI/ML-enabled M&O, NDTs can help mitigate those risks by generating 'what-if' scenarios for training, or for applying orchestration actions and observing their effects before carrying them out on the real environment.

Therefore, it is necessary to develop the tools and mechanisms that help generate an accurate NDT of the network infrastructure, which reflects its behaviour and properties, captures the intricacies of diverse topologies, and scales well to various network sizes. In this context, mechanisms for creating accurate NDT of the infrastructure have been developed, and the functions deployed while considering the impact of virtualization, in particular, on service performance. Indeed, while many mechanisms for NDTs have been provided in the literature, they do not consider the virtualization aspects for network infrastructures.

The proposed solution uses graph neural networks (GNNs) to generate the NDT due to their demonstrated performance in modelling communication networks (Suárez-Varela et al. 2023). The solution leverages attention mechanisms in GNNs and virtualisation-aware modelling, where the GNN model is trained using information on the impact that NFs have on each other when they are deployed on the same physical node, including central processing units (CPUs), memory, input/output (I/O) subsystems, and network interfaces. The training data is collected by performing extensive simulations where multiple NFs are deployed on the same nodes, and their performance is evaluated under different traffic rates by measuring resource usage and packet delay.

6.5.3 Multi-agent-Based Solutions for Distributed Services Orchestration

The increased complexity of network services and application graphs that could be deployed across the network continuum does not fit well with centralized approaches such as optimisation techniques or single-agent-based AI algorithms, whose decision space grows exponentially with the addition of services or links. Handling resources or tasks in a decentralized manner becomes more and more crucial for the satisfaction of the service level objectives (SLOs) defined in a set of available deployment options. The aim of this algorithmic approach is to provide a framework of responsibility distribution across the service or infrastructure domain, defining the agents and the corresponding items they manage towards performing adaptive scaling to optimize efficiency. This includes homogeneous agent groups such as agents for each service, computing or network node, as well as heterogeneous groups that allow different types of resources to efficiently work together to satisfy heterogeneous SLOs (e.g. service latency and CPU consumption).

For realizing such a scenario, a directed acyclic graph (DAG) – formed application graph with individual services communicating with each other – is deployed in a multi-provider setup. The traffic created by user requests is distributed across the graph according to the application's structure, and so, each service needs to handle different loads at each time point. The developed autoscaling mechanisms make the application adaptive to the variable incoming workload by spawning the necessary number of replicas for each service to handle the corresponding load, while considering the defined optimization objectives, i.e. latency and energy requirements, and the respective resource constraints.

For distributing decision-making across different entities, a multi-agent approach is considered, where different services are handled by their individual agents. For instance, the agents may share a common reward function with their main objective being the common end-to-end latency of a service chain. The algorithm can be targeted to optimize towards the total latency of the requests and the overall resource consumption, while each agent takes responsibility for the monitoring of each service's metrics and its deployment's configuration for satisfying the given constraints. Each agent may monitor the metrics for its own service and take its own actions based on its model. A collaborative approach may be selected for the training process, aiming to optimize towards a common reward. To implement such a collaborative approach, various multi-agent reinforcement learning (MARL) algorithms can be used. For instance, the QMIX variation is selected in the work in Rashid et al. (2018) for enforcing policies that enable global latency minimization instead of local optimization, which takes place in cases where collaboration strategies are not applicable (e.g. independent RL for managing agents for non-communicating services). In cases where there is not a global reward, but collaboration can still improve local rewards (e.g. for disaggregated services that use a common resource pool), decentralized/message-passing approaches (Chu et al. 2020) can be considered for enabling the understanding of the agents' working environment.

6.5.4 AI-Enabled Network Management

AI-enabled network management can offer service providers flexible, programmable, and scalable approaches to handle increasingly complex networking demands. The foundation of such an approach involves network programmability supported by key technologies like software-defined networking (SDN), APIs, and cloud-based network management platforms (Vilalta et al. 2021). Building on this programmable foundation, novel AI-driven capabilities, ranging from LLM-based intent management to end-to-end traffic forecasting and anomaly detection, can further automate and optimize network operations. In particular, NDT can play a significant role in mirroring and predicting network behaviour, enabling proactive planning and reduced operational risk. In this section, details on contributions to novel AI-enabled features for ETSI TeraFlowSDN controller are described (ETSI TeraFlowSDN 2025a,b). TeraFlowSDN has achieved an unprecedented level of AI integration within a cloud-native SDN framework. By embedding dedicated microservices for forecasting, anomaly detection, and LLM-based intent parsing, it enables more sophisticated network management and policy-driven network automation.

A primary AI-driven enhancement for network management involves LLM-based intent processing, where operators express high-level service requests in natural language. Through automated parsing and classification, these requests are translated into actionable network configurations, which reduce manual interventions. By automating the entire intent lifecycle, from creation and validation to deployment and explanation, LLM-based systems minimize human errors, shorten service delivery times, and increase network agility. An additional benefit is the potential for automated security policy enforcement: LLMs can generate access control lists (ACLs) and other context-specific rules on the fly, offloading time-consuming tasks from security teams while strengthening overall network protection.

The concept of NDT adds another valuable dimension to AI-enabled operations. These virtual replicas of physical networks use real-time analytics to forecast how changes, such as new features, QoS policies, or hardware upgrades, will perform before being deployed in production (ETSI TeraFlowSDN 2025a,b). By simulating various failure modes and capacity constraints, NDTs help identify potential issues early, enabling informed decision-making and proactive risk mitigation. This predictive capability supports a self-adaptive, self-healing infrastructure that reduces downtime and ensures a more consistent QoS.

Finally, AI-based forecasting and anomaly detection modules further improve operational efficiency by predicting traffic surges, maintenance needs, and potential failures. Through accurate capacity planning and real-time problem identification, operators can mitigate disruptions before they affect end users. This proactive approach leads to higher service reliability and lower operational costs, which are key advantages in an environment where demands and risks are continually evolving.

The combination of network management with AI-driven features offers powerful means to automate tasks, enhance security, and accelerate innovation. By integrating LLM-based intent management, NDTs, and predictive analytics, service providers can streamline network operations, minimize manual overhead, and deliver a more flexible and responsive infrastructure to meet emerging challenges.

6.5.5 Causal AI for Intent-Based Management

While intent-based systems excel in goal-oriented management, they are further empowered by cognitive features, which uncover the intricate dependencies among KPIs and configurations. The cognitive feature comes from *causal reasoning* capability by utilizing causal information in the network, which enables the network to learn the cause/effect among the network KPIs and configurations using the logic and facts collected from the network. Usually, causal reasoning is realized through a causal graph (Pearl 2009) where nodes in the graph represent a network KPI, metric, or configuration parameter, whereas edges represent a causal relation between the nodes. Building the relevant causal graph is a fundamental problem, and there are various research efforts in developing algorithms that help with the discovery of a causal graph with different assumptions. One can categorize those algorithms into two groups: (i) *Constraint-based algorithms:* these algorithms rely on statistical independence tests to construct a skeleton graph and orient edges based on conditional independence (Spirtes and Glymour 1991), and (ii) *Score-based algorithms:* these algorithms optimize a scoring function to iteratively add or remove edges, aligning the graph structure with observed data (Chickering 2002).

Once the causal graph is constructed, exemplary motivation for the usage and benefit of causal learning, particularly for a cognitive network point of view, can be given as follows: (i) *Root-cause analysis* seeks to identify the fundamental causes of problems rather than merely addressing their symptoms. By applying causal reasoning, the sequence of events can be traced and dependencies that led to an incident, enabling the prevention of recurrence by addressing the root causes directly. (ii) *Counterfactual reasoning* is a cognitive process that helps to understand the alternative possibilities scenarios (Pearl 2009). (iii) *Explainable AI*: Causality plays a pivotal role for XAI by providing a structured framework for understanding and clarifying the decision-making processes of AI systems. Causality helps to understand why a particular decision was made by bringing the causal factors and relationships that influenced the outcome, ultimately making AI systems more accessible and reliable (Wu et al. 2022).

The integration of IbM with causal reasoning fosters a holistic approach to network automation. Agents within intent managers use causal graphs to predict the downstream effects of proposed actions, ensuring that selected interventions align with both immediate and long-term network objectives (Karaca et al. 2024). The causal reasoning can be utilized for Proposal and Prediction agents within an intent management function (IMF), shown in Figure 6.12. When an issue is raised due to performance degradation (e.g. latency becomes unmet), the corresponding set of Proposal agents that are capable of proposing actions to mitigate this issue is triggered. The causal relations among the actions and KPIs can be effectively used at the Proposal agent to find out the root cause leading to an issue (i.e. unmet KPI expectation or fault). Like the typical fault management procedures, the Prediction agent, whose responsibility is to predict the impact of a proposed action on all relevant KPIs, can subscribe to the causal graph service to learn the actual causal features so that the developed prediction model can be more robust and perform better.

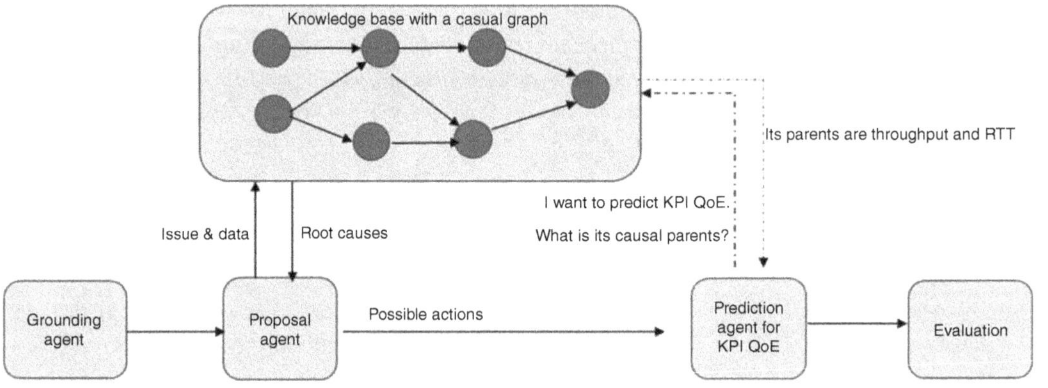

Figure 6.12 Integration of causal graph in IMF.

6.6 AI/ML and Trustworthiness for 6G

The role of AI/ML related to the trustworthiness of 6G is twofold, as depicted in Figure 6.13. On one hand, AI/ML can be used as a technique for improving the trustworthiness in 6G networks and the relevant key technological enablers in 6G. On the other hand, the trustworthiness of AI/ML techniques needs to be improved, and it is necessary to ensure the use of robust AI/ML mechanisms as they are applied in all network segments in the 6G end-to-end system. This section provides an overview of both of these aspects related to AI/ML and the trustworthiness of 6G.

6.6.1 AI/ML for Trustworthiness

AI and ML play an integral role in enhancing the trustworthiness of 6G networks in many ways, serving as technological enablers or tools for security and privacy-related activities in cellular networks, as well as complementary enablers to secure the operations of other key enablers. This may vary from deploying the AI/ML algorithms to detect or predict security vulnerabilities to using them as tools to mitigate or take remediation actions to prevent any harm to normal network operations. For instance, the classification algorithms, e.g. deep neural networks (DNNs), or the learning algorithms, e.g. RL, can be utilized for the detection of potential eavesdroppers, anomalies, or attempts to intrusions (Hexa-X-II 2025) AI/ML are powerful tools for particularly enhancing physical layer security (PLS) for detecting eavesdroppers or anomalies by analysing wireless link features, e.g. CSI

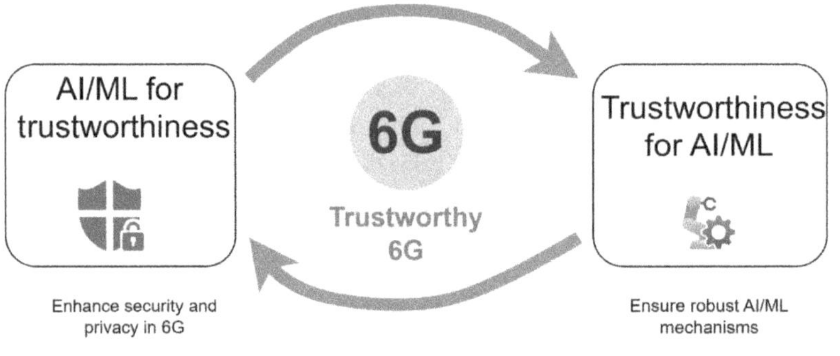

Figure 6.13 Illustration of the role of AI/ML related to the trustworthiness of 6G.

or SNR, and for secret key generation. Moreover, the use of distributed learning algorithms like FL or SL for distributed data processing and model training will assure privacy-enhanced data management in 6G networks. Mainly four types of security enablers are discussed below, which entail AI/ML tools to support their operations towards achieving trustworthiness in 6G networks. These include the *physical context awareness*, *physical anomaly detection*, *intrusion detection in the RAN*, and *physical layer deception* (Porambage 2023).

Physical context-awareness enabler in wireless systems utilizes the dynamic characteristics of the environment, such as physical obstacles, movement, and channel conditions, to generate unique keys based on the current state of the wireless channel. This adaptability impacts the key generation rate (KGR) through various factors like environment type, obstacles, movement, and other physical parameters. Recognizing the condition of the wireless channel allows the system to adjust the KGR appropriately. ML algorithms (e.g. Logistic Regression, decision trees, random forest, support vector machine (SVM), k-nearest neighbour (kNN), and CNN can be applied to classify environments based on channel impulse responses (CIRs), e.g. by extracting two types of features: momentary features, which capture individual CIR properties (e.g. amplitude, phase differences, and statistical measures), and temporal changeable features, which track variations across consecutive CIRs (e.g. cross-correlation and rates of change).

Physical anomaly detection has been studied under two pathways (Alemany 2024). First is the application of FL for anomaly detection in disaggregated RAN scenarios that allows distributed model training across edge devices. The federated client models within disaggregated RAN nodes are trained using locally available data, and model vectors are communicated to the relevant first-level aggregators hosted in RICs. Following aggregation, the calculated aggregator model vector is sent back to each local trainer in the same cluster, completing a cluster round. After a predefined number of cluster rounds, first-level aggregators transmit their model vectors to the second-level aggregator located in a centralized RIC. Second is the use of a DT of the radio environment for spectrum anomaly detection in licensed bands, which includes any type of interfering signals such as jamming signals. Measurement data regarding the spectrum are compared against expected data from a DT to identify deviations. Unsupervised learning algorithms for binary classification can be performed to identify the anomalies, considering the difference between the real world and DT data (e.g. expected received wideband power).

Intrusion detection systems (IDS) are crucial for monitoring network activities and detecting suspicious behaviours that may potentially lead to denial of service (DoS) or distributed denial of service (DDoS) attacks. The advancements in AI/ML methods have enhanced the capabilities of IDS while leading their role towards the security automation process in 6G networks. An XAI-based IDS framework has been proposed in Alemany (2024) to strengthen its defence against DDoS attacks that use adversarial techniques to evade detection. Here XAI is used to identify features altered by attackers, whereas the system gains the ability to adapt and optimize its responses in real time without the need for human intervention.

Physical layer deception can be applied to enhance security in wireless channels by manipulating signal characteristics (e.g. waveforms or beam patterns) to confuse eavesdroppers while ensuring the intended receiver decodes the message correctly. For example, it has been demonstrated in Alemany (2024) how DNNs can be used to further optimize these processes by adapting to adversarial behaviour and designing secure, semantically meaningful signals. Thereby, this approach improves communication efficiency, security, and robustness against attacks, particularly in semantic communication systems.

6.6.2 Trustworthy AI/ML for 6G

As moving towards 6G, AI/ML is becoming an essential component, directly affecting the security, privacy, and resilience (SPR) of the system. However, with the increased flow of data, ensuring the security and privacy of these AI systems is critical. Incorrect decisions made by AI/ML can impact performance, energy consumption, and the detection of security attacks. Additionally, since AI/ML involves data processing, potential privacy considerations must also be addressed. Therefore, the SPR of AI/ML is also important to create strong security measures and transparent strategies to protect user data and build trust, while also utilizing AI to enhance the efficiency and reliability of 6G networks. This aspect is generally referred to as the trustworthiness of AI/ML, encompassing elements like the robustness of AI/ML against security attacks (such as poisoning attacks), privacy and data confidentiality, transparency, and explainability. Privacy aspects in AI/ML were also discussed in Sections 6.3.1 and 6.3.2, focusing on privacy-enhanced aggregation, privacy levels, and hierarchical FL. In this section, addressing both the security and privacy aspects in AI/ML, utilization of XAI for robustness of AI/ML, and a focused look at the trustworthy AI/ML from the network M&O perspective are investigated.

Increasing security and privacy in federated learning: FL is an important tool for collaborative ML that can be utilized in different domains of 6G networks, so addressing security and privacy requirements in FL is essential to secure the 6G network and protect sensitive and privacy-related information. Increasing only privacy without paying attention to security can make the FL and potentially 6G system vulnerable to security attacks, which can be a threat to the reliability of the system and can introduce new security and privacy risks. Thus, it is important, depending on the threat model and trust assumptions, to address the security and privacy requirements simultaneously. However, addressing these two main aspects simultaneously in FL is a challenging problem because enhancing privacy makes security analysis a hard computation for the FL server. Although there are some solutions available in the literature, they have some considerable drawbacks in terms of computation or communication overhead or in terms of assumptions such as the need for having two non-colluding servers which help the aggregation operation but should not collaborate (e.g. share messages received from clients) with each other. To have a practical solution with reasonable assumptions, the FASIL framework was introduced in Karakoç et al. (2024). It prevents information leakage about the training data of the clients and simultaneously enables the FL server to execute security attack detection mechanisms to prevent security attacks on the FL process. It is foreseen that the developed FASIL framework can be utilized in FL applications in the 6G network when the privacy and security requirements in the FL are needed, depending on the trust assumptions and threat models. Some example scenarios where security and privacy aspects are required for the FL application can be an FL setup having UEs as clients, while a network element within RAN or core acts as the server; or NWDAFs in different network slices assume the role of FL clients and a single NWDAF, becoming the FL server.

Explainability-enhanced robust intrusion detection: Although increased connectivity from 6G provides significant benefits, it also introduces significant security challenges due to the large volume of network traffic. To address these challenges, IDSs are essential to monitor traffic, identify potential threats, and respond to threats in real time. When enhanced with ML, these systems can effectively detect unusual activities. However, some ML models might be vulnerable to adversarial attacks, which can risk not only IDS control but also sensitive data and crucial network processes. Attackers exploiting these vulnerabilities may launch several attacks to overcome and disable defence mechanisms. Thus, it is crucial to develop advanced IDSs with ML and robust security measures. Addressing this complex issue requires a comprehensive framework that leverages ML for adversarial attacks, managing sophisticated attack vectors, and real-time applicability. These measures might include regularly updating the systems and training them with a variety of data. In addition to current approaches, the incorporation of XAI to improve understanding of how decisions are made and, more generally, to detect malicious behaviours has been proven to be effective in the proposed adversarial attack detection framework (Paltun and Fuladi 2024). In the context of adversarial attacks, XAI is used to recognize characteristics of adversarial modification patterns by extracting and comparing the most important features. In this way, the system enables more accurate and efficient identification of unusual network behaviour by identifying specific vulnerabilities.

AI/ML security and privacy in network management and orchestration: AI/ML technologies will play a big role in enabling the ultra-high-performance capabilities, automation, and intelligence that 6G networks promise to offer. AI/ML-driven networks will be able to automatically configure, optimize, and repair themselves with minimal human intervention. Although these technologies are useful for the intelligence of 6G networks, they introduce new security and privacy vulnerabilities that need to be carefully managed. Security threats mostly target the robustness of the network against adversarial attacks; however, the privacy threat mostly focuses on the disclosure of sensitive information. There are potential security risks, such as poisoning attacks (i.e. data poisoning and model poisoning) and evasion attacks, as well as privacy attacks such as model inversion and membership inference attacks. In security attacks, the attacker aims to manipulate the input data at the training phase or inference time, leading the model to produce incorrect predictions. This can compromise the accuracy and reliability of an AI-driven system. On the other hand, in model inversion attacks, the malicious actor could reverse-engineer an AI model to gain access to sensitive data. Also, in membership inference attack, the attacker may attempt to determine whether a specific individual's data was used in training a particular model or not, which can lead to privacy violations.

In Hexa-X II, the vulnerability of the AI-driven IbM system is analysed from a security perspective. The aim of IbM is to oversee and control telecommunication systems by using high-level, abstract definitions of objectives and requirements through intents. By leveraging AI/ML techniques, IbM systems can autonomously orchestrate and optimize network operations, enhancing automation and improving overall efficiency in network management. Adversarial attacks targeting IbM systems can undermine the security of network management systems. These attacks may exploit vulnerabilities in the algorithms and decision-making processes that are used to define and implement network intents (goals or requirements). By manipulating intents, adversaries can cause misconfiguration or faulty optimizations, leading to network instability. As a result, the quality and reliability of services are compromised. Additionally, such attacks can corrupt the intent-handling process, causing AI-driven agents to fail in executing tasks as intended. This failure may prevent the network from achieving its goals, such as ensuring optimal bandwidth allocation or meeting user-specific requirements. The IMF is responsible for overseeing and processing intents. It coordinates a set of agents that make decisions, take actions, and ultimately work towards achieving the desired outcomes within the network. These agents can be enhanced with AI/ML models to deliver more optimized and efficient decision-making, improving the overall performance of the system. However, to ensure the robustness and reliability of IbM systems, it is required to ensure that the AI/ML models used by these agents are secure against adversarial attacks.

Another aspect of trustworthiness in 5G and beyond is privacy. In the context of 5G, the NWDAF in the core network and the management data analytics functions (MDAF) in the M&O layer play a pivotal role in improving network performance. These functions collect data from various

NFs and historical data to generate analytics and insights that help optimize resource allocation and enhance overall network efficiency. However, the reliance on large-scale data collection raises significant privacy concerns. Ensuring that sensitive network data is handled securely and in compliance with privacy regulations is fundamental to upholding trust among users, service providers, and regulators. For example, personal and location data, as well as usage patterns, may be leveraged for analytics. If mishandled or exposed, this data could violate user privacy or lead to unintended disclosure of sensitive information. To mitigate these risks, a comprehensive protection system tailored for diverse analytics functions is needed. This system must ensure that the integrity and confidentiality of data are preserved throughout the analysis process, while still enabling effective decision-making and network optimization. Multiple analytic functions can operate in a network, especially in a modern, dynamic, and multi-vendor network environment. FL techniques, which are a privacy-preserved distributed learning mechanisms, can be used to enable effective data analysis while ensuring that privacy is maintained. This approach allows data to be aggregated or analysed without exposing local data at each analytics function. However, model updates which are shared by each analytic function may contain sensitive information. In Hexa-X II, a privacy protection framework for distributed data analytics in M&O is provided. In this approach, while FL is used among multiple analytics functions, two functions (i.e. privacy management function and privacy operation function) are introduced which are responsible for orchestrating the privacy operation, generation and distribution of key pairs for privacy operations between multiple analytics functions, as well as applying decided privacy operation to protect local model updates at each analytics function.

6.7 An Overview of AI/ML Standardizations

Different standardization bodies have been undertaking the task of specifying support for AI/ML in network architectures. In particular, multiple Working Groups (GWs) within 3GPP have been studying the integration of AI/ML both in the RAN and the 5G core to support various functionalities as described in the following.

6.7.1 AI/ML Standardization in 3GPP SA2

The 3GPP SA2 WG specifies the core network architecture and functionalities of 3GPP systems. To support AI/ML use within the 5G system, the 3GPP standards define the NWDAF in the 5G architecture, which is in charge of training ML models based on data collected from different functions in the architecture, and providing data analytics (3GPP TS 23.288 2024). Examples of supported analytics include UE mobility analytics, QoS sustainability analytics, abnormal behaviour detection, or application detection.

The NWDAF can be split into two logical functions: The analytics logical function (AnLF), which is in charge of performing inference and exposing analytics and predictions, and the model training logical function (MTLF), which trains ML models and provides them to the AnLF. To improve the correctness of NWDAF analytics, the AnLF can provide accuracy feedback to the MTLF.

The 3GPP 5G architecture allows the NWDAF to collect data from and provide analytics to any 5G Core NF via the service-based interfaces, and multiple NWDAF instances can be deployed in a layered hierarchy or cooperate to perform an FL task where the NWDAFs can be an FL client or FL server. Additionally, analytics exposure for roaming use cases is supported, where the NWDAFs of *visited PLMN* and the *home PLMN* exchange information that facilitates operations for roaming UEs, such as Network Slice selection or QoS control.

On the other hand, the 5G system also supports a set of procedures to enable AI/ML applications. In particular, the 5G system supports the transmission of application AI/ML traffic between endpoints, optionally with specific QoS for the AI/ML flows, and can provide assistance to applications for selecting UE members for FL operations.

In Release 19, further work was performed to progress AI/ML support in 5G Advanced with the addition of support for vertical FL, AI/ML-based positioning, and NWDAF assistance for policy control and QoS enhancements, as well as network abnormal behaviours detection, mitigation, and prevention. Alignment work with the RAN groups for cross-domain (i.e. UE, RAN, 5GC, OAM, and AF) coordination is also under discussion.

6.7.2 AI/ML Standardization in 3GPP SA5

The 3GPP SA5 WG focuses on management, orchestration, and charging for 3GPP systems. Since Release 18, the WG has started specifying the AI/ML management capabilities in 5GS (3GPP TS 28.105 2024; 3GPP TR 28.858 2024), where management functionalities are defined to support the AI/ML operational workflow (i.e. training, deployment, and inference) while considering the requirements for different types of AI/ML such as FL, RL, online and offline training, distributed learning, and GenAI. Additionally, sustainability and trustworthiness aspects have been taken into account by introducing procedures for evaluating the energy consumption and efficiency impact for all operational phases of the AI/ML solutions, and by including trustworthiness indicators for the AI/ML functionalities.

To enable further automation of the M&O of 5G systems, in Release 19, the WG has started working on identifying use cases and providing solutions to integrate NDT for facilitating network planning, operation, and management (3GPP TR 28.915 2023). In particular, the WG aims to investigate the potential enhancements for the network M&O procedures based on input from the NDTs, and on the other hand, to determine the architecture and procedures for managing the NDTs, potentially using AI/ML.

6.7.3 AI/ML Standardization in 3GPP SA6

3GPP SA6 specifies the application enablement capabilities and APIs that facilitate the interaction between the 3GPP network layer and third-party applications, such as the SEAL and CAPIF for exposure. To support AI/ML applications, the following frameworks have been specified:

- **Application data and analytics enablement (ADAE):** This enablement layer provides mechanisms for collecting, storing, and managing data from various sources within the network, such as the NWDAF in the 5G core, or the vertical application layer (VAL) (3GPP TS 23.482 2024). The ADAE server can then process and combine the data and expose the analytics to application enablement clients.
- **AI/ML enablement (AIMLE):** This enablement layer defines the architecture, APIs, and protocols for 5G system support of AI/ML applications. Different APIs and procedures are specified to manage the AI/ML operations, such as AI/ML model training, deployment and update, storage and retrieval, or AI/ML client registration and discovery to perform AI/ML operations. Additional procedures are also defined to support FL, transfer learning, and SL (3GPP TS 23.482 2024, 3GPP TS 23.436 2024).

6.7.4 AI/ML Standardization for Air Interface in 3GPP RAN1/RAN2

3GPP initiated a study item on AI/ML for the 5G NR air interface in Release 18, aiming to establish a framework for leveraging AI/ML to enhance air interface performance. This study examined various aspects, including AI/ML LCM, and assessed the potential benefits of AI/ML compared to traditional methods across three key use cases:

- **CSI feedback:** Using AI/ML to reduce CSI acquisition overhead, improve feedback accuracy, and enable CSI prediction.
- **Beam management:** Using AI/ML to minimize overhead and latency for initial beam selection while enhancing beam selection accuracy.
- **Positioning:** Using AI/ML to improve positioning accuracy.

Building on the insights gained from Release 18, Release 19 is set to define a standardized framework for AI/ML integration in the air interface, with a focus on positioning and beam management. Additionally, Release 19 explores AI/ML applications for other emerging use cases. These developments by 3GPP are expected to accelerate AI/ML adoption in the telecom industry, paving the way for its broader implementation as the industry transitions toward 6G.

6.7.5 AI/ML Standardization for Air Interface in O-RAN

The O-RAN ALLIANCE is advancing RANs with two key principles: *intelligence* and *openness*. O-RAN introduces two primary approaches for managing network infrastructure through AI/ML-driven optimization:

1) **Service management and orchestration (SMO) framework:** This includes the Non-RT RIC and rApps, which facilitate intelligent RAN optimization in Non-RT (beyond 1 second) using data analytics and AI/ML.
2) **Near-real-time RIC and xApps:** These components enable the control and optimization of the centralized unit (CU) and distributed unit (DU) with near-real-time control loops (ranging from 10 ms to 1 s).

By leveraging AI/ML through rApps and xApps, O-RAN enables real-time prediction and adaptation of network parameters, particularly at the physical layer, to optimize performance. This AI-driven flexibility enhances RAN capabilities for various tasks, including MIMO beamforming, traffic steering, resource allocation, and power-saving strategies (e.g. dynamically switching off radio antennas based on traffic patterns). These AI-powered optimizations are among the potential 6G use cases currently under discussion within the O-RAN ALLIANCE, paving the way for more adaptive and efficient RAN operations.

6.8 Conclusion

AI/ML is envisioned to play a pivotal role in 6G systems. On the one hand, 6G is expected to benefit from integrated AI/ML functionalities across multiple layers of the network – for example, through AI-driven air interfaces, AI-powered M&O, and AI/ML techniques that enhance 6G trustworthiness. On the other hand, 6G aims to enable the large-scale deployment of AI/ML-based services across various industry sectors and societal domains, for instance, through AIaaS. There are strong motivations for the integration of AI/ML into 6G from multiple perspectives, as outlined in this chapter. AI/ML must be incorporated into 6G from the outset, following the AI/ML-empowered 6G system blueprint presented here. The anticipated performance benefits are demonstrated through a range of use cases – such as AI-driven air interfaces and intelligent M&O – which are also discussed in this chapter. These use cases will be supported by architectural enablers and the necessary signalling and protocols, and they will require standardization efforts to ensure seamless integration of AI/ML in 6G systems wherever applicable.

Acknowledgment

The authors acknowledge the contributions from Hexa-X-II project team to this chapter.

Acronyms and Abbreviations

ACL	Access control list
ADAE	Application data and analytics enablement
AF	Application function
AI	Artificial Intelligence
AIaaS	AI-as-a-Service
AIaaSF	AIaaS Function
AIMLE	AI/Ml enablement
AnLF	Analytics logical function
AP	Access point
API	Application programming interface
AR	Augmented reality
CaaS	Compute-as-a-Service
CAPIF	Common API framework
CNN	Convolutional neural network
CSI	Channel state information
CSP	Communication service providers
CU	Centralized unit
DAG	Directed acyclic graph
DataF	Data function
DDoS	Distributed denial of service
DDPG	Deep deterministic policy gradient
DF	Differential privacy
DN	Data network
DNN	Deep neural network
DoS	Denial of service
DRL	Deep reinforcement learning
DT	Digital twin
DU	Distributed unit
FEDaaS	Federation as a Service
FL	Federated learning
FLaaS	Federated Learning as a Service
GDPR	General data protection regulation
GenAI	Generative AI
GH	Group head
GNN	Graph neural network

IBM	Intent-based management
IDS	Intrusion detection system
IETF	Internet engineering task force
IMF	Intent management function
ISAC	Integrated sensing and communication
KGR	Key generation rate
kNN	k-nearest neighbour
LCM	Life cycle management
LLM	Large language model
M&O	Management and orchestration
MARL	Multi-agent reinforcement learning
MBL	Model-based learning
MDF	Management data analytics function
MIMO	Multi-input multi-output
ML	Machine learning
MLOp	ML operation
MLOpsF	MLOps function
MTLF	Model training logical function
NDT	Network digital twins
NEF	Network exposure function
NF	Network function
NFE	Network exposure function
NN	Neural network
NWDAF	Network data analytics function
PA	Power amplifier
PLS	Physical layer security
QoS	Quality of service
RAN	Radio access network
RF	Radio frequency
RIC	RAN intelligent controller
RL	Reinforcement learning
SDN	Software-defined networking
SEAL	Service enabler architecture layer
SL	Split learning
SLA	Service level agreement
SLO	Service level objective
SMO	Service management and orchestration
SNR	Signal-to-noise ratio
SON	Self-optimizing network
SPR	Security, privacy and resilience
SVM	Support vector machine
UE	User equipment
ULA	Uniform linear array
VAL	Vertical application layer
VR	Virtual reality
WG	Working group
XAI	Explainable AI

References

3GPP TR 23.700-82 (2024). Technical Specification Group Services and System Aspects; Study on application layer support for AI/ML services. V19.0.0, June 2024.

3GPP TR 28.915 (2023). Study on management aspect of Network Digital Twin. 3GPP TR 28.915, version 18.0.0. Sophia Antipolis, France: 3rd Generation Partnership Project (3GPP).

3GPP TR 38.843 (2023). Study on artificial intelligence (AI)/machine learning (ML) for NR air interface, v0.1.0, June 2023.

3GPP TS 23.222 (2024). Technical Specification Group Services and System Aspects; Functional architecture and information flows to support Common API Framework for 3GPP Northbound APIs. V19.2.0, June 2024.

3GPP TS 23.288 (2024). Architecture enhancements for 5G System (5GS) to support network data analytics services, v19.1.0, December 2024.

3GPP TS 23.436 (2024). Functional architecture and information flows for Application Data Analytics Enablement Service. 3GPP TS 23.436, version 18.3.0. Sophia Antipolis, France: 3rd Generation Partnership Project (3GPP).

3GPP TS 23.482 (2024). Functional architecture and information flows for AI/ML Enablement Service. 3GPP TS 23.482, version 18.3.0. Sophia Antipolis, France: 3rd Generation Partnership Project (3GPP).

Akgul O., Ericson, M. P. Botsinis, et al. (eds.) (2025). Hexa-X-II (2025) D3.5 Final architectural framework and analysis. Available at: https://zenodo.org/records/15774984 (accessed 1 September 2025).

Alemany, P. and B. Ojaghi Kahjogh (eds.) (2024). Hexa-X-II (2024). D2.4 End-to-end system evaluation results from the interim overall 6G system, September 2024. https://zenodo.org/records/15796347 (accessed 1 September 2025).

Arulkumaran, K., Deisenroth, M.P., Brundage, M., and Bharath, A.A. (2017). Deep reinforcement learning: a brief survey. *IEEE Signal Processing Magazine* 34 (6): 26–38. https://doi.org/10.1109/MSP.2017.2743240.

Bhat, N.N., Berkvens, R., and Famaey, J. (2023a). Gesture recognition with mmWave Wi-Fi access points: lessons learned. *2023 IEEE 24th International Symposium on a World of Wireless, Mobile and Multimedia Networks (WoWMoM)*, June, pp. 127–136. IEEE.

Bhat, N.N., Sameri, J., Struye, J. et al. (2023b). Multi-modal pose estimation in XR applications leveraging integrated sensing and communication. *Proceedings of the 1st ACM Workshop on Mobile Immersive Computing, Networking, and Systems*, October, pp. 261–267.

Boneza, B.M.I. (2023). A*n Integration Architecture of AIaaS: Interoperability and Functional Suitability*. Thesis for the Degree of Master of Science in Computer Science – Software Engineering, Mälardalen University.

Ericson, M., Botsinis, P., Saimler, M., et al. (eds.) (2024). Hexa-X-II (2024) D3.3 Initial analysis of architectural enablers and framework. Available at: https://zenodo.org/records/16995133 (accessed 1 September 2025).

Camara Project (2024). The Linux Foundation. Available at: https://camaraproject.org. https://zenodo.org/records/16995133 (accessed 1 September 2025).

Charatsaris, P., Diamanti, M., and Papavassiliou, S. (2024). Joint user association and resource allocation for hierarchical federated learning based on games in satisfaction form. *IEEE Open Journal of the Communications Society* 5: 457–471.

Chickering, D.M. (2002). Optimal structure identification with greedy search. *Journal of Machine Learning Research* 3 (November): 507–554.

Chu, T., Chinchali, S., and Katti, S. (2020). Multi-agent reinforcement learning for networked system control. *International Conference on Learning Representations (ICLR) 2020*. doi: https://doi.org/10.48550/arXiv.2004.01339.

Corrigan-Gibbs, H. and Boneh, D. (2017). Prio: private, robust, and scalable computation of aggregate statistics. *14th USENIX Symposium on Networked Systems Design and Implementation (NSDI 17)*.

Deng, J., Roh, U., Bao, J. et al. (2020). 5G and AI integrated high performance mobile SoC process-design co-development and production with 7nm EUV FinFET Technology. *2020 IEEE Symposium on VLSI Technology*, Honolulu, HI, USA, 2020.

Ericson, M. (ed.) (2023). Hexa-X-II (2023) D3.2 Initial architectural enablers. Available at: https://hexa-x-ii.eu/wp-content/uploads/2023/11/Hexa-X-II_D3.2_v1.0.pdf https://zenodo.org/records/16995133 (accessed 1 September 2025).

ETSI (2025a). TeraFlowSDN. Available at: https://tfs.etsi.org (accessed 26 March 2025).

ETSI (2025b). TeraFlowSDN contributions from HEXA-X-II. Available at: https://labs.etsi.org/rep/search?group_id=96&project_id=74&repository_ref=master&scope=issues&search=HEXA-X-II (accessed 26 March 2025).

European Commission (2021). Proposal for a regulation of the European parliament and of the council laying down harmonised rules on artificial intelligence (artificial intelligence act). [Online]. Available at: https://ec.europa.eu/info/business-economy-euro/banking-and-finance/financial-regulation-and-supervision_en (accessed 28 April 2025).

Farhadi, H. and Sundberg, M. (2020). Machine learning empowered context-aware receiver for high-band transmission. *2020 IEEE Globecom Workshops (GC Wkshps)*, Taipei, Taiwan, pp. 1–6. doi: https://doi.org/10.1109/GCWkshps50303.2020.9367518.

Farhadi, H., Haraldson, J., and Sundberg, M. (2023). A deep learning receiver for non-linear transmitter. *IEEE Access* 11: 2796–2803. https://doi.org/10.1109/ACCESS.2023.3234501.

Farhadi, H., Banerjee, B., Berknevs, R. et al. (2025a). 6G AI-Driven Air Interface — Hexa-X-II View. *IEEE Communications Magazine*, vol. 63, no. 10, pp. 118–125, October 2025.

Farhadi, H., Gowgi, P., and Sandberg, D. (2025b). AI/ML-based beamforming with imperfect channel state information for multi-user MIMO transmission. *IEEE International Conference on Acoustics, Speech and Signal Processing (ICASSP)*, 2025.

Geoghegan, T., Patton, C., Pitman, B. et al. (2024). Distributed aggregation protocol for privacy preserving measurement, 21 May. Available at: https://www.ietf.org/archive/id/draft-ietf-ppm-dap-11.html (accessed 1 September 2025).

Hexa-X-II (2025). AI/ML as a key enabler of 6G networks: methodology, approach, and AI-mechanisms in the SNS JU. White Paper, January 2025. Available at: ai_ml_white-paper-sns_tb_v1.0.pdf (accessed 1 September 2025).

Nimr, A., Atzeni, I., Tejerina, G., et al. (eds.) Hexa-X-II (2025). D4.5 – final results of 6G radio key enablers. Available at: https://zenodo.org/records/15773214 (accessed 1 September 2025).

Honkala, M., Korpi, D., and Huttunen, J.M.J. (2021). DeepRx: fully convolutional deep learning receiver. *IEEE Transactions on Wireless Communications* 20 (6): 3925–3940.

Ickin, S., Roeland, D., and Hall, G. (2023). Independent split model inference at operator network for network performance estimation. *NOMS 2023-2023 IEEE/IFIP Network Operations and Management Symposium*, Miami, FL, USA, pp. 1–9.

Karaca, M., Sadasivan, J., Baktir, A.C. et al. (2024). Utilizing causal learning for cognitive management of 6G networks. *2024 IEEE International Conference on Machine Learning for Communication and Networking (ICMLCN)*, Stockholm, Sweden, pp. 234–239. doi: https://doi.org/10.1109/ICMLCN59089.2024.10624867.

Karakoç, F., Paltun, B.G., Karaçay, L. et al. (2024). FASIL: a challenge-based framework for secure and privacy-preserving federated learning. *2024 8th Cyber Security in Networking Conference (CSNet)*, pp. 182–189.

Korpi, D., Honkala, M., and Huttunen, J.M.J. (2023). Deep learning-based pilotless spatial multiplexing. *2023 57th Asilomar Conference on Signals, Systems, and Computers*, Pacific Grove, CA, USA, pp. 1025–1029.

KServe (2025). KServe documentation. Available at: https://kserve.github.io/website/latest/ (accessed April 2025).

Lillicrap, T.P., Hunt, J.J., Pritzel, A. et al. (2016). Continuous control with deep reinforcement learning. *4th International Conference on Learning Representations (ICLR 2016), San Juan, Puerto Rico*, May 2–4, 2016.

Mateos-Ramos, J.M., Chatelier, B., Häger, C. et al. (2024). Semi-supervised end-to-end learning for integrated sensing and communications. *IEEE International Conference on Machine Learning for Communication and Networking (ICMLCN)*, Stockholm, pp. 132–138.

Mateos-Ramos, J.M., Häger, C., Keskin, M.F. et al. (2025). Model-based end-to-end learning for multi-target integrated sensing and communication under hardware impairments. *IEEE Transactions on Wireless Communications* 24 (3): 2574–2589.

Merluzzi, M., Borsos, T., Rajatheva, N. et al. (2023). The Hexa-X project vision on artificial intelligence and machine learning-driven communication and computation co-design for 6G. *IEEE Access* 11: 65620–65648. https://doi.org/10.1109/ACCESS.2023.3287939.

MLflow (2025). MLflow. Available at: https://mlflow.org/ (accessed April 2025).

Moore, S.K. (2024). Nvidia unveils Blackwell, its next GPU. *IEEE Spectrum* March 2024. https://spectrum.ieee.org/nvidia-blackwell.

OpenFL (2025). OpenFL. Available at: https://openfl.io/ (accessed April 2025).

O-RAN Alliance (2021). O-RAN non-RT RIC functional architecture. [Online]. Available at: https://www.o-ran.org/specifications (accessed 28 April 2025).

Paltun, B.G. and Fuladi, R. (2024). Introspective intrusion detection system through explainable AI. *2024 8th Cyber Security in Networking Conference (CSNet)*, pp. 28–32. IEEE.

Pearl, J. (2009). *Causality*. Cambridge: Cambridge University Press.

Porambage, P. (Ed.) (2023). Hexa-X-II (2023) D2.2 Foundation of overall 6G system design and preliminary evaluation results, December 2023. Available at: https://zenodo.org/records/16995047 (accessed 1 September 2025).

Prefect (2025). Prefect. Available at: https://www.prefect.io/ (accessed April 2025).

Rashid, T., Samvelyan, M., Schroeder de Witt, C., et al. (2018). QMIX: Monotonic value function factorization for deep multi-agent reinforcement learning. *International Conference on Machine Learning (ICML)*, 2018.

Rydén, H., Farhadi, H., Palaios, A. et al. (2023). Next generation mobile networks' enablers: machine learning-assisted mobility, traffic, and radio channel prediction. *IEEE Communications Magazine* 61 (10): 94–98. https://doi.org/10.1109/MCOM.001.2200592.

Saimler, M., İckin, S., Bernini, G. et al. (2024). The role of AI enablers in overcoming impairments in 6G networks. *2024 Joint European Conference on Networks and Communications & 6G Summit (EuCNC/6G Summit)*, Antwerp, Belgium, pp. 913–918.

Seldon (2025). Seldon core documentation. Available at: https://docs.seldon.io/projects/seldon-core/en/v2/index.html (accessed April 2025).

Spirtes, P. and Glymour, C. (1991). An algorithm for fast recovery of sparse causal graphs. *Social Science Computer Review* 9 (1): 62–72.

Suárez-Varela, J., Almasan, P., Ferriol-Galmés, M. et al. (2023). Graph neural networks for communication networks: context, use cases and opportunities. *IEEE Network* 37 (3): 146–153. https://doi.org/10.1109/MNET.123.2100773.

Tamburri, D.A. (2020). Sustainable MLOps: trends and challenges. *2020 22nd International Symposium on Symbolic and Numeric Algorithms for Scientific Computing (SYNASC)*, Timisoara, Romania, pp. 17–23.

Toumi, N. and Dimitrovski, T. (2024). AI-native architecture for 6G networks and services with model dependencies. *2024 Joint European Conference on Networks and Communications & 6G Summit (EuCNC/6G Summit)*, Antwerp, Belgium, pp. 901–906. doi: https://doi.org/10.1109/EuCNC/6GSummit60053.2024.10597080.

Uusitalo, M.A., Farhadi, H., Boldi, M. et al. (2025). Toward 6G – Hexa-X project's key findings. *IEEE Communications Magazine* 63 (5): 142–148. https://doi.org/10.1109/MCOM.003.2300735.

Vilalta, R., Muñoz, R., Casellas, R. et al. (2021). Teraflow: Secured autonomic traffic management for a tera of SDN flows. *Joint European Conference on Networks and Communications & 6G Summit (EuCNC/6G Summit)*, 8 June, pp. 377–382. IEEE.

Wu, Y., Lin, G., and Ge, J. (2022). Knowledge-powered explainable artificial intelligence for network automation toward 6G. *IEEE Network* 36 (3): 16–23. https://doi.org/10.1109/MNET.005.2100541.

Xiao, C., Han, D., Ma, Y., and Qin, Z. (2019). CsiGAN: robust channel state information-based activity recognition with GANs. *IEEE Internet of Things Journal* 6 (6): 10191–10204.

3GPP TS 28.105, (2024). 3rd Generation Partnership Project (3GPP) (2024) *TS 28.105: Management and orchestration; Artificial Intelligence/Machine Learning (AI/ML) management*. Technical Specification, Release 17.

3GPP TR 28.858, (2024). 3rd Generation Partnership Project (3GPP) (2024) *TR 28.858: Study on Artificial Intelligence/Machine Learning (AI/ML) management phase 2*. Technical Report, Release 19.

7

Integrated Sensing and Communication

Henk Wymeersch[1], Sami Mekki[8], Stefan Wänstedt[12], Athanasios Stavridis[12], Kawon Han[10], George C. Alexandropoulos[7], Benoît Denis[5], Sharief Saleh[1], Hui Chen[1], Yu Ge[1], Sokratis Barmpounakis[4], Vasileios Tsekenis[4], Rreze Halili[11], Rafael Berkvens[11], Pablo Picazo-Martínez[6], Giyyarpuram Madhusudan[9], Henry Blue[2], Nhan Thanh Nguyen[3], Markku Juntti[3], Musa Furkan Keskin[1], Francesca Costanzo[5], and Christos Masouros[10]

[1] *Chalmers University of Technology, Gothenburg, Sweden*
[2] *VTT Technical Research Centre of Finland, Espoo, Finland*
[3] *University of Oulu, Oulu, Finland*
[4] *WINGS ICT Solutions, Athens, Greece*
[5] *Université Grenoble Alpes and CEA-Leti, Grenoble, France*
[6] *University Carlos III of Madrid, Leganes, Madrid, Spain*
[7] *National and Kapodistrian University of Athens, Athens, Greece*
[8] *Nokia Networks France, Massy, France*
[9] *Orange, Meylan, France*
[10] *University College London, London, UK*
[11] *IMEC, Antwerp, Belgium*
[12] *Ericsson Research, Lund, Sweden*

As 6G networks transition from conceptualization to realization, integrated sensing and communication (ISAC) has emerged as a foundational paradigm (Tan et al. 2021). This integration represents a significant leap forward in wireless technology, offering a unified platform that merges wireless sensing and communication. ISAC is a groundbreaking technology that redefines the capabilities of wireless networks by blending these two traditionally distinct functionalities (Dong et al. 2024). Historically, radar-like sensing and communication used to operate independently, each relying on distinct hardware and spectrum allocations (Liu et al. 2020). This separation leads to inefficiencies and limitations in system performance and resource utilization when both communication and sensing functions are required. ISAC disrupts this traditional separation by enabling devices to sense their environment and transmit data. This dual functionality boosts network efficiency, responsiveness, and adaptability, creating a more cohesive and integrated system. The motivation for ISAC lies in its ability to support novel use cases, address stringent performance requirements, and advance the technological frontiers of 6G. This integration is pivotal for emerging applications like autonomous systems, smart infrastructure, and next-generation networks (Meng et al. 2024).

To appreciate ISAC's transformative potential, it is essential to explore its foundational principles, operational mechanisms, and diverse applications. ISAC operates on the principle of shared resources, where the same hardware and spectrum are utilized for both sensing and communication tasks. This approach not only optimizes resource usage but can also enhance the overall system performance. The operational mechanics of ISAC involve sophisticated signal processing techniques that allow for the simultaneous execution of sensing and communication

6G to Build a Sustainable Future, First Edition. Edited by Mikko A. Uusitalo, Patrik Rugeland, Mauro R. Boldi, and Ahmad Nimr.

functions. These techniques include joint waveform design, integrated signal processing, and dynamic resource allocation. By leveraging these advanced methods, ISAC systems can achieve high levels of efficiency and reliability.

The diverse applications of ISAC span various domains, each benefiting from the unique capabilities of this integrated approach. In autonomous systems, ISAC can enable precise navigation and collision avoidance by providing real-time environmental data. Smart infrastructure applications, such as intelligent transportation systems and smart cities, can leverage ISAC for enhanced monitoring and control. Next-generation networks can benefit from ISAC through improved connectivity, reduced latency, and support for advanced applications like holographic communication and immersive extended reality (XR). Despite its transformative potential, ISAC faces several challenges that must be addressed to unlock its full capabilities. Technical challenges include the need for advanced hardware designs, efficient signal processing algorithms, and robust interference management strategies. Security-related challenges involve ensuring data privacy, protecting against jamming and spoofing attacks, and developing secure authentication mechanisms. Regulatory challenges encompass spectrum allocation policies, compliance with safety standards, and adherence to privacy regulations.

ISAC represents a novel approach that will fundamentally transform the capabilities of future 6G networks. Continued research, development, and collaboration among industry stakeholders, researchers, and regulatory bodies will be important in overcoming these challenges and ensuring the successful deployment of ISAC in future wireless networks.

This chapter is organized to provide a comprehensive exploration of ISAC in 6G. Following an introduction on use cases and foundations of ISAC in Section 7.1, Section 7.2 considers the ISAC architecture, addressing its key components, centralized and distributed processing paradigms. Section 7.3 focuses on the ISAC physical layer, examining waveform design and signal processing techniques. Section 7.4 discusses the hardware considerations for ISAC. Finally, Section 7.5 outlines the next steps for ISAC in 6G.

7.1 The Role of ISAC in 6G

7.1.1 ISAC Use Cases

ISAC's dual-functional nature is expected to support a plethora of applications (see Figure 7.1) where communication and sensing are seamlessly intertwined, creating opportunities that are harder to attain with separate systems (Tan et al. 2021; Liu et al. 2020). One prominent use case is in **autonomous transportation and connected vehicles** (related to Network-Assisted Mobility from Chapter 2). Modern vehicles require not only the ability to communicate with other vehicles and infrastructure but also precise sensing capabilities to navigate complex environments safely. ISAC can facilitate this by combining vehicle-to-everything (V2X) communication with radar-based sensing, enabling functions such as obstacle detection, collision avoidance, and cooperative lane merging. By reducing hardware redundancy, ISAC can make it feasible to deploy these capabilities at scale in future intelligent transportation systems. Another critical use case is in **industrial automation and smart manufacturing** (related to Cooperating Mobile Robots from Chapter 2). In the Industry 4.0 era, factories relied on interconnected robots and sensors to optimize production. ISAC can enable robots to communicate while simultaneously sensing their surroundings, ensuring efficient coordination and reducing the risk of accidents. For example, an ISAC-enabled robot can transmit its position data to a central controller while

Figure 7.1 ISAC use cases.

using its radar capabilities to detect nearby objects or humans, making real-time adjustments to its operations. This capability is indispensable for achieving the high precision and safety standards required in automated manufacturing environments. In **healthcare and assisted living** (related to Human-centric Services from Chapter 2), ISAC promises to revolutionize patient care by integrating communication and sensing into a unified framework. Wearable devices and smart home systems equipped with ISAC functionality can monitor patients' vital signs and physical activities while transmitting real-time data to healthcare providers. For instance, a single ISAC device could detect a patient's fall using radar sensing and immediately notify emergency responders through communication networks, reducing response times and improving outcomes. **Smart cities and infrastructure** (related to Smart Intersections from Chapter 2) are another domain where ISAC can play a transformative role. By enabling real-time environmental sensing and communication, ISAC can support applications such as traffic management, public safety monitoring, and pollution detection. For example, an ISAC-equipped streetlight could detect traffic flow using radar and adjust its illumination or transmit the data to a central management system to optimize urban mobility. Such applications exemplify ISAC's potential to make cities more sustainable and efficient. Finally, ISAC can enhance **immersive experiences in XR and gaming** (related to Seamless Immersive Reality from Chapter 2), as these applications demand ultra-low-latency communication to synchronize user movements with virtual environments, as well as precise sensing to track user gestures and positions. By integrating these functions, ISAC can ensure seamless and responsive XR interactions, paving the way for next-generation entertainment and training platforms.

6G communication itself is also a use case of ISAC (related to Real-time Digital Twins from Chapter 2), enhancing **spectral and energy efficiency** by integrating sensing and communication into a unified framework (Liu et al. 2022). This integration reduces hardware redundancy, conserves spectrum, and enables spectral coexistence in congested bands. ISAC can also support **context-aware communication**, allowing networks to adapt dynamically through beamforming, resource allocation, and interference management. For example, an ISAC-enabled base station (BS) can use radar sensing to track blockers and adjust beams, improving signal quality and efficiency. Additionally, ISAC can enhance **positioning and localization**, achieving centimetre-level accuracy crucial for applications like AR, autonomous driving, and logistics. Lastly, ISAC can strengthen **network resilience and security**, enabling real-time detection of threats like jamming and spoofing. By sensing anomalies and adjusting communication strategies, ISAC can enhance the robustness and reliability of 6G networks.

7.1.2 ISAC Requirements and Metrics

ISAC systems must meet requirements that arise from the diverse use cases they support, covering both communication and sensing key performance indicators (KPIs) (Farhadi et al. 2023). One fundamental requirement is **high spectral efficiency**. Since ISAC systems share spectrum resources between sensing and communication, efficient spectrum utilization is critical to avoid interference and maximize performance. This is particularly challenging at higher frequencies, such as millimetre-wave (mmWave) bands, where ISAC operations must address severe path loss and limited coverage. Another critical requirement is **accurate, low-latency, and high-reliability sensing**. Applications such as autonomous driving and industrial automation rely on precise range, velocity, and angle measurements to make real-time decisions. ISAC systems must ensure high-resolution sensing capabilities, even in dynamic and multipath-rich environments. Advanced signal processing techniques, such as coherent integration and Doppler estimation, are essential to meet this requirement. Accuracy, low latency, and high reliability are equally important, particularly for mission-critical applications like vehicle safety systems and remote surgery. ISAC systems must deliver sensing and communication results within tight time constraints while maintaining robust performance under varying network conditions. Achieving this requires advanced resource allocation strategies and low-latency protocols that optimize the trade-off between sensing and communication tasks. Another requirement is the ability to support **massive connectivity**. Future 6G networks will connect billions of devices, ranging from smartphones to sensors and vehicles. ISAC systems must be scalable enough to handle this density while maintaining quality of service (QoS) for both sensing and communication functions. Technologies such as massive multiple input, multiple output (MIMO) and reconfigurable intelligent surfaces (RIS) may play an important role in meeting this challenge by enhancing spatial resolution and coverage. Finally, ISAC systems must be **energy-efficient** to support sustainable operations. By sharing hardware and spectrum, ISAC already reduces power consumption compared to separate systems. However, further advancements in low-power hardware design and energy-aware signal processing are necessary to meet the energy constraints of battery-operated devices and green communication networks.

7.1.3 Global View on ISAC

ISAC has received considerable attention in academia, industry, and standardization bodies, while the first standardized version is expected in 6G Release 21. Below is a brief overview of the status of ISAC in different standardization bodies across the globe.

ISAC is presented as a key feature of future wireless networks by the International Mobile Telecommunications (IMT) 2030 framework, which is being published by the International Telecommunication Union (ITU) as the vision for 6G wireless communication. It aims to surpass 5G (IMT-2020) by incorporating new technologies, enhanced connectivity, and improved network efficiency. Among the 6G usage scenarios, ISAC is listed together with the need for higher data rates (potentially terabit speeds), ultra-low latency (for applications like holographic communication), massive connectivity (supporting billions of devices in IoT ecosystems), ubiquitous connectivity, and artificial intelligence (AI)-driven networks (enhancing automation and efficiency) (ITU 2023). ISAC is expected to become a key enabler for a wide range of use cases such as AI-driven sensing, enhanced situational awareness, secure and distributed data sharing for specialized services, precise positioning, real-time 3D mapping, digital twins, industrial automation, gesture recognition, health monitoring, sports analytics, environmental sensing, and material inspection. The US National Institute of Standards and Technology (NIST) has been actively researching ISAC, focusing on its enabling techniques, applications, and standardization efforts. A comprehensive study draws attention on the design paradigms of ISAC systems, exploring aspects like transmit waveform design, environment modelling, sensing source, signal processing, and data processing. The study also presents some emergent smart-world applications that could benefit from ISAC, as it highlights challenges and future research directions in the field (Wang et al. 2022). Furthermore, the Alliance for Telecommunications Industry Solutions (ATIS) accredited by the American National Standards Institute (ANSI), a partner of the Third-Generation Partnership Project (3GPP), and a member of the ITU, has published a study on ISAC channel models covering midbands, mmWave, and sub-THz spectrum bands (Alliance for Telecommunications Industry Solutions 2024). The study highlights modifications to channel models developed in the 3GPP TR 38.901 (3GPP 2022) for 7–24 GHz based on new measurements. The 3GPP TR 22.837 has listed a variety of novel use cases and scenarios where ISAC could be applied. This includes object and intruder detection for smart homes, highways, railways, factories, and predefined secure areas around critical infrastructure. Additionally, ISAC would play a role in collision avoidance and trajectory tracking of unmanned aerial vehicles (UAVs), vehicles, automated ground vehicles (AGVs), as well as automotive manoeuvring and navigation. Other applications include public safety search and rescue, rainfall monitoring and flooding, and health and sports monitoring (3GPP 2023). In 2023, the European Telecommunications Standards Institute (ETSI) approved the Industry Specification Group on ISAC (ISG ISAC), which has established the technical foundations for ISAC technology development and standardization in 6G. The group has defined pre-standard research efforts on ISAC, with a focus on use cases, channel models through measurement campaigns, KPIs, evaluation assumptions, the radio access network (RAN) architecture frameworks, privacy and security aspects, and impact on the United Nations (UN) sustainability development goals (ETSI 2024). Finally, the Institute of Electrical and Electronics Engineers (IEEE) Task Group IEEE 802.11bf as well, was formed to develop an amendment to the Wi-Fi IEEE 802.11 standard and will enhance its ability to support Wi-Fi sensing and applications such as user presence detection, environment monitoring in smart buildings, and remote wellness monitoring (Ropitault et al. 2024).

7.1.4 Foundations of ISAC

Central to ISAC's foundation are the diverse sensing configurations – monostatic, bistatic, multistatic, and passive sensing – that define how signals are transmitted and received for environmental awareness (González-Prelcic et al. 2024). These configurations closely relate to positioning and are thus key technologies for autonomous navigation and real-time spatial intelligence. This section explores these foundational aspects, shedding light on their roles and potential in the context of next-generation wireless networks.

7.1.4.1 Sensing Configurations

Sensing techniques can be categorized based on how signals are transmitted and received, each offering distinct advantages and challenges. Monostatic, bistatic, and multistatic sensing differ in device configuration and synchronization requirements, impacting their spatial coverage and detection capabilities. Meanwhile, passive sensing leverages existing ambient signals (IEEE 802.11 working group, C.S. 2019). These approaches are briefly explored (see Figure 7.2), highlighting their key characteristics, applications, and limitations.

Monostatic sensing involves a single device performing both transmission and reception of signals, making it a prevalent choice in radar systems. A monostatic radar transmits a signal that reflects off an object and returns to the same device for analysis. This setup offers simplicity in terms of synchronization since all operations occur at a single location, at the cost of requiring full-duplex operation. However, monostatic systems have limitations in spatial coverage and detection range, as the single-point reception may struggle to pick up weak or distant signals. Applications of monostatic sensing are common in traditional radar systems used for weather monitoring, air traffic control, and basic surveillance tasks.

Bistatic sensing separates the transmitting and receiving devices, which are positioned at distinct locations. This separation offers a broader perspective and potentially improved detection capabilities. The bistatic configuration is particularly advantageous in scenarios requiring stealth or flexibility, such as military surveillance or terrain mapping. However, before starting the procedure, the devices need to negotiate to establish parameters such as timing and synchronization. Additionally, the physical separation of the devices necessitates precise synchronization, introducing additional complexity in system deployment and operation.

Multistatic sensing extends the concept further by employing multiple transmitters (TXs) and receivers (RXs) distributed across a region. This distributed configuration provides diverse perspectives, enhancing spatial resolution and enabling three-dimensional mapping of the environment.

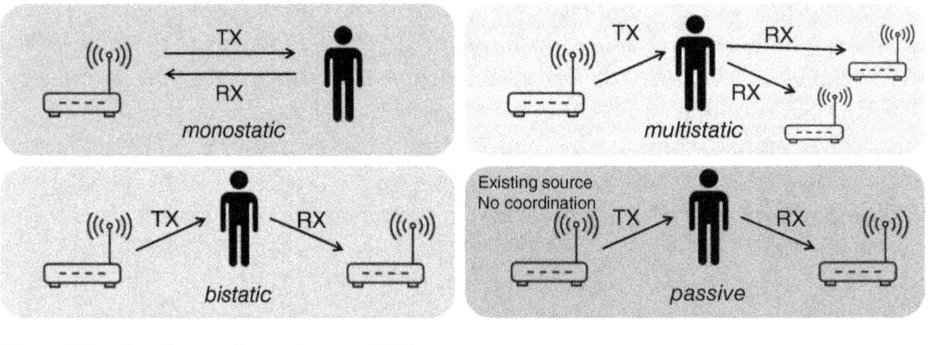

Figure 7.2 Sensing configurations in ISAC.

The redundancy of multistatic systems improves robustness, making them less susceptible to signal obstructions or device failures. These attributes make multistatic sensing suitable for applications such as urban traffic monitoring, environmental mapping, and advanced surveillance systems. Nevertheless, the high cost and complexity of deploying and synchronizing multiple devices pose significant challenges.

Passive sensing, in contrast, relies on ambient signals present in the environment, such as Wi-Fi or cellular transmissions, instead of actively generating signals. Thus, this procedure is initiated by a single device and does not require negotiation with the transmitting device. This approach takes advantage of existing infrastructure, making it cost-effective and energy efficient. Passive sensing is particularly useful in applications like crowd monitoring or detecting unauthorized activities, where non-intrusive observation is desirable. However, its performance depends heavily on the availability and characteristics of ambient signals, which may vary across environments, limiting its resolution and accuracy.

Positioning: User devices can act as TXs or RXs in sensing configurations (Figure 7.2), requiring accurate user positioning via radio signals (Behravan et al. 2022). This close link between positioning and sensing enables simultaneous localization and mapping (SLAM), required for autonomous navigation (Picazo et al. 2023). SLAM combines sensor data to map unknown environments while tracking the device position. mmWave sensing enhances SLAM, offering high-resolution obstacle detection and mapping, particularly in cluttered indoor spaces (Ge et al. 2024). However, handling non-line-of-sight (NLOS) and multipath effects remains a challenge. Multimodal sensor fusion, combining Wi-Fi, mmWave, visual, and inertial data, enhances SLAM accuracy by compensating for individual sensor limitations (Amjad et al. 2023).

7.1.4.2 Integrating Sensing into Communication

ISAC integration occurs at different levels, ranging from external sensor deployment to full resource sharing. The simplest form involves adding auxiliary sensors, like cameras or radar, to BSs or user devices to assist communication functions. Deeper integration ensures that sensing and communication share spectral resources for efficient coexistence. Hardware integration optimizes radio components, such as oscillators and converters, for both functions. Further, using the same waveform, like orthogonal frequency division multiplexing (OFDM), enables joint optimization of sensing and communication signals. The highest integration level allocates identical time and frequency resources to both tasks, maximizing efficiency, particularly in monostatic sensing but also feasible in bistatic and multi-static setups. Hence, at the waveform level, there is *joint transmission* and *separate transmission* (see Figure 7.3). In joint transmission, the same data signals are used for both communication and sensing. This is efficient in terms of spectral and hardware usage but introduces fundamental trade-offs: signals optimized for communication (e.g. beamforming towards a user) may not be ideal for sensing, making sensing opportunistic. Conversely, optimizing the signal for sensing (e.g. sweeping beams or optimizing a Cramér-Rao bound) degrades communication performance (Xiong et al. 2024). In a separate transmission, dedicated pilots are allocated for sensing, ensuring a predictable sensing QoS. However, this reduces the available time-frequency resources for communication, again leading to a trade-off. The extent of this trade-off depends on waveform design and signal processing, including interference coordination, synchronization, and resource allocation. New waveforms like orthogonal time frequency space (OTFS) may offer good time-frequency localization, balancing sensing resolution with communication reliability. However, managing interference and ensuring seamless operation in distributed ISAC deployments, such as multistatic configurations, remains a key challenge.

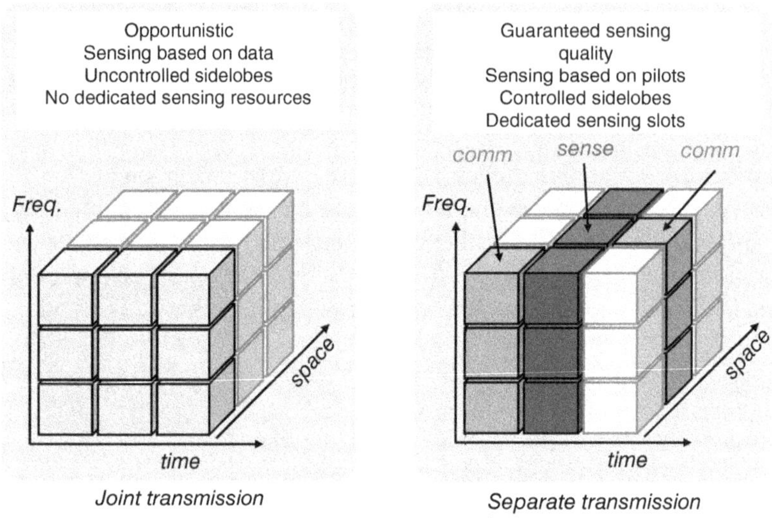

Figure 7.3 Joint versus separate transmissions of communication and sensing signals.

7.2 The ISAC Architecture

The 3GPP network architecture currently offers support for a variety of communication requirements and positioning/localization of active communication devices (i.e. user equipment (UEs)). However, sensing, which deals with the detection and parameter estimation of passive targets, necessitates significant changes to provide integrated sensing functionalities. Based on the architecture design principles elaborated in Chapter 3 and (Kerboeuf et al. 2024), several challenges can be identified, as listed in Table 7.1.

Integrating sensing into 6G requires fundamental architectural adaptations to efficiently allocate resources and ensure scalability. A key component is the new functions, which coordinate operations and separate sensing control from processing to maintain modularity. Another challenge is balancing centralized and distributed processing, as each approach affects latency, scalability, and accuracy. Reliable sensing also depends on tracking and handover of passive targets. Finally,

Table 7.1 Challenges for sensing in the 6G architecture.

Design principle	Challenges for sensing architecture
Flexibility to different network scenarios	Sensing service requirements vary widely for different use cases. The architecture should be adaptable to support these requirements efficiently. Support of data from multiple sensing modalities (e.g. LiDAR, cameras, and radio sensed devices)
Scalability	Many sensing use cases are spread over large areas and extended timeframes. Sensing service continuity needs to be ensured in this context.
Separation of concerns of network functions	Separate control and management of sensing nodes from the sensing data processing and fusion of sensing data from multiple sensing nodes
Minimizing environmental footprint and enabling sustainable use cases	Meet the needs of sensing service requirements in an efficient manner (e.g. intelligent resource allocation and compact data representation)

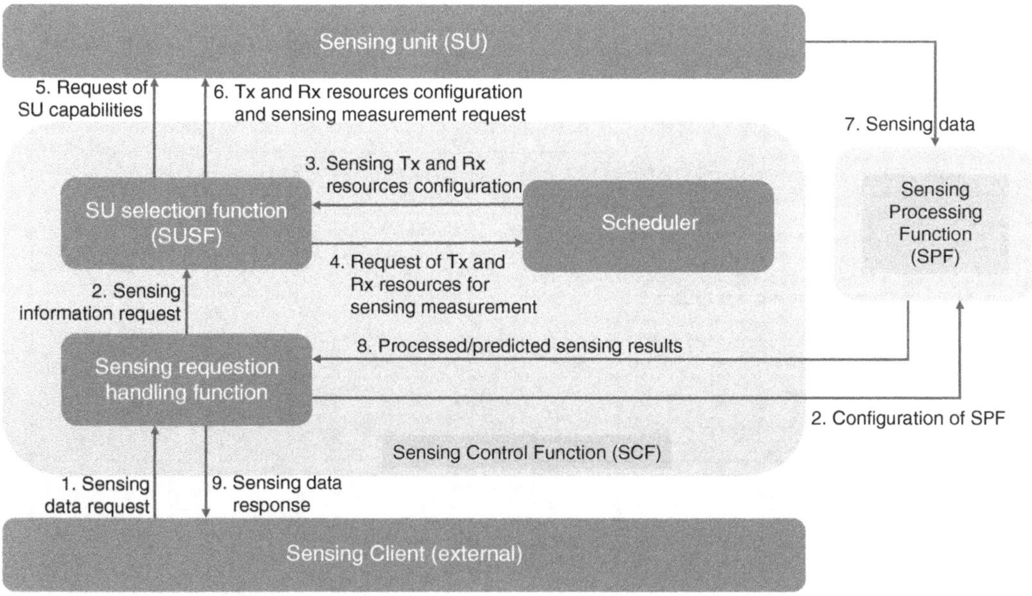

Figure 7.4 Sensing Network Functions.

machine learning (ML), AI, and semantic communication further enhance ISAC by optimizing resource allocation, target detection, and data fusion. These architectural elements address some of the key challenges in Table 7.1, providing a structured approach to integrating sensing into future networks.

7.2.1 Sensing Network Functions and Procedures

ISAC services require incorporating a new *sensing management function* (SeMF), which coordinates and manages sensing processes to ensure the service operates successfully. In line with the design principle of separation of concerns, the control function and the processing functions of the SeMF are separated as two entities, as illustrated in Figure 7.4:

- *Sensing control function* (SCF), which manages control and configuration aspects.
- *Sensing processing function* (SPF), which handles data processing and fusion.

Privacy and security are other primary functions of the SeMF, ensuring the ethical handling of personally identifiable information in sensing data during all stages of the sensing processes. Although privacy is a crucial aspect of ISAC services, its detailed treatment is beyond the scope of this chapter. Instead, the focus is on how the sensing functions interact to handle requests – from initiating the measurement process to processing and delivering results. The acronyms used in Figure 7.4 are intended to provide a clear representation of the concepts and entities involved. The main functions and principles described are widely recognized and will remain consistent across different terminologies that may evolve.

The *sensing procedure* begins with a request from a sensing client to collect information on one or more passive targets. Figure 7.5 presents a complete process flow for one such sensing procedure. The client may be a vehicle detecting pedestrians or a network application requiring sensing data. The request includes authentication details, the area of interest, time requirements, and required sensing performance.

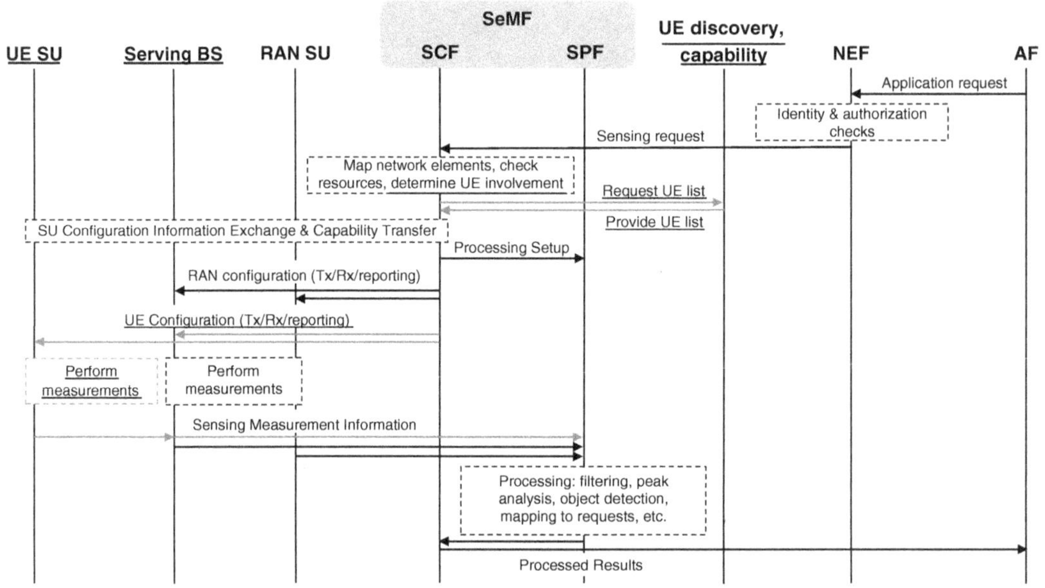

Figure 7.5 Setup and execution of sensing in the ISAC architecture. The underlined actions are needed when UEs are involved in sensing.

The application function (AF) forwards the sensing request to the network exposure function (NEF) for authorization and privacy checks. If valid, it proceeds to the SCF, where a sensing request handler enriches it with relevant contexts, such as KPI requirements and known environmental details. The request is then passed to the sensing unit selection function (SUSF), which determines the sensing units involved. At this stage, UEs and serving BSs, acting as sensing units (SUs), may also be included. The SUSF polls candidate sensing units for configuration details, such as reference signals, timing, supported sensing methods, and measurement capabilities, ensuring the request meets KPI requirements. Next, the scheduler reserves and configures radio resources, ensuring synchronization across sensing units. Once set, transmitting sensing units broadcast while receiving units forward data to the SPF. If a UE acts as a sensing RX, it may require significant uplink resources, potentially impacting battery life. The SPF then processes and fuses the measurements, especially in multistatic sensing, before returning results to the sensing client. The SCF and SPF can be deployed *centrally or in a distributed fashion* – SCF may reside in the core network, managing multiple cells, while SPF can be distributed between the core network and edge computing resources near the RAN. The next section will further explore the trade-offs between centralized and distributed processing.

7.2.2 Centralized and Distributed Processing

In ISAC systems, choosing between centralized and distributed processing approaches significantly affects key performance aspects such as latency, scalability, accuracy, and fault tolerance. A centralized approach typically aggregates raw or minimally processed data at a core node, enabling a comprehensive view of the environment. In contrast, a distributed approach leverages local processing at sensor nodes or edge devices, which can reduce latency and distribute computational load. Table 7.2 outlines the primary trade-offs between these processing approaches, illustrating the benefits and challenges of each.

Table 7.2 Trade-offs between centralized and distributed processing for ISAC.

Aspect	Centralized processing	Distributed processing
Latency	Longer delays due to data transport to a central fusion centre. May prevent real-time responses in time-critical applications.	Local pre-processing reduces response time. Ideal for real-time applications like obstacle detection in industrial automation.
Scalability	A single central node may become a bottleneck as the number of sensors increases.	Spreads the computational load across multiple nodes, supporting large-scale deployments.
Accuracy	Access to complete raw data can enable fine-grained fusion and higher-resolution detection.	Local processing may involve data compression or quantization; however, algorithm design (e.g. leveraging spatial diversity in D-MIMO) can still maintain high accuracy.
Fault tolerance	Centralized systems create a single point of failure – if the core node fails, the entire system is affected.	Local processing units operate independently; failure of one node has minimal impact on overall system performance.

As an example (shown in Figure 7.6), distributed MIMO (D-MIMO) deployment in an industrial environment highlights the trade-offs in distributed processing. Multiple radio units (RUs) are positioned throughout a factory floor to sense objects such as workers, robots, and equipment. Each RU performs local signal processing tasks, such as feature extraction or initial classification, before sending condensed information to a distributed unit (DU) via high-speed links. The DU then aggregates these inputs and forwards them to an edge SPF at a central unit (CU) for more complex sensor fusion, while the SCF in the 5G core network manages connectivity and resource orchestration. This architecture enables rapid local decision-making at the RUs, crucial for real-time safety checks and collision avoidance, while also supporting higher-level fusion at the SPF for tasks requiring a broader environmental perspective. The distributed nature of the setup reduces the risk of a single point of failure and allows for scalable network expansion as additional sensing nodes or functionalities are introduced. By balancing localized processing at the RUs with global data fusion at the SPF, the system achieves low-latency responses and high sensing accuracy, essential for dynamic, safety-critical industrial environments.

7.2.2.1 Data Representation and Sharing

While the above paragraph deals with where the sensing processing takes place, a related question is that of the data representation used between the sensing processing taking place in the distributed nodes (RU/DU) and the SPF at the edge or core network. Figure 7.7 depicts some of the data representation options. The choice of data representation is based on the performance required by the sensing service and the capabilities and limitations of the participating sensing nodes. These representations include:

- **Data Quantization** directly describes the raw signals, leading to large data volumes, which can be reduced under a few bit-quantization, at a cost of risking loss of important details.
- **Range-Angle-Doppler (RAD) Tensors** (Range-Angle and Range-Doppler maps) structure spatial and velocity information for radar sensing, but require large data volumes, high computational resources, and struggle with noise and real-time processing in multi-target environments.

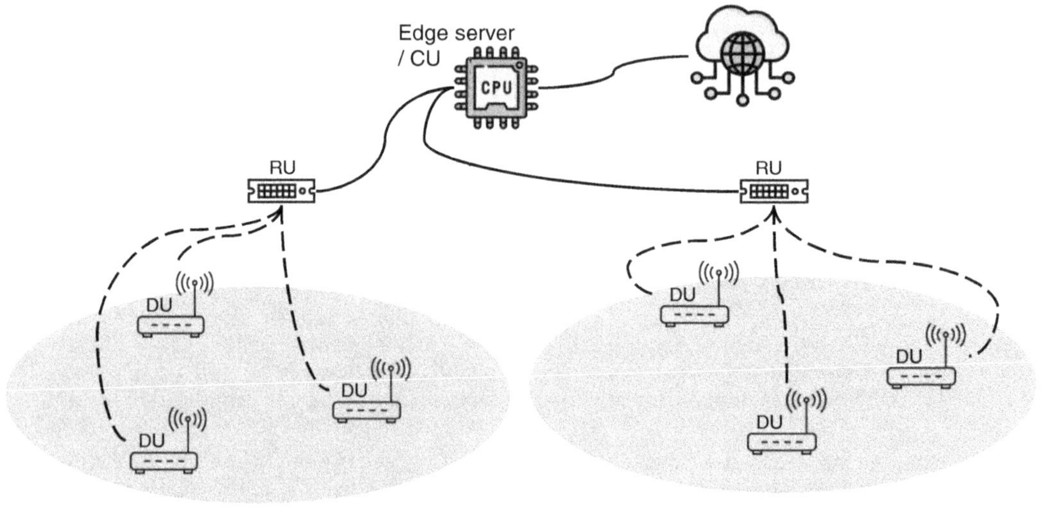

Figure 7.6 Distributed ISAC in a D-MIMO system.

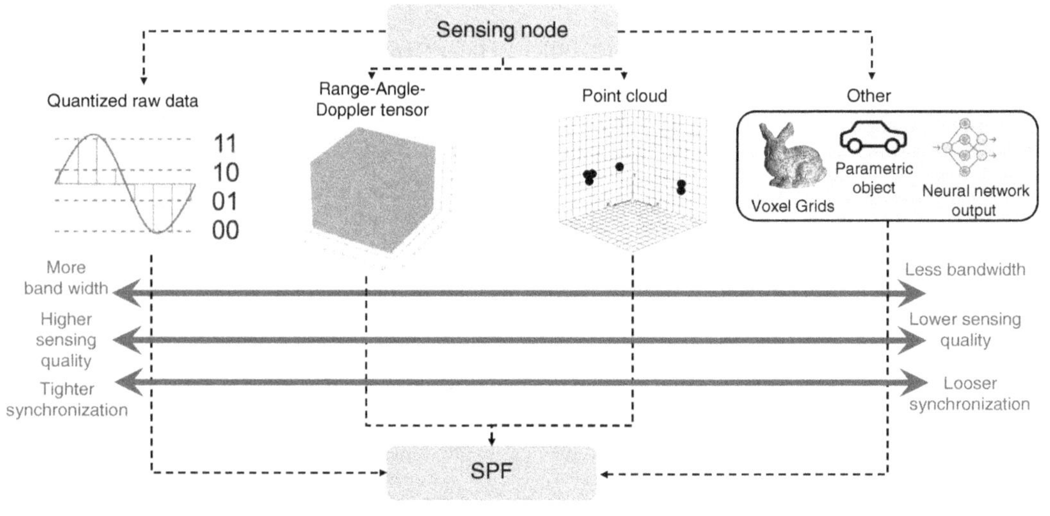

Figure 7.7 Data representations and their benefits and drawbacks.

- **Point Clouds** capture discrete 3D target positions, including range, velocity, azimuth, and elevation. They offer compact data representation and precise tracking but require filtering to handle noise, clutter, and segmentation challenges.
- **Other Object Representations** reduce point cloud data by clustering objects and creating compact models that preserve shape and structure, improving efficiency. Moreover, neural-network-based representations use deep learning to extract high-level features, enhancing sensing accuracy and robustness. While effective for 3D recognition and vision tasks, they have high computational costs, require large datasets, and lack interpretability.

7.2.2.2 Tracking and Handover of Passive Targets

Many sensing use cases span large areas and extended timeframes. In this context, it is essential to ensure the continuity of sensing services for passive targets, as visualized in Figure 7.8. The

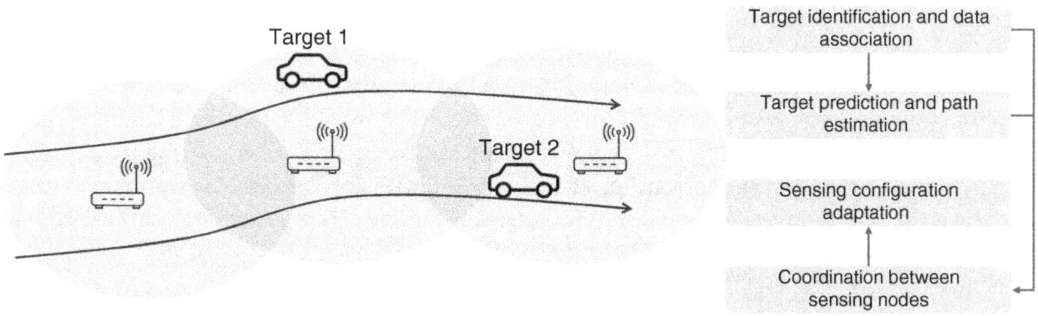

Figure 7.8 Multi-target tracking and handover pipeline.

ISAC framework should enable seamless tracking of moving targets across time and space, extending beyond the field of view (FoV) of individual sensors or even groups of bistatic or multi-static sensors. This distributed approach provides two key advantages: First, it ensures target identity maintenance, allowing a target's assigned identity to be consistently recognized across multiple sensors, facilitating smooth target handover. Second, it enables optimized sensing, where sensors can adjust their resource usage (e.g. radio resources, processing complexity, and data representation) based on predicted target movements, even before the target enters their FoVs, enhancing efficiency and accuracy. A procedure for target tracking and handover can be envisioned as follows.

- **Coordination between sensing nodes** – Smooth tracking requires coordination among sensors to ensure seamless handover as targets move between FoVs. The SeMF facilitates this process.
- **Target identification and data association** – All sensing nodes must recognize and associate target identifiers to maintain tracking consistency. The SeMF assigns and communicates target identities, ensuring each sensor correctly links new detections to existing tracks.
- **Target prediction and path estimation** – Accurate prediction of target movement enables proactive handover. Sensors anticipate when a target will exit their FoV and determine which nodes will track it next. The SeMF may maintain a global target map for improved tracking accuracy.
- **Sensing configuration adaptation** – Sensors adjust resolution, radio resources, and detection parameters based on predicted movements. The SeMF optimizes configurations dynamically, ensuring effective target tracking and efficient resource use.

7.2.3 Role of AI/ML and Semantics in ISAC

Chapter 6 introduced the motivations for data-driven architecture and AI-driven methods for cellular communication networks. In this section, these methods are considered from the ISAC perspective.

7.2.3.1 AI/ML in ISAC

AI/ML approaches can be readily incorporated into the ISAC network stack to provide substantial benefits. First, learning-based approaches move most of the required computation during their offline training phase, such that during deployment, the computation requirements are minimal, assuming, of course, the deployment of hardware acceleration mechanisms. As a result, operations, including performing inference, making predictions, or deciding on actions that affect the system, may take place in time scales that are shorter than typical channel coherence frames, offering precise control of the wireless system (Alexandropoulos et al. 2022b). Additionally, the inherent pattern recognition mechanisms of learning-based approaches allow for the design of operations that are tailored to the particularities of the deployment environments. Since practical ML requires specialized hardware, the **centralized approach** involves collecting the signal(s) at the participating sensors and forwarding them to an SPF with appropriate hardware capabilities to produce the output of the ML model. When the sensing process only requires signals from a single entity, the accelerator SPF may be collocated with the device. However, this increases the deployment costs and energy requirements of the sensing devices. To that end, ML-specific SPFs could be located on edge servers, to which the sensor data are shared. Depending on the nature of the input data (e.g. high-resolution RAD tensors), sharing may come with increased latency and more data-intensive representations must be shared. The inherent trade-off between latency and computation is only exemplified in cases where the multisensory data from physically separated devices must be shared, e.g. when D-MIMO systems are considered.

For **distributed ML systems**, mitigating the considerable overheads in terms of control and feedback link capacities is crucial. To that end, distributed ML approaches endow each distributed RU with its own local version of the trainable model. Each model observes only local information and only controls its associated actuators (e.g. RXs and beamformers). As a result, each local ML model is typically associated with smaller computational requirements. Training such swarms of models, however, requires some considerations. From the perspective of reinforcement learning (RL), frameworks such as the asynchronous advantage actor critic (A3C) (Mnih et al. 2016) or distributed neuroevolution enable coordinated learning by transferring each local model periodically to the CU for fusion and update to the previous model, so that all RUs receive an updated copy of the global model afterwards, on timescales depending on the application. Federated learning operates under a similar notion of model sharing instead of data sharing, where all model parameters are transmitted and averaged at the CU, an operation that may also take place over-the-air, by exploiting the superimposition properties of the wireless signals (Xiao et al. 2024; Stylianopoulos et al. 2025). While model sharing does not always lead to reduced data transmissions compared to input sharing of centralized ML, such exchanges take place only during training, whereas input sharing and fusion take place at every inference instance at the deployment system. Another approach is to endow the models with messaging exchange capabilities. In such cases, parts of the outputs of the RUs are broadcast over the network for the purpose of giving information to the cooperating nodes (Stamatelis et al. 2025). This approach necessitates information exchange both during training and inference; however, the messages are typically few-bit representations of each model's state and intention that can be transmitted with relatively negligible requirements.

The final consideration towards the architectural support of ML for ISAC is the dedicated need for model re-training and re-deployment. Since the statistics of the sensory data may change over time, as the system evolves, periodic re-training or substitution of deployment models with ones that are better trained for the current particularities of the environment is necessary. While data collection and training of a new model may take place without the disruption of the operation of the existing model, the architecture must be designed to accommodate such data transfers over its control links, where a central SPF is responsible for retraining. Finally, the new model must be transferred to the deployed devices, which calls for the sporadic utilization of high-capacity duplex links.

7.2.3.2 Semantic Communication in ISAC

Semantic communication focuses on transmitting only the meaningful information, rather than the raw data itself, that can be analysed to extract only useful information. The semantic meaning behind the data symbols becomes more important than simple reconstruction, enabling efficient data processing based on situational context, environmental information and goal selection. Besides the huge reduction of traffic load, semantic communication together with semantic joint source and channel coding can also enable transmissions highly robust to the dynamic wireless environment. Semantic sensing focuses on the simultaneous implementation of sensing and compression, where only task-related signals are extracted, reducing the amount of data acquired at the sensor level, to alleviate the computational and sampling costs of sensors (Zhang et al. 2023a). In the ISAC context, leveraging several semantic and goal-oriented methodologies can enhance performance by making it more scalable and adaptable to real-world applications. Multi-modal data fusion is one of the main challenges for the ISAC framework. Cross-modal sensed information can be efficiently integrated by leveraging AI technologies to enable semantics, such as those described below:

- **Semantic fusion** can combine different types of data from multiple types of sensors, leveraging the contextual meaning and improving environment recognition while reducing the need for high-bandwidth data transmission and becoming more resilient to noise, environmental interference, and channel fading. In Wang et al. (2025), the potential of generative AI (GenAI) is explored in solutions for radio frequency (RF) environment reconstruction, localization, and imaging, where multi-modal sensing makes a connected sensor network system more adaptive to dynamic environments and significantly improves overall efficiency.
- **Semantic segmentation** provides object-level understanding classifying each pixel of an image into different meaningful regions. Once an image, or sensor data, are segmented, the identified objects or areas can be composed into more complex structures, recognizing relationships between them.
- **Semantic composition** builds on segmentation to form context-aware, higher-level representations of a scene, combining components into structured representations, and discovering how individual segments relate to a meaningful whole.
- **Semantic alignment** can ensure that different modalities (e.g. images and point clouds) are properly synchronized and refer to the same environment, improving the accuracy and robustness of the sensing. In the multi-modal sensing case, this method aligns information from different sensors to contribute coherently to a shared understanding, helping with (semantic) fusion. In Cheng et al. (2025), multimodal large language models (MLLMs) are introduced as ISAC enablers in different use cases (intelligent transportation, drone swarms, and smart factory), showing how multimodal alignment learning and other AI methodologies can enhance communication and sensing performance.
- **Semantic compression** improves efficiency by transmitting only extracted features instead of raw data, reducing bandwidth demands and ensuring *sensing continuity* even in constrained networks. Additionally, semantic multi-modal fusion enhances reliability by mitigating sensor failures and adapting to environmental changes. Goal-oriented communication further optimizes performance by prioritizing contextually relevant information, reducing unnecessary transmissions while ensuring that critical updates remain available (Merluzzi et al. 2024).
- **Semantic resource allocation** optimizes network resources by dynamically adjusting sensing and communication priorities based on extracted semantic information. A major advantage of semantics in ISAC is, thus, *adaptive resource allocation*. AI-driven approaches, such as edge computing, are a key enabler, particularly in distributed ISAC, allowing sensor nodes to locally refine data into structured representations before transmission. This reduces latency and computational overhead while enhancing real-time decision-making for applications such as autonomous driving and smart surveillance.

7.3 The ISAC Physical Layer

The physical layer is central to the ISAC radio system, enabling dual-purpose waveforms, beamforming, and signal processing for both communication and sensing. This section covers waveform design, beamforming strategies, and signal processing methods, including OFDM optimization as well as new waveform paradigms for enhanced efficiency. ML for beam prediction and multi-modal sensing is explored, along with techniques for RAD processing, target tracking, and sensing optimization. Finally, the section addresses deployment challenges in massive MIMO, full-duplex, and RIS-aided systems, highlighting ISAC's adaptability to diverse network architectures.

7.3.1 Waveforms for ISAC

The design of ISAC waveforms plays a crucial role in enabling joint communication and sensing functionalities. Unlike conventional communication waveforms that optimize spectral efficiency and data rates, ISAC waveforms must also support high-resolution sensing and target localization, while minimizing interference between the two functions. This section examines waveform design strategies for monostatic, bistatic, and multi-static ISAC, explores beamforming and multi-modal sensing-assisted communication, and discusses emerging waveform paradigms such as OTFS.

7.3.1.1 MIMO-OFDM for ISAC

MIMO-OFDM serves as the foundation for modern wireless communication, and its use for ISAC has the potential to accelerate the adoption and commercialization of the new functionality of ISAC in future networks. In ISAC, the waveform must balance high-data-rate transmission with the ability to extract information regarding the surrounding environment from the received signals, such as precise range, Doppler, and angular information. Compared to monostatic sensing, in distributed ISAC (bistatic or multi-static), in addition to the previous balance, the coordination and synchronization of the involved sensing nodes pose additional challenges and opportunities. The opportunities mainly originate from the plethora of nodes in the modern wireless network that can act as sensors.

- **Degrees of freedom in time, frequency, and space** play an essential role in optimizing sensing performance and communication. From the sensing perspective, the choice of the OFDM symbol duration, cyclic prefix (CP) length, bandwidth, and time allocation, along with the periodicity and number of pulses inside a pulse-train, determine the fundamental quality metrics of radar sensing, e.g. range and Doppler resolution, and unambiguous range and Doppler. Note that the previous design parameters need to be optimized jointly or at least efficiently to satisfy both the sensing and the communication requirements. Beamforming strategies further enhance the performance of MIMO-OFDM-based ISAC and enable the coexistence/integration of sensing and communication. Digital beamforming enables per-user and per-target optimization, while hybrid analogue/digital beamforming reduces hardware complexity in large-scale MIMO deployments. Optimizing precoders in the space-frequency-time domain is a key challenge, requiring trade-offs between spectral efficiency and sensing precision. The adoption of joint optimization approaches that balance these factors remains an active area of research for the next-generation ISAC networks.
- **Waveform design for monostatic ISAC:** In monostatic sensing, the TX and RX are co-located, simplifying certain aspects of system design and reducing synchronization challenges, but monostatic-based systems are limited in detecting certain targets at extreme angles (Gavras and

Alexandropoulos 2024c). Full-duplex sensing expands on monostatic operation by allowing a single node to transmit and receive simultaneously over the same frequency band. Generally, full-duplex enhances spectral efficiency; however, it also introduces self-interference (SI) at the receiving end, necessitating advanced SI mitigation techniques (Gavras et al. 2023). While for bistatic ISAC, dedicated sensing pilots are preferred to ensure high sensing performance among distributed and unsynchronized nodes, monostatic and full-duplex sensing can rely on position data or channel information extracted from the received communication signal at the RX. For example, the monostatic full-duplex systems discussed in Alexandropoulos and Gavras (2025) and Gavras and Alexandropoulos (2024a,b) leverage the reflected downlink (DL) communication signals to extract position information, which can then be used to optimize an ISAC-oriented system. A hybrid approach balances both methods by dividing the transmitted waveform between data communication symbols and sensing probing streams. While this trade-off allows for flexible system optimization across all configurations, it also introduces added complexity, as both communication- and sensing-based beamforming must be considered, while the sensing probing streams may cause interference at the RX (Alexandropoulos et al. 2022a).

- **Waveform design for distributed ISAC:** Both communication data signals and dedicated pilot signals can be used in distributed ISAC (Strinati et al. 2025). Pilots and data signals in this context can be shared or exchanged between the distributed nodes to synchronize the sensing process and improve spatial resolution. However, distributed systems also present challenges in terms of *coordination, synchronization, and interference management* across the various nodes, which must be considered when optimizing the ISAC waveforms. To overcome these challenges, advanced algorithms for distributed signal processing, localization, and cooperative beamforming are essential to ensure seamless communication and sensing performance (Miridakis et al. 2017).

7.3.1.2 Beamforming for ISAC

As discussed above, the ISAC waveforms must serve both communication and sensing purposes. For communication, the waveform carries data intended for communication users, while for sensing, it also consists of probing signals directed at the directions of the potential sensing targets. To ensure that data signals reach communication users while probing signals are transmitted to the intended directions with minimal interference, proper ISAC waveform design in the spatial domain is essential.

There are three primary principles for the integration of radar sensing and communications in ISAC systems based on the focused functionalities:

- **In radar-communication (radcom) system**, the communication function is built on top of an existing sensing system by integrating digital messages into radar waveforms through sidelobe control, waveform diversity or index modulation (IM) (Hassanien et al. 2016).
- **Communication-radar (comrad)** typically employs conventional communication signals for probing the environment. This approach is advantageous for communication since it avoids additional interference from the sensing subsystem.
- **Under functional-coexistence ISAC** (concurrent sensing and communications on a shared waveform), a composite waveform with superimposed data and probe signals is formed, each using a dedicated precoder; representing the third design alternative. These precoders are optimized to balance the two functions and associated objectives, though sensing signals can introduce additional interference to the communication system in addition to the inter-user interference (Zhang et al. 2021).

Based on their design priorities, ISAC beamforming/waveform designs are classified as radar-centric, communication-centric, or joint designs.

- **Radar-centric designs** prioritize radar performance (e.g. beampattern shaping, Cramér–Rao lower bound minimization, or detection probability maximization) while ensuring communication QoS constraints (e.g. rate or signal-to-interference-plus-noise ratio (SINR)).
- **Communication-centric designs** maximize QoS while maintaining the sensing performance requirements.
- **Joint designs** seek to balance both functionalities by optimizing an objective function that combines communication and sensing metrics in a weighted sum formulation.

ISAC waveforms are typically shaped by the beamformer but can also be designed directly as the overall transmit vector of the antenna array. In this case, the transmit vector contains both communication data and probing signals, influencing the interference management capabilities of the communication subsystem. As a result, waveform design problems are often formulated as multiuser interference mitigation tasks subject to power constraints (Liu et al. 2018). These constraints can be *average power constraints*, which simplify the design and optimization algorithms, or more practical *constant modulus constraints* to enable energy-efficient transmission by avoiding signal distortion in low-cost and nonlinear power amplifiers (PAs). However, constant modulus constraints introduce nonconvexity, making the optimization problem significantly more challenging. More generally, beamforming design in ISAC systems presents inherent challenges from both communication and sensing perspectives. In particular, the complexity increases when massive MIMO and hybrid beamforming techniques are employed to achieve large beamforming gains while maintaining reasonable hardware costs and power consumption. These challenges often necessitate iterative numerical algorithms based on convex optimization, which can be computationally expensive. To address this, data-driven approaches leverage deep learning models to generate dual-functional waveforms/beamformers with reduced computational complexity. For instance, convolutional neural networks (CNNs) can estimate radar target directions and generate hybrid precoders efficiently (Elbir et al. 2021). However, deep learning methods conventionally developed for image and natural language processing have black-box characteristics, a lack of interpretability, and the need for extensive training data. An alternative is explainable and interpretable model-based ML, which integrates learning techniques with iterative numerical optimization (Nguyen et al. 2024a,b). This hybrid approach enhances traditional optimizers by ensuring high performance within a fixed number of layers, leading to predictable and manageable computational complexity and runtime. Multi-modal sensing can reduce overheads in large-scale ISAC networks, where massive antenna arrays make channel state information (CSI) acquisition and beamforming challenging. Data from distributed cameras, LiDAR, and radar can enhance communication processes by providing insights into spatial covariance, line-of-sight (LoS) conditions, and blockage status (Ali et al. 2020; Alkhateeb et al. 2023). ML enables mapping of environmental data to optimize beamforming and waveform design. However, challenges remain in latency, scalability, and data privacy. The need for real-time sensor data exchange increases fronthaul demands, while privacy concerns arise from sensitive user information. Further research is needed to balance these trade-offs.

7.3.1.3 Emerging Waveforms for ISAC: OTFS and Beyond

While OFDM remains the dominant waveform for communication, its application to ISAC faces challenges such as range-Doppler ambiguity, high peak-to-average power ratio (PAPR), and inter-symbol/inter-carrier interference. To address these challenges, several waveform designs are being explored.

- **OTFS** modulates information in the delay-Doppler domain rather than the time-frequency domain used in OFDM. OTFS may offer superior resilience against Doppler shifts and multipath fading, making it potentially well-suited for high-mobility ISAC applications (Wei et al. 2021).
- **IM** embeds communication data within sensing waveforms. IM-based ISAC designs have demonstrated the ability to lower PAPR and enhance data throughput without sacrificing sensing accuracy (Cheng et al. 2018).
- **Single-carrier frequency division multiple access (SC-FDMA)** offers another alternative by spreading the signal spectrum, reducing PAPR, and improving energy efficiency. This waveform is particularly attractive for ISAC applications with stringent power constraints, such as IoT and edge sensing.
- **Frequency-modulated continuous wave (FMCW)** waveforms, such as phase-coded FMCW or OFDM-FMCW have been explored as well for ISAC systems. This approach aims to efficiently utilize spectral resources and enhance system performance by allowing simultaneous data transmission and environmental sensing (Boddapati et al. 2023). While it is a widely adopted waveform for high-resolution radar sensing due to its superior range and Doppler estimation capabilities, it is particularly useful in automotive radar applications. Its simple hardware implementation and robustness in detecting objects at varying distances further contribute to its popularity. However, its limited spectral efficiency restricts data transmission capacity, making it less suitable for high-throughput applications.
- **Modulation on conjugate-reciprocal zeros (MOCZ)** is a novel waveform design that has been proposed for ISAC applications. MOCZ leverages optimal auto-correlation properties, thereby supporting reasonable data rates while maintaining effective sensing performance (Dehkordi et al. 2023).

Selecting the optimal waveform for ISAC depends on multiple factors, including mobility, interference resilience, and computational complexity. Ongoing research aims to refine waveform selection criteria to maximize ISAC performance across diverse application scenarios.

7.3.2 Signal Processing for Sensing

Signal processing plays a fundamental role in ISAC by enabling the extraction of both communication data and sensing information from shared wireless signals. Unlike conventional systems that focus solely on data transmission, ISAC requires additional processing to derive environmental awareness while maintaining high communication efficiency. This section outlines the key signal processing techniques used in ISAC, before zooming in on specific examples.

7.3.2.1 Generic ISAC Signal Processing

This section provides the following: an overview of the RX chain, highlighting the common processing steps and the additional refinements required for sensing. Next, the section delves into RAD measurement generation, distinguishing between radar-based and CSI-based sensing methods. The following part focuses on processing RAD measurements, detailing how range, angle, and Doppler data are converted into target detections, tracked over time, and used for resource optimization. Finally, the section explores multi-modal fusion, demonstrating how integrating RF sensing with complementary data sources – such as LiDAR and cameras – enhances robustness and accuracy.

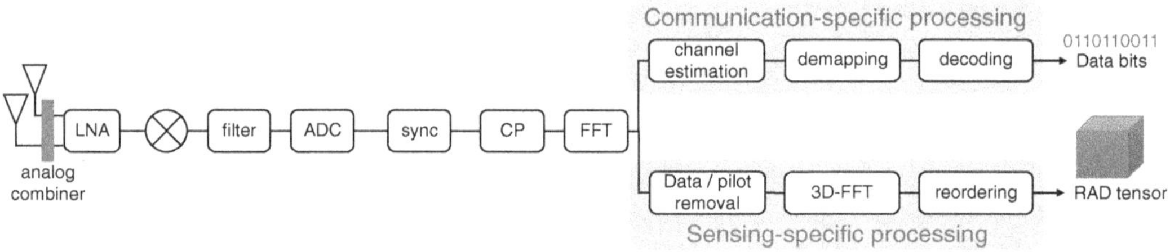

Figure 7.9 Example signal processing chain for communication and sensing.

Measurement generation: The signal processing pipeline for ISAC largely mirrors that of conventional communication systems (see Figure 7.9), with additional steps for sensing. It begins with analogue combining across antennas, followed by low-noise amplification, filtering, and down-conversion to baseband. The signal is then digitized via an analogue-to-digital converter (ADC) and processed with digital filtering. Synchronization aligns the RX with the transmitted signal, after which the CP is removed, and an FFT transforms the signal into the frequency domain. Channel estimation extracts key characteristics for data detection and decoding. For sensing, the process remains similar up to the FFT stage, but data symbols are removed to isolate sensing-related components. Coherent processing across frequency, time, and antennas is then applied to extract range, Doppler, and angle information, forming a RAD tensor or data cube. Methods can be enhanced with spectrum estimation techniques, such as Multiple Signal Classification (MUSIC), for improved resolution.

Measurement processing: RAD data must be converted into actionable detections and position estimates, involving three key stages: location conversion, tracking and data association, and feedback for resource optimization. In monostatic sensing, RAD measurements directly map to positions and radial velocities, forming a 4D point cloud. Bistatic and multistatic setups introduce additional complexity due to separate TX-RX reference frames, requiring synchronization and geometric calibration. Probabilistic models and statistical filtering techniques, such as Kalman filters or particle filters, refine noisy measurements and provide uncertainty estimates. Tracking and data association involve predicting target motion and linking new detections with existing tracks. Motion models like constant velocity (CV) or constant acceleration (CA) improve prediction accuracy, while data association techniques – such as nearest-neighbour (NN), joint probabilistic data association (JPDA), and multi-hypothesis tracking (MHT) – help resolve ambiguity in dense environments. Semantic processing further classifies targets based on motion patterns or radar cross-section signatures. Sensing feedback enhances ISAC efficiency by dynamically allocating resources. Slow-moving or well-localized targets may require fewer sensing updates, whereas unpredictable targets might necessitate increased sensing resources. Reconfigurable antennas or RIS can refine beamforming based on tracking insights, while ML techniques predict future movements, optimizing beam alignment and reducing latency. Distributed ISAC nodes can also coordinate to prioritize the most advantageous sensing perspectives, minimizing redundancy.

Measurement fusion: Integrating multiple sensing modalities – such as RF signals, LiDAR, and cameras – enhances ISAC robustness by compensating for individual sensor limitations. Radar provides velocity and range accuracy, cameras offer rich visual data, and LiDAR excels in 3D mapping. Fusing these heterogeneous data streams improves object detection, localization, and situational awareness. AI-driven fusion techniques, including deep learning and cross-modal attention mechanisms, dynamically adjust sensor weighting based on environmental conditions and task priorities. This fusion can be performed centrally at an edge/cloud server or in a distributed fashion at sensor nodes.

7.3.2.2 Examples Deployment Scenarios

In this section, the emphasis is put on sensing signal processing approaches that are suited to more specific network configurations and deployment scenarios, presented as illustrative study cases (see Figure 7.10).

1) **Sensing with hybrid analogue/digital beamforming:** Hybrid analogue/digital beamforming (HBF) enables ISAC in higher-frequency bands by balancing high data rates with radar-like sensing. Large antenna arrays employing phase shifters, switches, metasurfaces, and vector modulators optimize resource use while reducing RF chain requirements (Nguyen et al. 2024a,b). However, HBF introduces performance trade-offs, such as beam squint in wideband systems, which can improve search efficiency but degrade estimation accuracy (Luo et al. 2024). Constraints from analogue components, like constant-modulus phase shifters, further complicate optimization, making traditional communication-focused beamforming methods inadequate for ISAC (Nguyen et al. 2023).
2) **Massive MIMO and cell-free MIMO sensing:** Massive MIMO enhances ISAC by leveraging large antenna arrays for simultaneous communication and sensing. Mono-static ISAC suffers from added interference but mitigates these effects as antenna counts increase, leading to interference-free communication (Nguyen et al. 2024a,b). In cell-free MIMO, multiple access points (APs) collaborate, improving both functions as AP density grows (Nguyen et al. 2025). Larger antenna arrays also enhance sensing accuracy, with key metrics such as main lobe-to-average-side lobe ratio (MASR) scaling linearly and the Cramér-Rao lower bound decreasing inversely (Liao et al. 2024).
3) **Full-duplex sensing:** Full-duplex ISAC nodes enable simultaneous DL communication and uplink sensing of radar targets (Gavras and Alexandropoulos 2024a,b; Alexandropoulos and Gavras 2025). Since communication signals are known at the RX, super-resolution methods can estimate target positions without dedicated pilots. This reduces overhead and enables target tracking via reflection channel evolution (Gavras and Alexandropoulos 2024a). However, SI from transmitted signals distorts reflected signals, necessitating advanced cancellation techniques (Alexandropoulos and Gavras 2025).
4) **Sensing with extremely large antenna arrays (ELAAs) and RIS:** Extremely large antenna arrays (ELAAs) introduce near-field properties due to their large aperture, allowing beam focusing for energy-efficient communication and high-resolution sensing (Gavriilidis and Alexandropoulos 2025; Delbari et al. 2024). The transition from planar to spherical wavefronts improves directionality and target resolution (Alexandropoulos and Gavras 2025; Gavras and Alexandropoulos 2024a,b). Near-field capabilities also enhance wireless power transfer and imaging applications (Cardoso et al. 2025). Hybrid RISs further optimize ISAC by reflecting DL signals for both communication and bistatic sensing (Gavras and Alexandropoulos 2024a).
5) **RIS-enabled positioning:** RIS improves non-terrestrial network (NTN) positioning by providing additional RAD measurements. A single low earth orbit (LEO= satellite typically struggles with poor geometric dilution of precision (GDoP), but an RIS can enable accurate snapshot positioning by introducing angle-based measurements. Moreover, through the added RAD measurements, the clock bias and carrier frequency offset (CFO) between the satellite and the UE can be resolved. However, the power degradation in RIS paths remains a key bottleneck, which can be leveraged for simplified channel estimation (Saleh et al. 2024).

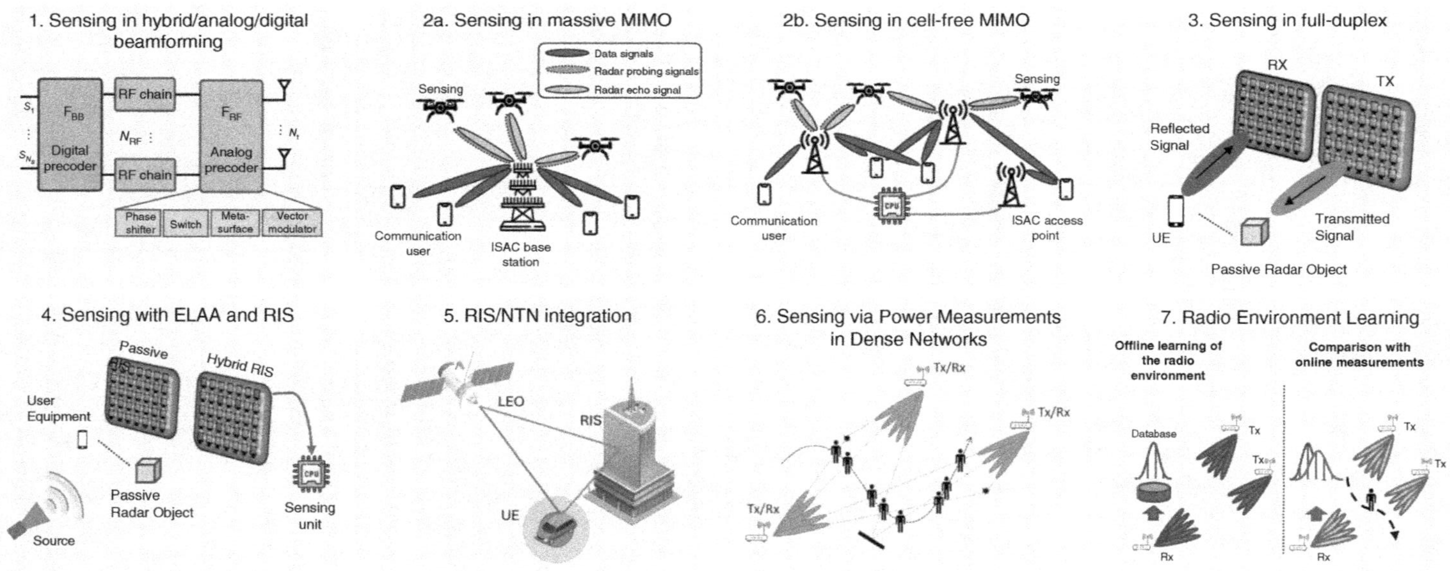

Figure 7.10 Examples of network deployment scenarios requiring specific sensing signal processing.

6) **Sensing via power measurements:** Device-free localization tracks passive objects through transient shadowing effects on communication links (Schmidhammer et al. 2021). In dense mmWave networks, interference power across sensing nodes reveals motion patterns, enabling blocker detection via spatiotemporal power analysis (Sana et al. 2023). Network densification in 6G is expected to enhance detection and tracking continuity.
7) **Radio environment learning:** Environmental sensing can leverage pre-learned site-specific radio maps, detecting anomalies in new signal observations (Devoti et al. 2020; Vaca-Rubio et al. 2023). Beamforming alignment and uplink activity tracking help establish baseline models, while statistical or image-based outlier detection identifies new targets, improving adaptive situational awareness.

7.4 ISAC Hardware Considerations

The hardware layer of ISAC plays a crucial role in ensuring reliable communication and sensing by addressing the challenges introduced by real-world hardware imperfections. Unlike theoretical models that assume ideal components, practical systems face various hardware impairments (HWIs) such as phase noise (PN), mutual coupling, PA non-linearity, and in-phase/quadrature imbalance. This section explores the impact of HWIs on ISAC, discussing their implications on communication and sensing accuracy, and reviewing state-of-the-art calibration strategies. Additionally, it examines SI suppression for full-duplex ISAC, where preserving sensing signals while eliminating interference is a key challenge. The final part of the section focuses on synchronization requirements across monostatic, bistatic, and distributed ISAC systems, highlighting how synchronization errors affect system performance and how they can be mitigated through advanced techniques.

7.4.1 Hardware Impairments and Calibration

In practical wireless communication and sensing systems, ideal hardware assumptions rarely hold due to the presence of HWIs. These impairments arise from imperfections in analogue and digital hardware components, manufacturing tolerances, and environmental factors.

7.4.1.1 Hardware Impairments

In classical wireless systems, key HWIs include PN caused by imperfections in local oscillators (LOs), CFO due to mismatch between TX and RX frequencies, and mutual coupling resulting from electromagnetic interactions between closely spaced antennas (Chen et al. 2023). Additionally, PA non-linearity introduces distortions, array gain errors cause inconsistencies in antenna gains, and antenna displacement errors alter expected signal propagation paths. In-phase and quadrature (IQ) imbalance further degrades system performance due to imperfect signal mixing in IQ modulators. While these impairments exist in both communication and sensing systems, their effects vary depending on frequency, bandwidth, and application requirements. At lower frequencies, HWIs often have a negligible impact on system performance, as classical signal processing techniques can compensate for them. However, in emerging mmWave and sub-THz systems, these impairments become more pronounced due to narrow beamwidths, shorter wavelengths, and higher sensitivity to hardware imperfections. This makes effective modelling and mitigation of HWIs crucial for reliable operation.

With the advent of 6G, new technologies such as RIS and ELAA (as illustrated in Figure 7.11) introduce additional HWIs beyond those seen in conventional systems.

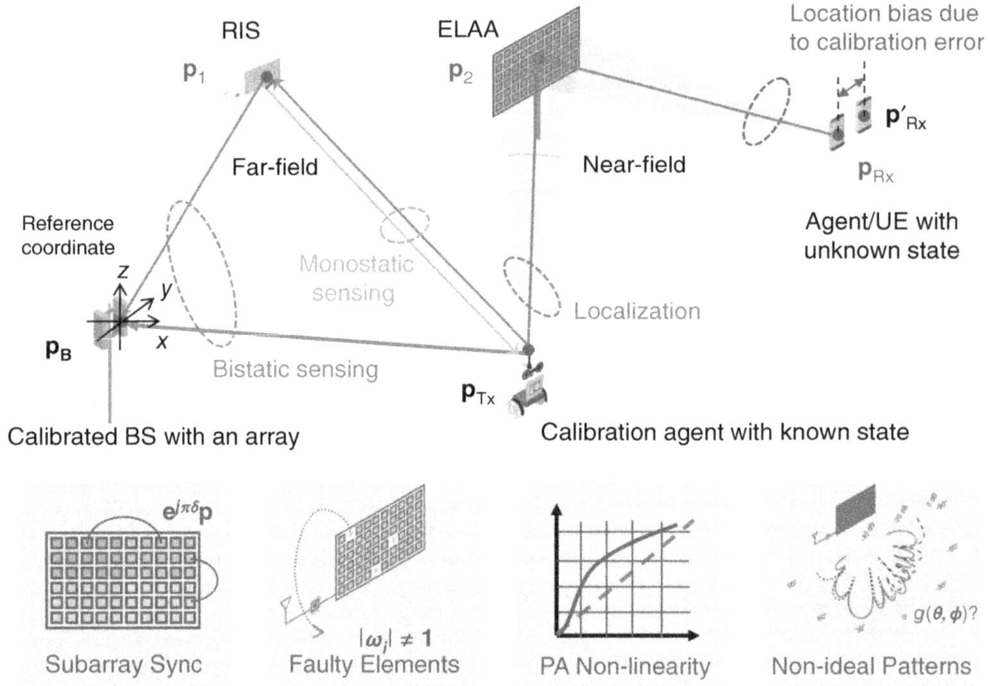

Figure 7.11 Illustration of RIS- and ELAA-aided localization and sensing scenarios and several types of hardware impairments.

- **RIS-specific impairments:** RIS units, composed of numerous tuneable elements, are subject to pixel failures, where individual elements become non-functional due to hardware defects or environmental factors like dust or moisture. Mutual coupling is particularly problematic in ultra-dense RIS implementations, distorting the expected phase shifts. Furthermore, phase-dependent amplitude variations occur in RIS elements, violating idealized unit-modulus constraints. Additional RIS impairments include power non-linearity in active RIS, PN in hybrid RIS, and reflection coefficient inaccuracies caused by non-ideal impedance tuning (Ghazalian et al. 2024).
- **ELAA-specific impairments:** ELAA face significant spatial correlation issues due to closely packed antennas, amplifying the impact of mutual coupling. Distributed PN becomes an issue when multiple sub-arrays operate asynchronously. Moreover, array misalignment errors in ELAA introduce signal distortions that affect both communication and sensing accuracy.

In both RIS- and ELAA-based systems, traditional compensation methods designed for small-scale MIMO systems are inadequate, necessitating more advanced calibration techniques, as will be discussed in this section.

7.4.1.2 Impact of Hardware Impairments

Impact of HWIs on communication: HWIs degrade communication performance in multiple ways. PN and CFO introduce frequency shifts and inter-carrier interference, significantly affecting OFDM systems. Mutual coupling and IQ imbalance distort the signal constellation, increasing the bit error rate and reducing spectral efficiency. PA non-linearity leads to spectral regrowth and in-band distortions, impacting both data transmission and adjacent channel interference. For 6G, where high-frequency, large-scale MIMO and intelligent surfaces are integral, HWIs pose a significant challenge to maintaining high reliability and low latency. Impairments like RIS phase errors or ELAA misalignment can lead to beam misdirection, reducing beamforming gain and degrading throughput. Although modern equalization techniques and adaptive algorithms can mitigate some HWIs, their effectiveness diminishes at extreme frequencies.

Impact of HWIs on sensing: Unlike communication, where error correction techniques can compensate for many HWIs, sensing is inherently more sensitive to hardware imperfections. Localization and radar sensing rely heavily on precise phase, time, and angle measurements, making them vulnerable to even minor hardware imperfections. For instance, PN and CFO distort time-of-flight and Doppler estimations, degrading range and velocity accuracy, respectively. Mutual coupling introduces errors in angle-of-arrival (AoA) and angle-of-departure (AoD) estimation, affecting direction-finding accuracy. PA non-linearity introduces unpredictable distortions in radar backscatter, reducing the reliability of target detection. RIS and ELAA impairments amplify sensing errors. RIS phase mismatches introduce location bias, while pixel failures degrade angular resolution. Studies have shown that for 6G localization and tracking applications, hardware imperfections can lead to estimation errors exceeding 10x compared to ideal conditions (Chen et al. 2023). Thus, calibration is a necessity, not an option, for high-accuracy sensing applications.

HWIs severity: While HWIs impact both communication and sensing, the latter is more susceptible to their effects. In communication, impairments such as PN, CFO, or IQ imbalance often lead to increased error rates but can be mitigated with equalization, coding, and diversity techniques. However, in sensing, these same impairments directly degrade fundamental measurements such as range, angle, and velocity, leading to incorrect interpretations of the environment. For example, a small PN variance might be tolerable for a communication system using error correction, but in sensing, it can bias angle and range estimation, leading to incorrect positioning. Similarly, mutual coupling can be ignored in communication if the beamforming gain is sufficiently high, but in sensing, it distorts spatial signatures, reducing target detection accuracy. RIS miscalibration might degrade throughput in communication, but can lead to incorrect object tracking and localization in sensing applications. Thus, *more sophisticated calibration techniques are required for sensing* applications to ensure robust performance.

7.4.1.3 Mitigation of Hardware Impairments

Given the increasing complexity of ISAC systems, various calibration techniques have been proposed to mitigate HWIs. These can be broken down along different axes.

First of all, offline calibration involves pre-deployment hardware calibration using known reference signals, often in a controlled environment, such as an anechoic chamber. It is effective for static impairments like mutual coupling and RIS phase errors but cannot address time-varying issues like PN. Also, this type of calibration is time-consuming and costly. On the other hand, online calibration methods compensate for HWIs dynamically during operation, using feedback mechanisms. For instance, adaptive beam training can realign RIS phases in real time. Another breakdown is in terms of the applied methodology. The model-based calibration uses analytical models of HWIs to estimate and correct impairments. For example, mutual coupling matrices can be precomputed and applied as compensation filters. On the other hand, data-driven calibration applies ML techniques, such as deep learning, to estimate and mitigate hardware-induced distortions. This is particularly useful in RIS and ELAA, where complex impairments make model-based approaches infeasible.

In some cases, joint calibration and sensing are feasible. Emerging 6G systems also explore the possibility of jointly calibrating HWIs and performing sensing (Ozturk et al. 2024), having performed joint pixel failure, and localization to identify the malfunction pixels. Additionally, the antenna displacement is tackled using a model-driven deep learning approach (Mateos-Ramos et al. 2025). The mutual coupling has also been considered and calibrated together with the localization tasks by Wen et al. (2020). These methods utilize the same ISAC framework used for environment sensing and self-calibration, leading to reduced overhead and improved accuracy. However, most of the existing works consider only a few types of impairments, and more comprehensive and practical solutions are needed.

7.4.2 Self-Interference Suppression

The SI channel between the RX and TX of an in-band full-duplex (IBFD) node consists of both the direct TX-RX path and any relevant multipath components. As a result, it can be represented as a multipath channel characterized by an exponential power delay profile. This profile includes a quasi-static component influenced by the TX/RX antenna structures, as well as a time-varying component caused by reflections from the surrounding environment. Considering an IBFD ISAC-oriented massive MIMO node, which features an analogue and digital (A/D) hybrid beamforming architecture at both the multi-antenna TX and multi-antenna RX ends (Alexandropoulos and Gavras 2025), a full duplex (FD) node operating in the presence of multiple radar targets is considered hereinafter, where the DL signal from the node is reflected by all passive and active (i.e. connected users) radar targets in the environment, as shown in Figure 7.12. Consequently, the SI channel can be mathematically expressed as the summation of the *end-to-end reflection channel* matrix between the FD node and the radar targets, as well as the *direct SI channel* matrix between the TX and RX antennas of the FD transceiver (Smida et al. 2024). This characterization of the overall SI channel reveals that all sensing information is embedded within the end-to-end reflection signal. However, in IBFD communications, it is crucial to fully eliminate the direct SI

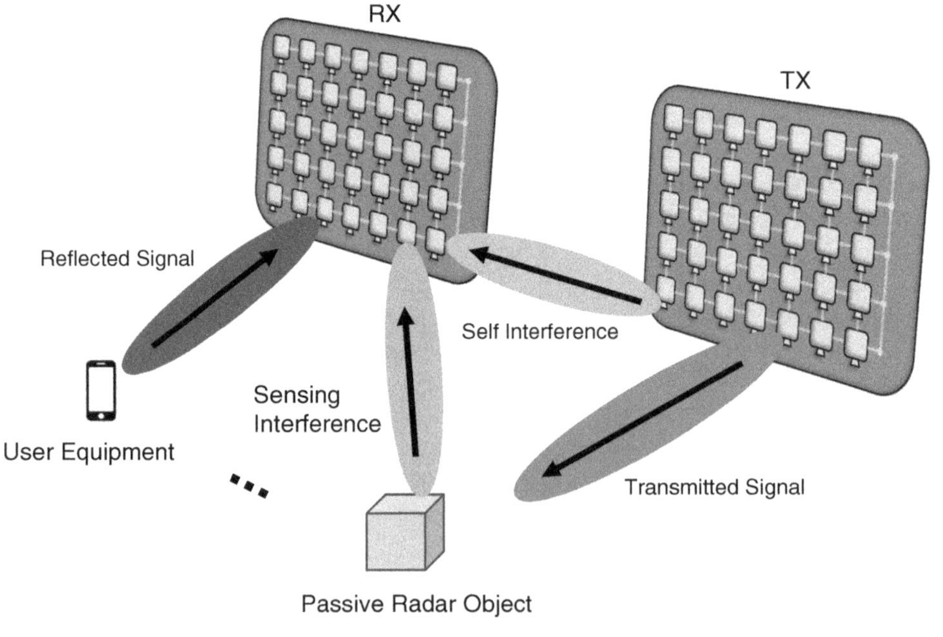

Figure 7.12 Illustration of a generalized FD-enabled ISAC system design, where an FD node operates near multiple active and passive radar targets, simultaneously supporting communication and sensing functions.

signal to ensure proper operation at the RX side (e.g. uplink communications and estimation via the received signal).

7.4.2.1 SI Mitigation Mechanisms

For a successful FD-ISAC system, it is essential to suppress the direct SI channel to ensure sufficient TX/RX isolation at the FD node's RX, while preserving the received reflections from targets. In IBFD MIMO transceivers, SI removal typically occurs in three stages, categorized by their domain of operation: **propagation, analogue, and digital**. Effective implementation of SI suppression mechanisms across these domains is crucial for enabling reliable FD-ISAC communication.

- **Space and propagation domain:** Beamforming in the propagation domain reduces SI power by leveraging propagation losses, antenna element directionality, and relevant signal processing techniques. By controlling the amplitude and phase of each antenna element, beamforming enables precise signal shaping to either enhance or suppress transmission and reception at specific locations. In FD-MIMO-ISAC, the transmit beamformer is optimized with respect to an ISAC-focused performance metric while minimizing SI leakage, with the receive beamformer further suppressing the residual SI while enhancing signal reception in the direction of the passive (unknown) and active (can also be unknown) radar targets. Beamforming can be implemented in several ways: (i) in the digital domain, where each antenna is connected to a dedicated RF chain, providing maximum flexibility; (ii) in the analogue domain, where a single RF chain is shared across all antenna elements for greater power efficiency; and (iii) a hybrid combination of both, where a reduced number of RF chains balances flexibility in beamforming patterns with power efficiency.
- **Analogue domain:** In the analogue domain, the FD node estimates and subtracts the SI signal from the received signal. Since the RX side is aware of the transmitted signal, it can reconstruct and remove its own interference after accounting for the direct TX-RX channel effects. Analogue cancellers typically use a limited number of cancellation taps, each consisting of a fixed delay, a variable phase shifter, and an attenuator (Alexandropoulos et al. 2022a). These taps can be connected through multiplexers (MUXs) and demultiplexers (DEMUXs) for efficient routing between TX and RX RF chains, thus keeping only the necessary TX-RX connections. Two reduced-complexity hardware architectures for IBFD MIMO analogue SI cancellation were proposed in (Alexandropoulos et al. 2022a) one based on analogue taps and another utilizing auxiliary (AUX) TXs. Unlike conventional designs, these architectures jointly optimize tap values, AUX TX configurations, and digital TX/RX beamforming matrices to meet performance objectives. The tap-based analogue canceller was later extended to wideband scenarios in Islam et al. (2022). Other works (Iimori et al. 2019) explored IBFD MIMO HWIs and imperfect channel knowledge. For lower-complexity analogue SI cancellation, prioritizing the suppression of the line-of-sight SI component is often sufficient. After analogue cancellation, the remaining SI signal must be within acceptable RF saturation limits at the RX, otherwise, further cancellation is needed.
- **Digital domain:** Digital-domain SI suppression operates on processed signals after passing through the Variable Gain Amplifier (VGA) and ADC. Following beamforming and analogue cancellation, the residual SI consists of IQ imbalances, nonlinear PA distortions, and the RX's noise figure. To mitigate these distortions, a combination of maximum likelihood methods, deep neural networks, transfer learning, and traditional signal processing techniques can be used (Muranov et al. 2021). In FD-MIMO-ISAC systems, a key challenge is distinguishing between target echoes and direct SI in the presence of noise. ML-based classification techniques, such as independent component analysis, can help differentiate between moving target

reflections and static close-proximity reflectors. However, as the number of antennas increases, the severity of SI leakage from the TX to the RX in IBFD systems also increases, making SI cancellation more complex. To address this, a unified IBFD MIMO architecture can integrate both analogue/digital (A/D) TX/RX beamforming and A/D SI cancellation. By designing these components jointly, the system can optimize performance while balancing complexity and cost (Alexandropoulos et al. 2022a). This unified approach distributes processing demands strategically across space/propagation, analogue, and digital domains, ensuring efficient SI suppression in large-scale FD-MIMO-ISAC deployments.

7.4.2.2 Simultaneous DL Data Communication and Monostatic Sensing

Consider the FD-MIMO-ISAC node depicted in Figure 7.12, featuring hybrid A/D TX/RX beamforming and A/D SI cancellation (Alexandropoulos et al. 2022a), enabling simultaneous data transmission and monostatic sensing, in a deployment scenario where the FD node serves one or more users in the DL direction while concurrently receiving and processing reflections from radar targets, in the node's vicinity, to estimate their parameters. Meanwhile, uplink communication remains possible and may occur concurrently. This FD-MIMO-enabled system has been extensively studied in the literature within ISAC-focused scenarios (Gavras et al. 2023; Gavras and Alexandropoulos 2024a, b; Alexandropoulos and Gavras 2025). These studies explore the co-design of A/D TX and RX beamformers in conjunction with the A/D canceller to balance communication and sensing objectives. Notably, an ISAC-focused design can be framed as an optimization problem, where the system's adjustable parameters are optimized to: (i) maximize a communication-driven objective while meeting a sensing constraint, (ii) maximize a sensing-driven objective while adhering to a communication constraint, or (iii) alternatively, it may aim for a balanced trade-off, jointly optimizing both sensing and communication objectives simultaneously. Obviously, this co-design must account for the impinging SI signal while adhering to hardware constraints, which vary between the analogue and digital domains depending on the chosen cancellation architecture, as previously described.

Joint beamforming and direct SI cancellation design: The co-design of A/D SI cancellation with TX and RX beamforming in an FD MIMO system for ISAC presents unique challenges. Effective SI suppression must ensure minimal SI leakage in the RX-end of the FD node, while preserving echoes critical for sensing. The key to achieving this balance lies in the numerous beamforming degrees of freedom, in the joint optimization of the A/D beamforming TX/RX and SI cancellation matrices across the space/propagation, analogue, and digital domains. In the space/propagation domain, leveraging spatial characteristics allows TX and RX beamforming to be designed to minimize SI while maximizing directional gain towards targets. Directional nulling can be employed to place deep nulls in the direction of SI while ensuring maximum energy is transmitted towards intended sensing targets (e.g. an intended region of interest) or data-requesting users. Additionally, antenna placement plays a significant role in mitigating SI at the hardware level. By spacing antennas appropriately and leveraging polarization techniques, SI power can be significantly reduced before reaching the RX.

The SI management challenge in FD-MIMO-ISAC: Implementing adaptive analogue SI cancellers enhances system flexibility. A combination of tuneable phase shifters, variable attenuators, and AUX TX paths can be used to cancel SI at the RF level before down-conversion (Alexandropoulos et al. 2022a), while wideband analogue SI cancellation requires dynamic adaptation due to frequency-selective SI effects (Islam et al. 2022). Frequency-dependent cancellation taps can be optimized dynamically based on real-time channel conditions. Residual SI that persists after analogue cancellation can be mitigated digitally by estimating the direct SI channel matrix between the TX and RX antennas of the FD transceiver. Deep learning-based approaches can be employed to model nonlinear SI effects and improve cancellation efficiency in real time. This is particularly beneficial in wideband systems where traditional linear filters may be insufficient. Adaptive filtering and nonlinear regression techniques help mitigate PA distortions and in-phase/quadrature imbalances that arise from HWIs. In ISAC systems, a major challenge is preserving the reflections from targets while cancelling only the unwanted SI. Unlike communication systems, where total SI removal is desirable, sensing applications rely on reflected signals to extract spatial and Doppler information about targets. Traditional SI cancellation techniques risk removing reflections from targets, thereby degrading sensing performance. To overcome this, advanced signal processing methods must differentiate between direct SI and target-induced reflections. ML classifiers can be trained to distinguish between static and dynamic reflections, preserving only useful echo components. Sparsity-based reconstruction techniques, such as compressive sensing, can be applied to extract useful reflections while nulling direct SI. Additionally, frequency-domain techniques, including Doppler filtering, can help isolate moving target echoes from static SI components.

A robust FD-MIMO-ISAC system must fully leverage all available beamforming degrees of freedom to optimize ISAC-specific objectives and constraints, while striking a balance between efficient SI cancellation for communication and target-reflection preservation for sensing. Jointly optimizing TX/RX beamforming and A/D SI cancellation is critical for achieving high spectral efficiency while maintaining accurate target detection. By leveraging ML-based classifiers, hybrid beamforming techniques, and dynamic filtering, ISAC systems can operate effectively in complex environments, ensuring seamless integration of communication and radar functionalities.

7.4.3 Synchronization Requirements for ISAC

In wireless communication, time, frequency, and phase synchronization errors between the BS and communication users impact the reliability and efficiency of the system by degrading the bit error rate (BER), spectral efficiency, and communication rate. Various technical solutions have been extensively investigated to mitigate these errors, including pilot or training sequences, channel equalization, and the use of CP or guard interval (Morelli et al. 2007). Unlike conventional wireless communication systems, ISAC imposes additional synchronization requirements for sensing functionality, which cannot be fully addressed using synchronization techniques designed solely for communication. Therefore, the impact of synchronization on ISAC must be carefully evaluated based on its specific configuration, and appropriate mitigation techniques should be applied accordingly, as shown in Figure 7.13.

7.4.3.1 Impact of Synchronization on Monostatic ISAC

In a monostatic ISAC system, where signals are transmitted and received at a co-located transceiver for sensing, the impact of time, frequency, and phase synchronization is relatively minimal. This is because synchronizing the TX and RX is straightforward when they share a common reference or LO signal source. The hardware configuration that utilizes a shared LO effectively eliminates time offset (TO) and CFO between TX and RX, making synchronization-related impairments negligible for monostatic sensing performance (Koivunen et al. 2024).

As discussed earlier regarding HWIs, PN can still impact monostatic sensing performance. Nevertheless, in systems where the LO is shared between TX and RX, this effect remains insignificant. This is because the PN associated with the received echo, reflected from a sensing target, is highly correlated with that of the reference signal source used for downconverting the received signal. As a result, the residual PN is significantly suppressed compared to the original signal source. This

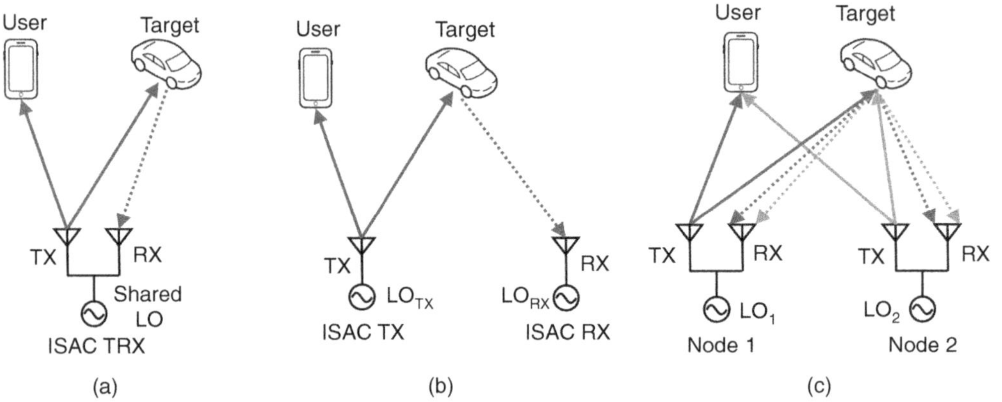

Figure 7.13 Illustration of different ISAC configurations and their respective signal sources: (a) monostatic ISAC, where a single transceiver shares the same signal source; (b) bistatic ISAC, where the transmitter and receiver operate with independent signal source; and (c) distributed ISAC, involved several transmitters and receivers, each with independent signal sources.

phenomenon, known as the *range-correlation effect* in radar, allows monostatic sensing to maintain reliable performance despite the presence of PN (Gerstmair et al. 2019).

However, the effectiveness of the range-correlation effect diminishes as the target range increases. As the distance grows, the differential PN between the received and reference signals becomes less correlated, leading to degraded range and Doppler estimation performance in monostatic sensing. Interestingly, such an effect can be exploited to improve ranging performance by capitalizing on the additional range information conveyed by the range-dependency of the differential PN statistics, effectively turning PN from foe to friend (Keskin et al. 2023).

7.4.3.2 Impact of Synchronization on Bistatic ISAC

Unlike the monostatic ISAC system, bistatic ISAC is highly vulnerable to synchronization errors, including time, frequency, and phase offsets. Since TX and RX are deployed at physically separate locations, a common LO cannot be shared, making it impossible to maintain perfect frequency and phase coherence. This inherently leads to degradation in sensing performance due to the use of independent signal sources (Wu et al. 2024).

In terms of time synchronization, TO between TX and RX in bistatic ISAC introduces bias in range estimation. If TO is smaller than the CP length or guard interval, the bias in range estimation is proportional to the TO value, necessitating precise calibration techniques to correct the induced error. If TO exceeds the CP length, inter-symbol interference (ISI) occurs, significantly degrading both range estimation and target detection performance, leading to poor bistatic ISAC sensing capabilities. CFO in bistatic ISAC is also critical to range-Doppler estimation. Similar to TO, CFO causes Doppler estimation errors, shifting the detected Doppler frequency by the magnitude of the frequency mismatch between TX and RX. With conventional ISAC signalling, if residual CFO is comparable to the subcarrier spacing, it disrupts subcarrier orthogonality, resulting in ICI. ICI not only increases range sidelobe levels but also degrades Doppler sidelobe levels, further impacting target detection and estimation accuracy (Keskin et al. 2021). Without proper synchronization correction, TO and CFO critically degrade bistatic ISAC performance, affecting range-Doppler resolution, target detection, and overall sensing reliability. Therefore, advanced calibration techniques, such as precise time synchronization methods and CFO compensation algorithms, are essential to mitigate these effects and ensure robust bistatic ISAC operation (Pegoraro et al. 2024).

In terms of phase synchronization, a constant phase offset does not directly impact target detection performance or the estimation of target parameters such as range, Doppler, and AoA. However, varying phase offsets or PN can significantly degrade these performances. Unlike monostatic ISAC, bistatic ISAC introduces uncorrelated PN, where the PN of TX and RX is completely independent. This makes bistatic sensing more sensitive to PN effects and prevents it from benefiting from the range correlation effect that monostatic sensing exploits. The impact of PN in bistatic ISAC behaves similarly to CFO effects, leading to degradation in both range and Doppler estimation and ultimately deteriorating target detection performance.

7.4.3.3 Impact of Synchronization on Distributed ISAC

The multi-node configuration of distributed ISAC introduces more complex synchronization requirements across ISAC nodes, and its impact on ISAC performance varies depending on the level of cooperation.

Distributed ISAC can be categorized into two different operational systems based on the level of phase synchronization: noncoherent distributed ISAC and coherent distributed ISAC. In *noncoherent distributed ISAC*, multiple ISAC nodes operate without phase-level synchronization, aiming to achieve noncoherent cooperation gains through distributed configurations (Meng et al. 2025). For sensing, noncoherent distributed ISAC improves target localization accuracy over co-located ISAC by leveraging the geometric diversity of ISAC nodes and targets. In communication, it utilizes coordinated beamforming to mitigate inter-node interference (INI) or employs noncoherent joint transmission cooperative multi-point (CoMP) to enhance power gain across multiple BSs. In contrast, *coherent distributed ISAC* enables greater cooperation gains in both sensing and communication by maintaining phase synchronization across ISAC nodes. For sensing, it utilizes coherent D-MIMO radar, leveraging phase information to enhance detection and parameter estimation accuracy (Sakhnini et al. 2022). For communication, coherent joint transmission CoMP provides signal combining gain, improving transmission efficiency and reliability. Although coherent distributed ISAC outperforms noncoherent distributed ISAC in both sensing and communication performance, practical limitations constrain its achievable gains. Factors such as PN, imperfections in CIS, and drifts in TO and CFO introduce synchronization errors that degrade coherent system performance.

Even in noncoherent distributed ISAC, TO and CFO significantly impact overall system performance, particularly target localization and velocity estimation (Yang et al. 2024). The distributed ISAC configuration benefits from spatial diversity by integrating multiple monostatic and bistatic sensing links. Since it combines both monostatic and bistatic sensing for target localization, the effects of TO and CFO differ from those in standalone monostatic or bistatic ISAC systems. Synchronization requirements in distributed ISAC vary based on the processing approach, which can be categorized into *decentralized* and *centralized* processing. In *decentralized processing*, each ISAC node locally estimates the target location using its monostatic link and multiple bistatic links from other TX nodes. In this case, TOs between nodes directly influence target localization accuracy, as bistatic sensing is highly sensitive to TO, as previously discussed. The localization error increases proportionally with TO misalignment, leading to degraded sensing performance. In *centralized processing*, the system aggregates data from multiple ISAC nodes to perform joint target estimation (Yang et al. 2023). Since this approach relies more on monostatic measurements, which are unaffected by TO and CFO, it is inherently more robust to synchronization errors compared to decentralized processing. However, centralized processing requires higher backhaul capacity to collect and process raw received data from all ISAC nodes, imposing additional system constraints (Zou et al. 2024).

7.5 Conclusion

ISAC is expected to be a key enabler of 6G networks, transforming wireless infrastructure into a dual-purpose system that seamlessly integrates communication and environmental awareness. This chapter has explored the fundamental components of ISAC, covering waveform and beamforming strategies, signal processing techniques, hardware challenges, and the role of AI and semantic communication in optimizing performance. This chapter examines how ISAC can leverage existing wireless technologies like MIMO-OFDM while also exploring new paradigms such as multi-modal sensing and distributed architectures. Despite the substantial progress in ISAC research, significant technical and practical challenges remain, requiring further investigation and refinement before large-scale deployment in 6G networks.

Many open challenges must be addressed to fully realize the potential of ISAC. One of the most pressing issues is resilience and security, as ISAC introduces new vulnerabilities to jamming and spoofing attacks. Malicious actors could exploit sensed data to mislead network decisions, requiring adaptive security strategies tailored to ISAC's unique risks. Another critical challenge is privacy and data protection, particularly in urban environments where continuous sensing could unintentionally expose personal information. Mobile network operators (MNOs) should develop stringent privacy safeguards while balancing the operational benefits of ISAC-enabled infrastructures. HWIs also pose challenges, as ISAC places additional demands on RF components, calibration, and synchronization. These issues become more pronounced in mmWave and sub-THz bands, where small deviations in phase or power can significantly degrade sensing accuracy. The energy efficiency of ISAC must also be considered, as sensing operations could introduce high computational and power demands. The trade-off between ISAC's benefits and its energy consumption, particularly for applications like NLOS detection, must be carefully evaluated.

Despite these challenges, ISAC remains a cornerstone of 6G, enabling intelligent, context-aware networks that optimize beamforming, enhance interference mitigation, and support emerging applications such as holographic communication and smart infrastructure. To ensure successful deployment, a unified framework for ISAC is needed, addressing interoperability, security, privacy, and energy constraints. Standardization bodies like ITU and 3GPP are already investigating ISAC's integration into 6G, but further industry-wide collaboration is necessary. The future of ISAC will be driven by advances in AI, signal processing, and hardware design. ML will play a key role in optimizing sensing techniques, improving target tracking, and enhancing security. Research efforts should focus on refining sensing algorithms, reducing hardware limitations, and developing efficient, scalable solutions that enable ISAC to operate reliably in real-world 6G deployments. As ISAC continues to evolve, it has the potential to fundamentally reshape wireless networks, providing a new level of intelligence and adaptability in next-generation communications.

Acknowledgement

The authors would like to acknowledge that certain paragraphs and figures were generated with support from ChatGPT 4.0.

Acronyms and Abbreviations

ADC	Analog-to-digital converter
ADE	Antenna displacement errors
AF	Application function
AGE	Array gain errors
AI	Artificial intelligence
ANSI	American National Standards Institute
AP	Access point
ATIS	Alliance for Telecommunications Industry Solutions
AUX	Auxiliary
BER	Bit error rate
BS	Base station
CA	Constant acceleration
CFO	Carrier frequency offset
CP	Cyclic prefix
CU	Central unit
CV	Constant velocity
DL	Downlink
DU	Distributed unit
ELAA	Extra large antenna array
ETSI	European Telecommunications Standards Institute
FMCW	Frequency modulated continuous wave
HBF	Hybrid beamforming
IBFD	In-band full-duplex
IEEE	Institute of Electrical and Electronics Engineers
IM	Index modulation
IMT	International Mobile Telecommunications
INI	Inter-node interference
IQI	Phase and quadrature imbalance
ISAC	Integrated sensing and communication
ITU	International Telecommunication Union
JPDA	Joint probabilistic data association
LO	Local oscillator
MAC	Medium access control
MASR	Main lobe-to-average-side lobe ratio
MC	Mutual coupling
MHT	Multi-hypothesis tracking
MIMO	Distributed MIMO
ML	Machine learning
MOCZ	Modulation on conjugate-reciprocal zeros
MUSIC	Multiple Signal Classification

NEF	Network exposure function
NIST	National Institute of Standards and Technology
NLOS	Non-line-of-sight
NN	Nearest neighbour
NTN	Non-terrestrial network
OFDM	Orthogonal frequency division multiplexing
OFDMA	Orthogonal frequency division multiple access
OTFS	Orthogonal time frequency space
PA	Power amplifier
PAN	PA nonlinearity
PAPR	Peak-to-average power ratio
PHY	Physical layer
PN	Phase noise
RAD	Range-angle-doppler
RAN	Radio access network
RL	Reinforcement learning
RX	Receiver
SAR	Synthetic aperture radar
SCF	Sensing control function
SI	Self-interference
SINR	Signal-to-interference-plus-noise ratio
SLAM	Simultaneous localization and mapping
SPF	Sensing processing function
STAP	Space-time adaptive processing
SUSF	Sensing unit selection function
TO	The hardware configuration that utilizes a shared LO effectively eliminates time offset
TX	Transmitter
UN	United Nations
VGA	Variable gain amplifier
XR	Extended reality

References

3GPP (2022). TR. 38.901, Study on channel model for frequencies from 0.5 to 100 GHz, V17.0.0.

3GPP (2023). SA1 TR 22.837 v100, Feasibility Study on Integrated Sensing and Communication (Release 19).

Alexandropoulos, G.C. and Gavras, I. (2025). Extremely large full duplex MIMO for simultaneous downlink communications and monostatic sensing at sub-THz frequencies. *arXiv preprint*, arXiv:2502.10693.

Alexandropoulos, G.C., Islam, M.A., and Smida, B. (2022a). Full-duplex massive multiple-input, multiple-output architectures: recent advances, applications, and future directions. *IEEE Vehicular Technology Magazine* 17 (4): 83–91.

Alexandropoulos, G.C., Stylianopoulos, K., Huang, C. et al. (2022b). Pervasive machine learning for smart radio environments enabled by reconfigurable intelligent surfaces. *Proceedings of the IEEE* 110 (9): 1494–1525.

Ali, A., Gonzalez-Prelcic, N., Heath, R.W., and Ghosh, A. (2020). Leveraging sensing at the infrastructure for mmWave communication. *IEEE Communications Magazine* 58 (7): 84–89.

Alkhateeb, A., Charan, G., Osman, T. et al. (2023). DeepSense 6G: a large-scale real-world multi-modal sensing and communication dataset. *IEEE Communications Magazine* 61 (9): 122–128.

Alliance for Telecommunications Industry Solutions (2024). Next G Alliance Report Channel Measurements and Modeling for Joint/Integrated Communication and Sensing, as well as 7-24 GHz Communication.

Amjad, B., Ahmed, Q.Z., Lazaridis, P.I. et al. (2023). Radio SLAM: a review on radio-based simultaneous localization and mapping. *IEEE Access* 11: 9260–9278.

Behravan, A., Yajnanarayana, V., Keskin, M.F. et al. (2022). Positioning and sensing in 6G: gaps, challenges, and opportunities. *IEEE Vehicular Technology Magazine* 18 (1): 40–48.

Boddapati, H.K., Kumar, K., Chavva, A.K.R., and Khan, M.S. (2023). Joint communication and sensing for MIMO systems with overlapped OFDM and FMCW. *2023 IEEE 98th Vehicular Technology Conference (VTC2023-Fall)* (October 2023) (pp. 1–5). IEEE.

Cardoso, F., Matos, S., Pessoa, L.M., and Alexandropoulos, G.C. (2025). Indoor channel characterization with extremely large reconfigurable intelligent surfaces at 300 GHz. *19th European Conference on Antennas and Propagation (Eucap)* (pp. 1–5). IEEE.

Chen, H., Keskin, M.F., Aghdam, S.R. et al. (2023). Modeling and analysis of OFDM-based 5G/6G localization under hardware impairments. *IEEE Transactions on Wireless Communications* 23 (7): 7319–7333.

Cheng, X., Zhang, M., Wen, M., and Yang, L. (2018). Index modulation for 5G: striving to do more with less. *IEEE Wireless Communications* 25 (2): 126–132.

Cheng, L., Zhang, H., Di, B. et al. (2025). Large language models empower multimodal integrated sensing and communication. *IEEE Communications Magazine* 63: 190–197.

Dehkordi, S.K., Jung, P., Walk, P. et al. (2023). Integrated sensing and communication with MOCZ waveform. *arXiv preprint*, arXiv:2307.01760.

Delbari, M., Alexandropoulos, G.C., Schober, R., and Jamali, V. (2024). Far-versus near-field RIS modeling and beam design. *arXiv preprint*, arXiv:2401.08237.

Devoti, F., Sciancalepore, V., Filippini, I., and Costa-Perez, X. (2020). PASID: Exploiting indoor mmWave deployments for passive intrusion detection. *IEEE INFOCOM 2020-IEEE Conference on Computer Communications* (July 2020) (pp. 1479–1488). IEEE.

Dong, F., Liu, F., Cui, Y. et al. (2024). Sensing as a service in 6G perceptive mobile networks: architecture, advances, and the road ahead. *IEEE Network* 38 (2): 87–96.

Elbir, A.M., Mishra, K.V., and Chatzinotas, S. (2021). Terahertz-band joint ultra-massive MIMO radar-communication: model-based and model-free hybrid beamforming. *IEEE Journal of Selected Topics in Signal Processing* 15 (6): 1468–1483.

ETSI (2024). ETSI Launches a New Group for Integrated Sensing and Communications, a Candidate Technology for 6G. https://www.etsi.org/newsroom/press-releases/2291-etsi-launches-a-new-group-for-integrated-sensing-and-communications-a-candidate-technology-for-6g (accessed 23 February 2025).

Farhadi, H., Nimr, A., Mahmood N.H., et al. (Eds.) (2023). Hexa-X-II Deliverable 4.2 Radio Design and Spectrum Access Requirements and Key Enablers for 6G Evolution. [Online]: Available at: https://zenodo.org/records/15773276 (accessed 1 September 2025).

Gavras, I. and Alexandropoulos, G.C. (2024a). Simultaneous near-field THz communications and sensing with full duplex metasurface transceivers. *2024 IEEE 25th International Workshop on Signal Processing Advances in Wireless Communications (SPAWC)* (September 2024) (pp. 126–130). IEEE.

Gavras, I. and Alexandropoulos, G.C. (2024b). Joint near-field target tracking and communications with full duplex holographic MIMO. *ICASSP 2024-2024 IEEE International Conference on Acoustics, Speech and Signal Processing (ICASSP)* (April 2024) (pp. 13301–13305). IEEE.

Gavras, I. and Alexandropoulos, G.C. (2024c). Simultaneous communications and sensing with hybrid reconfigurable intelligent surfaces. *19th European Conference on Antennas and Propagation (EuCAP)* (pp. 1–5). IEEE.

Gavras, I., Islam, M.A., Smida, B. and Alexandropoulos, G.C. (2023). Full duplex holographic MIMO for near-field integrated sensing and communications. *2023 31st European Signal Processing Conference (EUSIPCO)* (September 2023) (pp. 700–704). IEEE.

Gavriilidis, P. and Alexandropoulos, G.C. (2025). Near-field beam tracking with extremely large dynamic metasurface antennas. *IEEE Transactions on Wireless Communications* 24: 6257–6272.

Ge, Y., Khosravi, H., Jiang, F. et al. (2024). Experimental validation of single BS 5G mmWave positioning and mapping for intelligent transport. *IEEE Transactions on Vehicular Technology* 73: 16744–16757.

Gerstmair, M., Melzer, A., Onic, A., and Huemer, M. (2019). On the safe road toward autonomous driving: phase noise monitoring in radar sensors for functional safety compliance. *IEEE Signal Processing Magazine* 36 (5): 60–70.

Ghazalian, R., Zheng, P., Chen, H. et al. (2024). Calibration in RIS-aided Integrated Sensing, Localization and Communication Systems. *arXiv preprint*, arXiv:2409.16931.

González-Prelcic, N., Keskin, M.F., Kaltiokallio, O. et al. (2024). The integrated sensing and communication revolution for 6G: vision, techniques, and applications. *Proceedings of the IEEE* 112: 676.

Hassanien, M., Amin, M.G., Zhang, Y.D., and Ahmad, F. (2016). Dual-function radar-communication: Information embedding using sidelobe control and waveform diversity. *IEEE Transactions on Signal Processing* 64 (8): 2168–2181.

IEEE 802.11 working group, C.S (2019). IEEE draft standard for information technology-telecommunications and information exchange between systems local and metropolitan area networks-specific requirements part 11: Wireless LAN medium access control (MAC) and physical layer (PHY) specifications-amendment: Enhanced throughput for operation in license-exempt bands above 45 GHz. *IEEE P802. 11ay D, 3*, p. 0.

Iimori, H., de Abreu, G.T.F., and Alexandropoulos, G.C. (2019). MIMO beamforming schemes for hybrid SIC FD radios with imperfect hardware and CSI. *IEEE Transactions on Wireless Communications* 18 (10): 4816–4830.

Islam, M.A., Alexandropoulos, G.C., and Smida, B. (2022). Joint analog and digital transceiver design for wideband full duplex MIMO systems. *IEEE Transactions on Wireless Communications* 21 (11): 9729–9743.

ITU (2023). IMT towards 2030 and beyond (IMT-2030).

Kerboeuf, S., Porambage, P., Jain, A. et al. (2024). Design methodology for 6G end-to-end system: Hexa-X-II perspective. *IEEE Open Journal of the Communications Society* 5: 3368–3394.

Keskin, M.F., Wymeersch, H., and Koivunen, V. (2021). MIMO-OFDM joint radar-communications: is ICI friend or foe? *IEEE Journal of Selected Topics in Signal Processing* 15 (6): 1393–1408.

Keskin, M.F., Wymeersch, H., and Koivunen, V. (2023). Monostatic sensing with OFDM under phase noise: from mitigation to exploitation. *IEEE Transactions on Signal Processing* 71: 1363–1378.

Koivunen, V., Keskin, M.F., Wymeersch, H. et al. (2024). Multicarrier ISAC: advances in waveform design, signal processing, and learning under nonidealities [Special Issue on Signal Processing for the Integrated Sensing and Communications Revolution]. *IEEE Signal Processing Magazine* 41 (5): 17–30.

Liao, B., Ngo, H.Q., Matthaiou, M., and Smith, P.J. (2024). Power allocation for massive MIMO-ISAC systems. *IEEE Transactions on Wireless Communications* 23 (10): 14232–14248.

Liu, F., Zhou, L., Masouros, C. et al. (2018). Toward dual-functional radar-communication systems: optimal waveform design. *IEEE Transactions on Signal Processing* 66 (16): 4264–4279.

Liu, F., Masouros, C., Petropulu; A.P. et al. (2020). Joint radar and communication design: applications, state-of-the-art, and the road ahead. *IEEE Transactions on Communications* 68 (6): 3834–3862.

Liu, F., Cui, Y., Masouros, C. et al. (2022). Integrated sensing and communications: toward dual-functional wireless networks for 6G and beyond. *IEEE Journal on Selected Areas in Communications* 40 (6): 1728–1767.

Luo, H., Gao, F., Yuan, W., and Zhang, S. (2024). Beam squint assisted user localization in near-field integrated sensing and communications systems. *IEEE Transactions on Wireless Communications* 23 (5): 4504–4517.

Mateos-Ramos, J.M., Häger, C., Keskin, M.F. et al. (2025). Model-based end-to-end learning for multi-target integrated sensing and communication under hardware impairments. *IEEE Transactions on Wireless Communications* 24: 2574–2589.

Meng, K., Masouros, C., Petropulu, A.P., and Hanzo, L. (2024). Cooperative ISAC networks: opportunities and challenges. *IEEE Wireless Communications* 32: 212–219.

Meng, K., Masouros, C., Petropulu, A.P., and Hanzo, L. (2025). Cooperative ISAC networks: performance analysis, scaling laws, and optimization. *IEEE Transactions on Wireless Communications* 24 (2): 877–892.

Merluzzi, M., Filippou, M.C., Gomes Baltar, L. et al. (2024). 6G goal-oriented communications: how to coexist with legacy systems? *Telecom* 5 (1): 65–97.

Miridakis, N.I., Tsiftsis, T.A., Alexandropoulos, G.C., and Debbah, M. (2017). Simultaneous spectrum sensing and data reception for cognitive spatial multiplexing distributed systems. *IEEE Transactions on Wireless Communications* 16 (5): 3313–3327.

Mnih, V., Badia, A.P., Mirza, M. et al. (2016). Asynchronous methods for deep reinforcement learning. *International Conference on Machine Learning*, pp. 1928–1937.

Morelli, M., Kuo, C.C.J., and Pun, M.O. (2007). Synchronization techniques for orthogonal frequency division multiple access (OFDMA): a tutorial review. *Proceedings of the IEEE* 95 (7): 1394–1427.

Muranov, K., Islam, M.A., Smida, B., and Devroye, N. (2021). On deep learning assisted self-interference estimation in a full-duplex relay link. *IEEE Wireless Communications Letters* 10 (12): 2762–2766.

Nguyen, N.T., Shlezinger, N., Eldar, Y.C., and Juntti, M. (2023). Multiuser MIMO wideband joint communications and sensing system with subcarrier allocation. *IEEE Transactions on Signal Processing* 71: 2997–3013.

Nguyen, N.T., Nguyen, L.V., Shlezinger, N. et al. (2024a). Joint communications and sensing hybrid beamforming design via deep unfolding. *IEEE Journal of Selected Topics in Signal Processing* 18 (5): 901–916.

Nguyen, N.T., Nguyen, V.-D., Nguyen, H.V. et al. (2024b). Massive MIMO joint communications and sensing with MRT beamforming. *IEEE Radar Conference (RadarConf24)*, Denver, CO, USA.

Nguyen, N.T., Fang, T., Ngo, H.Q. and Juntti, M. (2025). Multi-static cell-free massive MIMO ISAC: performance analysis and optimization. *Asilomar Conference on Signals, Systems, and Computers (ASILOMAR)*.

Ozturk, C., Keskin, M.F., Sciancalepore, V. et al. (2024). RIS-aided localization under pixel failures. *IEEE Transactions on Wireless Communications* 23 (8): 8314–8329.

Pegoraro, J., Lacruz, J.O., Azzino, T. et al. (2024). JUMP: Joint communication and sensing with Unsynchronized transceivers Made Practical. *IEEE Transactions on Wireless Communications* 23: 9759–9775.

Picazo, P., Groshev, M., Blanco, A. et al. (2023). waveSLAM: empowering accurate indoor mapping using off-the-shelf millimeter-wave self-sensing. *2023 IEEE 98th Vehicular Technology Conference (VTC2023-Fall)* (October 2023) (pp. 1–7). IEEE.

Ropitault, T., da Silva, C.R., Blandino, S. et al. (2024). IEEE 802.11 bf WLAN sensing procedure: enabling the widespread adoption of WiFi sensing. *IEEE Communications Standards Magazine* 8 (1): 58–64.

Sakhnini, A., De Bast, S., Guenach, M. et al. (2022). Near-field coherent radar sensing using a massive MIMO communication testbed. *IEEE Transactions on Wireless Communications* 21 (8): 6256–6270.

Saleh, S., Keskin, M.F., Priyanto, B. et al (2024). 6G RIS-aided single-LEO localization with slow and fast Doppler effects. *2024 IEEE Global Communications Conference (GLOBECOM).*

Sana, M., Dakdouk, H., and Denis, B. (2023). Sensing of side lobes interference for blockage prediction in dense mmWave networks. *2023 IEEE Annual International Symposium on Personal, Indoor and Mobile Radio Communications (PIMRC)* (September 2023) (pp. 1–6). IEEE.

Schmidhammer, M., Gentner, C., Sand, S., and Fiebig, U.C. (2021). Multipath-enhanced device-free localization in wideband wireless networks. *IEEE Antennas and Wireless Propagation Letters* 20 (4): 453–457.

Smida, B., Alexandropoulos, G.C., and Riihonen, and Islam, M.A. (2024). In-band full-duplex MIMO systems for simultaneous communications and sensing: challenges, methods, and future perspectives. *IEEE Signal Processing Magazine* 41 (5): 8–16.

Stamatelis, G., Stylianopoulos, K., and Alexandropoulos, G.C. (2025). Evolving multi-branch attention convolutional neural networks for online RIS configuration. *IEEE Transactions on Cognitive Radio and Networking* available in early access: https://ieeexplore.ieee.org/document/10930892.

Strinati, E.C., Alexandropoulos, G.C., Amani, N. et al. (2025). Toward distributed and intelligent integrated sensing and communications for 6G networks. *IEEE Wireless Communications* 32 (1): 60–67.

Stylianopoulos, K., Di Lorenzo, P., and Alexandropoulos, G.C. (2025). Over-the-air edge inference via end-to-end metasurfaces-integrated artificial neural networks, *arXiv preprint*, arXiv:2504.00233.

Tan, D.K.P., He, J., Li, Y. et al. (2021). Integrated sensing and communication in 6G: motivations, use cases, requirements, challenges and future directions. *2021 1st IEEE International Online Symposium on Joint Communications & Sensing (JC&S)* (February 2021) (pp. 1–6). IEEE.

Vaca-Rubio, C. J., Ramirez-Espinosa, P., Kansanen, K., Tan, Z. H., and De Carvalho, E. (2023). Radio sensing with large intelligent surface for 6G. *2023 IEEE International Conference on Acoustics, Speech and Signal Processing (ICASSP)* (June 2023) (pp. 1–5). IEEE.

Wang, J., Varshney, N., Gentile, C. et al. (2022). Integrated sensing and communication: enabling techniques, applications, tools and data sets, standardization, and future directions. *IEEE Internet of Things Journal* 9 (23): 23416–23440.

Wang, L., Zhang, C., Zhao, Q. et al. (2025). Generative AI for RF sensing in IoT systems. *IEEE Internet of Things Magazine* 8 (2): 112–120.

Wei, Z., Yuan, W., Li, S. et al. (2021). Orthogonal time-frequency space modulation: a promising next-generation waveform. *IEEE Wireless Communications* 28 (4): 136–144.

Wen, F., Wang, J., Shi, J., and Gui, G. (2020). Auxiliary vehicle positioning based on robust DOA estimation with unknown mutual coupling. *IEEE Internet of Things Journal* 7 (6): 5521–5532.

Wu, K., Pegoraro, J., Meneghello, F. et al. (ed.) (2024). Sensing in bistatic ISAC systems with clock asynchronism: a signal processing perspective. *IEEE Signal Processing Magazine* 41 (5): 31–43.

Xiao, B., Yu, X., Ni, W. et al. (2024). Over-the-air federated learning: status quo, open challenges, and future directions. *Fundamental Research* 5: 1710–1724.

Xiong, Y., Liu, F., Wan, K. et al. (2024). From torch to projector: fundamental tradeoff of integrated sensing and communications. *IEEE BITS the Information Theory Magazine* 4: 73–90.

Yang, S., Jakobsson, A., and Yi, W. (2023). Moving target detection using a distributed MIMO radar system with synchronization errors. *IEEE Transactions on Geoscience and Remote Sensing* 61: 1–17.

Yang, X., Wei, Z., Xu, J. et al. (2024). Coordinated transmit beamforming for networked ISAC with imperfect CSI and time synchronization. *IEEE Transactions on Wireless Communications* 23: 18019–18035.

Zhang, J.A., Liu, F., Masouros, C. et al. (2021). An overview of signal processing techniques for joint communication and radar sensing. *IEEE Journal of Selected Topics in Signal Processing* 15 (6): 1295–1315.

Zhang, B., Qin, Z., and Li, G.Y. (2023a). Semantic-aware image compressed sensing. *2023 IEEE 33rd International Workshop on Machine Learning for Signal Processing (MLSP)*, Rome, Italy, pp. 1–6, doi: 10.1109/MLSP55844.2023.10285864.

Zhang, Z., Jiang, T., and Yu, W. (2023b). Localization with reconfigurable intelligent surface: an active sensing approach. *IEEE Transactions on Wireless Communications* 23 (7): 7698–7711.

Zou, Q., Behdad, Z., Demir, Ö.T., and Cavdar, C. (2024). Distributed versus centralized sensing in cell-free massive MIMO. *IEEE Wireless Communications Letters* 13: 3345–3349.

8

Early Validation of 6G Concepts

Sokratis Barmpounakis[1], Vasileios Tsekenis[1], Vasiliki Lamprousi[1], Pietro G. Giardina[2], Ioannis Tzanettis[3], Grigorios Kakkavas[3], Anastasios Zafeiropoulos[3], Panagiotis Demestichas[1], Ricard Vilalta[4], Daniel Adanza[4], Raul Muñoz[4], Pol Alemany[4], and Rafael Pires[5]

[1] *WINGS ICT Solutions, Athens, Greece*
[2] *Nextworks, Pisa, Italy*
[3] *Institute of Communication and Computer Systems, National Technical University of Athens, Athens, Greece*
[4] *Centre Tecnològic de Telecomunicacions de Catalunya (CTTC-CERCA), Castelldefels (Barcelona), Spain*
[5] *VTT Technical Research Centre of Finland, Oulu, Finland*

This chapter explores the early validation of 6G concepts through the design and implementation of proof-of-concepts (PoCs) for applications leveraging 5G and 6G capabilities. As part of this validation process, a variety of 6G enablers, functionalities, and mechanisms have been evaluated, covering areas such as management and orchestration across the 6G continuum, network transformation, 6G devices, network control programmability and telemetry, radio protocols, and sensing.

The process of validating 6G networks unfolds in a step-by-step manner, beginning with foundational mechanisms and progressively incorporating more advanced features. This structured approach ensures that different components of the system evolve in a way that supports intelligent, adaptive, and efficient 6G infrastructures. The evaluation follows a natural progression from single-domain configurations to more complex multi-domain setups, culminating in the integration of novel radio and device capabilities.

In the first part of the chapter, the discussion begins by establishing the foundation of the validation process, outlining the key performance indicators (KPIs) and key value indicators (KVIs) that serve as the basis for assessing the system's effectiveness as far as performance and sustainability are concerned. The validation framework evaluates critical network KPIs, including throughput, end-to-end latency, scalability, and integration potential, ensuring that the developed solutions meet the stringent performance requirements of future 6G systems. In addition to traditional performance metrics, the evaluation adopts a value-driven approach by incorporating KVIs that measure the system's impact across the environmental, social, and economic sustainability axes. This holistic perspective ensures that the technological advancements not only enhance network capabilities but also align with broader societal and sustainability goals. The chapter then presents a detailed and incremental analysis of the PoC evaluations, demonstrating how these metrics translate into real-world system performance and long-term value creation.

The validation activities presented in this chapter focus on the collaborative robots (cobots) use case (Hexa-X-II 2024), with a specific emphasis on automated logistics operations within industrial environments. Collaborative robotics represents an ideal scenario to showcase the transformative

6G to Build a Sustainable Future, First Edition. Edited by Mikko A. Uusitalo, Patrik Rugeland, Mauro R. Boldi, and Ahmad Nimr.

potential of advanced functionalities, particularly within industrial logistics settings that demand robust, efficient, and real-time coordination among heterogeneous actors and systems. Such environments require precise orchestration of autonomous robots and intelligent systems that collaboratively execute complex logistical tasks. Communication among cobots, other machines, and humans is pivotal for ensuring operational safety, real-time cooperation, and enhanced environmental perception beyond individual sensory capabilities. Current 5G ultra-reliable and low-latency communication (URLLC) implementations face substantial challenges in achieving the targeted levels of latency and reliability in real-world scenarios, thus limiting their practical applicability in demanding industrial use cases such as automated logistics (Maghsoudnia et al. 2024). More specifically, two scenarios are leveraged to showcase the system-level validation activities, both under the collaborative robots UC, that is, the cobot-powered warehouse inventory management scenario and the zero-touch cobot-powered video surveillance scenario.

The first stage of the system validation focuses on building a foundational single-domain setup, designed to assess the early-stage components of the automated warehouse inventory management and video surveillance systems. This environment consists of autonomous mobile robots (AMRs) and unmanned aerial vehicles (UAVs) operating within controlled industrial (e.g. warehouse) settings. At this stage, the primary objectives are to establish advanced management and orchestration (M&O) mechanisms and optimize functionality allocation for efficient workload distribution across the system's computing resources.

To achieve this, an energy-aware allocation strategy is employed, ensuring that compute-intensive tasks such as object detection, navigation, and quality inspection are executed optimally. A metaheuristic algorithm dynamically distributes workloads across available compute nodes, considering factors such as energy consumption, processing capacity, and system trustworthiness. In parallel, trust evaluation functions (TEFs) are introduced to assess system reliability in real time, guiding resource placement decisions to maximize resilience. With these fundamental orchestration and allocation mechanisms in place, the system is capable of autonomously handling the business operations with a higher degree of efficiency and precision. This initial phase serves as the foundation for extending the system to more complex multi-domain environments.

Building on the successes of the single-domain setup, the second phase introduces a more flexible and adaptable architecture, expanding beyond a centralized control framework. The inclusion of flexible topologies plays a key role in this phase, allowing the system to integrate temporary communication and compute resources as needed. This enhancement improves system adaptability, ensuring that resources can be allocated dynamically based on demand fluctuations. At this stage, the integration of beyond communication services (BCS) transforms the system into a more intelligent and resource-efficient ecosystem. The compute-as-a-service (CaaS) and AI-as-a-service (AIaaS) architectural enablers are integrated as part of the validation activities, enabling real-time offloading of computationally intensive tasks to distributed compute nodes. These enhancements ensure that latency-sensitive applications, such as high-definition video streaming for real-time industrial operations, work smoothly even in dynamic conditions.

The second phase also incorporates more advanced orchestration and management strategies, further refining intent-driven resource allocation and closed-loop automation. Through continuous monitoring, the system dynamically adjusts resource allocation to ensure that operational demands are met while optimizing energy consumption. As the system expands across multiple domains, the importance of security and privacy also increases, leading to the integration of advanced trust evaluation mechanisms that safeguard data and resource-sharing processes. By the end of this phase, the system is capable of seamlessly managing multi-domain interactions, balancing efficiency, security, and adaptability in real time.

In the final phase of system validation, the focus shifts towards enabling multi-domain, multi-site synergetic monitoring and orchestration scenarios. As 6G systems expand beyond isolated deployments, they must seamlessly operate across geographically distributed sites while efficiently managing heterogeneous resources in real time. This phase introduces an advanced, federated approach to monitoring, orchestration, and network automation, ensuring that distributed compute, network, and device resources can be jointly optimized across multiple locations, leveraging integration fabrics and management exposure mechanisms, as well as intent-based networking capabilities.

8.1 PoC Components in the E2E 6G System Blueprint

The 6G E2E system blueprint defined by Hexa-X-II (2025) and presented in Chapter 3 is built upon an iterative methodology focusing on the 6G enablers, which are further analysed and selected for their integration into the E2E 6G system design, considering the system requirements defined from the 6G use cases. In alignment with the 6G E2E system blueprint, the PoC architecture integrates a structured set of components (represented by the dark-coloured rectangles in Figure 8.1), operating as part of all the different layers of the system: infrastructure, network functions, application enablement platform, and application, as well as pervasive functionalities. These PoC components provide a systematic approach to validating key 6G functionalities and enabling advanced capabilities across applications such as cobots, intelligent automation, and real-time decision-making.

Primary observations comprise the multiple application components (as part of the application layer at the top) consuming exposed capabilities and interfaces available via the northbound interfaces of the application enablement platform layer. In a nutshell:

At the application layer, the PoCs provide mechanisms for real-time data processing and user interaction. The trigger/UI modules enable intuitive system interaction, ensuring that users can engage with network functionalities efficiently. Swarm planning and management components implement cobots' operations and mission-specific planning and task allocation. Video streaming and sensor data exposure allow real-time access to critical information, including visual feeds, processed sensor data, and operational device conditions. These elements are essential for maintaining situational awareness and enabling automated decision-making. Additionally, the PoCs incorporate aggregated compute resources and pre-processed sensing data, improving the availability and utilization of computational and sensory inputs. This ensures that latency-sensitive applications receive the necessary processing capacity to maintain performance standards.

Another layer of the system blueprint is the application enablement platform layer, which provides a structured environment for service exposure, programmability, and interaction with network intelligence. This platform facilitates the integration of new applications, APIs, aggregation frameworks, and developer tools, ensuring that services can dynamically interact with the 6G system. By enabling service programmability, the application enablement platform supports intent-based automation and real-time service adaptation, allowing applications to request and adjust network resources according to specific operational requirements.

Within the network layer, the architecture integrates AI-driven decision-making frameworks and flexible network management mechanisms. The AIaaS framework provides managed AI/ML services, including pre-built models, on-demand training, inference, and lifecycle management (LCM) to enable automated network optimization by dynamically adjusting system parameters based on real-time conditions. Predictive analytics facilitates proactive resource allocation, reducing inefficiencies and improving overall system responsiveness. The system also employs adaptive

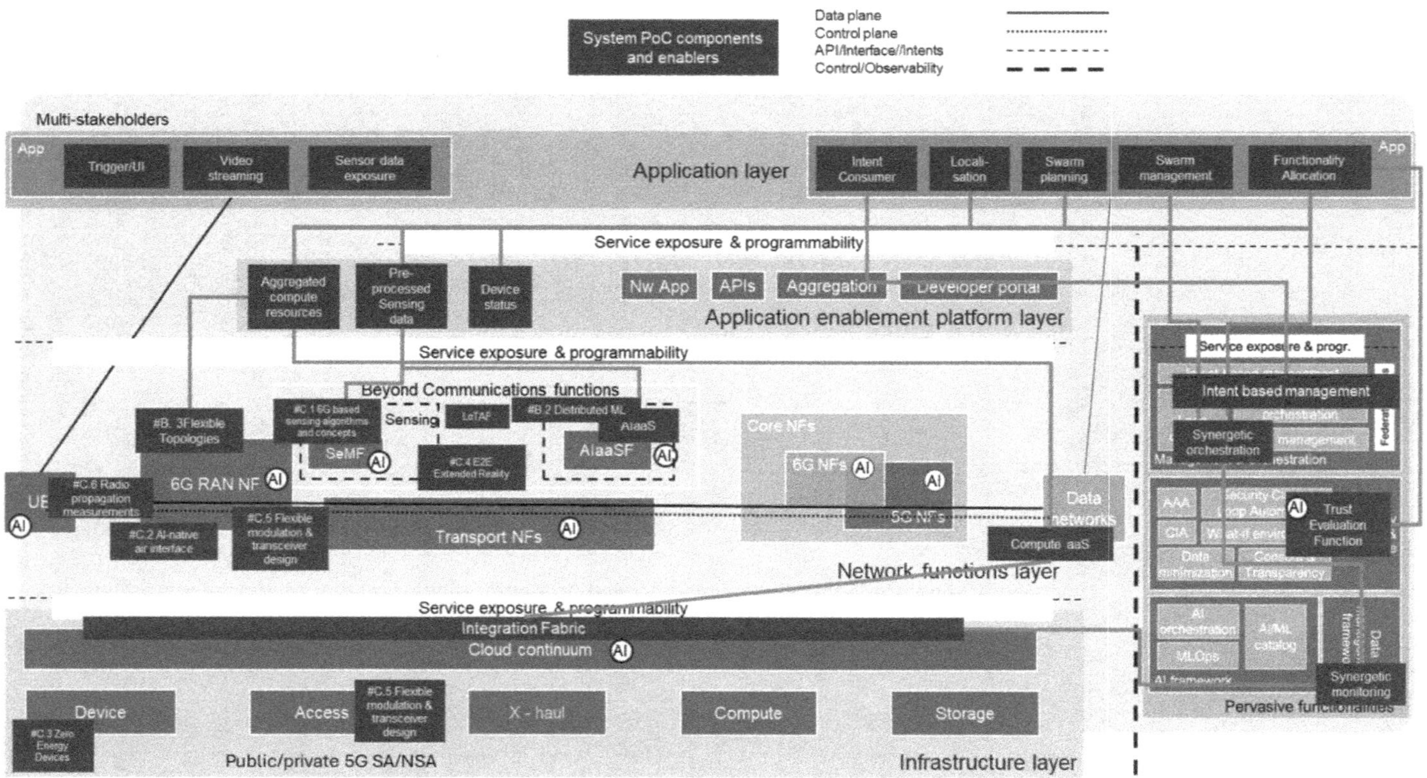

Figure 8.1 PoC components mapped as part of the E2E 6G system blueprint.

workload distribution, ensuring that computational resources are allocated efficiently according to current demands. Network functionalities such as flexible topologies, AI-native air interfaces, and optimized transceiver designs contribute to enhanced communication efficiency, supporting diverse deployment scenarios and maintaining high spectral utilization. Additional technical enhancements include AI-based sensing algorithms, flexible modulation techniques, and precise radio propagation measurements, all of which improve radio and network efficiency.

The integration fabric serves as a central component in connecting network, compute, and storage resources. It ensures seamless device interoperability, enabling efficient communication across heterogeneous hardware and software environments. Dynamic resource orchestration continuously adjusts the allocation of computing and network resources, optimizing performance according to system needs. Automated management and AI-driven closed-loop (CL) mechanisms maintain network stability by continuously monitoring and optimizing operations. A TEF is integrated within the system to assess security and ensure the resilience and reliability of deployed services.

8.2 KPIs Related to the Validation Activities

As already stated, the validation activities of this chapter concentrate on the collaborative robots use case, having a specific emphasis on automated logistics processes within industrial environments. Deploying collaborative mobile robots in these scenarios demands a network architecture capable of supporting highly localized, task-specific interactions, ensuring strict reliability and ultra-low latency. Such stringent requirements are essential for enabling real-time coordination among robots, humans, and other machinery, where even minor communication disruptions can lead to substantial financial and material losses.

The system-level PoC assesses KPIs that ensure the reliability, efficiency, and robustness of the deployed solutions. These indicators measure not only the overall system performance across network, compute, and application layers, supporting complex operations such as orchestration, real-time decision-making, teleoperation, and swarm management, but also aspects that highlight environmental, economic, and societal sustainability.

Among the critical technical KPIs, bitrate, E2E latency, service provisioning time, recovery time, and service reliability play a fundamental role. Bitrate and latency, in particular, have a direct impact on compute-heavy applications such as computer vision and intelligent automation tasks. The interaction between robotic platforms, infrastructure sensors, and backend processing systems is highly dependent on low-latency communication, network throughput, and efficient resource orchestration.

Orchestration KPIs are particularly relevant in intent-based management, ensuring fast response times for provisioning, deployment, and recovery actions. Intent deployment latency captures the time required for an intent-based service request to be fully operational, while intent conflict resolution latency measures how quickly conflicting service requests are resolved to maintain system stability. Similarly, scaling time evaluates the speed at which the system can adjust resources dynamically in response to demand.

In scenarios where network and compute resources are constrained, the system's ability to adapt through intelligent orchestration becomes critical. Service provisioning time and termination time provide insights into how efficiently the infrastructure can allocate and release resources, optimizing service continuity. Additionally, API performance metrics, such as request processing time, error rates, and concurrent tenant handling, assess the responsiveness of key orchestration functions.

8.3 Social, Environmental, and Economic Sustainability Aspects

Beyond technical performance metrics, sustainability KVIs are central to evaluating the system-level PoC. Environmental, social, and economic sustainability are key dimensions, ensuring that the system operates efficiently, securely, and with long-term viability.

The energy-efficient operation of both the networking and computing infrastructure plays a crucial role in reducing the system's overall environmental footprint. Optimized resource allocation and orchestration within the 6G continuum enhance energy efficiency, maximizing operational lifetimes in resource-constrained environments such as industrial and manufacturing settings. The deployment of collaborative robotic (cobot) systems with intelligent coordination mechanisms further contributes to sustainability by reducing unnecessary computational loads and extending device longevity.

Additionally, dynamic reallocation of functionality and up/down-scaling of compute resources ensure adaptive energy utilization, minimizing waste and optimizing efficiency across heterogeneous network and compute environments. By enforcing energy-efficient orchestration policies, overall energy consumption across the infrastructure is significantly reduced, supporting the broader sustainability goals of next-generation networks.

Trustworthiness and resilience are fundamental social sustainability indicators, ensuring that the system remains secure, reliable, and fault-tolerant under dynamic operational conditions. The ability to seamlessly hand over operations in case of failures, combined with distributed intelligence and automation, enhances system self-healing capabilities and service continuity.

Moreover, network reliability – measured through uptime, fault tolerance, and failure recovery speed – is a critical KPI in minimizing service disruptions. Automated failure recovery mechanisms further contribute to system resilience, ensuring that services remain operational even in adverse conditions. As 6G networks evolve, embedding these principles into orchestration frameworks strengthens end-to-end security, privacy, and system integrity, which are essential for trust in automated and AI-driven infrastructures.

From an economic sustainability perspective, maintaining high resilience and minimizing downtime is key to ensuring continuous and reliable service delivery. Reducing system failures and optimizing energy efficiency directly lowers operational costs (OPEX) associated with power consumption. Additionally, flexible and adaptive network topologies contribute to a cost-effective infrastructure, reducing capital expenditure (CAPEX) by enabling more efficient deployment of resources.

Together, these elements create a robust economic foundation, enhancing the long-term viability and competitiveness of the 6G ecosystem. The ability to dynamically allocate and optimize resources across heterogeneous infrastructures ensures that 6G technologies remain scalable, sustainable, and adaptable to future demands.

By integrating environmental, social, and economic sustainability into system-level PoC evaluations, 6G networks can achieve a balance between high-performance service delivery, energy-efficient operations, and long-term economic feasibility. These sustainability dimensions will not only enhance network and compute efficiency but also foster trust, resilience, and adaptability – crucial factors for the widespread adoption and success of future 6G architectures.

The system-level PoC integrates these KPIs and KVIs to demonstrate the feasibility of scalable, sustainable, and resilient next-generation networks. By analysing network performance, orchestration efficiency, and sustainability impacts, the system PoC offers a comprehensive evaluation framework for 6G-driven automation, teleoperation, and intelligent decision-making processes.

8.4 Validation Use Cases

8.4.1 The Cobot-Powered Warehouse Inventory Management Use Case

The warehouse inventory management use case, implemented under the project's system PoC activities, exemplifies how 6G-powered automation and intelligent connectivity can optimize industrial workflows. Warehouse operations are becoming increasingly complex, requiring highly efficient, real-time, and automated inventory management to reduce errors, optimize logistics, and increase operational flexibility. This scenario explores how collaborative robots (cobots), AI-driven decision-making, and real-time data analytics can enhance logistics and inventory tracking in smart warehouses. The focus of the scenario is on leveraging advanced Management & Orchestration (M&O) mechanisms, integrating BCS, as well as novel mechanisms such as trustworthy flexible topologies towards ensuring a seamless and adaptive system beyond current system capabilities in terms of performance, resilience, and trust.

The cobot-powered automated inventory management solution (Figure 8.2) leverages a cobot-powered warehouse inventory management scenario to demonstrate the system's capabilities. This setup integrates advanced computer vision and sensor data fusion for real-time navigation, object identification, counting, and localization, along with a dynamic translation system that maps symbolic warehouse locations to 3D geometric coordinates. This enables smooth navigation for UAVs/drones and AMRs while ensuring precise inventory location.

The key application components supporting the automated logistics use case encompass the following essential functionalities:

Intent management interface: An intuitive user interface component enabling warehouse managers or operational users to articulate high-level requirements, performance expectations, operational timeframes, and resource specifications through intent-based commands.

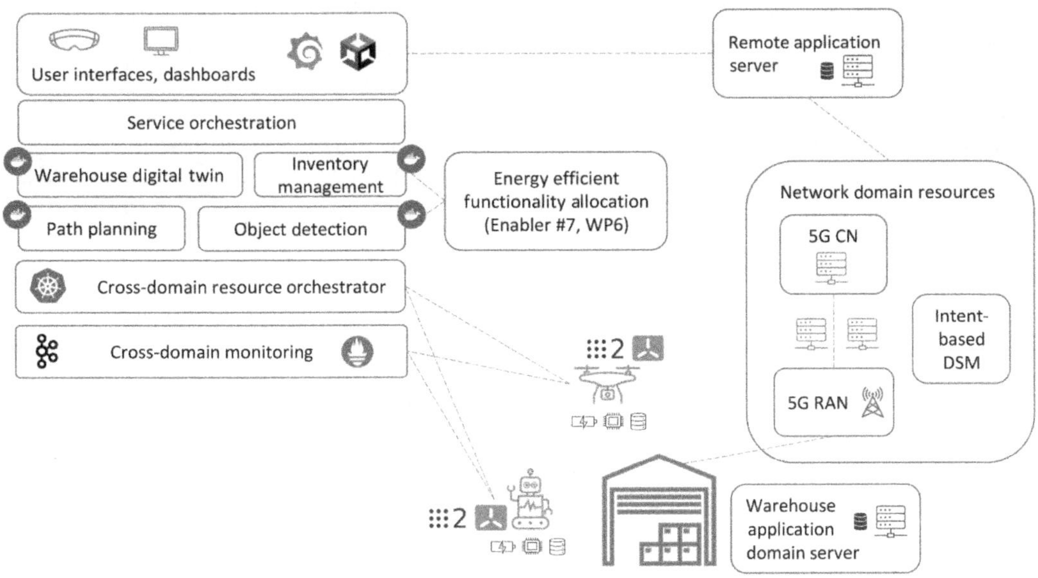

Figure 8.2 Cobot-powered inventory management orchestration components, platforms, and tools overview.

This simplifies and accelerates the translation of business needs into precise system actions and resource allocations.

Inventory realization module: Responsible for advanced identification, classification, and tracking of warehouse items. It incorporates sophisticated techniques, including computer vision and machine learning, ensuring real-time, accurate, and efficient inventory management operations.

Computer vision processing components: Encompassing advanced algorithms and software responsible for object detection, localization, and real-time analysis to guide AMRs and UAVs. These components support critical operations such as inventory audits, navigation, obstacle avoidance, and automated inspections, significantly enhancing operational precision and reducing human oversight.

High-resolution video streaming: Facilitates the continuous and real-time transmission of visual data from robotic devices (AMRs and UAVs) to centralized processing nodes or edge computing infrastructure. It ensures that visual analytics processes receive high-quality data streams, thus enabling timely and informed decision-making.

Sensor data exposure platform: Provides standardized mechanisms and interfaces for securely and efficiently sharing sensor data collected by robots and infrastructure equipment. This accessibility supports various downstream applications, enabling enhanced situational awareness and informed real-time decision-making.

Localization services: Delivers precise positioning information necessary for effective swarm coordination, robot navigation, inventory location identification, and spatial task execution. Accurate localization is critical to ensuring seamless, synchronized operations within dynamic warehouse environments.

Swarm planning and deployment: An intelligent orchestrator tasked with coordinating robotic group activities, including formation planning, dynamic resource distribution, and cooperative task execution. It ensures optimal group behaviour, resource efficiency, and rapid adaptability to changing operational scenarios.

Sensor data exposure interface: A core component of the sensor data exposure platform. This interface provides standardized access to sensor-derived data from AMRs, UAVs, and warehouse infrastructure, enabling seamless integration and reuse of sensor information across multiple application components.

Trust evaluation function: Analyses real-time and historical performance metrics from the system's compute and network nodes to establish trust indices. These indices guide workload distribution, ensuring tasks are allocated to the most reliable nodes, thereby enhancing system robustness and overall reliability.

Synergetic monitoring and orchestration: Provides comprehensive, real-time visibility of infrastructure, device, and network performance metrics across multiple domains and facilities. This multi-domain orchestration framework dynamically monitors, analyses, and optimizes resources, ensuring optimal operational performance, proactive fault management, and efficient system-wide resource utilization.

A single-domain network setup was initially assumed for the System PoC and later extended to multi-domain scenarios. An integration fabric at the capability operator level facilitates the connection of various application domain resources, AI domain resources, and service and cloud domain managers.

The resource domain includes AMRs and UAVs, serving as a testbed for advanced automated inventory management. Key components include observability features and data fusion techniques utilizing computer vision and sensor data for accurate object identification, counting, and tracking.

A central aspect of the scenario is related to the optimal allocation of functionality, focused on optimizing the placement of inventory management services across the continuum, that is, the robots, the edge, and the cloud compute resources. This involves strategic assignment of warehouse sectors to devices, defining the role of the robots, and managing computationally intensive workloads such as real-time computer vision tasks. The placement strategy prioritized trustworthiness, workload distribution, energy availability, hardware capabilities, and physical environment constraints. Services include object detection, path planning, quality inspection, and warehouse digital twin creation.

The high-level system conceptual architecture (Figure 8.2) integrates orchestration and monitoring components, AI-powered functionality allocation, and trust- and energy-efficient optimization strategies to enhance system reliability and performance. The infrastructure monitoring component continuously tracks physical and virtual resource metrics, triggering alerts when thresholds are exceeded or a device fails.

The intent-based API server receives high-level requests from users or external sources, coordinating with the functionality allocation mechanism to optimize resource distribution. This mechanism gathers real-time data on compute node capabilities, task requirements, and trust indices to generate an optimized workload placement. The API server verifies feasibility before the orchestration enforcer executes the placement, ensuring efficient task execution. The described workflow is shown in Figure 8.3.

8.4.2 Zero-Touch Cobot-Based Video Surveillance Use Case

The zero-touch cobot video surveillance scenario demonstrates the deployment of service intents and their LCM. This view represents a single domain, with potential to establish connections to other domains with the integration fabric, becoming a multi-domain scenario (Pires et al. 2024).

The use case is demonstrated by monitoring the life cycle of the intent instantiation and deployment and testing the resilience of the system stack due to changes which may affect operations. The main events that describe the use case scenario are as follows:

1) **Declaring the business intent:** The process begins with defining the business intent using a solution called the intent-based networking – intent management entity (IBN-IME) platform in a human-readable format. This platform is based on the digital service management – intent management entity (DSM-IME) functional architecture described in Chapter 6, interprets the intent, identifies the necessary services, and determines the configuration changes required for deployment in the testbed.
2) **Deployment by the CL Platform:** The CL Platform receives the processed intent from the IBN-IME and manages its implementation. It uses internal modules to deploy the video surveillance application, activate the cobot state machine (SM) to start patrols, and continuously monitor the testbed, executing corrective actions when needed to ensure service continuity.
3) **Cobot navigation and task execution:** The cobot SM, under the CL Platform's guidance, issues commands to subordinate SMs that control individual cobots. While one cobot actively performs surveillance tasks, another remains idle, recharging its battery in preparation for the next operational cycle.
4) **Seamless task transition for continuous operation:** When the active cobot's battery depletes, the system ensures uninterrupted surveillance by proactively managing the transition. The CL Platform anticipates battery depletion, seamlessly migrates required functions to the charged cobot, and instructs the transition, maintaining zero service downtime.

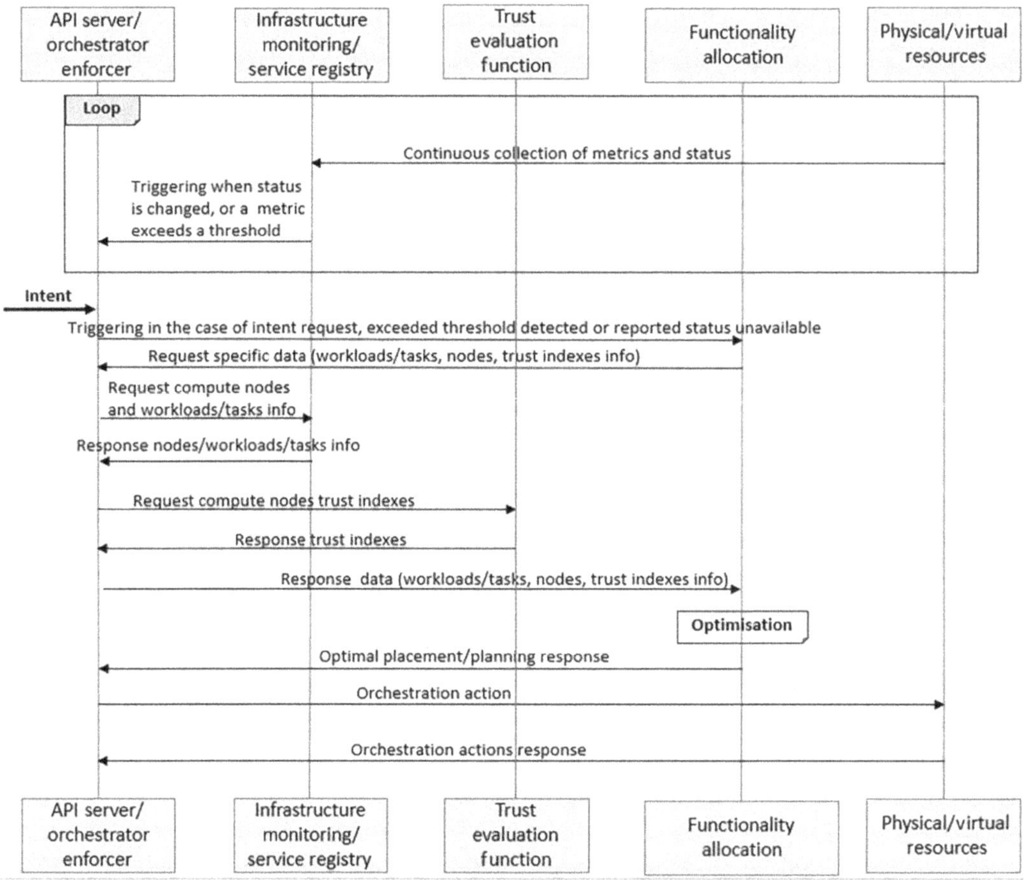

Figure 8.3 Message sequence diagram of the cobot-powered warehouse inventory management PoC.

Figure 8.4 presents the testbed architecture where a zero-touch cobot video surveillance scenario is evaluated. It integrates 5G connectivity, ROS-based SM control, and Kubernetes for application orchestration, 6G enablers: CL Platform and IBN-IME Hexa-X-II (2024).

8.4.2.1 Edge Computing and Orchestration Layer

The edge servers run multiple Kubernetes-based services, managing video processing, data storage, and MQTT messaging for event communication. Additionally, CL machine learning services are deployed in the cluster to monitor and orchestrate any corrective actions.

A separate workstation runs virtual machines for each enabler. Additionally, a set of Docker containers deploys ROS SMs that ensure coordinated operation, switching between active patrol and battery charging, as shown in Figure 8.5. The figure shows a bird's-eye view of the surveillance premises for each cobot (Cobot 1 on the top left and Cobot 2 on the top right), and below the SMs of each cobot (bottom left and right, respectively) and the robot manager service (RMS) that synchronizes both cobots (bottom centre).

The testbed is a multistakeholder environment. From the cobot warehouse perspective, the CL Platform and IBN-IME enabler modules are designated as third-party components. This means these components are not authorized to interact with other processes that do not concern the cobot operations. Therefore, several security measures must be considered to ensure the use case pipeline is completely isolated from other processes in the warehouse (use case premises).

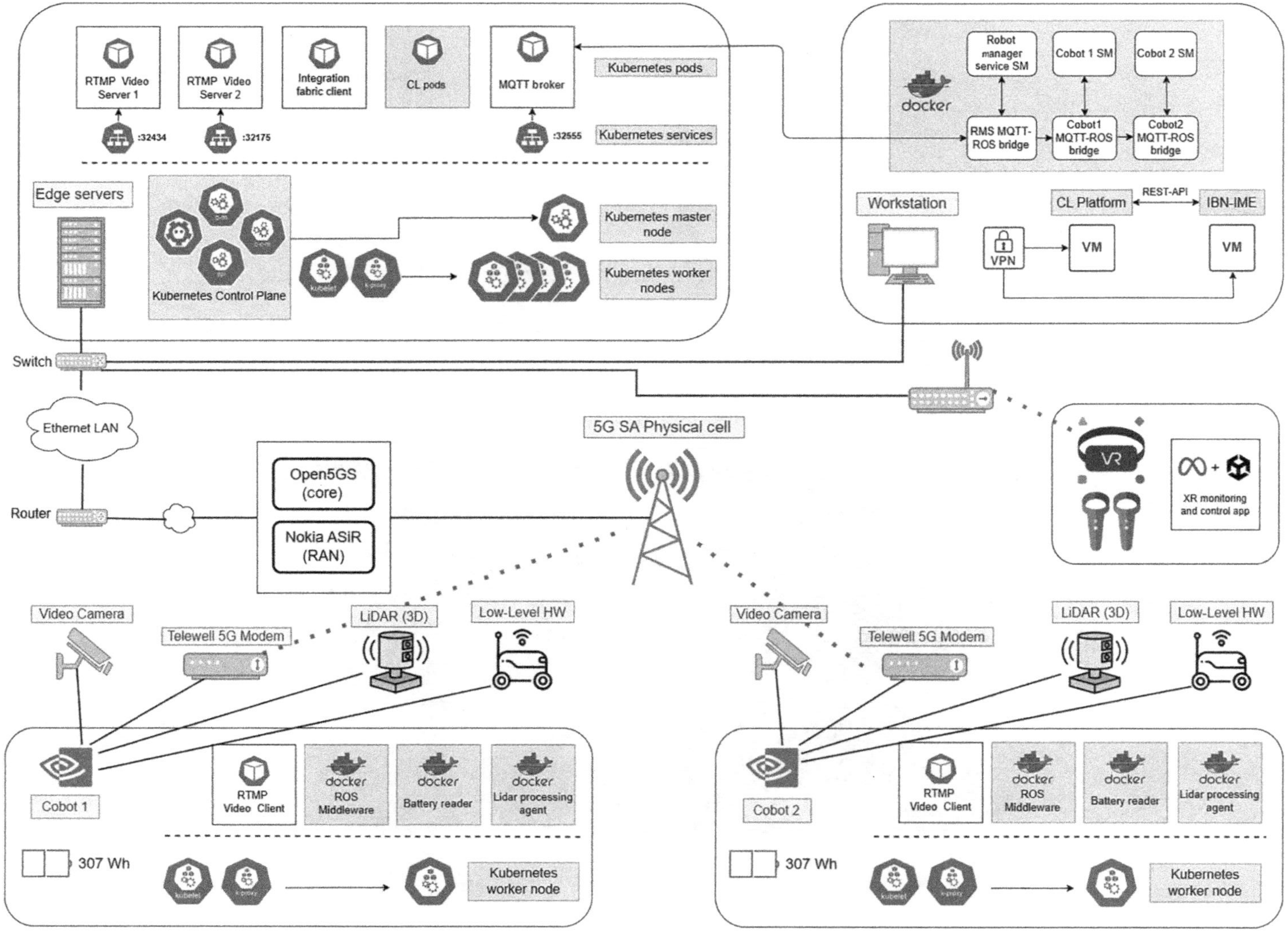

Figure 8.4 The zero-touch cobot video surveillance scenario.

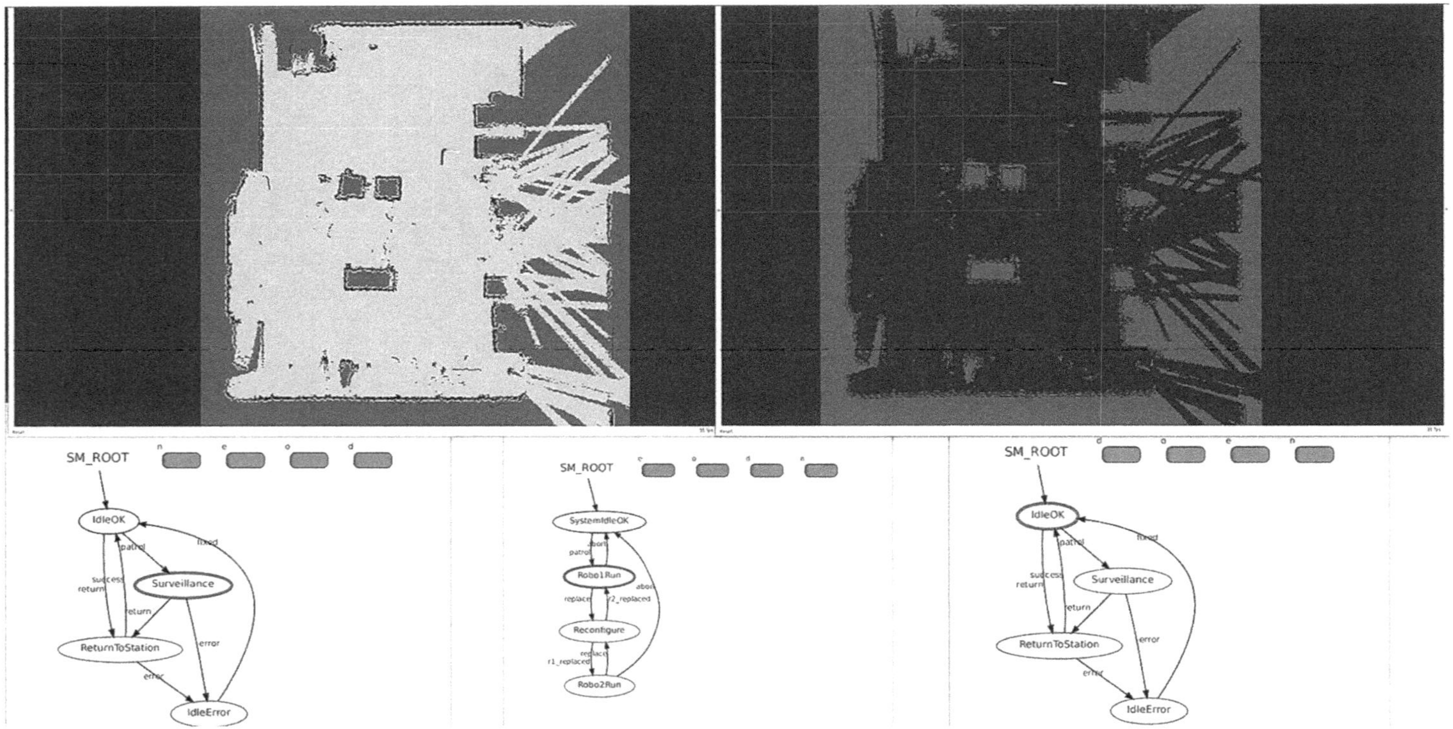

Figure 8.5 ROS cobot navigation and state machines.

Access to the testbed is restricted through a VPN, and port forwarding rules are applied to allow enabler-operators (CL-Platform and IBN-IME) to reach the virtual machines where enablers are deployed. Additionally, access control lists (ACLs) regulate and restrict the IP addresses the enablers can reach, ensuring controlled and secure access.

Furthermore, Kubernetes Role-Based Access Control (RBAC) is applied to define user privileges within the cluster. This ensures that the CL Platform operates only with the components, preventing it from interfering with other deployed services. By implementing these layered security measures, the testbed maintains system integrity, minimizes unintended interactions between services, and preserves data privacy.

8.4.2.2 Cobot Surveillance System (Robot Layer)

The surveillance system relies on two autonomous cobots, Cobot 1 and Cobot 2, which execute patrol tasks within this scenario. Each cobot is equipped with LIDAR sensors and cameras for real-time navigation and environmental monitoring, along with a battery monitoring system that enables energy management. To ensure seamless communication and low-latency control, both cobots are integrated with a Telewell network modem, supporting 5G Standalone (SA) connectivity; see Figure 8.6. The cobot hardware includes LiDAR sensors, a video camera, a 5G modem, and a chassis and motor for movement.

Figure 8.6 The cobot hardware includes LiDAR sensors, a video camera, a 5G modem, and a chassis and motor for movement.

All computational processes are handled by an Nvidia Jetson Orin Nano, which is configured as a Kubernetes worker node. This setup allows the CL Platform to dynamically schedule and deploy surveillance applications when a cobot is actively patrolling. Additionally, the cobots continuously publish low-level battery metrics to the MQTT broker, enabling the CL Platform to monitor battery status and predict each unit's operational lifespan.

8.4.2.3 VR Control and Monitoring Layer

The VR-based control interface provides operators with an immersive and interactive environment to monitor and manage the cobot surveillance system. Developed using Meta Quest 2 and Unity, this interface enables users to visualize and interact with the cobots in real-time through a digital twin representation.

In this virtual environment, operators can observe and control the robots as they navigate their patrol routes. The digital twin mirrors the cobots' physical state, displaying real-time positional data, sensor feedback, and operational status. One of the key features of the interface is the LIDAR sensor visualization, which allows operators to perceive the cobots' surroundings exactly as they do, using point cloud data rendered in 3D space, as shown in Figure 8.7.

Beyond passive monitoring, the VR interface supports manual intervention and control, allowing operators to override autonomous navigation if necessary. Through intuitive VR-based commands, an operator can redirect a cobot and modify its patrol path.

By leveraging 5G connectivity, the VR interface ensures low-latency, real-time synchronization between operator commands and sensor feedback, enabling a fluid and responsive interaction with the cobots. The system is designed to provide a harmonious balance between automation and human control, ensuring that manual interventions enhance rather than disrupt ongoing operations and provide situational awareness to stakeholders. Although the scenario is designed to provision a service with minimal intervention, a surveillance use case eventually needs human supervision. The cobot can push a notification when an anomaly is detected, and a human operator can operate the robot through the VR digital twin, where the cobot sensor data can be best visualized. This synergy between the zero-touch service management (ZSM) system and human operators is critical for achieving a robust, adaptable, and intelligent surveillance system.

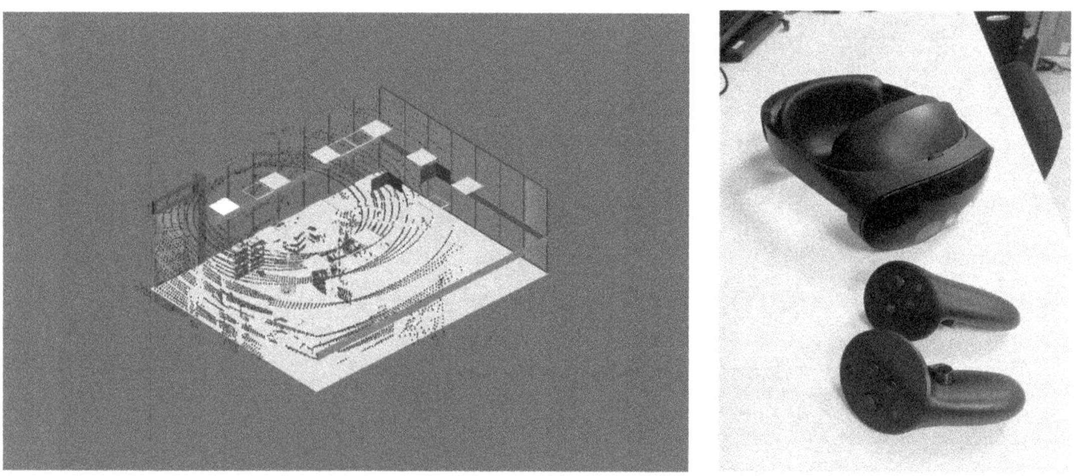

Figure 8.7 Left: A digital twin model with real-time LiDAR projection of the measurement premises. Right: A Meta Quest XR headset for immersive control of cobots.

8.5 System PoC Enablers

The following sections present a set of system-level enablers that underpin the PoC implementations introduced earlier, particularly those described in the two collaborative robotics scenarios. These enablers have been instantiated to support various aspects of the use cases – including orchestration, flexible topologies and beyond communication mechanisms, trust management, energy efficiency, zero-touch automation, and AI-driven service management – thus enabling the realization of robust, adaptive, and sustainable 6G-powered systems.

8.5.1 Advanced Management and Orchestration Mechanisms: Trust- and Energy-Driven Functionality Allocation

Efficient management and orchestration of services and workloads in future networks are essential for minimizing energy consumption, especially as wireless environments continue to experience significant growth in device numbers and complexity, each with distinct and dynamic operational requirements. The ability to dynamically and intelligently allocate workloads becomes particularly critical to address these challenges effectively. The proposed approach focuses specifically on optimizing energy efficiency by carefully distributing computational tasks across available network and compute nodes. Energy efficiency optimization is achieved by minimizing both computing and transmission energy usage, while also leveraging potential energy sources such as RF energy harvesting or renewable solutions (e.g. solar panels) where applicable. This balanced approach ensures optimal resource utilization, significantly reducing the network's overall energy footprint and thus aligning with the sustainability objectives of next-generation wireless infrastructures.

8.5.1.1 Implementation (Energy-Driven Optimization)

The computational workloads placement problem (M&O placement model) studies energy efficiency and trustworthiness when computational workloads and tasks are distributed to the various compute nodes of a system. The input to this algorithm is the computational workloads/tasks $\{1, \ldots, i, \ldots, n\}$ with their computation and physical requirements, as well as the data size ds_i and the location of data generation corresponding to each workload. Also, the compute nodes $\{1, \ldots j, \ldots, m\}$ are considered, with their capabilities (e.g. available CPUs $cpuN_{j'}$, memory), power consumption when idle and when fully loaded $\left(W^j_{\text{idle}}, W^j_{\text{max}}\right)$, the battery level, if applicable and the trust indexes $trustN_j$ assessed by the TEF of enabler 4.3. The network topology graph consisting of the nodes and the communication channels (with their throughput $cap_{j,j'}$) is also input to the algorithm. The aim is to minimize the objective function:

$$\min_{x,z} \left(a_1 E + a_2 L - a_3 T\right)$$

which is weighted (a_1, a_2, a_3) and consists of the energy consumption (processing, transmission, RF harvester-generated energy) term:

$$E = \sum_{i=1}^{n} \sum_{j=1}^{m} \sum_{j'=1}^{m} \left(\left(W^{j'}_{\text{max}} - W^{j'}_{\text{idle}} \right) U^{j'}_{\text{CPU}} \left(x_{i,j'} \right) + W^{j'}_{\text{idle}} \right) \frac{ds_i}{cpuN_{j'}} z_{i,j,j'}$$
$$+ \sum_{i=1}^{n} \sum_{j=1}^{m} \sum_{j'=1}^{m} z_{i,j,j'} p_{j,j'} \left(\frac{ds_i}{cap_{j,j'}} \right) - E_{\text{harvest}}$$

The end-to-end (E2E) latency (processing and transmission):

$$L = \sum_{i=1}^{n}\sum_{j=1}^{m}\sum_{j'=1}^{m} z_{i,j,j'} \left(\frac{ds_i}{cpuN_{j'}} + \frac{ds_i}{cap_{j,j'}} \right)$$

The trustworthiness term (this term is maximized):

$$T = \sum_{i=1}^{n}\sum_{j=1}^{m} trustN_j\, x_{i,j}$$

The $U^j_{\mathrm{CPU}}(x_{i,j})$ term denotes the CPU utilization rate of the compute node j which is a function of the binary decision variable $x_{i,j}$ which shows if the workload i is placed in the node j. The $p_{j,j'}$ denotes the power consumption in the communication between nodes j and j' and $z_{i,j,j'}$ is the binary variable showing if workload i using data produced in node j is placed in node j'. The constraints that should be respected are that each workload should be allocated to only one node, the available resources of the nodes should be respected, and the functional requirements of the tasks should also be respected if applicable. The output is the near-optimal placement of the workloads to the available nodes.

For solving this NP-hard problem, a metaheuristic algorithm was developed based on the genetic algorithm paradigm. The flowchart of this algorithm is shown in Figure 8.8. In particular, the algorithm initializes a population of possible solutions, which are called chromosomes. Each chromosome consists of a series of nodes. Each node is the proposed node for each workload to be placed. The fitness score of each chromosome is calculated based on the objective function described before. The parent selection, crossover, and mutation operations are, respectively, the tournament selection, uniform crossover, and adaptive mutation and are responsible for improving the overall fitness of the population at each generation. The number of generations varies since it is determined by a dynamic stopping criterion.

The functionality allocation algorithm (Figure 8.8) is designed to allocate compute resources efficiently across the continuum, being use case-agnostic. Effective management of services and workloads in future networks involves carefully allocating tasks and roles to available physical nodes to maximize operational efficiency and energy savings. Central to this process is the intelligent assignment of physical tasks based on detailed requirements, including specific robotic capabilities such as cameras, wheels, propellers, and mobility constraints. Additionally, spatial considerations, such as the physical distance to task locations, are integral inputs to this allocation mechanism. The physical nodes considered for executing these tasks typically comprise robotic platforms, including AGVs and UAVs, each characterized by their hardware capabilities and functional limitations, battery capacities, and energy harvesting capabilities. By intelligently matching task requirements with available robotic units' capabilities and locations, the system optimizes resource utilization, significantly enhancing overall productivity, adaptability, and energy efficiency within automated logistics environments.

The objective function to be minimized incorporates several critical cost factors, including the energy consumption associated with operating each physical node, the total travel distance required for task execution, and the duration of the longest individual robot tour. Constraints taken into account include the maximum allowable energy consumption per node, respecting each robotic unit's battery capacity; the fulfilment of task-specific functional requirements (e.g. capabilities such as cameras, wheels, or propellers); and ensuring the compatibility between robotic capabilities and task demands. Additionally, task allocation must respect each robot's available resources (CPU capacity, memory, etc.) and ensure that the robots' specific functional capabilities – such as

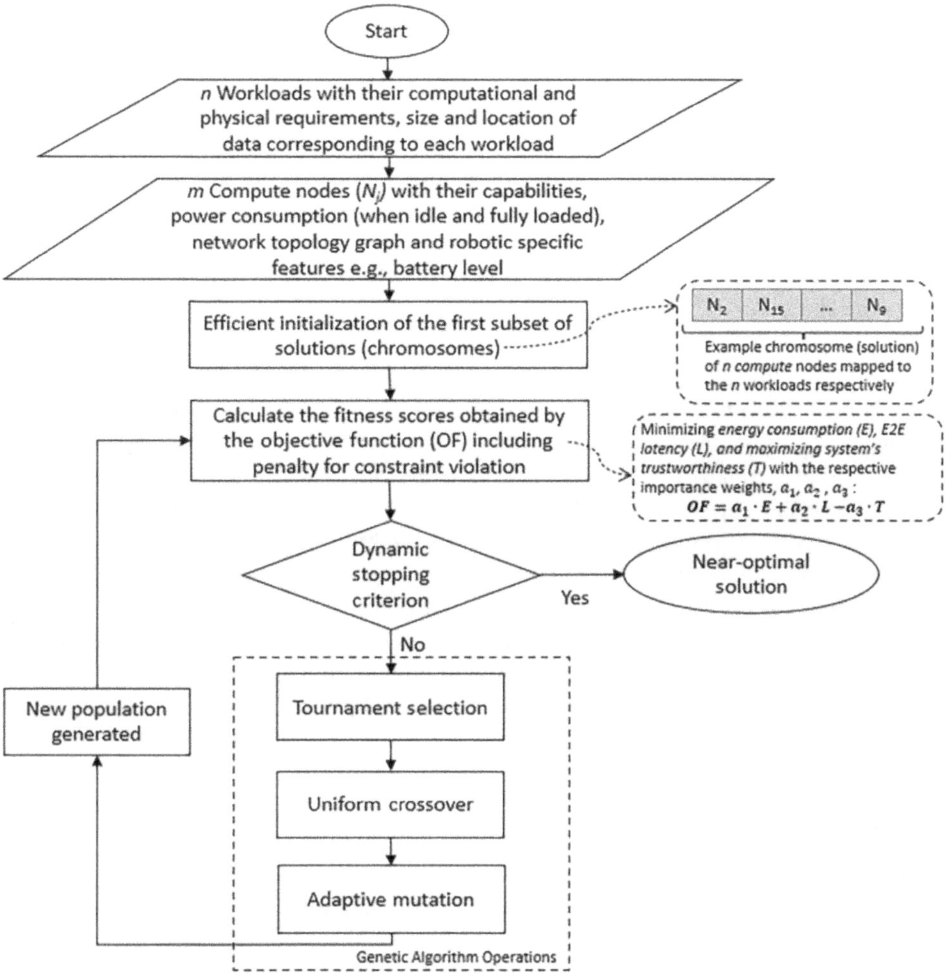

Figure 8.8 Flowchart of functionality allocation algorithm.

the presence of cameras, wheels, or propellers – are appropriately matched to task requirements. The output of this optimization is an optimal task assignment plan, detailing the precise sequence of tasks allocated to each robotic unit. At this stage, task sequences are defined in such a way that workload distribution is balanced, functional constraints are satisfied, and energy efficiency and operational effectiveness are maximized.

8.5.1.2 Implementation (Trust-Driven Optimization)

6G-enabled networks in industrial settings must accommodate diverse devices with varying characteristics that continuously exchange information, often through communication channels that can be unstable or inherently unreliable. Ensuring secure communication in such dynamic and potentially untrusted conditions requires a multi-layered security approach. At the infrastructure level, mechanisms like VPN access restrictions, ACLs, and Kubernetes RBAC can be used to enforce baseline security and isolation, as demonstrated in the video surveillance use case. These mechanisms regulate access to physical and virtual resources and restrict communication pathways, helping to maintain system integrity and data privacy. However, to further enhance security at the orchestration and management layers, particularly in dynamic and intent-driven environments, a flexible, lightweight, and adaptive trust evaluation mechanism is also needed. This is where the TEF comes into play, enabling real-time, context-aware decisions for secure communications among heterogeneous and evolving industrial IoT ecosystems.

To meet these stringent security demands, the TEF has been proposed as a critical enabler within the broader management and orchestration architecture. TEF facilitates the establishment and maintenance of secure communication paths, particularly when sensitive workloads traverse the network, move between subnets, or interconnect components across distinct trust domains. Depending on the specific application scenario or use case, TEFs are designed to dynamically evaluate the trustworthiness of relevant system components, including infrastructure nodes, subnetworks, or entire systems.

For instance, in workload placement and orchestration scenarios, the TEF at the infrastructure layer assesses the trustworthiness of available computing nodes to determine optimal workload allocation. It ingests real-time telemetry and monitoring data from system components, evaluates trust indices based on predefined criteria or use-case-specific intents, and then provides actionable insights to the orchestration layer. Specifically, the orchestration engine utilizes these insights to prioritize task placements onto compute nodes exhibiting the highest trust scores, thereby maximizing overall system reliability and security.

Moreover, the application of TEFs extends to scenarios involving flexible network topologies and multi-domain communications. In such contexts, the TEF plays a pivotal role in establishing trustworthiness across dynamic network segments, subnets, and different trust domains, safeguarding data workloads during transit and ensuring secure connectivity between diverse system elements.

Ultimately, the orchestrator integrates TEF recommendations to enforce decisions on workload placement, dynamically balancing trustworthiness, reliability, and resource availability. A schematic representation of the high-level architecture of TEFs, illustrating these relationships and processes, is depicted in Figure 8.9.

The early-stage implementation outlined in this section includes a detailed description of how the trust indexes of the compute and infrastructure nodes were calculated and leveraged as inputs for the functionality allocation mechanism presented earlier. As described in Hexa-X (2023), trustworthiness has a wide scope and is related to security and privacy, as well as availability and reliability. Based on this, the TEF envisioned for the infrastructure layer, for now, considers the following aspects:

- Availability, A_j, which is initially approached as the percentage of a time window during which the compute node was not fully loaded or unavailable.
- Reliability, R_j, which is approached as the times the compute node succeeded in executing a task/workload within a time threshold.
- Security, S_j, which is related to secure communication and trusted computing, but for this preliminary implementation, it is envisioned as a binary variable indicating if Transport Layer Security (TLS) is available or not.
- Multi-connectivity capabilities, M_j, which spans between 0 and 1 depending on which capabilities (e.g. 4G/5G, Wi-Fi, NB-IoT, BT, interfaces – device/UE-based adaptive RAT selection) are supported.
- Battery level, B_j, of the battery-powered devices/nodes (e.g. robotic units).

Hence, the trust index $trustN_j$ of the compute node j is calculated based on the following formula:

$$trustN_j = w_1A_j + w_2R_j + w_3S_j + w_4M_j + w_5B_j$$

The weights, w_1, w_2, w_3, w_4, w_5 vary depending on the use case and the intent request.

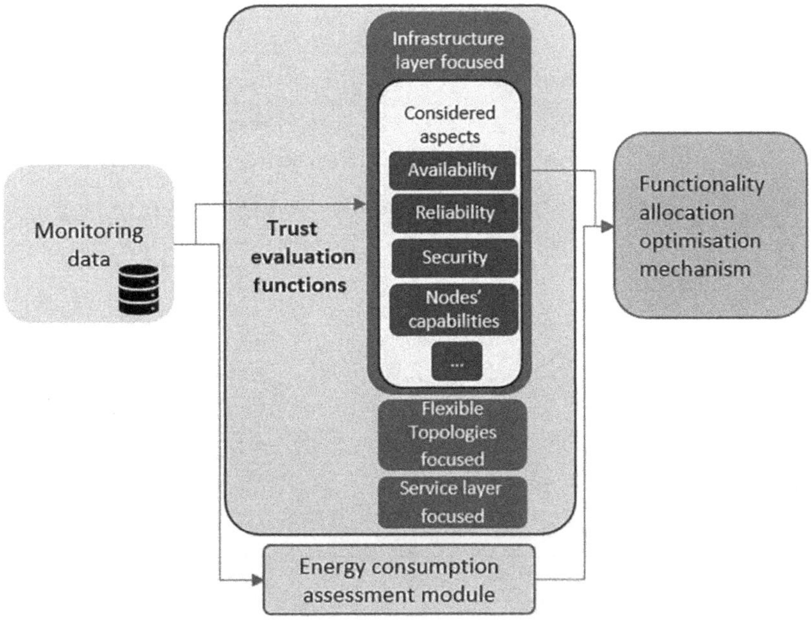

Figure 8.9 High-level architecture of the TEFs.

8.5.2 Advanced Management and Orchestration Mechanisms: Zero-Touch Automation

The management of vertical services implies the deployment of dedicated orchestration components also in charge of allocating required cloud resources offered by different providers across the cloud continuum. Furthermore, zero-touch automation mechanisms are also crucial to maintaining the QoS level agreed upon among the parties (e.g. in the SLA). This component PoC focuses on resource orchestration and zero-touch automation based on the concept of CL and is part of the scenario described in Section 8.5, *Zero-touch cobot-based video surveillance.*

The resource orchestrator collects and maintains information related to the different cloud platforms and selects the proper one to provision vertical service applications. It offers four macro-components implementing its functionalities (Figure 8.10):

- **Platform manager:** Encompasses a set of technology-specific drivers to interact with the different target cloud platforms supported, enabling the dynamic discovery and continuous monitoring of cloud resources in the continuum.
- **Resource manager:** Stores information related to the platforms and the devices, dynamically discovered through the platform manager. Such information can be retrieved by third parties (e.g. service orchestrators) through a dedicated REST interface.
- **Service deployer:** In charge of application vertical deployment on the target cloud platform.
- **Service repository:** Stores platform-specific orchestration templates, e.g. Helm charts for Kubernetes, retrieved by the service deployer at orchestration time.

CL-based automation is widely explored and applied in the context of mobile networks. Nonetheless, CLs are often and paradoxically built statically and hardly configurable at runtime. This represents a limit for the level of automation required by 6G systems.

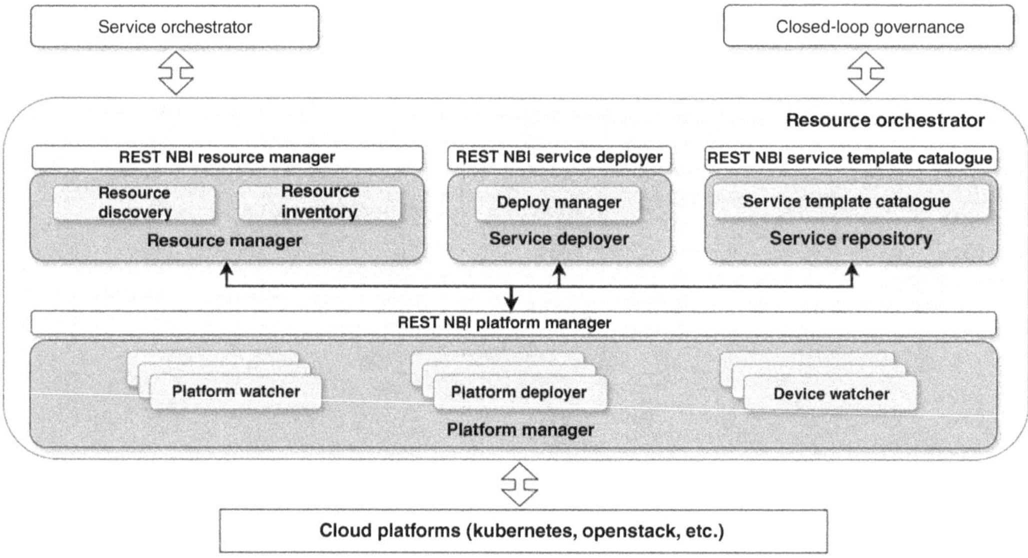

Figure 8.10 Resource orchestrator high-level architecture.

The last evolution in the design of the closed loops considers the different stages (monitor, analysis, decision, and execution) completely orchestrable and configurable at runtime and provides specific functions for their management and coordination: multiple coexisting closed loops are indeed possible, making the automation truly pervasive in the 6G systems. In this regard, ETSI ZSM (2021) introduces two specific functions known as (i) CL Governance, which provides interfaces and logic to manage the loop lifecycle, and (ii) CL Coordination, which is in charge of coordinating coexisting loops, avoiding race conditions and conflicting decisions that may lead to system instability. Forms of collaboration between loops are also considered. The component considered in this PoC is an implementation of the CL Governance, which includes several internal modules shown in Figure 8.11. The goal is to cover the governance functionalities identified by ETSI, as the standard does not provide any reference architecture.

- **Closed-loop descriptors catalogue:** Repository storing CL and CL functions descriptors. It allows CRUD operations on the descriptors through its REST NBI.
- **Resource allocation:** In charge of computing CL functions' placement by exploiting information retrieved from the REC-EXEC: given the CL requirements, the resource allocation selects the most suitable cloud platform among the available.
- **Closed-loop LCM:** Is the core module of the CL-Governance, responsible for the LCM of the CLs. It relies on three main sub-modules.
  - **Closed-loop LC manager:** Selects proper CL and CL functions' descriptors given the service requirements, provisions, and decommissions CLs
  - **CL records manager:** Tracks the existing instances of CL.
  - **Closed-loop runtime manager:** Performs runtime operation of existing CL instances, for example, configuration and start/stop.

To validate the service LCM and automation, a state-of-the-art service orchestrator (European Space Agency (ESA) 2023) has been integrated with the proposed resource orchestrator and CL Governance components.

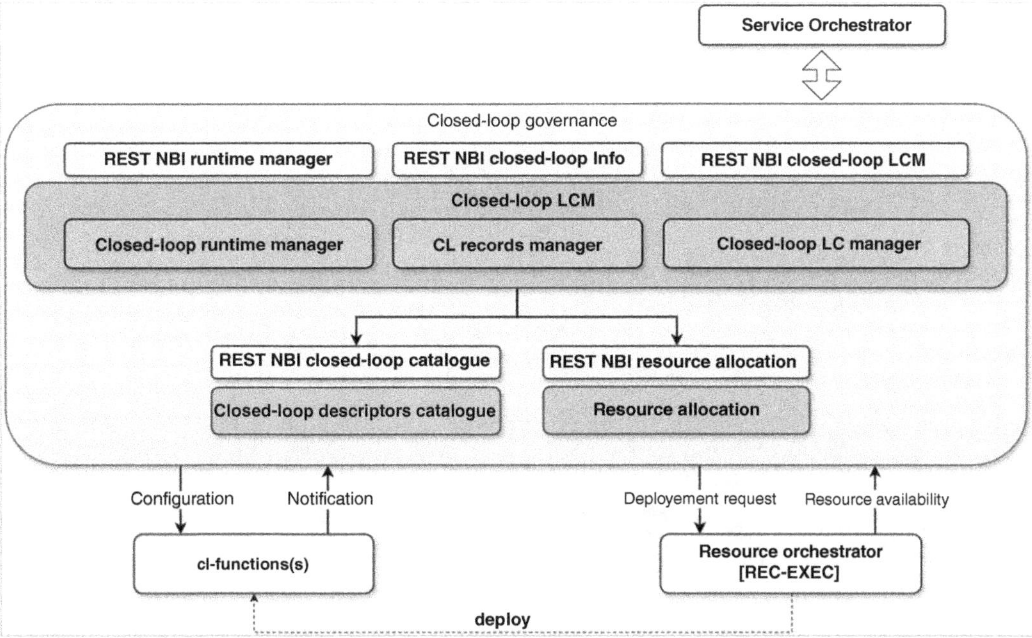

Figure 8.11 CL governance high-level architecture.

The interactions between the three entities are shown in Figure 8.12. The service orchestrator decomposes a given service provisioning request into sub-requests towards the resource orchestrator (1a) and the CL Governance (1b), respectively. In 2a, the resource orchestrator provisions the vertical application(s) into the target cloud environment. In parallel, the CL Governance identifies the proper set of functions to implement the automation loop and sends a request to the resource orchestrator (2b) for their deployment (3b).

Vertical application(s) and the CL are both part of the vertical service. Nonetheless, it is important to note that the CL deployment might not necessarily happen in the same cloud environment as the vertical application(s) but rather in the most convenient one to fulfil the CL goal (e.g. close to the monitoring data): each CL function can reside in a different cloud environment if required.

As mentioned, the service LCM and automation have been validated in the zero-touch cobot-based video surveillance use case described in Section 8.5.

Figure 8.13 shows the deployment of the different elements requested by the resource orchestrator in a Kubernetes environment belonging to the cobot management platform.

In particular, the deployment includes a video streaming application, required for video surveillance, and the functions representing the stages of the CL dedicated to the service: *cl-analysis* (Analysis), *cl-decision* (Decision), and *cl-execution* (Execution). A screenshot from the graphical user interface (GUI) exposed by the CL Governance with the details of the CL instance is shown in Figure 8.14.

A specific *CL-monitoring* function is missing in this case because the monitoring stage is realized through a programmable monitoring platform deployed outside the Kubernetes environment and not directly controlled by CL Governance. The service orchestrator programs the monitoring platform at orchestration time to collect the required data, which is consumed by *cl-analysis*.

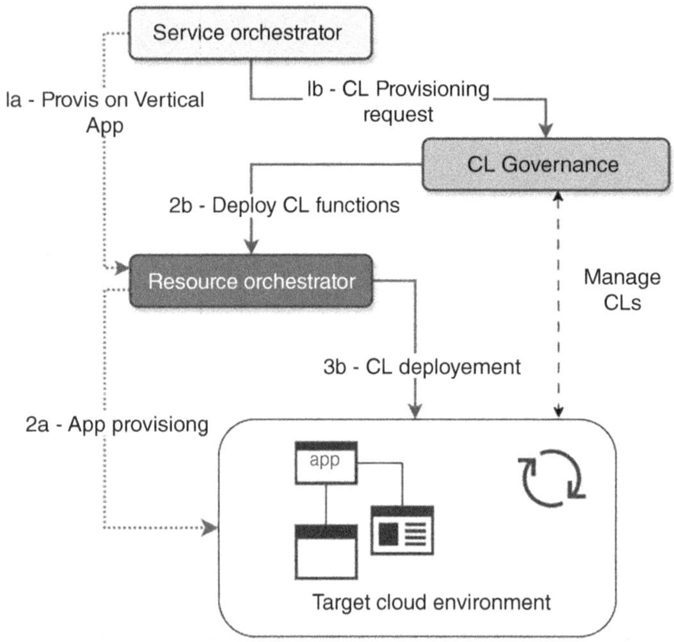

Figure 8.12 Orchestration and CL management components' interactions.

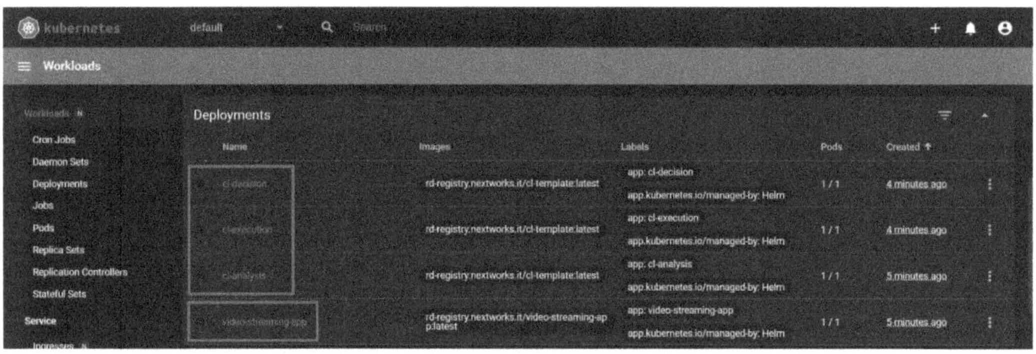

Figure 8.13 CL functions and vertical applications deployed by the resource orchestrator.

8.5.3 Trustworthy Flexible Topologies and Beyond Communication Aspects

Trustworthy, flexible topologies represent a significant evolution in 6G network design, driven by the exponential growth in data, devices, and emerging applications that demand dynamic and secure infrastructures. Traditional static networks, designed primarily for predictable, human-centric communications, are increasingly inadequate for managing the diverse and complex requirements of modern services such as extended reality, massive digital twinning, and real-time robotics coordination. In this new era, networks must process vast amounts of sensor data, localization inputs, and multimedia streams while ensuring that critical performance metrics such as latency, throughput, energy consumption, and node reliability, among others, are continuously monitored and integrated into adaptive reconfiguration processes. Therefore, in these advanced networks, dynamic reconfiguration is essential, particularly when connectivity may be

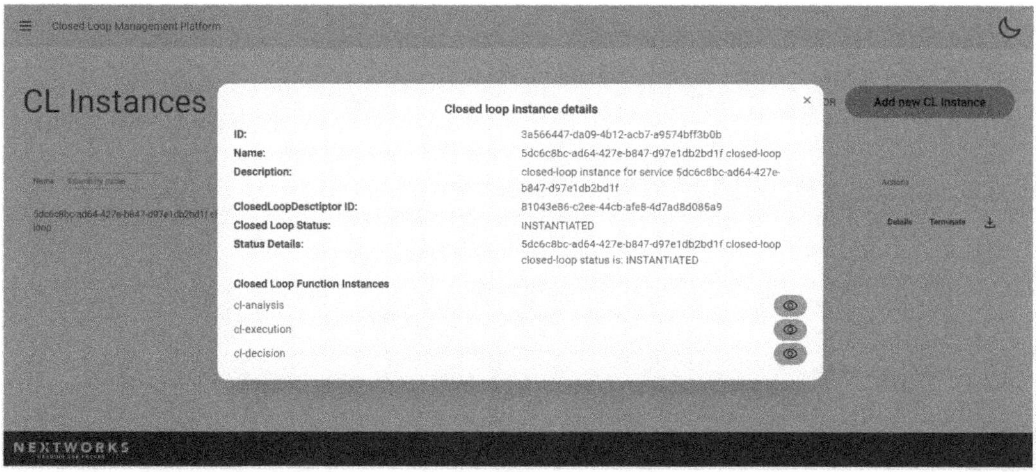

Figure 8.14 Details of the deployed CL instance displayed from the CL Governance GUI.

disrupted or when resource-constrained environments challenge conventional infrastructures. For this reason, flexible topologies employ continuous node discovery and comprehensive trust evaluation methods that assess each network entity, both terrestrial and aerial, using multiple criteria such as capacity, energy efficiency, performance, and operational cost (see Section 8.6.1 for an in-depth description of these trust metrics). Note that trust management has evolved from a security concern into a multidimensional approach ensuring only reliable nodes participate in critical functions, thereby enhancing overall network resilience. Therefore, when connectivity is lost, for instance, in the case where ground-based AMRs performing their predefined actions become isolated, the network autonomously needs to reconfigure itself. As shown in Figure 8.15, UAVs operate as ad hoc mobile access points, with optimal placement determined by advanced methods and techniques that ensure both operational efficiency and security.

Recent improvements in UAV navigation and autonomous operations have significantly enhanced network reliability and scalability. Innovations in three-dimensional navigation now allow aerial vehicles to dynamically avoid static obstacles during path planning, while refined trajectory tracking ensures precise navigation within predefined operational zones. These advancements are crucial for maintaining continuous network coverage and effectively responding to rapidly changing environmental conditions. In such environments, operational dynamics are orchestrated by a combination of centralized and distributed control mechanisms that coordinate various network entities. Ground-based units perform essential tasks on a planar level, whereas UAVs operating in three dimensions facilitate difficult tasks and extend connectivity in areas where traditional infrastructure may fall short. Distributed controllers embedded within robots and UAVs manage local tasks such as obstacle avoidance, link adaptation, and rapid fault recovery, while central controllers, deployed on edge or cloud platforms, continuously monitor device states and reallocate tasks dynamically based on real-time feedback. User interfaces present live updates on KPIs such as battery levels, video streams, and geographic positions. This integrated approach ensures that any disruptions are promptly identified and addressed, thereby sustaining overall system performance and resilience.

The evolution of flexible topologies is also closely linked to the broader concept of BCS, which extends the capabilities of networks to process and manage data for advanced applications. Modern 6G networks are increasingly required to support applications that demand not only high-speed

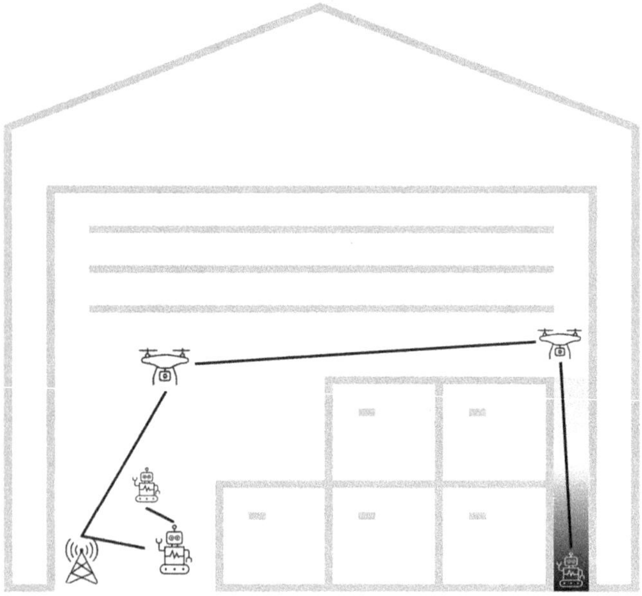

Figure 8.15 UAV-assisted flexible topology in the warehouse inventory scenario.

connectivity but also robust data processing and analytical capabilities. The integration of BCS within network architectures allows for the seamless exposure of network metrics via standardized, secure interfaces. These interfaces enable application functions to access critical insights, thereby facilitating intelligent decision-making and real-time responsiveness. In practice, this means that networks can dynamically adjust their configurations based on continuous feedback regarding network performance, environmental conditions, and user requirements. The convergence of dynamic reconfiguration and BCS leads to an environment where the network is not simply a passive means for data transmission, but an active participant in ensuring optimal performance, energy efficiency, and security across both network infrastructures and application services. A key driver of these advancements is the ability to perform intelligent resource allocation through AI and ML techniques. AI-driven algorithms analyse real-time data to determine the optimal distribution of computational and communication resources across the network. A numerical example (evaluation results) is presented in Section 8.6.2.

Operationally, the orchestration of these networks is complex and multifaceted. Central controllers, operating on edge or cloud servers, monitor real-time metrics across the network and reallocate tasks as conditions change. They rely on a continuous stream of data provided by both ground-based units and aerial vehicles, ensuring that every element of the network is functioning optimally. The information displayed on user interfaces is not limited to basic metrics; it includes detailed visualizations of network topology, real-time status updates, and performance indicators that enable operators to quickly grasp the state of the entire system. This comprehensive monitoring is essential for maintaining situational awareness in environments where both predictable and unpredictable factors may affect network performance.

Furthermore, when disruptions occur, automated recovery protocols are activated. These protocols ensure that if a node becomes isolated or if connectivity is compromised, the system rapidly reconfigures itself to restore service continuity, often by reassigning mobile access roles to aerial vehicles or by redirecting data flows through alternate pathways. To demonstrate their

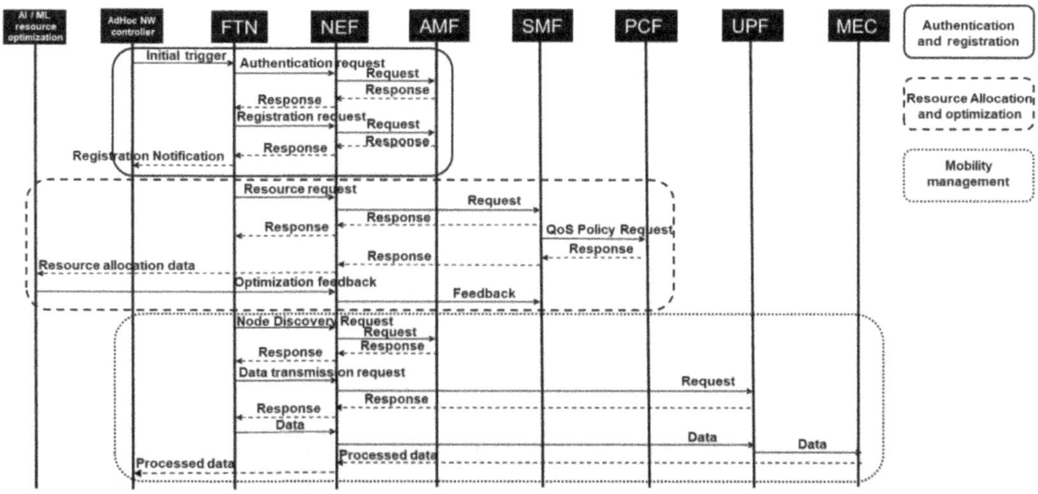

Figure 8.16 MSC for authentication, registration, resource allocation, and mobility management of a trustworthy, flexible, and unstructured network.

coordination, Figure 8.16 presents a high-level message sequence chart (MSC) that outlines the flexible topology network (FTN) orchestrating key operations, such as authentication, resource allocation, and mobility management, across various network functions. In this workflow, the FTN initiates authentication and registration procedures via the network exposure function (NEF) and access and mobility management function (AMF), each step validated by corresponding responses. Once registration is confirmed, resource requests traverse the NEF to the session management function (SMF) and policy control function (PCF), ensuring QoS adherence before the AI/ML resource optimization component adjusts allocations in real time. Meanwhile, the AMF and user plane function (UPF) handle node discovery and data flows to maintain seamless connectivity. Through these interactions, metrics such as latency, throughput, and energy consumption guide decision-making, reinforcing both reliability and performance within the network.

In summary, the evolution of trustworthy flexible topologies in 6G networks is driven by the dual constraints of processing vast amounts of diverse data and maintaining robust, adaptive connectivity in dynamic environments. The integration of continuous node discovery, multidimensional trust evaluation, and intelligent resource allocation creates a network architecture that is both resilient and efficient. Enhancements in UAV navigation and autonomous operations have further cemented the role of aerial units as critical components in extending network coverage and sustaining service continuity. By harmonizing centralized oversight with edge-level autonomy, the control architecture continuously adapts by reallocating resources and correcting faults on the fly to maintain seamless operation. This holistic approach lays a strong foundation for future innovations in network design, where the seamless integration of data processing, connectivity management, and real-time adaptive reconfiguration will be essential to meet the demands of an increasingly interconnected world.

8.5.4 AI-Assisted E2E Lifecycle Management of a 6G Latency-Sensitive Service

This PoC addresses the AI-assisted end-to-end LCM of a 6G latency-sensitive service across the compute continuum, operating on a multi-cluster and multi-domain configuration. Specifically, the considered service is a microservices-based application as part of a cobot-based video surveillance service in an industrial setup (see Section 8.4.2), where low-latency communication is crucial for coordinating robots and machinery, monitoring processes in real time, and ensuring safety through immediate responses to anomalies. The examined use case involves a stationary robotic arm that performs a designated manufacturing task. Meanwhile, a smaller robotic vehicle is remotely controlled by a human operator relying on a video feed from its onboard camera to surveil the workspace. Sampled frames from this video feed are provided as input to an ML-based object detector, which generates alerts that pause the robotic arm's operation when necessary. In the conducted experiments, this occurs whenever the presence of human(s) is detected in the room where the arm is located. The main aspects tested and validated in this PoC include the multi-cluster deployment of the application graph, the optimized application-level service component placement (e.g. video encoder and object detector), and various orchestration mechanisms such as reinforcement learning-based service autoscaling, computation offloading, and component migration.

The developed application consists of two service chains: one concerning the teleoperation of the robotic vehicle and another dedicated to generating alerts that safely pause and resume the arm's operation based on an ML-based object detection component analysing the vehicle's video stream. The teleoperation service chain includes three essential components. The *encoder* receives the native MJPEG video stream from the vehicle, encodes it to H.264, and forwards it to the *media server*, which manages and distributes the stream using WebRTC. The *remote operation centre* serves as the interface with the human operator, displaying the received video feed and accepting the remote-control input. These remote-control commands are relayed to the vehicle via the bridge component. The object detection service chain also consists of three main components. The *frame sampler* receives the same video feed from the vehicle's camera and samples frames at a dynamically adjustable rate. The *rescaler* then resizes these sampled frames to a suitable resolution and passes them to the *object detector*, which is built upon a pre-trained deep neural network (YOLOv8) suitable for real-time detection. If an anomaly is detected – specifically, one or more persons in the workspace – an alert is generated, and the robotic arm's operation is paused via the arm controller to ensure safety (Figure 8.17).

The deployment of the application and the real-time orchestration of its services take place in a multi-cluster environment, enabling the placement of services across the different layers of the compute continuum: device, edge, and cloud. This environment, along with its corresponding monitoring mechanisms, is built using state-of-the-art tools such as Karmada, Zipkin, and Prometheus,

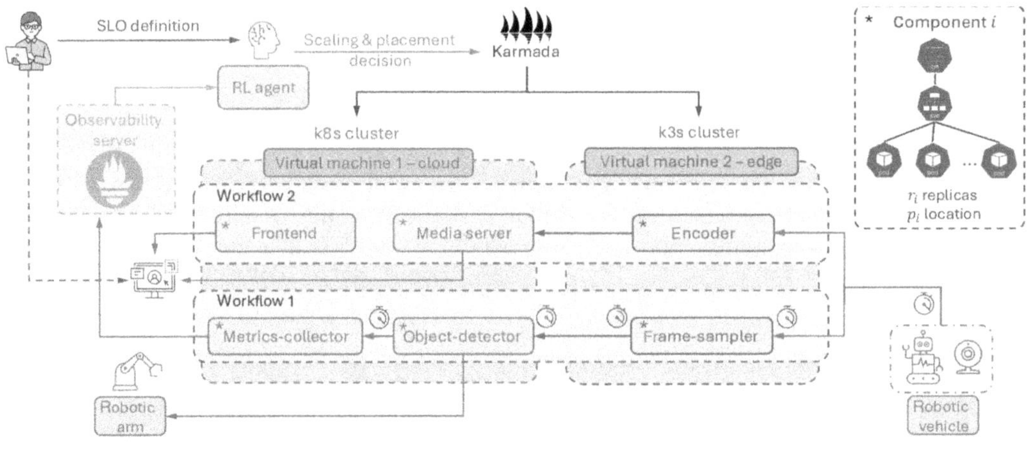

Figure 8.17 6G latency-sensitive service setup for AI-assisted orchestration mechanism evaluation.

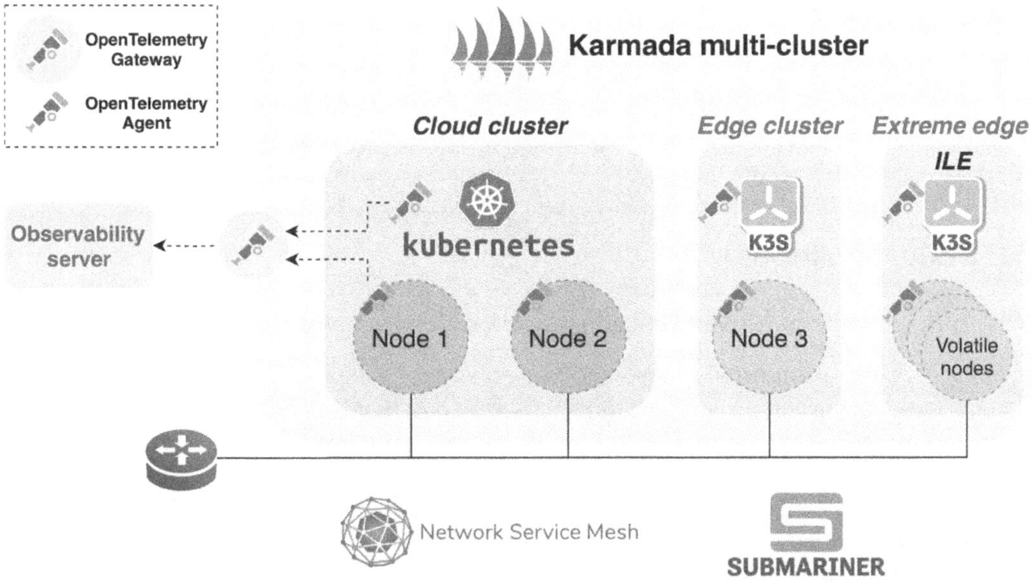

Figure 8.18 The 6G latency-sensitive service experimental set-up.

which enable the propagation of Kubernetes services across multiple clusters and the collection of operational data signals. More precisely, the experimental setup consists of two clusters: one deployed at the edge, closer to the robotic vehicle and arm, and another in the cloud with higher computational resources. An observability platform is responsible for collecting key metrics related to performance evaluation, including service computation and communication latency measured in milliseconds, CPU usage, and workload quantified as frames per second (Figure 8.18).

During the experiments, the object detection workflow is orchestrated using a set of autoscaling and computation offloading mechanisms across the two clusters. Object detection through the robot's surveillance camera operates continuously at a low sampling rate during normal operation. When a person is identified in the video stream, the object detector attempts to confirm the presence of a human by increasing the sampling rate and shifting the system into warning operation. If no human presence is confirmed, the system returns to normal operation. However, if a human continues to be detected in the incoming footage, an alert is triggered to stop the robotic arm, and the sampling rate is subsequently decreased, transitioning to danger operation. Once the human moves out of the camera's scope, the object detector resumes the arm's operation, restoring normal operation. The changes in the sampling rate directly impact the workload of the system since higher sampling rates generate a higher number of requests to the object detector, increasing computational demand.

The autoscaling and placement mechanisms continuously analyse collected data at regular intervals to determine the optimal placement of services while considering resource availability and migration constraints. Additionally, the number of replicas is dynamically adjusted to ensure compliance with end-to-end latency requirements, maintaining compliance with service-level objectives. Specifically, the ML-based object detector component is selected for horizontal scaling based on the measured workload and end-to-end latency at the application layer, with migration between the edge and cloud as required. A centralized joint autoscaling-placement mechanism is the intelligent agent responsible for managing the application. A reinforcement learning-based algorithm was developed to enable adaptive behaviour in response to the variable incoming workload. These orchestration decisions are made by evaluating real-time data regarding delays, which reflect both computation and communication latency, collected at specific monitoring points. For every interval, the reinforcement learning agent determines the optimal placement of services while considering resource constraints and migration limitations and adjusts the number of replicas as needed to ensure that end-to-end latency requirements remain within acceptable limits.

8.6 Evaluation Results

8.6.1 Trust- and Energy-Driven Functionality Allocation Evaluation Results

The implementation related to the trust- and energy-driven functionality allocation, flexible network topologies, and beyond-communication network aspects within a cobot-powered warehouse inventory management system involved monitoring various metrics. Figure 8.19 presents simulation results related to energy consumption (left graph) and trustworthiness (right graph) when using the functionality allocation algorithm based on a genetic algorithm approach, compared to the round-robin load balancing algorithm (baseline). Specifically, the right graph illustrates the percentage increase in trustworthiness achieved by the functionality allocation algorithm over the baseline, while the left graph shows the corresponding percentage decrease in energy consumption. The functionality allocation optimization mechanism seeks near-optimal placement of computational workloads on available compute nodes by minimizing the weighted objective function, $w_1E + w_2L - w_3T$, which includes energy consumption (E), E2E latency (L), and trustworthiness (T). The trustworthiness term is calculated as the sum of the trust indexes of the compute nodes used for workload placement. The trust index for each node j is derived from a weighted sum of several factors: availability, reliability, security, multi-connectivity capabilities, and battery level. The energy consumption term accounts for the processing, transmission, and energy generated by the RF harvester.

The conducted experiments evaluated how varying weights assigned to trustworthiness and energy efficiency influence workload allocation. Specifically, trustworthiness and energy consumption weights ranged from minimal influence to full prioritization relative to other terms within the optimization objective function. Trustworthiness was evaluated by adjusting its relative

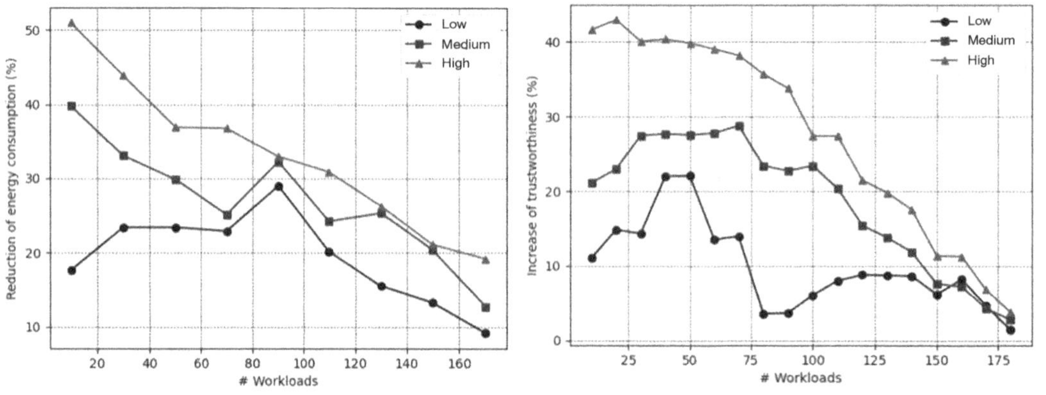

Figure 8.19 Energy consumption (left graph) and trustworthiness (right graph) gains of the functionality allocation mechanism compared to the round-robin load balancing algorithm (baseline) for different workload levels (low, medium, and high).

contribution within the optimization process through different weighting parameters, denoted as w_3, varying from minimal to maximal influence, alongside energy-related weights a_1. In this evaluation, fifty virtualized compute nodes with diverse capabilities were fixed, while the number of computational workloads, each with varying resource requirements, progressively increased. A detailed description of the specific parameter ranges and numerical values utilized for this analysis is available in the supplementary material.

As illustrated in the experimental results, an increase in the trustworthiness weight significantly influences system reliability improvements, although this influence diminishes as workload complexity grows. Specifically, the metaheuristic-based functionality allocation algorithm demonstrated a trustworthiness improvement of up to 43% over the baseline round-robin workload placement method. This observed limitation arises as the number of workloads grows, thereby reducing the solution space for optimal task placements, which consequently restricts the achievable improvements in trustworthiness.

Similarly, energy efficiency results revealed a direct relationship between the energy weighting factor a_1 and reductions in total energy consumption. Higher weighting factors yielded more significant energy savings, demonstrating the algorithm's capability to prioritize energy-efficient workload distributions effectively. Notably, the proposed functionality allocation approach reduced energy consumption by 9.9% to 50.9% relative to the baseline round-robin placement. However, this improvement margin also depends on workload complexity and system constraints, as fewer placement options are available when workload volume increases, thus inherently limiting achievable energy reductions.

Moreover, the evaluation of UAV power consumption across different scenarios, as shown in Table 8.1 and Figure 8.20, emphasizes the importance of strategically distributing tasks between UAVs and edge computing resources in 6G networks. Figure 8.20 presents UAV battery depletion under three configurations: moderate transmission load with low CPU utilization, high transmission load with low CPU utilization, and low transmission load with high CPU utilization. This analysis reveals that onboard processing of high-volume data, rather than relying on edge resources for data transmission, can extend UAV operational time. This flexibility is vital for ensuring sustainable 6G-powered operations, enabling UAVs to complete tasks for longer periods without frequent recharges. Additionally, 6G technology's advanced capabilities facilitate efficient UAV-to-edge communication, allowing UAVs to maintain performance without excessively draining the battery during moderate computational tasks. By intelligently distributing tasks based on UAV capabilities, the functionality allocation mechanism can optimize resource usage, reducing power consumption on compute nodes nearer to the cloud while ensuring high efficiency and performance.

Another set of experiments examined power consumption when AI/ML operations were allocated either (i) at the edge (through compute offloading) or (ii) at the robotic platform (using on-board processing). Figure 8.21 presents a comparison of battery usage over time for three processing scenarios: on-board processing with data transmission, on-device processing only, and

Table 8.1 Battery test scenarios for characteristics of UAV.

Conf. #	Setup
1	Transmission Load: 2 Mbps, CPU Utilization: 22%, CPU cores: 7, Battery Autonomy: 0.32 h
2	Transmission Load: 100 Mbps, CPU Utilization: 22%, CPU cores: 7, Battery Autonomy: 028 h
3	Transmission Load: 0 Mbps, CPU Utilization: 60%, CPU cores: 7, Battery Autonomy: 0.30 h

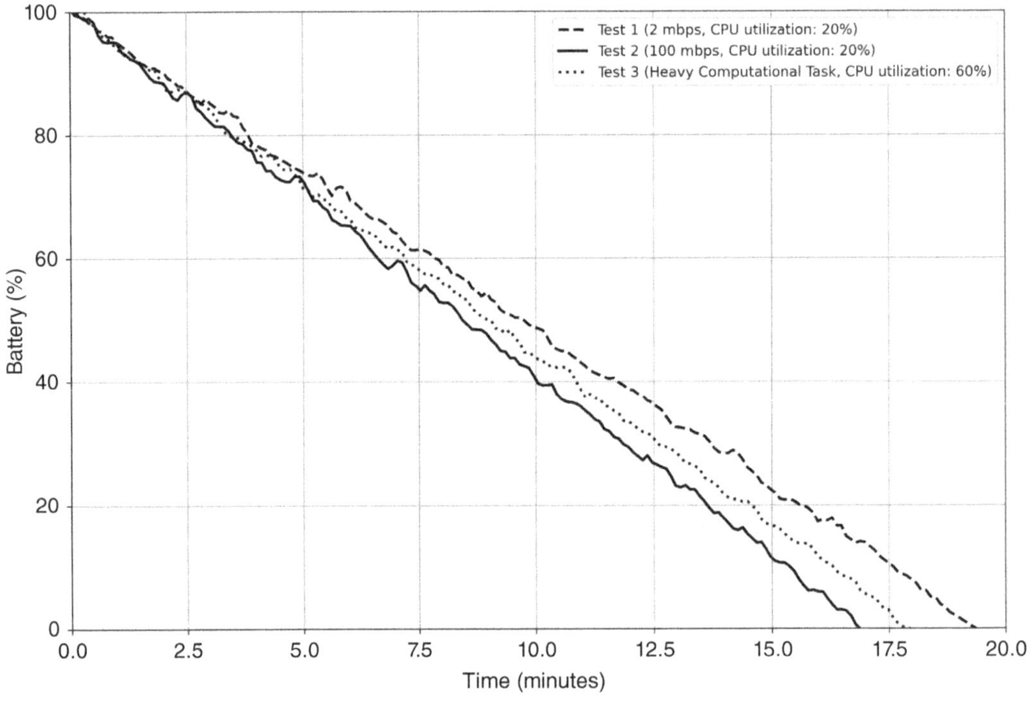

Figure 8.20 Battery depletion of UAV under different scenarios.

offloading processing to the edge (with no on-board processing). The dashed line indicates the scenario where both data processing and transmission occur on the device, showing a rapid decline in battery percentage, reflecting high power consumption. The only on-device processing line (dark line) shows a more gradual decline, indicating moderate power usage. In contrast, offloading tasks to edge computing via a 5G line (light grey line) shows the slowest battery depletion, highlighting the lowest power consumption. According to the chart, edge computing reduced battery depletion by approximately 20% under heavy ML operations compared to on-board processing.

For mobile robotics that prioritize battery efficiency, this approach has significant advantages. The reduced power consumption enables devices to run for longer periods without recharging, improving cost-effectiveness and operational efficiency. Offloading tasks to the edge also enhances device performance by preventing processor overload, leading to faster task completion and better overall efficiency. Additionally, leveraging edge resources allows devices to manage more demanding applications, optimizing resource utilization.

The impact of offloading ML operations was further analysed, focusing on CPU utilization and inference time for the ML model. For comparison, the model was deployed using CPU resources in both on-device processing and offloading scenarios (since the robots were not GPU-enabled). Figure 8.22 (left) illustrates the CPU utilization of the mobile robotic platform across three scenarios: on-device inference with data transmission, on-device inference only, and offloading inference to the edge. The light grey line representing the scenario where both inference and video streaming occur on the device and the dark line representing the on-device inference alone case are almost identical in the graph, leading to the highest CPU utilization, with nearly all processing cores engaged. The only on-device processing line (dark line) shows slightly lower CPU usage with less

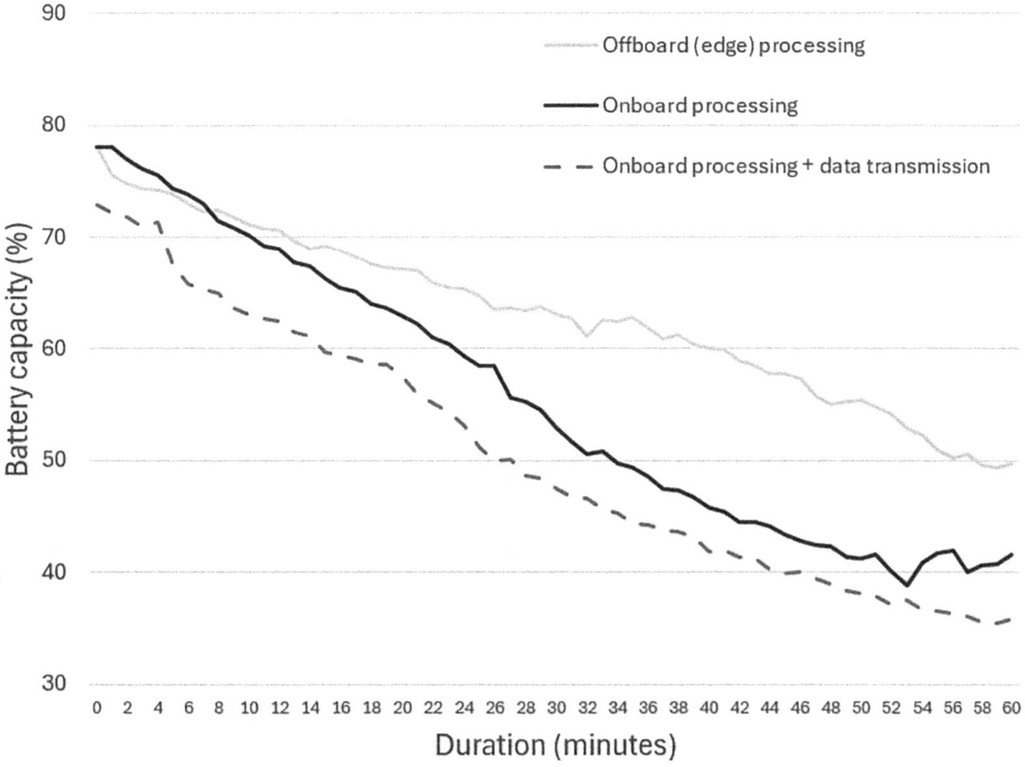

Figure 8.21 On-device versus edge computing setup power consumption.

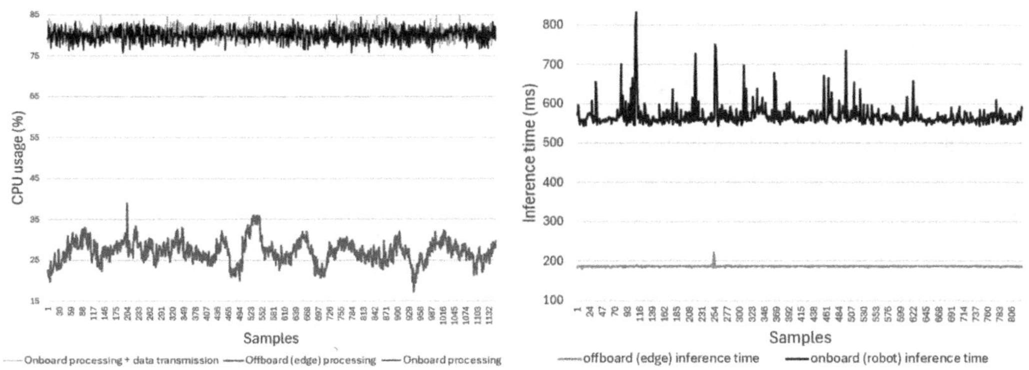

Figure 8.22 CPU utilization benefits of offloading computation via 5G (left graph) and inference time benefits of offloading computation via 5G (right graph).

fluctuation. The dark grey line, representing offloading inference to edge computing via 5G, shows a significant drop in CPU utilization, as the robot is merely publishing camera data through ROS2.

For a more detailed performance comparison, Figure 8.22 (right) visualizes the inference time of the model throughout the experiment. It is clear from the chart that edge computing (light grey line) provides faster and more stable inference performance, while on the robot (dark line), inference time increases nearly threefold, with more significant fluctuations in performance.

In summary, offloading computationally intensive tasks to edge or cloud nodes via B5G/6G greatly improves robot performance and efficiency. As shown in Figure 8.22 (left), CPU usage drops significantly from 80% to 30% when AI models are offloaded, while the inference time variability also decreases, with stable times around 100 ms compared to 400–800 ms on the robot. These improvements highlight that leveraging B5G/6G for cloud/edge computing enhances CPU efficiency, speeds up processing, and optimizes overall system performance.

8.6.2 Trustworthy Flexible Topologies and Beyond Communication Results

Building on the evolution of flexible topologies (see Section 8.6.3) and the integration of BCS capabilities, AI-driven orchestration enables 6G networks to intelligently allocate resources in real time, adapting to application-specific demands. This intelligent reconfiguration not only enhances responsiveness but also leads to measurable performance gains in latency, energy efficiency, and data exposure. When AI is employed to optimize the placement of application functions and to orchestrate the allocation of network resources, there can be reductions in end-to-end latency of over 10% compared to non-optimized configurations, as seen in Figure 8.23.

Similarly, network-wide energy consumption may be reduced by up to 20%, while data exposure is minimized by approximately 5–10%, enhancing both security and privacy. These performance improvements are critical in scenarios where networks must serve a multitude of applications simultaneously, each with its own unique set of requirements and constraints. The ability to reconfigure the network dynamically – ensuring that the right resources are allocated at the right time – has emerged as a basis of 6G network resilience and efficiency.

8.6.3 AI-Assisted E2E Lifecycle Management of a 6G Latency-Sensitive Service Evaluation Results

The orchestration actions of agent setups are decided based on the evaluation of real-time data regarding delays (reflecting computation and communication latency) collected at specific monitoring points. The autoscaling and placement mechanisms evaluate the collected data, and for every interval, the agents select where the optimal placement is (taking into consideration resource and migration constraints) and how many replicas are needed to guarantee service level objectives (SLOs) satisfaction in terms of end-to-end latency requirements.

Three different agent setups are used to demonstrate how different stakeholder groups can influence SLA satisfaction: (i) joint scaling-migration (JSM), (ii) independent scaling-migration (ISM), and (iii) mixed scaling-migration (MSM). The state S_t in all cases considers the *request arrival rate, replica number,* and *placement* for each of the deployed services. Thus:

$$S_t = \{W_t, R_t, P_t\}$$

where W_t is the workload (arrival rate), R_t is the number of replicas of the service, and P_t is the cluster on which the service is deployed.

The action space, on the other hand, for each approach is analysed as:

$$\text{JSM: } A_t = R_t \times P_t$$

$$\text{ISM: } A_{t,1} = R_t, \quad A_{t,2} = P_t$$

$$\text{MSM: } A_{t,1} = P_t, \quad A_{t,2} = P_t$$

JSM assumes a central DQN (Mnih et al. 2015) agent taking joint scaling (1–3 replicas) and migration (edge/cloud) decisions. Thus, the underlying network of DQN with 1 hidden layer is

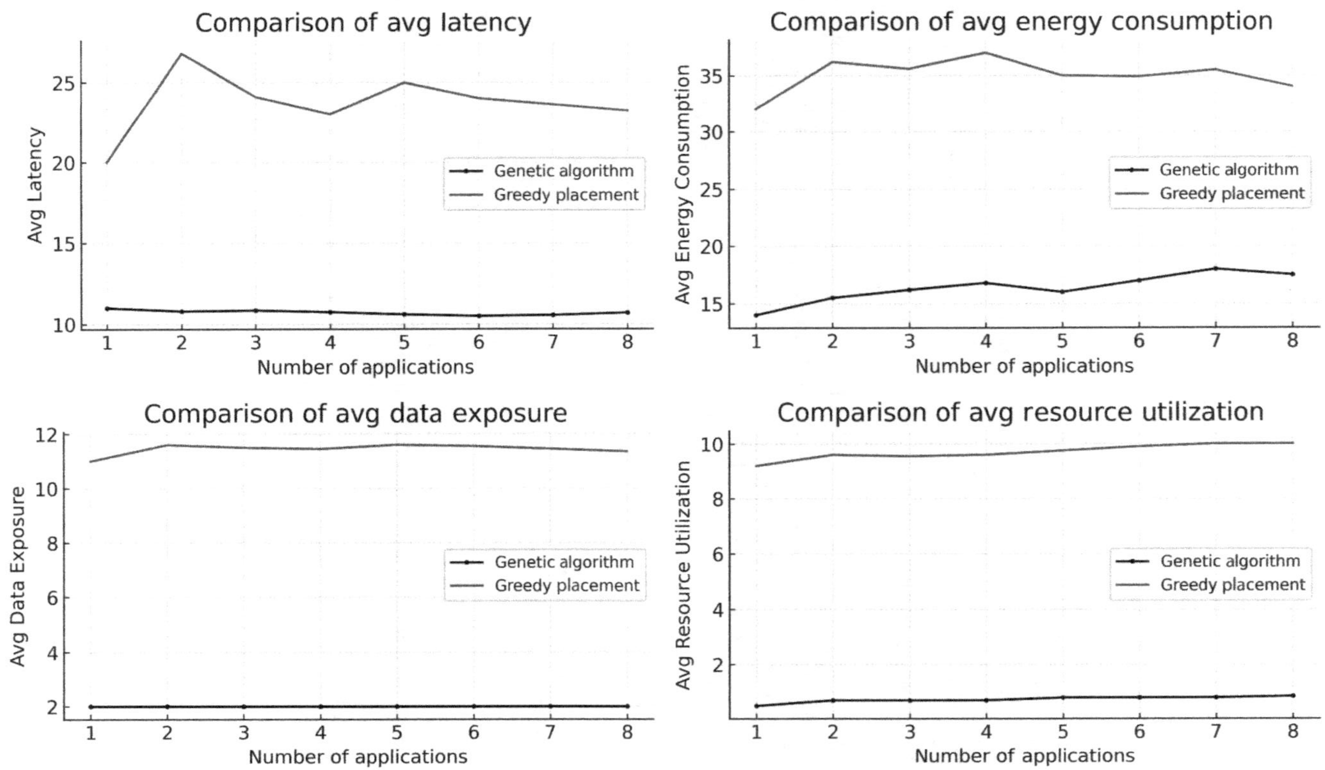

Figure 8.23 Performance comparison of Genetic Algorithm versus Greedy Placement for latency, energy consumption, data exposure, and resource utilization.

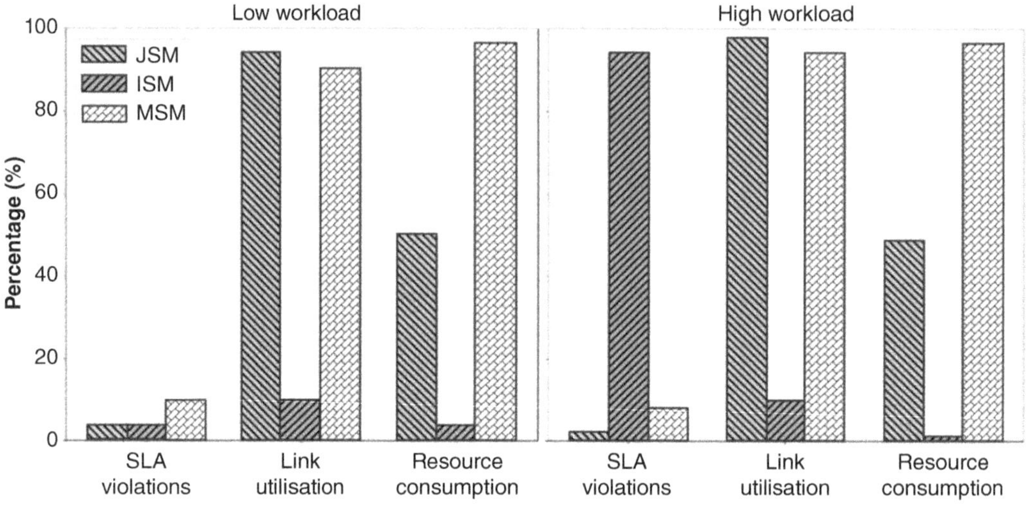

Figure 8.24 Multi-agent service orchestration mechanism evaluation.

of size $3 \times 64 \times 6$. ISM assumes independent scaling and migration by two DQN agents; thus, the two networks are of sizes $3 \times 64 \times 3$ and $3 \times 64 \times 2$. MSM tries a hybrid approach where agents act independently but consider global rewards as feedback, adopting the QMIX (Rashid et al. 2018) approach. The agent networks are similar to ISM, while the mixing network used for training is a 2-layered network with layers of sizes 2×64 and 64×1 accordingly. All weights and biases are provided by the corresponding hypernetworks.

Comparative measurements for the three approaches are illustrated in Figure 8.24. Specifically, the percentage of requests that violate the SLA (i.e. the latency threshold), the utilization of the link between the two clusters, and the resource consumption in terms of percentage of used replicas are depicted for the three different setups.

JSM meets the SLA requirements for both low (1 frame/s) and high (10 frames/s) workloads. However, this comes at the cost of high link utilization and medium to high resource consumption. On the other hand, ISM demonstrates minimal link utilization and resource consumption but fails to satisfy SLA requirements for high workloads, as its agents operate independently. MSM, in contrast to ISM, introduces collaboration between agents, reducing SLA violations but increasing link utilization and resource consumption, even for low workloads (1 frame/s). The comparison between ISM and MSM highlights how introducing collaboration directives can enable agents operating in a decentralized manner to achieve global objectives, albeit with higher resource costs. Furthermore, while JSM outperforms MSM in this relatively simple setup with only two orchestration tasks, MSM's advantage lies in its ability to scale more effectively to complex scenarios. Finally, centralized solutions like JSM may not always be feasible when multiple stakeholders are involved.

8.6.4 Service Migration-Related Evaluation Results

This implementation comes as an extension of the AI-assisted orchestration and E2E LCM of the 6G latency-sensitive service and is based on the deployment of the 6G latency-sensitive service. The service deployment for a multi-site synergetic monitoring and orchestration scenario includes three main components:

- **Infrastructure:** This includes two different sites, Site A and Site B, as described above.
- **Service components:** Two main components are considered for this scenario, 'Frame Sampler', which is deployed in Site B, and 'Object Detector', which is moved across the two sites depending on the availability of Site A's resources.
- **Observability:** Heterogeneous data is gathered from both components using a distributed monitoring data collection mechanism of the monitoring and telemetry framework.

The scenario's migration workflow is illustrated in Figure 8.25.

The intents coming from higher layers are translated into migration directives from Site A to Site B and vice versa. In the figure, the migration process of the 'Object Detector' component from Site A to Site B is described, taking into consideration the timestamps used as measurement points. After evaluation of the two active intents on Site A, the resources are assigned to another application running on Site A, and thus, the directive to move the service from Site A to Site B is initiated. The opposite process takes place when resources become available at Site A.

The migration overhead introduced was measured to identify service downtime due to the overlapping intents and is illustrated in Figures 8.26 and 8.27.

Messages to the integration fabric demonstrate average E2E latency performance close to 100 and 300 ms, and in the worst case, 800 ms. Service deployment duration is measured at around 550 ms for Site A, while service drop is measured as zero since the message to start the service at the local service is sent before actually dropping the service, so that the drop time is minimal. In a future extension of the scenario, a mechanism could be developed to ensure zero service downtime. For Site B, service start and drop demonstrate a similar duration and are smaller than Site A, as expected since it is a lightweight server.

Site A operations when deploying the service to the cloud seem to greatly increase downtime by 150%, and thus, such costs should be taken into account in cases where SLAs include downtime requirements.

8.6.5 Intent-Based Networking Evaluation Results

The tasks carried out in the process of synergetic monitoring and orchestration, and the results of the PoC in terms of execution time after testing the solution with a set of 20 intent requests (Vilalta et al. 2023).

The main functionalities of intent-based networking support for the PoC described in Section 8.6.4 were divided into five main tasks that occur sequentially and are illustrated in Figure 8.28.

1) **Interpret intent:** The workflow starts by extracting the most relevant attributes out of the input message received in the request. The algorithm first parses the received information into a different data structure (Adanza et al. 2024).
2) **Database storage:** The algorithm then stores this information in a database to keep track of each request.
3) **Feasibility check:** To ensure that the requirements included in the intent request can be accomplished, the algorithm performs a simplistic feasibility check by looking at the requested and the available resources.
4) **Integration with TeraFlowSDN (TFS):** The next natural step is to mount the internal intent descriptor and send a request to the TFS controller to create a layer 3 VPN.
5) **Updating status through Kafka:** To update the other involved entities in the project, a set of Kafka messages is exchanged to communicate that the process has been started and to update the final status of the request.

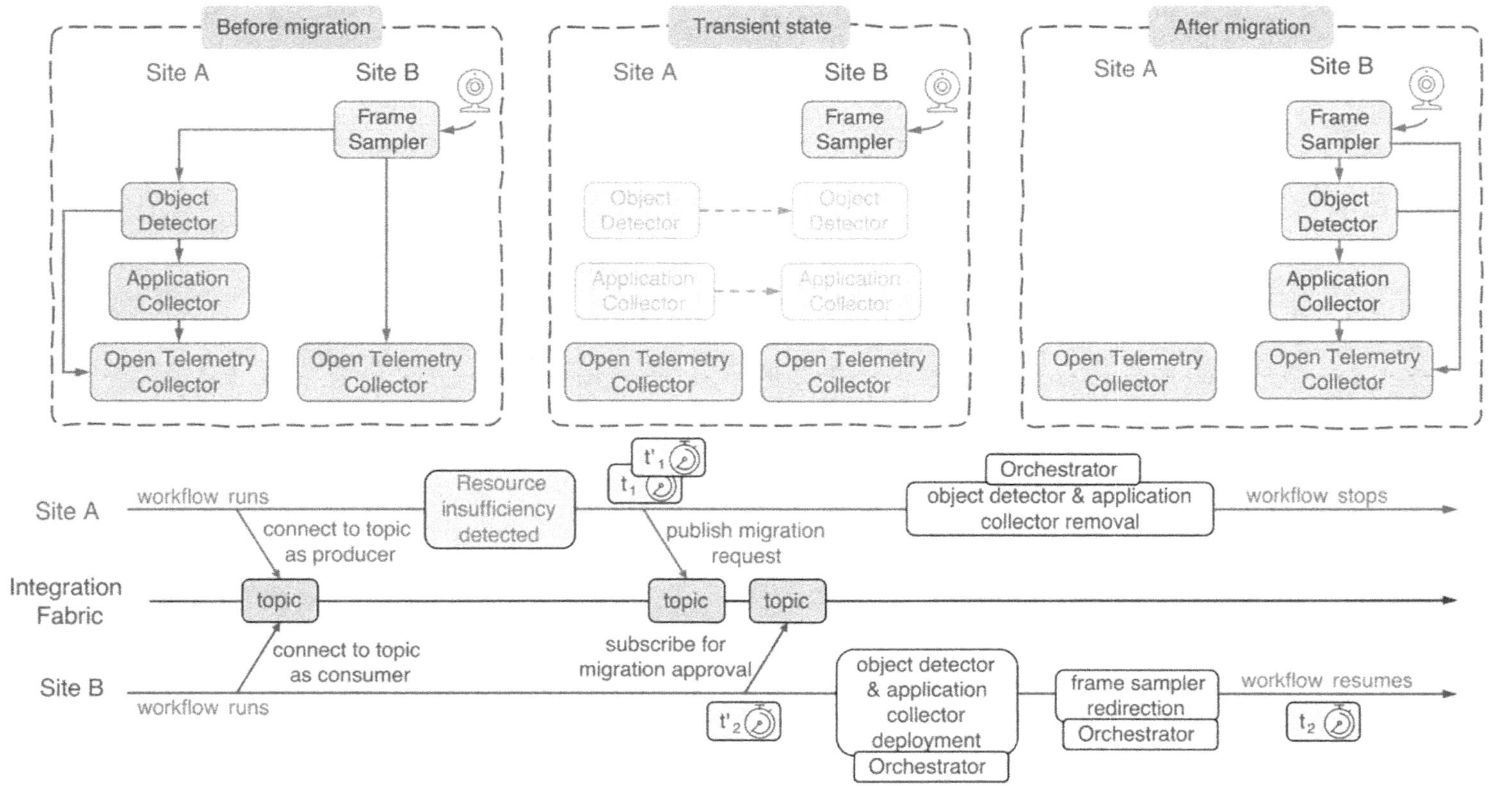

Figure 8.25 Synergetic scenario migration workflow.

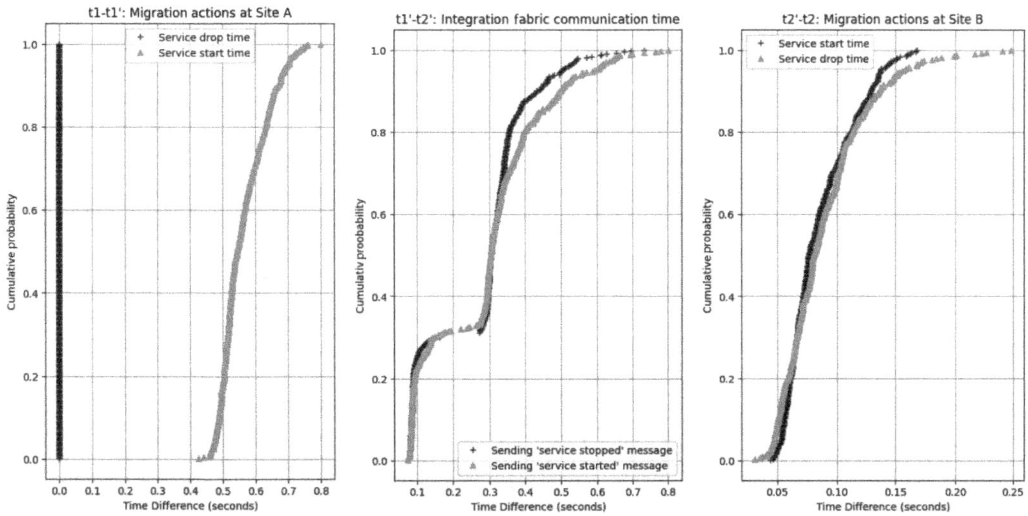

Figure 8.26 Migration delay for Site A/Site B operations.

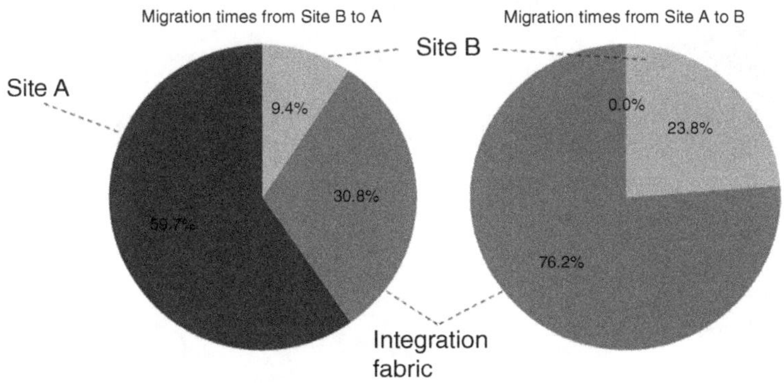

Figure 8.27 Migration delay for local/remote deployment.

Figure 8.28 Schema with the main process encompassed by the orchestrator.

To demonstrate that the solution can be executed in a real-time environment, the orchestrator has been tested with a set of 20 intents, and the execution times were gathered in Figure 8.29.

Figure 8.29 presents the cumulative distribution values (CDF) on the Y axis and the number of seconds taken to process an intent request on the X axis. It is possible to appreciate a low variability of the data samples with one outlier and a range of values from a minimum of 2.69 and a maximum of 4.16 seconds. The low variability of the samples can be attributed to the execution of actions like intent interpretation, database storage, and the feasibility check, which present a low variability in their respective tasks. However, the process of triggering the TFS call and the sending of the Kafka message, although they still occur in a short period of time, present a bigger variability since the execution of those tasks relies on external entities.

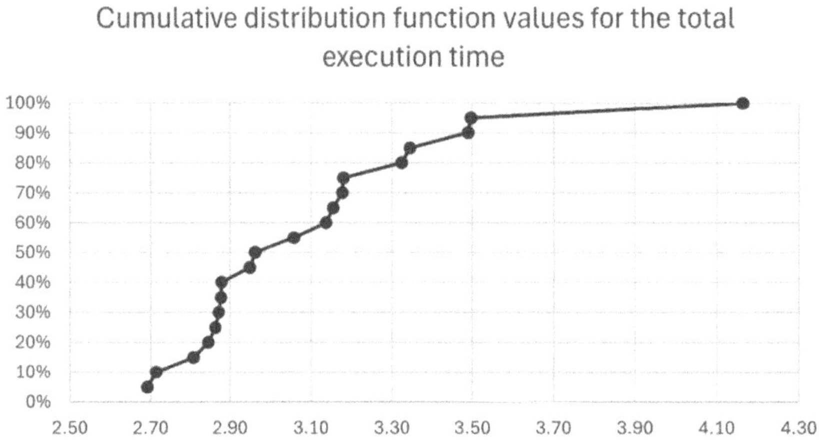

Figure 8.29 Cumulative distribution function values for the total execution time.

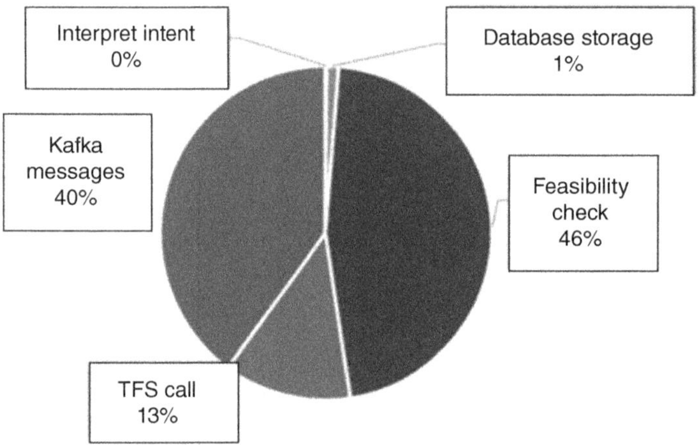

Figure 8.30 Relative values for the time taken in each task.

Alternatively, Figure 8.30 illustrates the average of the relative percentage values of each task. It is worth considering that the execution of the scenario occurs in a very short period of time (it took an average of 3.10 seconds to be executed), which means that a small change in terms of time implies a big change in the percentage distribution. However, it is also worth considering that the biggest execution time was in the tasks for performing the feasibility check and sending the Kafka message. That is because the algorithm relies on external entities, which slightly delay its performance time. On the other hand, the database storage and the intent interpretation took a marginal proportion of the execution time, followed by triggering the TFS request, which barely took an average of 13% of the time.

8.7 Conclusion

The early validation activities presented in this chapter have highlighted the transformative potential of 6G system components in enabling sophisticated, collaborative robotics applications within complex industrial logistics environments. Through targeted PoC implementations, the potential of advanced management and orchestration functionalities, including intent-based service provisioning and adaptive workload allocation mechanisms, was effectively demonstrated.

Specifically, intent-based management emerged as a pivotal component, translating high-level operational objectives seamlessly into precise system-level actions. Additionally, the introduction of advanced orchestration frameworks enabled dynamic, energy- and trust-aware allocation of both computational and physical robotic tasks, ensuring optimal utilization of available resources across distributed robotic and compute infrastructures. The incorporation of TEFs significantly contributed to system robustness, ensuring workloads are reliably executed by nodes exhibiting higher trust indices, thereby enhancing the overall reliability and security of industrial operations.

While certain enablers, such as ultra-reliable, low-latency communication, were directly validated in the PoCs, others, like precise network-assisted positioning and AI-enhanced sensing, are foreseen as critical complementary capabilities. These elements will play a key role in future system iterations, supporting more advanced robot coordination, navigation, and real-time decision-making in dynamic environments.

Furthermore, the integration of zero-touch orchestration and automation methods ensures rapid responsiveness to changing operational demands, thereby optimizing resource utilization and reducing operational overheads.

In conclusion, the validation activities provided clear insights into how the next generation of 6G technologies can holistically improve the performance, adaptability, and sustainability of automated logistics ecosystems. The integration of advanced orchestration, intent-driven management, and intelligent functionality allocation, along with trust-based security mechanisms, will be instrumental in shaping the industrial applications landscape of the future.

Acronyms and Abbreviations

ACL	Access control lists
AI	Artificial intelligence
AIaaS	AI-as-a-service
AMF	Access and mobility management function
AMR	Autonomous mobile robot
BCS	Beyond communication service
CaaS	Compute-as-a-service
CAPEX	Capital expenditure
CL	Closed loop
DSM	Digital service manager
E2E	End to end
ESA	European Space Agency
FTN	Flexible topology network
GUI	Graphical user interface
IBM	Intent-based management
IBN	Intent-based networking
IME	Intent management entity
ISM	Independent scaling-migration
JSM	Joint scaling-migration
KPI	Key performance indicator

KVI	Key value indicator
LCM	Lifecycle management
M&O	Management and orchestration
MSC	Message sequence chart
MSM	Mixed scaling-migration
NEF	Network exposure function
OPEX	Operational expenditure
PCF	Policy control function
PoC	Proof-of-concept
RAN	Radio access network
RBAC	Role-based access control
RMS	Robot manager service
ROS	Robot operating system
SA	Standalone
SM	State machine
SMF	Session management function
TEF	Trust evaluation function
TFS	TeraFlowSDN
TLS	Transport layer security
UAV	Unmanned aerial vehicle
UI	User interface
URLLC	Ultra-reliable and low-latency communications
ZSM	Zero-touch service management

References

Adanza, D., Natalino, C., Gifre, L. et al. (2024). , IntentLLM: an AI Chatbot to create, find, and explain slice intents in TeraFlowSDN. *2024 IEEE 10th International Conference on Network Softwarization.*

Alemant, P. and Ojaghi, B. (Eds.) Hexa-X-II (2024). End-to-end system evaluation results from the interim overall 6G system. Deliverable D2.4. [Online]. Available at: https://zenodo.org/records/15796347 (accessed 1 September 2025).

Bantouna, A. (Ed), Hexa-X-II (2024). 6G use cases and requirements. Deliverable D1.3. [Online]. Available at: https://zenodo.org/records/16780232 (accessed 1 September 2025).

ETSI GS ZSM (2021). *Zero-Touch Network and Service Management (ZSM); Closed-Loop Automation; Part 1: Enablers of Closed-Loop Automation.* Sophia Antipolis: European Telecommunications Standards Institute.

European Space Agency (ESA) (2023). 5G-HOSTS-SAT 5G hub over-the-air vertical segment validations. [Online]. Available at: https://connectivity.esa.int/projects/5ghostssat (accessed 10 May 2025).

Le Sauze, N. and Kerboeuf, S. (Eds.) Hexa-X-II (2025). Final overall 6G system design. Deliverable D2.5. [Online]. Available at: https://zenodo.org/records/15796361 (accessed 1 September 2025).

Maghsoudnia, A., Vlad, E., Gong, A. et al. (2024). Ultra-reliable low-latency in 5G: a close reality or a distant goal?*Proceedings of the 23rd ACM Workshop on Hot Topics in Networks*, November, pp. 111–120.

Masood Khorsandi, B., Uusitalo, M., Hamon, M.H, et al (Eds.) Hexa-X (2023). Hexa-X architecture for B5G/6G networks-final release. Deliverable 1.4. [Online]. Available at: https://hexa-x.eu/wp-content/uploads/2023/07/Hexa-X-D1.4-Final.pdf (accessed 1 September 2025).

Mnih, V., Kavukcuoglu, K., Silver, D., Rusu, A. A., Veness, J., Bellemare, M. G., Graves, A., Riedmiller, M., Fidjeland, A. K., Ostrovski, G., Petersen, S., Beattie, C., Sadik, A., Antonoglou, I., King, H., Kumaran, D., Wierstra, D., Legg, S., & Hassabis, D. (2015). Human-level control through deep reinforcement learning. *Nature*, 518(7540), 529–533. https://doi.org/10.1038/nature14236.

Pires, R., Blue, H., Malinen, J. et al. (2024). Closed-loop automation in 6G for minimum downtime task continuity in surveillance cobots. *2024 Joint European Conference on Networks and Communications & 6G Summit (EuCNC/6G Summit).*

Rashid, T., Samvelyan, M., Schroeder de Witt, C., Farquhar, G., Foerster, J., & Whiteson, S. (2018) QMIX: monotonic value function factorisation for deep multi-agent reinforcement learning, arXiv preprint. Available at: https://arxiv.org/abs/1803.11485 (accessed 10 March 2025).

Vilalta, R., Adanza, D., and Alemany, P. (2023). Demonstration of an intent-based networking for end-to-end packet optical cloud-native SDN orchestration. *European Conference of Optical Networks.*

9

Path Towards 6G

Ishita Mishra[1], Toon Norp[2], Per Hjalmar Lehne[3], Bahare M. Khorsandi[4], Ricard Vilalta[5], Alexandros Kaloxilos[6], Patrik Rugeland[7], and Mauro R. Boldi[8]

[1] *Institute PIIU, Rome, Italy*
[2] *TNO, The Hague, The Netherlands*
[3] *Telenor, Oslo, Norway*
[4] *Nokia, Munich, Germany*
[5] *Centre Tecnològic de Telecomunicacions de Catalunya (CTTC-CERCA), Castelldefels (Barcelona), Spain*
[6] *GIA, Brussels, Belgium*
[7] *Ericsson Research, Stockholm, Sweden*
[8] *Telecom Italia, Turin, Italy*

In the latest years, wireless communication networks have become an integral component of the fabric of modern society, fundamentally transforming how we connect and interact. In the decade of 2020s, the development of the sixth generation of telecommunications networks, or 6G, represents a further evolution from its predecessors, offering unprecedented opportunities while addressing new challenges. Building on the foundations laid by 5G, 6G is expected to evolve and transform connectivity with enhanced performance, more efficient spectrum usage, and innovative solutions that will improve various aspects of everyday life.

As seen in previous chapters, 6G promises transformative capabilities such as terahertz frequency communications, pervasive adoption of artificial intelligence (AI) / machine learning (ML), and integrated sensing, computation, and communication. These innovations will fuel the growth of autonomous systems, immersive augmented and virtual reality experiences, and cross-domain applications from healthcare to smart city infrastructures.

Implementing this innovative path will require overcoming significant technological, economic, and regulatory challenges. As research and development continue, the promise of 6G will gradually come to reality, offering transformative possibilities for industries, individuals, and societies at large.

This chapter presents the view on the current path towards the future introduction of 6G (Section 9.1), offering a review of what is going on worldwide (Section 9.2), with a specific insight on the industrial standpoint (Section 9.3). Finally, Section 9.4 presents the current status of the standardization efforts towards 6G.

9.1 Migration from 5G to 6G and Gaps

As the standardization of 6G recently started, with the ambition to see the first commercial deployments around 2030, it will be imperative to ensure a smooth transition from 5G to enable a large

6G to Build a Sustainable Future, First Edition. Edited by Mikko A. Uusitalo, Patrik Rugeland, Mauro R. Boldi, and Ahmad Nimr.

market uptake of 6G. To achieve this, 6G must provide a clear benefit compared to 5G from day 1, wherever it is deployed. In the transition from 4G to 5G, the rush to have an early launch of 5G led to the introduction of 5G non-standalone (NSA), which acted as a performance booster to 4G, as the 5GC was not ready when the initial 5G radio deployment started.

However, the downside was that 5G NSA, relying on the 4G evolved packet core (EPC), could not support any of the newer 5G features, such as network slicing or positioning. Thus, 5G NSA delayed the deployment of 5G standalone (SA) to such a degree that it is expected that around 40% of the subscriptions in 2030 will still rely on NSA (Ericsson 2024, November).

To avoid the same inconvenience in 6G, the 6G system should rely on an enhanced 5GC, which can more easily be upgraded to support the 6G services, instead of needing to replace all core network (CN) hardware with a new 6G core. For the radio access network (RAN), in order to benefit from the technological development and novel services intended, a new radio access technology (RAT) is envisioned to connect to the evolved 5GC. Although 5G introduced the dual connectivity (DC), initially standardized for 4G to support NSA and to provide data aggregation from multiple sites, in deployments, it has been shown to provide very little benefit while adding significant complexity. Furthermore, at the radio interface, the spectrum resources are scarce, with all low bands already allocated to legacy technology. Reallocation and refarming of spectra take significant time and would detrimentally impact the 5G technology if done too quickly. To overcome this, 6G needs to be designed to support so-called multi-RAT spectrum sharing (MRSS), where an operator can reallocate spectrum between 5G and 6G dynamically on the shortest timescale, i.e. single slots (sub-ms) based on the instantaneous needs. To support disaggregation of the RAN, the so-called lower-layer split (LLS), where only the radio units are distributed, with all other functionality centralized in the base station, is being developed and standardized by the O-RAN alliance.

Furthermore, as one of the emerging disruptive technologies, AI/ML is poised to fundamentally alter the interactions and operations of networks (see also Chapter 6). Even though it is too early to determine the impacts, its rapid development spurs the need to consider and evaluate the application and benefits of AI/ML in all parts of the network. For instance, many parts of the networks would benefit from automation and optimization provided by these tools. To facilitate this, the network will have to be designed in such a way that the ML agents can be deployed where needed, as well as collect and act upon requisite data and control and interact with the pertinent parts of the network. However, even though the AI/ML tools can be immensely powerful, a thorough cost-benefit analysis will be required for each scenario to determine whether their benefits outweigh their costs in terms of complexity, energy consumption, and TCO, as well as security and privacy.

Finally, as discussed in Chapters 2 and 3, the 6G networks will need to take sustainability aspects into account at all steps of the way, from the design to the development, deployment, and operation of the system, in terms of environmental, economic, and social sustainability. For instance, the need to reduce energy consumption spurs the innovation of solutions to improve the efficiency as well as reduce the energy consumption of the network, e.g. through more energy-efficient sleep modes or energy-neutral devices. In addition, the goal of social sustainability aims, among other things, to bridge the digital divide, i.e. to provide connectivity to the un- and under-connected through the incorporation of NTNs. On the other hand, economic sustainability involves fostering long-term economic growth without depleting resources or compromising the social or environmental well-being.

9.2 Global 6G Initiatives

As the global telecommunications industry progresses towards 6G, international alliances play a crucial role in shaping its development. Recognizing the importance of global collaboration, SNS-JU and 6G-IA have established Memorandums of Understanding (MoUs) with key 6G alliances worldwide. These partnerships facilitate knowledge exchange, technical alignment, and policy discussions, ensuring that 6G development remains globally harmonized and socially impactful.

This chapter highlights the major 6G alliances with which SNS-JU and 6G-IA have formal agreements, as depicted in Figure 9.1.

9.2.1 European Priorities for 6G Development

As the global race towards the development and standardization of 6G networks intensifies, Europe has established a clear and comprehensive vision to ensure its leadership in the next generation of telecommunications. Building on the foundations of 5G, Europe aims to position itself at the forefront of 6G Research and Development (R&D), focusing on technological sovereignty, sustainability, and global collaboration. These priorities reflect not only the technical ambitions of the continent but also its commitment to addressing broader social, economic, and environmental challenges.

The European 6G vision is shaped by the 6G-IA 6G Smart Networks and Services Industry Association (6G-IA 2024) and key initiatives such as the Smart Networks and Services Joint Undertaking (SNS-JU). These play a pivotal role in defining and guiding the European 6G priorities, with SNS flagship projects like Hexa-X-II serving as important contributors to the global 6G agenda. These initiatives emphasize the importance of aligning with global standards to foster a unified vision for 6G while addressing the unique needs of European stakeholders. Central to this effort is the goal of leveraging 6G technology to support social well-being, economic growth, and environmental sustainability.

From a technological perspective, Europe's priorities for 6G (6G-IA 2023) include advancements that will revolutionize mobile connectivity and introduce entirely new capabilities. Key innovations involve native support for artificial intelligence (AI) and machine learning (ML), which will enable real-time network intelligence, resource optimization, and enhanced security. Integrated sensing and communication (ISAC) represent another major area of development, offering the ability to merge communication networks with sensing capabilities for applications ranging from environmental monitoring to autonomous systems. Additionally, significant emphasis is placed on cybersecurity to ensure the resilience and trustworthiness of 6G networks, reflecting Europe's adherence to strict data protection and privacy standards. A detailed vision recently published by 6G-IA (2024) identifies the work items that 6G-IA and the SNS JU are considering as key priorities to meeting IMT-2030 targets (ITU-R 2023). These priorities are presented in Figure 9.2.

A defining feature of Europe's 6G strategy is its strong focus on sustainability. Recognizing the dual role of 6G in supporting environmental goals and enhancing sustainability across industries, Europe is committed to ensuring that 6G systems will provide solutions towards a green transition for Europe. This transition is one of the three strategic objectives for Horizon Europe's strategic plan. To evaluate the success of this transition, the notion of key value indicators (KVIs) pioneered

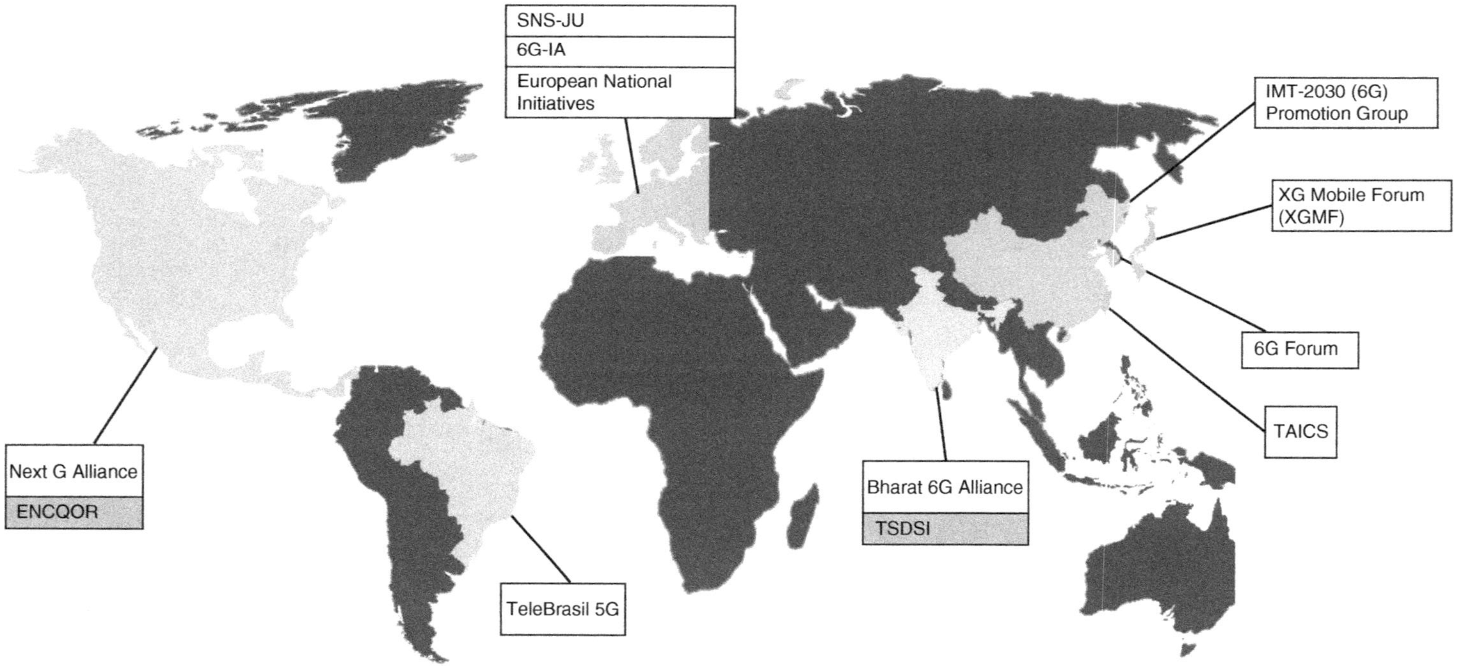

Figure 9.1 6G research and innovation alliances in Europe and other world areas.

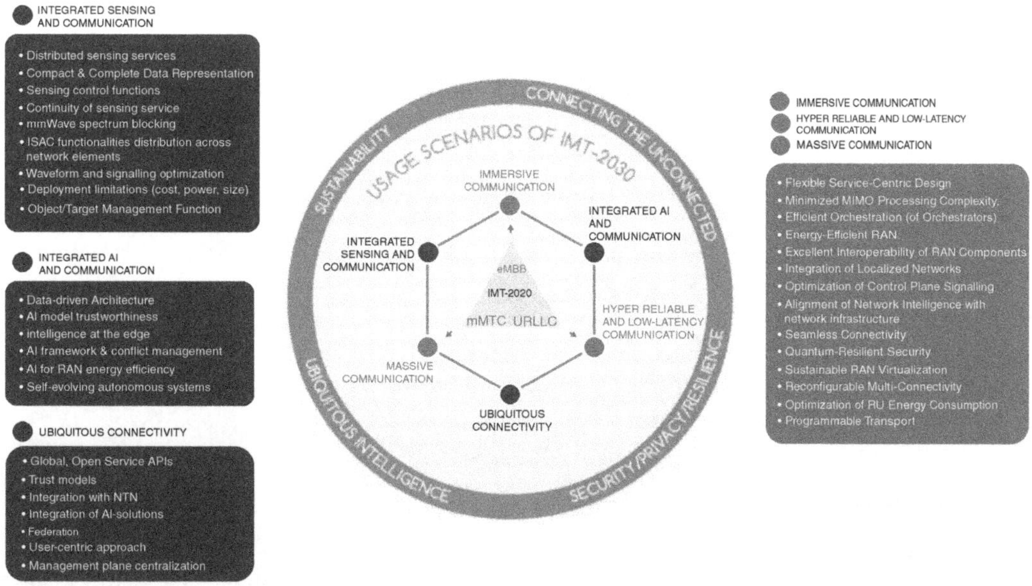

Figure 9.2 SNS JU/6G-IA work items to meet the IMT-2030 targets. *Source:* 6G-IA (2024) / 6G Smart Networks and Services Industry Association.

by SNS projects has been developed as presented in Section 2. Still, significant work remains to qualify/quantify these indicators such that they can be used in a standardization or certification context.

The European vision for 6G, apart from environmental sustainability (e.g. lower energy consumption), also emphasizes the need to address societal (e.g. digital inclusion, accessibility, and trustworthiness) and economic considerations (monetization of 6G services, reduction of the total cost of ownership – TCO), including digital inclusion, accessibility, and trustworthiness. By identifying and resolving technological and business gaps in 5G, particularly in areas such as support for vertical industries, Europe aims to create a more inclusive and comprehensive telecommunications ecosystem. This vision is exemplified by the development of innovative use cases as discussed in Chapter 2. Each use case family is underpinned by tailored key performance indicators (KPIs) such as data rates, latency, reliability, and positioning accuracy, reflecting Europe's commitment to meeting diverse application needs.

In addition to technical advancements, Europe places significant emphasis on operational innovations to ensure seamless and efficient 6G deployment. These include flexible network topologies, the integration of terrestrial and non-terrestrial networks, and spectrum sharing and optimization. The European vision emphasizes interoperability, interconnected smart networks, and the development of sustainable business ecosystems.

Technological sovereignty is another cornerstone of Europe's 6G strategy. To reduce technological dependence on non-European stakeholders and foster local innovation, Europe is investing in areas such as components and microelectronics, open telco cloud and distributed computing solutions, network intelligence, and security. By promoting open-source activities and ensuring interoperability, Europe aims to create robust and globally competitive solutions. The development of globally accepted AI/ML frameworks and secure, privacy-focused hardware and software further underscores Europe's commitment to technological independence and leadership.

Standardization and regulation play a vital role in shaping the future of 6G. Under the IMT-2030 framework, the International Telecommunication Union (ITU) is expected to define the technical requirements for 6G by 2025. Europe is actively contributing to this process, ensuring that its priorities, including ubiquitous connectivity, AI-driven communications, and integrated sensing, are well-represented. Compliance with regulatory frameworks such as the Radio Equipment Directive, the AI Act, and the Cybersecurity Act is crucial for creating secure, reliable, and socially responsible 6G systems.

In conclusion, Europe's priorities for 6G reflect a holistic approach that balances technological innovation with societal, economic, and environmental goals. By fostering collaboration, advancing R&D, and addressing key challenges, Europe aims to create a 6G ecosystem that is sustainable, inclusive, and globally competitive. This vision positions Europe as a leader in the next generation of telecommunications, ensuring that 6G serves as a transformative force for society while supporting global sustainability objectives.

9.2.2 European National Platforms View

While the Horizon Europe Smart Network and Services (SNS) programme has started, in several countries in Europe, large national initiatives have started aiming to make European 5G/6G research even stronger. The SNS ICE project, a CSA project funded by JU SNS, has collected information about these national initiatives based on publicly available information (e.g. based on presentations and websites).

Almost all EU member states have 5G/6G activities. These may be funded through, for example, the Horizon Europe programme, Connecting Europe Facility (CEF) or European Regional Development Funds (ERDF). In this section, however, the focus is on initiatives that are funded by the national governments of the member states. In several European countries, there are large national initiatives with budgets of some millions of euros: Figure 9.3 reports graphically these initiatives in the European countries, their names, and the allocated budget (Project 2023).

There are different ways that can be distinguished in how these large national initiatives are organized:

- Institute: here, a specific institute or university receives the funding, which is then used to fund researchers and projects (e.g. the University of Oulu with the 6GFlagship).
- Programmatic: here, one multi-year large programme is established, relatively independent from the government, in which research topics and planning are determined from the start. There is relatively little government steering on programme content for the duration of the programme (e.g. Restart in Italy).
- Portfolio of projects: Here, the government issues (regular) calls for proposals and funds multiple individual projects (e.g. España Digital 2026).

The Dutch national 6G programme called the Future Network Services (FNS 2024) programme is funded through the National Growth Fund, whereby the Dutch government will invest up to €203M in FNS. The programme aims to build a leading position in certain parts of 6G technology, which will strengthen the future earning power of the Dutch economy. The programme is shaped around two phases: Phase 1 runs from Q1 2024 up to the end of Q2 2026 with a budget of €61M, while Phase 2 of the programme, with €142M of funding, will run from Q3 2026 till Q3 2030.

The Spanish national initiative on 6G, called UNICO 6G R&D, is a part of the Digital Spain 2026, or the 'España Digital 2026' (ESD 2023) initiative started by the Ministry for Economic Affairs and

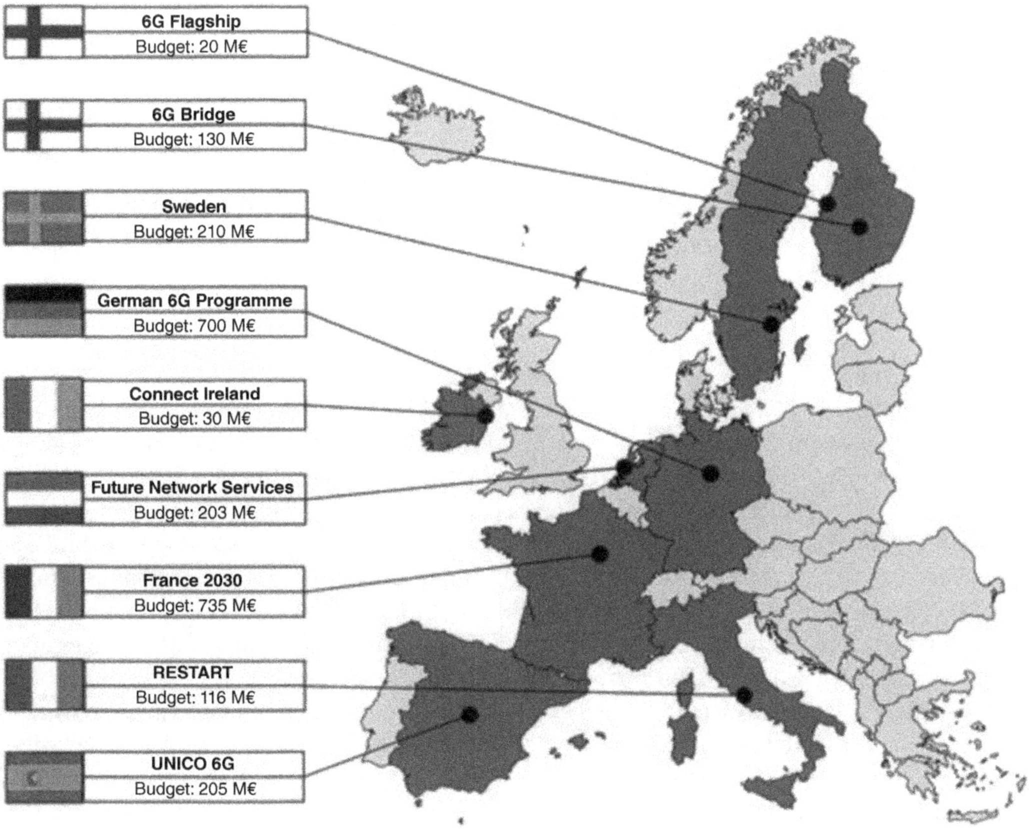

Figure 9.3 European National initiatives on 6G.

Digital Transformation as a part of the country's digital transformation roadmap. A total funding of 205 M€ was allocated to the UNICO 6G R&D programme and is distributed to the partners by means of three open calls, one in 2021, 2022, and 2023 each.

The national initiative from Italy is called RESTART (Restart 2023): 'RESearch and innovation on future Telecommunications systems and networks, to make Italy more smART'. The programme is planned to run from January 2023 to December 2025, with a funding of 116 M€, and comprising a total of 25 partners. The programme includes the following activities: fundamental and applied research; technology transfer and exploitation of research results, including dissemination activities; support for the creation and development of start-ups and spin-offs from research, promoting the activities and services of incubation and venture capital funds; training led in synergy by universities and enterprises, with particular reference to SMEs (small and medium enterprises) to reduce the mismatch between the skills required by enterprises and those offered by universities; PhDs programmes.

The French National Initiative is a part of the 'France 2030' recovery plan, which has ecology, competitiveness, and cohesion as its three main themes (2030, 2023). Launched in 2021 by the French Ministry of Economy, Finances and Industrial and Digital Sovereignty, this initiative is coordinated at the national level by the Directorate-General for Enterprise/Ministry of Economy and Finance, in collaboration with other competent ministries, government agencies, and the national telecoms regulatory authority (ARCEP). France 2030 has 10 objectives and 6 conditions for success, which make up the plan. The French national initiative started in 2021 and is funded with € 735M from the General Secretariat of the Prime Minister (SGPI), managed by Direction Générale des Entreprises (DGE) of the Ministry of Economy and the Ministry of Research.

In Finland, there are two NIs, 6G Flagship and 6G Bridge. The 6G Bridge (2023) programme aims to make Finland the global leader in providing new value with 5G Advanced and 6G technologies for sustainable industries and societies, e.g. in smart cities, smart energy, smart ports, and smart factories with different ecosystem players. While the 6G Flagship is focused primarily on fundamental research, the 6G bridge reduces the gaps between fundamental research and applied research. The two national programmes do not compete with each other but rather offer complementary services and insights. The 6G Flagship (2021) is a part of the Finnish government's national research spearhead programme from 2018 to 2026. The goal is to create the essential 6G technological components, the tools, and the equipment to build a 6G Test Network, develop chosen vertical applications for 6G to accelerate societal digitization, and continue to be a recognized vision leader and sought-after research partner in worldwide 6G research. The initiative is run by the Centre for Wireless Communications (CWC) at the University of Oulu in Finland.

The German 6G Programme comprises different kinds of projects. The '6G Platform Germany', or more precisely the 'Platform for Future Communication Technologies and 6G' (Platform 2023), is the so-called 'umbrella organization' of the German 6G Programme. It is intended both to make scientific contributions to the content design of 6G and to ensure the scientific-organizational support for the processes that are necessary for the successful implementation of the German 6G programme. Within the German national 6G initiative, four 6G research hubs were started, which comprise about 160 research groups at 21 universities and 15 research institutes overall. The use case scenarios and application areas of the programme are campus networks (automation and campus logistics), medical scenarios (hospitals, emergency, and operation theatre), mobility (automotive, commercial vehicles, and drones), and global coverage (satellites, rural areas, ...). The programme also focuses on several societal and policy aspects such as sustainability, sovereignty, privacy, and data protection, EM compatibility, scientific communication and gaining acceptance, trustworthiness, and gender equality (women in 6G).

In Sweden, the Advanced Digitalization programme has received funding of €210M from the Swedish Government for the period of 2023 to 2027 (Sweden 2023). The budget and the official representation of Sweden are handled by the innovation agency, Vinnova. The programme runs on a partnership basis whereby partners of the projects will, at the very least, match the funding provided by the government, hence bringing the total budget of Advanced Digitalization to €420M. The programme office for this initiative is financed by the industry in addition to Vinnova.

In Ireland, there is about €30M in core funding available from Research Ireland to CONNECT Ireland. This funding has been allocated for a period of six years, starting in 2021. The programme has a target to add funding from industry collaborations, and non-exchequer/non-commercial sources, including in-kind contributions, which should bring the total budget to about €75M.

9.2.3 North America: Strengthening 6G Innovation Through Strategic Alliances

The **Next G Alliance** (2025), spearheaded by the **Alliance for Telecommunications Industry Solutions (ATIS)**, stands out as North America's flagship initiative for 6G research and innovation. Emphasizing research, development, and standards, the Alliance promotes innovation and global influence. Established in 2020, it involves leading academic institutions, government agencies, and telecommunications firms like Verizon, AT&T, and Qualcomm. The alliance values key areas such as AI integration, sustainability, and advanced manufacturing, while addressing societal implications like digital equity and workforce readiness. Its ultimate goal is to ensure North American leadership in wireless innovation. With an MoU signed between 6G-IA and Next G Alliance, the partnership aims to align technical priorities, spectrum strategies, and policy frameworks between Europe and the United States. The collaboration fosters joint discussions on AI-driven networks, sustainability, and the integration of emerging technologies into 6G ecosystems.

Canada's **ENCQOR** (Evolution of Networked Services through a Corridor in Quebec and Ontario for Research and Innovation) (ENCQOR 2025) initiative, launched in 2017 and currently ongoing, bridges public and private sector collaborations to advance 6G research. The MoU between 6G-IA and ENCQOR in 2022 enables cross-border cooperation on AI-driven networking, new spectrum applications, and pre-commercial 6G testbeds, ensuring Canada remains at the forefront of next-generation wireless advancements.

9.2.4 South America: Advancing 6G for Emerging Markets

TeleBrasil 5G (TeleBrasil 2025) continues its long-standing collaboration with SNS-JU under an MoU signed in September 2022, focusing on 6G network evolution, regulatory policies, and the adoption of sustainable telecom infrastructure. This partnership strengthens joint efforts in spectrum harmonization, AI-based network management, and digital inclusion, reinforcing Brazil's role in global 6G discussions.

9.2.5 Asia-Pacific: Key Contributions to 6G Development

China's **IMT-2030 (6G) Promotion Group** (IMTChina 2025), established in June 2019, acts as a coordination body, uniting academic, governmental, and industrial stakeholders. Supported by the China Academy of Information and Communications Technology (CAICT) and with an MoU in place between 6G-IA and IMT-2030, the partnership facilitates knowledge-sharing on AI-native networks, spectrum harmonization, and space-terrestrial integration, ensuring China's contributions align with global 6G objectives.

The **XG Mobile Forum (XGMF)** (2025) is Japan's principal 6G research and policy initiative, established in June 2024 following the merger of the Beyond 5G Promotion Consortium (B5GPC) and the Fifth Generation Mobile Communications Promotion Forum (5GMF). This consolidation aimed to streamline Japan's efforts in next-generation mobile communications and enhance collaboration with international partners. With an MoU signed between 6G-IA and XGMF, the partnership is built on fostering technical exchanges, aligning research roadmaps, and contributing jointly to international standardization bodies such as the 3rd Generation Partnership Project (3GPP) and ITU-R. The collaboration focuses on sustainable network architectures, AI-driven automation, and advanced frequency band utilization. These efforts ensure that Japan's contributions to 6G development are well integrated into the global ecosystem, promoting innovation and interoperability across international markets.

In South Korea, the **6G Forum** (2025), formed in May 2021 and continuing the legacy of its predecessor, the 5G Forum, spearheads projects such as terahertz spectrum research, holographic communications, and intelligent urban systems. Through its MoU with 6G-IA, the collaboration enables joint studies on future wireless architectures, open RAN frameworks, and sustainable telecom policies, supporting South Korea's active engagement in global 6G standardization efforts.

The **Taiwan Association of Information and Communication Standards (TAICS)** (2025) plays a crucial role in aligning Taiwan's telecommunications research with global 6G standards.

Through its MoU with 6G-IA, TAICS engages in joint initiatives on spectrum policy, AI-native networks, and interoperability standards, contributing actively to international 6G cooperation and harmonized development.

The **Bharat 6G Alliance** (2025), launched in July 2023, and **Telecommunications Standards Development Society, India (TSDSI)** (2025), established in January 2014, have formalized MoUs with 6G-IA, reinforcing India's active role in global 6G development. These agreements enable cooperation on cost-effective 6G solutions, spectrum strategies, and standardization efforts, with a strong focus on rural connectivity, digital inclusion, and AI-driven telecom innovations.

9.2.6 Initiatives Towards Openness

As 6G networks evolve, openness and interoperability emerge as key principles driving their development and deployment. The transition towards open and software-defined architectures, such as O-RAN, TelcoCloud, or GSMA Operator Platform, reflects a broader industry shift towards open innovation, ensuring flexibility, competition, multi-provider interoperability, and resilience in network infrastructure.

As detailed in the previous section, Europe's approach to 6G emphasizes a collaborative computing ecosystem that integrates key industry players, academia, and policymakers (6G-IA 2024). Initiatives like the Smart Networks and Services Joint Undertaking (SNS JU) under Horizon Europe and the European Alliance for Industrial Data, Edge, and Cloud are fostering synergies across different sectors, including telecommunications, edge computing, and cloud services. This collaboration enables an integrated approach to developing open and federated infrastructures, which are critical for scaling innovation.

Open-source initiatives are central to ensuring the interoperability of 6G networks. The EU's strategy includes promoting open interfaces, cloud-native deployments, and vendor-neutral architectures. The focus is on leveraging open standards and fostering interoperability between cloud, edge, and network infrastructure providers. This approach also ensures that 6G solutions align with sustainability goals by optimizing resource usage across distributed computing environments. Several initiatives are being hosted by Linux Foundation Europe (LFE) or European Telecommunications Standards Institute (ETSI) Software Development Groups (SDG). Figure 9.4 highlights five major pillars of openness in 6G: Open RAN deployment, cloud-native networks, interoperability, security and sovereignty, and sustainability. The higher importance of interoperability and security underscores the need for robust, standards-driven networks that balance openness with resilience. Meanwhile, Open RAN and cloud-native implementations represent the industry's growing shift towards software-defined and disaggregated infrastructures.

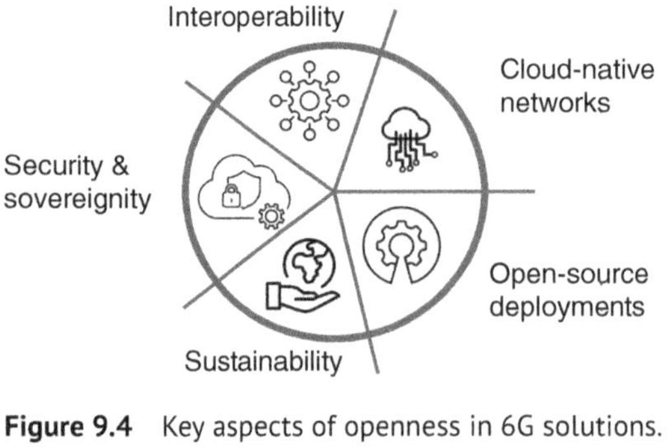

Figure 9.4 Key aspects of openness in 6G solutions.

With the move towards openness, ensuring network security and digital sovereignty remains a priority. The European Commission has emphasized the need for secure, open, and transparent supply chains to safeguard critical digital infrastructure. The development of quantum-safe encryption, federated data management, and AI-driven network orchestration are key aspects of the EU's 6G roadmap.

To sustain and expand the openness of 6G, Europe aims to establish large-scale pilot projects that integrate open 6G infrastructures with emerging applications, such as autonomous systems, smart mobility, and industrial IoT. Future regulatory frameworks will also encourage open business models that support fair competition and vendor diversity, ensuring that open 6G solutions remain a cornerstone of Europe's digital future.

9.3 6G Industrial Exploitation and Market Aspects

The efforts towards 6G represent huge investments from both the public and private sectors in developing a range of new technologies. For the society to be able to benefit from this effort, it must be possible to transform these results into products which can be profitable for the industry as well as beneficial for the society and the end users.

Both 5G and 6G technologies have potential for disruption. 5G has opened substantial improvements in technical performance and opened up for entirely new use cases, but this has not yet been monetized. 6G is expected on the market by 2030 with capabilities surpassing those of 5G regarding, e.g. speed, capacity, and latency. The performance may potentially lead to a seamless blending among the digital, physical, and human worlds.

A recent case study Rossi (2024) also indicates that EU industrial policy is changing towards being more technology specific as a response to the worry of unsatisfactory growth and competitiveness compared to the United States. This shift, together with other regions' industrial policy, will influence the uptake and possible industrial success of 6G. The following sections contain a discussion about the industry position and interest, the estimated impact and exploitation potential, as well as some reflections on the market prospects for 6G.

9.3.1 Industry's Main Interests in 6G

Both telecom and the adjacent industry are engaging in setting the direction and in the development of 6G technology through research, technology proof-of-concepts, experiments and pilots, and standardization. The end goal is usually to be in the position to develop, market and sell products, solutions, and services.

In Chapter 2, the value concept of 6G has been explained. For 5G and some previous generations, design goals were improved technical performance with a strong demand for increased bitrates. This demand still exists; however, it may be decreasing, and the value-based approach, where sustainability plays a pivotal role in developing future networks, is getting more attention.

The value-based approach will be important for the industry when assessing the exploitation potential. Sustainability is not a peripheral concern but a foundational principle encompassing environmental, social, and economic dimensions, going beyond mere energy efficiency improvements. The economic sustainability dimension has always been important for any exploitation of new technological advances, but now industry must also consider environmental and social sustainability.

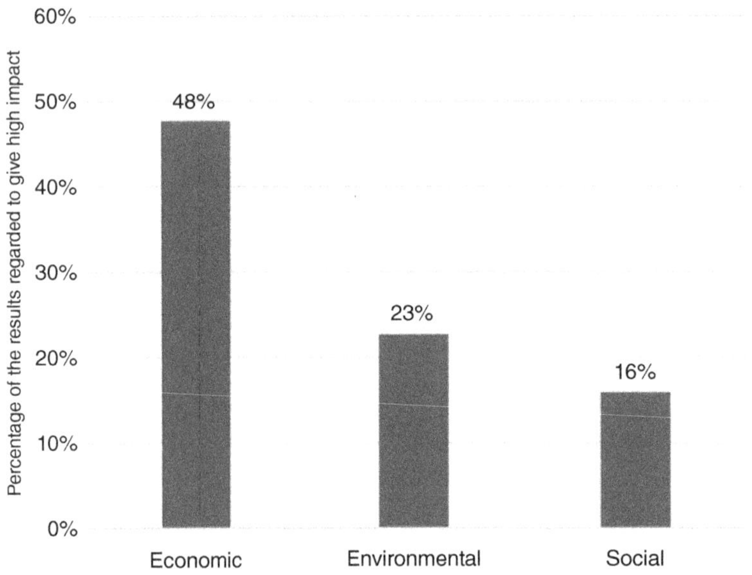

Figure 9.5 Estimated impact from 6G research and innovation results, as reported by the industrial partners of Hexa-X-II. The graph shows the percentage of the results that the partners regard as having a high impact in the respective areas.

Both the telecom industry (typically network operators and vendors) and vertical/adjacent industries are active in the shaping of 6G through participation in research, development, and piloting of 6G technologies. In this way, industry is seeking to make 6G viable from a business point of view.

9.3.2 Anticipated Impact from European R&I on 6G

Impact from technology can be estimated in three areas corresponding to the value-based approach focusing on the environmental, social, and economic dimensions. Figure 9.5 shows the estimated impact from 6G R&I results as reported by the industrial partners in a survey conducted amongst all the partners in the Hexa-X-II project. It shows clearly that the highest impact of 6G technologies is expected to be in the economic area.

Strictly, it does not tell anything about whether the impact is positive or negative; however, it is likely to assume that the responses reflect a belief in a positive impact. For the time being, there is less belief in the environmental and social impact, but since 6G development is only halfway to the target in 2030, this can change depending on what 6G will eventually look like.

Further, Figure 9.6 shows how the industrial partners expect the economic impact to be in terms of capex, opex, profit, and revenue from exploiting 6G technologies.

9.3.3 Investigating the Exploitation Potential

6G technology is still taking shape, and there are still many open issues; however, industry is already evaluating the exploitation potential of the results produced from 6G R&I. It is not surprising that 'Further research' is the area in which the highest exploitation potential is considered

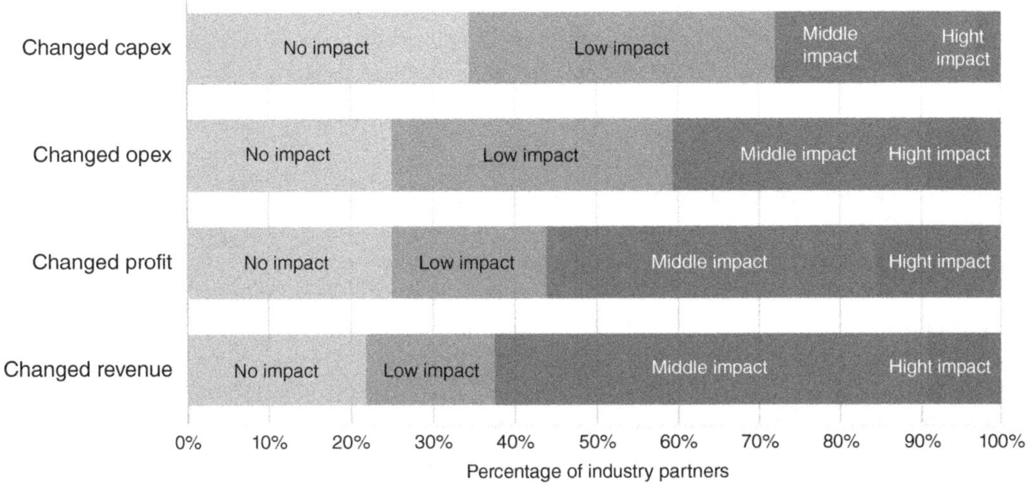

Figure 9.6 Estimated economic impact from 6G research and innovation results, as reported by the industrial partners of Hexa-X-II.

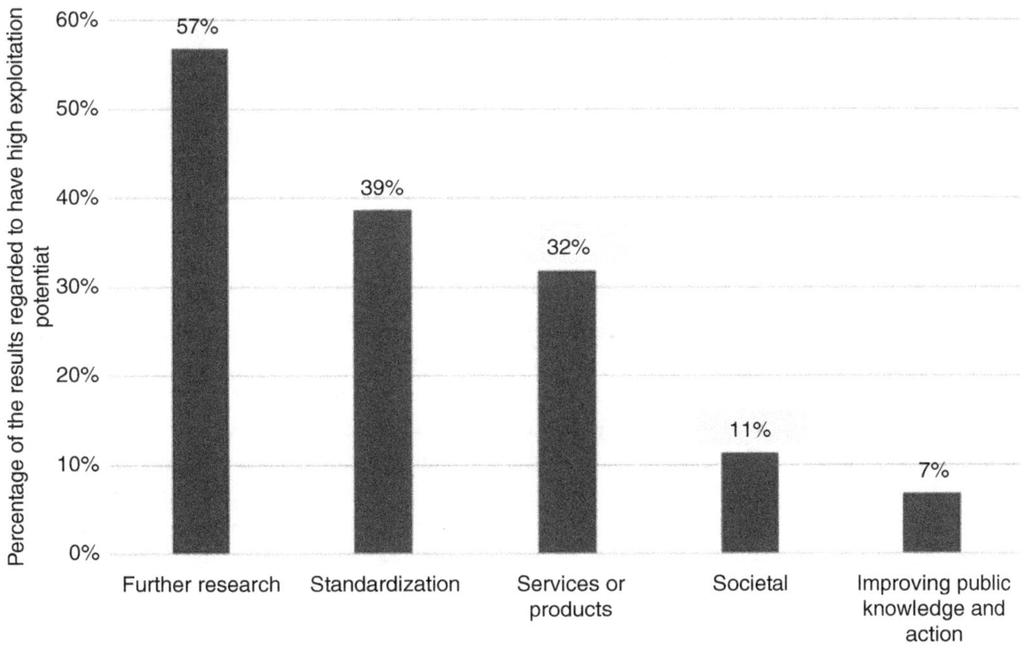

Figure 9.7 Areas of possible exploitation, as reported by industrial partners in Hexa-X-II. The graph shows the percentage of the results that the partners regard as having high exploitation potential in the respective areas.

at the moment, as shown in Figure 9.7. The other important area is on standardization, in which industry is heavily engaged through 3GPP, ETSI, and other SDOs, which is discussed in more detail in Section 9.4.

Many of the results stemming from 6G R&I are technology neutral, and industry is already considering that there is a significant potential for developing products and solutions already at this stage in the development towards 6G technologies, systems, and solutions.

9.3.4 6G Market Prospects

Annual surveys conducted in the SNS JU programme provide important insights into the market prospects for 6G technology (OPS 2025). Assessing the commercial viability of 6G is commonly done using techno-economic analysis and market evaluations. A critical view of the potential widespread adoption of innovations is believed to be important. This includes the quantitative modelling of the cost side, the identification of cost items, and the total cost of ownership.

Open and disaggregated solutions and AI-based solutions are anticipated to be the main drivers in the telecommunications market in the coming years, alongside MBB and FWA. Other technologies and innovations which are expected to play important roles are localization and sensing, and sustainability and energy efficiency.

Learning from the 5G uptake, major obstacles are the lack of demand for unique 6G services not doable using 5G technology. Too many deployment options have also been a hurdle. For network operators, the deployment cost is a major concern, and a smooth migration from 5G to 6G is seen as a prerequisite by many players in the emerging 6G ecosystem. Another factor which is believed to play a significant role is security and privacy concerns, which also points back to the value-based approach discussed above, meeting the expectations of developing a socially sustainable technology. There are also concerns related to a lack of trust in AI solutions.

For 6G to be successful, there is a consensus that 6G must be able to transform the current ecosystem. Most of the novel markets indicated are within AI, enhanced communication services, and the widespread adoption of AR and VR, but not underestimating the importance of the social impact of 6G. The most impacted vertical sector is expected to be Industry 4.0/manufacturing, followed by smart cities and automotive (OPS 2025).

Expected key results from the European 6G R&I cover a wide area of technologies. Many of these are not uniquely 6G specific, like AI and ML integration, advanced network management, security and privacy, communication and sensing, and energy efficiency.

Many of the innovations developed towards 6G are expected to be of interest for the main mobile network operators (MNOs) and vendors. Yet, these new business opportunities could also be exploited by existing and emerging SMEs and start-ups. Novel applications could encourage entrepreneurial activities in a wide range of fields. SMEs can broaden their product and service portfolios, thereby increasing their competitiveness. SMEs can access new markets, including verticals such as automotive or aerospace and niche markets such as cybersecurity, privacy and security, or cloud services.

9.4 Next Steps for 6G Standardization

The 6G standardization roadmap marks a strategic phase in the evolution of wireless communication, building upon the foundations laid by 5G. As global demands for connectivity, speed, and innovation continue to grow, 6G is poised to revolutionize industries, enabling transformative technologies like AI, holographic communications, and ultra-reliable low-latency networks. This section reports the strategic milestones and ongoing initiatives shaping the 6G standardization process. By understanding these efforts, insights are gained into how the future of wireless connectivity will be structured and the timeline for realizing its full potential.

9.4.1 Standardization and Industrial Fora

Global collaboration is pivotal in shaping the future of 6G technologies, with a diverse ecosystem of organizations contributing in different ways. Some bodies develop technical specifications and harmonize global standards, while others focus on implementing software frameworks and fostering industry-wide collaboration.

Standardization bodies such as **3GPP** and **ITU-R** remain central to 6G development, ensuring the alignment of technical specifications, spectrum allocation, and regulatory policies. ITU's **IMT-2030** initiative builds on insights from 5G, setting performance and architectural guidelines for the next generation, including spectrum frameworks, network capabilities, and technical benchmarks crucial for 6G.

On the **industrial collaboration side**, organizations like the **GSM Association (GSMA)** and **Next Generation Mobile Networks Alliance (NGMN)** play a crucial role in driving industry adoption, ecosystem development, and business strategies for 6G. GSMA, representing mobile network operators worldwide, works on policy advocacy, spectrum coordination, and industry-wide initiatives, ensuring interoperability and economic viability of next-generation networks. NGMN, composed of leading telecom operators and vendors, provides technical guidance and operator-driven requirements to shape 6G network evolution.

The **O-RAN Alliance** (2025), though primarily concerned with open and interoperable architectures for 5G RAN, plays a foundational role in promoting modular and flexible network architectures that may influence 6G. By advancing open interfaces and disaggregated RAN components, it contributes to the evolution of next-generation mobile networks, ensuring vendor neutrality and adaptability.

The **6G Forum** (2025) serves as an international multi-stakeholder platform distinct from the South Korean 6G Forum. While the latter focuses on South Korea's domestic R&D and standardization efforts, the global 6G Forum brings together stakeholders to align policies, share technological insights, and discuss the social, ethical, and legal implications of 6G adoption. It facilitates broader discussions to ensure that 6G development aligns with societal and regulatory considerations.

Additionally, the **Trade and Technology Council (EU-US)** reflects strong transatlantic partnerships aimed at harmonizing research priorities and regulatory strategies for next-generation wireless technologies, including 6G, ensuring unified global approaches in tackling the challenges of the evolving digital ecosystem.

By bringing together technical standardization, industry-driven initiatives, and policy harmonization, these organizations collectively ensure that 6G is not only technologically robust but also globally interoperable, sustainable, and commercially viable.

9.4.2 Roadmap and Next Steps

With the initial 6G deployment anticipated around 2030, the standardization process for 6G is already underway, involving multiple stakeholders, including international standards bodies like 3GPP, regulatory bodies, technology companies, and governments. The 6G standardization will be done by 3GPP, which has a long history of defining the technical specifications for cellular networks, including 3G, 4G, and 5G. This collaborative effort ensures a global, interoperable standard for 6G, facilitating seamless communication across different regions and devices. In order to achieve the multi-dimensional, comprehensive standardization for the next generation of mobile networks, the following remarks need to be considered.

The ITU and other industry bodies have since the year 2020 been working to define a vision and requirements for 6G. This involves identifying use cases, performance metrics, and societal needs that 6G should address, such as sustainability, digital inclusion, and ultra-low latency communication. Alongside ITU, major technology companies and academic institutions have published white papers to propose visions for 6G. These documents serve as foundational inputs for standardization discussions.

After setting the foundation for the 6G vision, identifying suitable frequency bands for 6G is a critical step. It is imperative that existing frequency bands for mobile communications can also be leveraged by 6G, as well as identifying new bands in the GHz range that can be utilized by 6G.

Although not within the scope of the first release, for specific short-range communication services, frequencies above 100 GHz, including terahertz waves, are being explored to support ultra-high data rates and capacity; however, this is still a research topic. The ITU-R's WRC meetings play a crucial role in identifying future harmonized spectrum. The 2027 WRC will be particularly important for identifying additional frequency bands in time for the commercial 6G deployments.

3GPP plays a pivotal role in the development of global 6G standards. Building on its expertise in all mobile network generations, 3GPP's work involves defining the architecture, interfaces, technical framework, and protocols for 6G. As of 2023, 3GPP has committed to developing the global 6G standards.

It is noteworthy to mention that while 3GPP is developing 6G, the evolution of 5G Advanced will continue in 3GPP for several releases, standardizing new technology that will serve the global market until 2030 and beyond. The work on requirements in 3GPP kicked off in May 2024 to start defining the 6G services and service requirements. Representatives from different regions presented their views on novel aspects to be considered for 6G. The 3GPP study on 6G services was approved in September 2024. The main task for 3GPP, in parallel to maintaining 5G advanced standardization, will be to design the 6G system going towards 2026. The technical work in the working groups will start in Q3 2025. The initial design of the 6G system will be done in two phases over two releases. The first is a study phase where different aspects of 6G are studied in 3GPP release 20. The second phase will go into detailed technical design and specification, culminating in the fully implementable 6G standard in Release 21. The study will begin in Q3 2025 and will be 21 months long. This will be parallel to normative work on 5G Advanced evolution during Release 20, which will have a duration of 18 months only. During the technical study and the work on requirements, the 3GPP community will get a clearer picture of the time needed to finalize 6G (Figure 9.8).

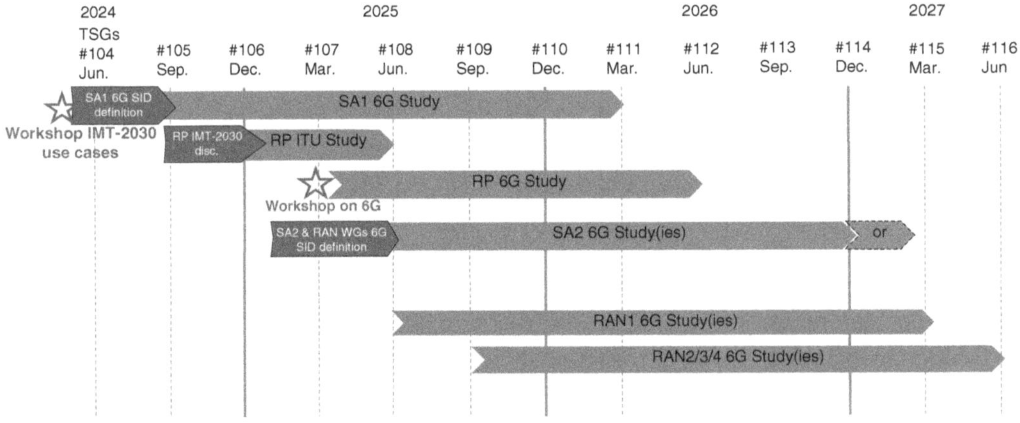

Figure 9.8 Standardization roadmap in the next years.

The journey to 6G standardization is a collaborative and iterative process, requiring input from diverse stakeholders across the globe. By focusing on a shared vision, robust research, and coordinated policy development, the industry aims to deliver transformative service enablement and connectivity that meets the demands of the 2030s and beyond.

9.5 Key Challenges and Future Directions

The global advancement of 6G is marked by both opportunities and challenges. Geopolitical tensions and fragmented regulatory frameworks could potentially hinder the establishment of unified global standards, while the deployment of 6G networks could raise environmental sustainability concerns, despite the actions to make 6G more energy efficient than previous generations. Overcoming these hurdles requires enhanced international collaboration to align research priorities and technical frameworks. Public-private partnerships and cross-border collaborations, described in this chapter, will be crucial for navigating technological and regulatory complexities.

By converging regional expertise and aligning technological advancements with societal needs, the global pursuit of 6G reflects a shared vision for a more sustainable and inclusive future. These efforts promise not only to advance telecommunications but also to redefine the role of connectivity in addressing pressing global challenges, from healthcare to education and urban infrastructure.

This chapter has presented an overview of the path towards 6G, as per the status of the works in standardization and in research and innovation at the time of publishing this book. First, a view on the global initiatives has been introduced, in Europe and elsewhere. Then, a snapshot of the status of the industrial expectations around 6G has been offered. Finally, a view on the current standardization initiatives concludes the chapter.

Acronyms and Abbreviations

3GPP	3rd generation partnership project
5GMF	Fifth-Generation Mobile Communications Promotion Forum
5GC	5G Core
6G	Sixth Generation
6G-IA	6G Smart Networks and Services Industry Association
AI	Artificial intelligence
ARCEP	Autorité de Régulation des Communications Électroniques et des Postes (French telecom regulator)
ATIS	Alliance for Telecommunications Industry Solutions
B5GPC	Beyond 5G Promotion Consortium
CAICT	China Academy of Information and Communications Technology
CEF	Connecting Europe Facility
CTTC	Centre Tecnològic de Telecomunicacions de Catalunya
CWC	Centre for Wireless Communications
DC	Dual connectivity
DGE	Direction Générale des Entreprises (French Ministry of Economy)
EPC	Evolved Packet Core
ERDF	European Regional Development Funds
ETSI	European Telecommunications Standards Institute
GSMA	GSM Association
ISAC	Integrated sensing and communication

ITU	International Telecommunication Union
LFE	Linux Foundation Europe
LLS	Lower-layer split
ML	Machine learning
MNO	Mobile network operator
MRSS	Multi-RAT spectrum sharing
NGMN	Next generation mobile networks alliance
NSA	Non-standalone
R&D	Research and development
RAN	Radio access network
RAT	Radio access technology
SA	Standalone
SDG	Software development groups
SDO	Standards development organisation
SGPI	General Secretariat of the Prime Minister (France)
SME	Small and medium enterprise
SNS	Smart networks and services
SNS-JU	Smart networks and services joint undertaking
TAICS	Taiwan Association of Information and Communication Standards
TSDSI	Telecommunications Standards Development Society, India
XGMF	XG Mobile Forum

References

6G Bridge (2023). 6G Bridge. [Online]. Available at: https://www.businessfinland.fi/en/for-finnish-customers/services/programs/6g-bridge (accessed August 2025).

6G Flagship (2021). 6G Flagship. [Online]. Available at: https://www.6gflagship.com/ (accessed August 2025).

6G Forum (2025). 6G Forum. [Online]. Available at: https://6gforum.or.kr/html/ (accessed August 2025).

6G-IA (2023). Key strategies for SNS, [Online]. Available at: https://6g-ia.eu/wp-content/uploads/2023/10/6g-ia-position-paper_2023_final.pdf (accessed 1 September 2025).

6G-IA (2024). European vision for the 6G network ecosystem. [Online]. Available at: https://6g-ia.eu/wp-content/uploads/2024/11/european-vision-for-the-6g-network-ecosystem.pdf (accessed 1 September 2025).

Bharat (2025). Bharat 6G Alliance. [Online]. Available at: https://bharat6galliance.com/bharat6G/ (accessed August 2025).

ENCQOR (2025). ENCQOR. [Online]. Available at: https://encqor.ca/en/home/ (accessed August 2025).

Ericsson (2024). Ericsson Mobility Report November 2024. [Online]. Available at: https://www.ericsson.com/en/reports-and-papers/mobility-report/reports/november-2024 (accessed 1 September 2025).

ESD (2023). Ministry of Economic Affairs and Digital Transformation, "Espana digital". [Online]. Available at: https://espanadigital.gob.es/en (accessed August 2025).

FNS (2024). Future Network Services. [Online]. Available at: https://futurenetworkservices.nl/ (accessed August 2025).

G. F (2023). Gouvernement. [Online]. Available at: https://www.gouvernement.fr/france-2030 (accessed August 2025).

IMTChina (2025). IMTChina. [Online]. Available at: https://www.imt2030.org.cn/ (accessed August 2025).

ITU-R (2023). *ITU-R Recommendation M.2160, IMT-2030 Framework*. s.l: ITU-R.
Next G Alliance (2025). Next G Alliance. [Online]. Available at: https://nextgalliance.org (accessed August 2025).
ORAN Alliance (2025). ORAN Alliance. [Online]. Available at: https://www.o-ran.org (accessed August 2025).
Platform, 6. G (2023). Germany. [Online]. Available at: https://www.6g-platform.com/ (accessed August 2025).
Project, S. I (2023). D2.2 Findings and trends from European 6G R&I initiatives, s.l.: s.n.
Restart (2023). Fondazione Restart. [Online]. Available at: https://www.fondazione-restart.it/ (accessed August 2025).
Rossi, M.A. (2024). EU technology-specific industrial policy. The case of 5G and 6G. *Telecommunications Policy* 48 (2): 102639, ISSN 0308-5961. https://doi.org/10.1016/j.telpol.2023.102639.
SNS OPS (2025). SNS OPS Deliverable 1.4. "Second period assessment and planning report", s.l.: s.n.
Sweden (2023). Regeringskansliet, "Regeringen förstärker innovations-och forskningsprogram som ska bidra till framtidens digitala lösningar". [Online]. Available at: https://www.regeringen.se/pressmeddelanden/2023/03/regeringen-forstarker-innovations--och-forskningsprogram-som-ska-bidra-till-framtidens-digitala-losningar-/ (accessed August 2025).
TAICS (2025). Taiwan Association of Information and Communication Standards. [Online]. Available at: https://www.taics.org.tw/eng/ (accessed August 2025).
TeleBrasil (2025). TeleBrasil. [Online]. Available at: https://www.telebrasil.org.br (accessed August 2025).
TSDSI (2025). Telecommunications Standards Development Society. [Online]. Available at: https://tsdsi.in (accessed August 2025).
XGMF (2025). XGMF. [Online]. Available at: https://xgmf.jp/en/ (accessed August 2025).

10

Conclusion

Patrik Rugeland[1], Mikko A. Uusitalo[2], and Mauro R. Boldi[3]

[1] *Ericsson Research, Stockholm, Sweden*
[2] *Nokia Bell Labs, Espoo, Finland*
[3] *Telecom Italia, Turin, Italy*

With the standardization of 6G ramping up during 2025, the 6G research conducted by the European SNS-JU projects, in particular the flagship projects Hexa-X and Hexa-X-II in the past five years, have given a solid foundation to build upon. With the immense environmental challenges facing our societies, the sustainability aspects have been lifted to the forefront, attempting to balance their esoteric requirements against the traditional performance requirements. However, it is recognized that it is not sufficient to solely consider environmental sustainability, as social and economic sustainability dictates whether any solution will be viable. As the legacy mobile communication systems are immensely capable, significant effort was placed on identifying and developing differentiating features and usages of 6G, where a wide range of use cases were identified that were grouped into use-case families: *immersive experiences, collaborative robots, physical awareness, digital twins, fully connected world*, and *trusted environments*. These have been analysed based on requirements and sustainability, as well as enablers for technical realization. Conversely, the main research focus has not been on the legacy technology, which will form the bedrock of 6G as these aspects are well understood from decades of research.

One of the main differentiating factors is that the 6G system will become a platform, opening up possibilities for exposing capabilities to allow more observability and control of the network. This will require many innovations in the architecture, incorporating cloud functionalities and exploring network modularization. To achieve the desired network coverage, improved network flexibility, e.g. with improved carrier aggregation and incorporation of NTN will be crucial. Similarly, enhancing the reliability of the network will enable deterministic communication necessary for many of the envisioned use cases for 6G.

Furthermore, to achieve the stringent requirements, the design of the radio interface and hardware has been evaluated, exploring novel deployment alternatives such as D-MIMO or RIS, and supports for energy-neutral devices as well as modelling and performance evaluation of sub-THz communication. The viability of these solutions will depend on both their marginal benefit compared to other solutions and their overall cost to implement and deploy.

Another important aspect is the emergence of AI/ML, which has the potential to significantly alter the operation and performance of the mobile networks. For instance, employing AI/ML on the air interface can potentially increase the bitrate, reduce the energy consumption, and boost the reliability through predictive management and bespoke configurations. On the other hand,

6G to Build a Sustainable Future, First Edition. Edited by Mikko A. Uusitalo, Patrik Rugeland, Mauro R. Boldi, and Ahmad Nimr.

employing AI/ML on the management level can introduce more autonomous intent-based management systems that are easier to manage and optimize.

The 6G platform is also envisioned to extend the operation beyond communication, incorporating novel functionality such as sensing or compute offloading. For ISAC the prospect of reusing parts of the radio network for sensing the environment at a low marginal cost can foster many previously inconceivable usages. For computational offloading, by moving some of the functionality from the edge devices to the network, more lightweight and simpler devices can be envisioned with a smaller form-factor and longer battery operation.

To showcase the potential of 6G, a system-level proof-of-concept incorporating an autonomous robot, integrating ISAC IBM, and XR has been developed. Although this PoC only includes a sliver of the potential of 6G, it nevertheless hints at its potential.

Finally, as the standardization of 6G is ramping up, targeting initial commercial deployments around 2030, the groundwork conducted by Hexa-X-II and the other European projects have shown the necessary considerations needed for 6G such as sustainability considerations and proper migration strategies from 5G to 6G, as well as the potential technological steps that can be taken to realize the 6G system.

Index

Note: Page numbers in *italic* refer to figures and **bold** refer to tables.

6G to Build a Sustainable Future, First Edition. Edited by Mikko A. Uusitalo, Patrik Rugeland, Mauro R. Boldi, and Ahmad Nimr.

e

f